AF615914

BD 026887901 7

The Target of Penicillin

The Target of Penicillin

The Murein Sacculus of Bacterial Cell Walls Architecture and Growth

Proceedings
International FEMS Symposium
Berlin (West), Germany, March 13–18, 1983

Editors
R. Hakenbeck · J.-V. Höltje · H. Labischinski

Walter de Gruyter · Berlin · New York 1983

Editors

Regine Hakenbeck
Max-Planck-Institut für Molekulare Genetik
Ihnestrasse 63–73, D-1000 Berlin 33, F. R. of Germany

Joachim-Volker Höltje
Max-Planck-Institut für Virusforschung, Abteilung Biochemie
Spemannstrasse 35/II, D-7400 Tübingen, F. R. of Germany

Harald Labischinski
Robert-Koch-Institut des Bundesgesundheitsamtes,
Abteilung Cytologie
Nordufer 20, D-1000 Berlin 65, F. R. of Germany

Library of Congress Cataloging in Publication Data

The Target of penicillin.

Based on papers presented at the International Symposium on the Murein Sacculus of Bacterial Cell Walls, Mar. 12–18 in West Berlin and sponsored by the Senat der Stadt Berlin (West) … [et al.]
Bibliography: p.
Includes indexes.
1. Peptidoglycans-Congresses. 2. Penicillin-Congresses.
I. Hakenbeck, R. (Regine), 1948 – II. Höltje, J.-V. (Joachim Volker), 1941 – III. Labischinski, H. (Harald), 1948 – IV. International Symposium on the Murein Sacculus of Bacterial Cell Walls (1983: Berlin, Germany) – V. Berlin (Germany: West). Senat.
QR77.3.T37 1983 616'.014 83-15215
ISBN 3-11-009705-2

CIP-Kurztitelaufnahme der Deutschen Bibliothek

The target of penicillin: the murein sacculus of bacterial cell walls, architecture and growth; proceedings, internat. FEMS symposium, Berlin (West), Germany, March 13–18, 1983/ ed. R. Hakenbeck … – Berlin; New York: de Gruyter, 1983.
ISBN 3-11-009805-2

NE: Hakenbeck, Regeine [Hrsg.]; Federation of European Micro-biological Societies

Printing: Karl Gerike, Berlin. – Binding: Dieter Mikolai, Berlin. – Printed in Germany.

Preface

This book is based on the papers read at the International Symposium on the Murein Sacculus of Bacterial Cell Walls (March 13-18 in West Berlin). The aim of the symposium was to present the current state of knowledge on the subject of the murein sacculus. The contributions to this book, therefore, include very different approaches, ranging from physical studies on the three-dimensional architecture of this unique biopolymer, the biological processes responsible for cellular growth to medical implications of the bacterial cell wall. We intentionally allowed equal space for poster presentations and talks, in order to cover as many experimental results as possible. It is up to the reader to interpret the papers and draw his own conclusions. We hope that this collection will provide a comprehensive up-to-date review of the many recent advances in this field. If it does this, then its success is mainly due to the excellent contributions by the participants at the symposium. For this, we would like to thank all authors.

We would like to express our gratitude to the Senat der Stadt Berlin (West), the Max-Planck-Gesellschaft, the Deutsche Forschungsgemeinschaft, and the Federation of European Microbiological Societies for sponsoring the symposium, and Bayer AG, the Bundesgesundheitsamt, the Paul Martini Stiftung and Schering AG for financial support. In the end, it was their aid which also enabled us to present this book. Thanks are also due to the publisher for publishing these proceedings so rapidly. Finally, we thank all our colleagues, especially I. Baxivanelis, A. Brauer, H. Lange, B. Nürnberg and E. Vorgel for their enthusiastic and devoted help in the organization of the symposium and/or in the realization of the proceedings.

Berlin (West) and Tübingen
May 1983

R. Hakenbeck
J.-V. Höltje
H. Labischinski

CONTENTS

VI. BIOSYNTHESIS OF MUREIN

CONTRIBUTORS AND PARTICIPANTS

The page number of the contribution is in brackets.

Juan A. Ayala (475)
Instituto de Bioquimica de Macromoleculas, Centro de Biologia Mol.
Universidad Autonoma de Madrid
Canto Blanco
Madrid-34 Spain

Vidar Bakken
Dept. of Biochemistry
University of Bergen
Arstadveien 19
N-5000 Bergen
Norway

Gerhard Barnickel (61)
Institut für Kristallographie
Freie Universität Berlin
Takustr. 6
D-1000 Berlin 33
Germany

Jose Berenguer (211)
Instituto de Bioquimica de Macromoleculas, Centro de Biologia Mol.
Universidad Autonoma de Madrid
Canto Blanco
Madrid-34 Spain

Terry J. Beveridge (35)
Department of Microbiology
College of Biological Science
University of Guelph
Guelph, Ontario
Canada N1G2W1

Peter Blümel (323)
Robert Koch-Institut des Bundesgesundheitsamtes
Nordufer 20
D-1000 Berlin 65
Germany

Guiseppe Botta (445)
Institute of Microbiology
University of Genova
Viale Benedetto XV/10
16132 Genova
Italy

Cyrille Brandt (223)
Institut de genetique et de biologie microbiennes
Rue Cesar-Roux 19
CH-1005 Lausanne
Switzerland

Thomas Briese (173)
Max-Planck-Institut f. Molekulare Genetik, Abt. Trautner
Ihnestr. 63-73
D-1000 Berlin 33
Germany

Jenny Broome-Smith (471)
Microbial Genetics Group
School of Biological Sciences
The University of Sussex
Brighton, BN1 9QG
England

Derek F.J. Brown (537)
Clinical Microbiology and Public Health Laboratories
New Addenbrooke's Hospital
Cambridge
England

Christine Buchanan (451)
Dept. of Biology
Southern Methodist University
Dallas, Texas 75275
U.S.A.

Petra Burghaus (317)
Institut f. Kristallographie
Freie Universität Berlin
Takustr. 60
D-1000 Berlin 65
Germany

Lars G. Burman
Dept. of Clinical Bacteriology
University of Umea
S-901 85 Umea
Sweden

Pietro Canepari (141)
Institute of Microbiology
University of Genova
Viale Benedetto XV, 10
16132 Genova
Italy

Paulette Charlier
Service de Cristallographie
Institut de Physique B5
Universite de Liege
Sart-Tilman
B-4000 Liege Belgium

Alessandro Costa
Institute of Microbiology
University of Genova
Viale Benedetto XV/10
16132 Genova
Italy

Jacques Coyette (523)
Service de Microbiologie
Institut de Chimie B6
Universite de Liege
Sart Tilman
B-4000 Liege Belgium

Lolita Daneo-Moore (493)
Dept. of Microbiology and
Immunology
Temple University School of Med.
3400 N Broad St.
Philadelphia, Pa. 19140 U.S.A.

Miguel Angel De Pedro (465)
Instituto de Bioquimica de Macro-
moleculas, Centro de Biologia Mol.
Universidad Autonoma de Madrid
Canto Blanco
Madrid-34 Spain

Ralf Dreher (643)
Mikrobiologie II
Universität Tübingen
Auf der Morgenstelle 28
D-7400 Tübingen
Germany

Heinz Ellerbrok (421)
Max-Planck-Institut f. Molekulare
Genetik, Abt. Trautner
Ihnestr. 63-73
D-1000 Berlin 33
Germany

Franz Fiedler
Lehrstuhl f. Mikrobiologie
Universität München
Maria-Ward-Str. 1a
D-8000 München 19
Germany

Erwin Fischer (73)
Institut f. Kristallographie
Freie Universität Berlin
Takustr. 6
D-1000 Berlin 33
Germany

Helene Fischer (625)
The Rockefeller University
1230 York Avenue
New York, N.Y. 10021
U.S.A.

Werner Fischer (179)
Institut f. Physiologische
Chemie
Universität Erlangen
Fahrstr. 17
D-8520 Erlangen Germany

Roberta Fontana (531)
Institute of Microbiology
Medical School
University of Padua
Via Gabelli 63
35100 Padua Italy

Helmut Formanek (55)
Botanisches Institut der
Universität München
Menzingerstr. 67
D-8000 München 19
Germany

Norbert Franken
Lehrstuhl f. Mikrobiologie
Technische Universität München
Arcisstr. 21
D-8000 München 2
Germany

Jean-Marie Ghuysen (369)
Service de Microbiologie
Institut de Chimie
Universite de Liege
Sart Tilman
B-4000 Liege Belgium

Peter Giesbrecht (243)
Robert Koch-Institut des
Bundesgesundheitsamtes
Nordufer 20
D-1000 Berlin 65
Germany

Isaac Ginsburg (341)
Dept. of Oral Biology
Hebrew University-Hadassah
Faculty of Dental Medicine
Jerusalem 91010
Israel

Bernd Glauner (29)
Max-Planck-Institut f. Virus-
forschung, Abt. Biochemie
Spemannstr. 35
D-7400 Tübingen
Germany

Jobst Gmeiner (613)
Institut f. Mikrobiologie
Techn. Hochschule Darmstadt
Schnittspahnstr. 9
D-6100 Darmstadt
Germany

Ernest W. Goodell (129)
Natural Science Dept.
SUNY College of Technology
811 Court St.
Utica, N.Y. 13502
U.S.A.

Laurent Gutmann (229)
Laboratoire de Microbiologie
Institut Biomedical des
Cordeliers
15, rue de l'Ecole de Medicine
75270 Paris, Cedex 06 France

Regine Hakenbeck (415)
Max-Planck-Institut f. Molekulare
Genetik, Abt. Trautner
Ihnestr. 63-73
D-1000 Berlin 33
Germany

Walter P. Hammes (577)
Institut f. Lebensmittel-
technologie der Universität
Hohenheim
Garbenstr. 25
D-7000 Stuttgart 70 Germany

Charles R. Harrington (607)
Dept. of Biochemistry
University of Cambridge
Tennis Court Road
Cambridge CB2 1QW
England

Jean van Heijenoort (191)
Centre National de la Recherche
Scientific, Inst. de Biochimie
Batiment 432
Universite de Paris-Sud
91405 Orsay France

Susanne Henning (571)
Institut f. Lebensmittel-
technologie der Universität
Hohenheim
Garbenstr. 25
D-7000 Stuttgart 70 Germany

Michael L. Higgins (113)
Department of Microbiology
Temple University School of Med.
3400 N Broad St.
Philadelphia, Pa. 19140
U.S.A.

Yukinori Hirota (393)
National Institute of Genetics
Yata 1, 111
Mishima, 411
Japan

Joachim-Volker Höltje (185)
Max-Planck-Institut f. Virus-
forschung, Abt. Biochemie II
Spemannstr. 35
D-7400 Tübingen
Germany

Jerzy Hrebenda
Instytut Mikrobiologii
Uniwersytetu Warszawskiego
Nowy Swiat 67
00-046 Warsaw
Poland

Gerhard Huber
Mikrobiologie H 780
Hoechst AG
Postfach 80 03 20
D-6230 Frankfurt (M) 80
Germany

Edward Ishiguro (631)
Dept. of Biochemistry
and Microbiology
University of Victoria
Victoria, B.C.
Canada V8W 2Y2

Harald B. Jensen
Dept. of Biochemistry
University of Bergen
Arstadveien 19
N-5000 Bergen
Norway

Lars Johannsen (261)
Robert Koch-Institut des
Bundesgesundheitsamtes
Nordufer 20
D-1000 Berlin 65
Germany

Uwe Jürgens (19)
Institut f. Biologie II
Albert-Ludwigs-Universität
Schänzlestr. 1
D-7800 Freiburg i.Br.
Germany

Dimitri Karamata (285)
Institut de genetique et de
biologie microbiennes
Rue Cesar-Roux 19
CH-1005 Lausanne
Switzerland

Judith A. Kelly (387)
Biochemistry and Biophysics Sec.
Biological Sciences Group
University of Connecticut
Storrs, Ct. 06268
U.S.A.

Thomas Kersten (335)
Robert Koch-Institut des
Bundesgesundheitsamtes
Nordufer 20
D-1000 Berlin 65
Germany

Gunnar Kleppe
Rogaland distriktshogskole
P.O. 2540 Ullandhaug
N 4001 Stavanger
Norway

Arthur L. Koch (99)
Department of Biology
Indiana University
Jordan Hall 138
Bloomington, Ind. 47405
U.S.A.

Hans U. Koch
Institut f. Physiologische
Chemie
Universität Erlangen
Fahrstr. 17
D-8520 Erlangen Germany

Mira Korat
Max-Planck-Institut f.
Virusforschung
Spemannstr. 35
D-7400 Tübingen
Germany

Werner Kraus (511)
Max-Planck-Institut f. Virus-
forschung, Abt. Biochemie
Spemannstr. 35
D-7400 Tübingen
Germany

Dominique Krüger (273)
Robert Koch-Institut des
Bundesgesundheitsamtes
Nordufer 20
D-1000 Berlin 65
Germany

Wolfgang Kusser
Lehrstuhl f. Mikrobiologie
Universität München
Maria-Ward-Str. 1a
D-8000 München 19
Germany

Harald Labischinski (49)
Robert Koch-Institut des
Bundesgesundheitsamtes
Nordufer 20
D-1000 Berlin 65
Germany

Aviva Lapidot (79)
Dept. of Isotope Research
Weizmann Institute of Science
Rehovot
Israel

Jose Laynez (91)
Thermochemistry Laboratory
Chemical Centre
University of Lund
S-22007 Lund
Sweden

Edgar Lederer (293)
Institut de Biochimie
Universite de Paris-Sud
Centre d'Orsay
91405 Orsay
France

Mireille Leduc
Institut de Biochimie
Universite de Paris-Sud
91405 Orsay
France

Bernd Leps (67)
Institut f. Kristallographie
Freie Universität Berlin
Takustr. 6
D-1000 Berlin 33
Germany

Melina Leyh-Bouille
Service de Microbiologie
Institut de Chimie B6
Universite de Liege
Sart-Tilman
B-4000 Liege Belgium

Werner Lubitz (205)
Fachbereich Biologie
Universität Kaiserslautern
Postfach 3049
D-6750 Kaiserslautern
Germany

Zdzislaw Markiewicz (25)
Instytut Mikrobiologii
Uniwersytetu Warszawskiego
Nowy Swiat 67
00-046 Warsaw
Poland

Robert E. Marquis (43)
Dept. of Microbiology
University of Rochester
601 Elmwood Avenue
Rochester, New York 14642
U.S.A.

Hans Herbert Martin
Fachbereich Biologie (10)
Techn. Hochschule Darmstadt
Schnittspahnstr. 9
D-6100 Darmstadt
Germany

Michio Matsuhashi (499)
Institute of Applied Micro-
biology
University of Tokyo
Bunkyo-ku
Tokyo, 113 Japan

Dominique Mengin-Lecreulx (559)
Institut de Biochimie
Universite de Paris-Sud
91405 Orsay
France

Helmut Mett (565)
Pharmaceuticals Division
Research Department
Ciba-Geigy AG
CH-4002 Basel
Switzerland

P. Diederik Meyer (237)
Laboratorium voor Microbiologie
Universiteit van Amsterdam
Nieuwe Achtergracht 127
1018 WS Amsterdam
The Netherlands

Stefan Mollner (305)
Institut f. Mikrobiologie
Universität Tübingen
Auf der Morgenstelle 28
D-7400 Tübingen 1
Germany

Juan-Carlos Montilla (409)
Instituto de Bioquimica de Macro-
moleculas, Centro de Biologia Mol.
Universidad Autonoma de Madrid
Canto Blanco
Madrid-34 Spain

Karl Heinz Müller
Max-Planck-Institut f. Virus-
forschung, Abt. Biochemie
Spemannstr. 35
D-7400 Tübingen
Germany

Nanne Nanninga (147)
Dept. of Electron Microscopy
and Molecular Cytology
University of Amsterdam
Plantage Muidergracht 14
1018 TV Amsterdam The Netherlands

Dieter Naumann (85)
Robert Koch-Institut des
Bundesgesundheitsamtes
Nordufer 20
D-1000 Berlin 65
Germany

Francis C. Neuhaus (589)
Dept. of Biochemistry, Molecular
Biology and Cell Biology
Northwestern University
Hogan 2-100
Evanston, Ill. 60201 U.S.A.

Roswitha Nöthling (267)
Max-Planck-Institut f. Molekulare
Genetik, Abt. Trautner
Ihnestr. 63-73
D-1000 Berlin 33
Germany

Martine Nguyen-Disteche
Service de Microbiologie
Institut de Chimie B6
Sart Tilman
B-4000 Liege
Belgium

James T. Park (119)
Dept. of Molecular Biology
and Microbiology
Tufts University School of Med.
136 Harrison Avenue
Boston, Mass. 02111 U.S.A.

Helmut Pelzer (105)
Dr. Karl Thomae AG
Abteilung Biochemie
D-7950 Biberach/ Riss
Germany

Harold R. Perkins (255)
Dept. of Microbiology
University of Liverpool
P.O. Box 147
Liverpool L69 3BX
England

Roland Plapp (197)
Fachbereich Biologie
Universität Kaiserslautern
Postfach 3049
D-6750 Kaiserslautern
Germany

Harold M. Pooley (279)
Institut de genetique et de
biologie microbiennes
Rue Cesar-Roux 19
CH-1005 Lausanne
Switzerland

Angelika Raible (249)
Dr. Karl Thomae AG
Abt. Biochemie
D-7950 Biberach/ Riss
Germany

V.S.R. Rao (359)
Molecular Biophysics Unit
Indian Institute of Science
Bangalore 560 012
India

Bernhard Reinicke
Robert Koch-Institut des
Bundesgesundheitsamtes
Nordufer 20
D-1000 Berlin 65
Germany

Victor M. Reusch, Jr. (601)
Dept. of Biological Sciences
University of Denver
Denver, Colorado 80208
U.S.A.

Peter E. Reynolds (517)
Dept. of Biochemistry
University of Cambridge
Tennis Court Road
Cambridge, CB2 1QW
England

Alfredo Rodriguez-Tebar (427)
Instituto de Bioquimica de Macro-
moleculas, Centro de Biologia Mol.
Universidad Autonoma de Madrid
Madrid-34 Spain

Howard J. Rogers (637)
National Institute for
Medical Research
Mill Hill
London NW7 1AA
England

Raoul S. Rosenthal (311)
Dept. of Microbiology and
Immunology
Indiana University Sch. Med.
Indianapolis, Ind. 46226
U.S.A.

Barry Ross
Hoechst AG
Postfach 80 03 20
D-6230 Frankfurt 80
Germany

Lawrence I. Rothfield
Dept. of Microbiology
University of Connecticut
School of Medicine
Farmington, Conn. 06032
U.S.A.

I. Mohamed Said (439)
Max-Planck-Institut f. Virus-
forschung, Abt. Biochemie
Spemannstr. 35
D-7400 Tübingen
Germany

Milton R.J. Salton (3)
Dept. of Microbiology
New York University School
of Medicine
550 First Avenue
New York, N.Y. 10016 U.S.A.

Guiseppe Satta (135)
Istituto di Microbiologia II
Universita' Degli Studi Di Cagliari
Via Procell, 12
09100 Cagliari
Italy

Karl H. Schleifer (11)
Lehrstuhl f. Mikrobiologie
Institut Botanik u. Mikrobiologie
Techn. Universität München
Arcisstr. 21
D-8000 München 2 Germany

Elmar Schrinner
Hoechst AG
Abt. Chemotherapie
Postfach 80 03 20
D-6230 Frankfurt 80
Germany

Schumacher
Hygieneinstitut der
Universität Köln
Goldenfeldsstr. 21
D-5000 Köln 41
Germany

Uli Schwarz
Max-Planck-Institut f. Virus-
forschung, Abt. Biochemie
Spemannstr. 35
D-7400 Tübingen
Germany

Karl Seeger (481)
Hoechst AG
H 811
Postfach 80 03 20
D-6230 Frankfurt 80
Germany

Peter H. Seidl (299)
Lehrstuhl f. Mikrobiologie
Techn. Universität München
Arcisstr. 21
D-8000 München 2
Germany

Gerald D. Shockman (165)
Dept. of Microbiology
Temple University School of
Medicine
Philadelphia, Pa. 19140
U.S.A.

Brian G. Spratt (403)
Microbial Genetics Group
School of Biological Sciences
University of Sussex
Brighton, BN1 9QG
England

Jack L. Strominger (349)
Dept. of Biochemistry and
Molecular Biology
Harvard University
Cambridge, Mass. 02138
U.S.A.

Hideho Suzuki (583)
Faculty of Science
Dept. of Biology
The University of Tokyo
Hongo, Tokyo 113
Japan

Rudolf Then
Hoffmann-la Roche AG
Pharmazeutische Forschungs-
abteilung
CH-4002 Basel
Switzerland

Alexander Tomasz (155)
The Rockefeller University
1230 York Avenue
New York, N.Y. 10021
U.S.A.

Shigeo Tomioka (505)
Institute of Applied Micro-
biology
University of Tokyo
Bunkyo-ku
Tokyo 113, Japan

Endre Vasstrand
Dept. of Biochemistry
University of Bergen
Arstadveien 19
N-5000 Bergen
Norway

David Vazquez (433)
Instituto de Bioquimica de Macro-
moleculas, Centro de Biologia Mol.
Universidad Autonoma de Madrid
Canto Blanco
Madrid-34 Spain

J. Barrie Ward (551)
Glaxo Group Research Limited
Greenford Road
Greenford, Middlesex, UB6 OHE
England

Jörg Wecke (329)
Robert Koch-Institut des
Bundesgesundheitsamtes
Nordufer 20
D-1000 Berlin 65
Germany

Frans Wientjes (459)
Dept. of Electron Microscopy
and Molecular Cytology
University of Amsterdam
Plantage Muidergracht 14
1018 TV Amsterdam The Netherlands

Brian J. Wilkinson (217)
Dept. of Biological Sciences
Illinois State University
Felmley Hall 206
Normal Ill. 61761
U.S.A.

Russell Williamson (487)
Laboratoire de Microbiologie
Institut Biomedical des
Cordeliers
15, rue de l'Ecole de Medicine
75270 Paris, Cedex 06 France

Gudrun Wolf (595)
Institut f. Lebensmittel-
technologie d. Universität
Hohenheim
Garbenstr. 25
D-7000 Stuttgart 70 Germany

Jan T.M. Wouters (619)
Laboratorium voor Microbiologie
Universiteit van Amsterdam
Nieuwe Achtergracht 127
1018 WS Amsterdam
The Netherlands

Anne Wyke (543)
National Institute of Medical
Research, Division of Micro-
biology
Mill Hill
London NW7 1AA England

INTRODUCTION

THE BACTERIAL CELL WALL - STRUCTURE AND FUNCTION

Milton R. J. Salton
Department of Microbiology
New York University School of Medicine
New York, NY 10016, USA

Introduction

The history of our current interest in bacterial cell walls can be traced all the way back to the birth of microbiology with Antony van Leeuwenhoek's description of the microscopic morphology of bacteria and other microorganisms in 1675. Leeuwenhoek accurately recorded the characteristic shapes of bacteria and obviously anticipated that surface structures were responsible for their shape, one of the main features of the cell wall still recognized today. The cell surface envelope structures are specially important for free-living bacteria as rigid protective barriers against the adverse environments they may encounter. Early attempts, prior to the turn of the century, to experimentally establish a cell wall in bacteria and to probe its chemical nature stemmed from two types of approach as exemplified by Fischer's (1891) "physiological" experiments on plasmolysis and attempts to chemically characterize the resistant rod-shaped structures of _Bacillus subtilis_ by Vincenzi (1887). The uncertainties of the traditional bacterial cytologists efforts to understand the chemical structure and function of the cell wall ended in the mid 1940's to early 1950's when a new era of the application of electron microscopy and biochemistry to subcellular fractionation ushered in the rapid advance of our knowledge of wall structure and function. In looking back over some of the major milestones in the devel-

The Target of Penicillin

Table I. Walls - Lysozyme - Penicillin

1. Vincenzi, 1887 - First attempt at cell wall chemistry
2. Fleming, 1922 - Lysozyme and bacteriolysis
3. Fleming, 1929 - First observations on penicillin
4. Epstein & Chain, 1940 - First chemical study of lysozyme substrate
5. Chain *et al.*, 1940 - Isolation of penicillin
6. Duguid, 1945 - Penicillin acts on walls
7. Park, 1952 - Discovery of Park nucleotides
8. Strange, 1956 - Muramic acid

opment of this field (Table I), studies of walls, lysozyme and penicillin became intertwined and finally integrated into a new unified understanding of wall structure and biosynthesis. At the forefront of this era, the outstanding and incisive contributions of Professor Wolf Weidel and his colleagues will long be remembered. It is fitting that we pay tribute to Wolf Weidel, who in collaboration with Pelzer developed the concept of the "Murein Sacculus" of bacterial walls as "Bagshaped Macromolecules." The significance of the chemical studies of walls in the mid 1950's was placed in perspective when Park and Strominger (1) proposed the biochemical basis for the mechanisms of action of penicillin and its selective toxicity and Weidel and Primosigh (2) drew attention to the biochemical parallels between phage and penicillin lysis. The exciting developments following this period form the solid basis on which the newer and equally exciting approaches of today are emerging to give us new insights into wall structure and biochemistry.

Wall Ultrastructure

Although we are often prone to think about cell walls and envelopes of Gram-negative bacteria as the flat collapsed structures we see in electron micrographs of isolated walls,they constitute the external compartment of the cell and functionally should perhaps be regarded more as an organelle rather than a simple rigid corset surrounding the more fragile plasma membrane. As seen in thin sections, the cell wall profile of many Gram-positive bacteria has a relatively thick, amorphous appearance. Occasional evidence of layered and fibrillar structures is seen in some. Profiles of the mycobacterial cell wall, however, are much more complex, showing distinct layering with the murein or peptidoglycan layer forming the innermost of four sheetlike structures (3). Fibrous, ropelike structures (peptidoglycolipid) although not resolved in thin-section profiles can be readily seen by negative-staining and freeze-fracture techniques and they appear to be a general ultrastructural feature of the walls of all species of mycobacteria (3). These surface components can be extracted with ethanolic-KOH revealing a residual smooth (murein) layer (3). These changes in the mycobacterial wall following extractions with ethanol-KOH or pyridine, contrast with the lack of any changes seen in the surface of the Gram-positive organism, *Listeria monocytogenes* following extraction of its lipopolysaccharide-like component with phenol (4). This suggests that the organization of this LPS-like macromolecule in the cell surface of *L. monocytogenes* differs from the arrangement of LPS in the outer membranes of Gram-negative species.

The surface topography of Gram-negative bacteria has been well documented by thin sections, showing under appropriate conditions the densely-stained profile of the peptidoglycan layer. Freeze-fracture of the envelopes of Gram-negative species also shows the complexity of their surface structures, but with the principal fracture faces passing through the plasma membrane (5).

Ultrastructural studies of isolated walls can also reveal further details of the organization of the peptidoglycan in cells as seen in recent experiments carried out with Dr.

K.-S. Kim on dividing Micrococcus agilis. The cross- wall septum possesses an annular region in the center with a fibrillar-like appearance, perhaps representing the centripetal growth and assembly of the peptidoglycan. Similar annular structures have also been previously documented in Staphylococcus aureus walls by Giesbrecht et al. (6).

Chemical Components of Bacterial Walls

Peptidoglycan or murein constitutes the major structural polymer of the walls of both Gram-positive and Gram-negative bacteria and the recently discovered pseudomurein of the Archaebacteria containing N-acetyltalosaminuronic acid instead of N-acetylmuramic acid in the glycan moiety forms the analogous structure in these organisms (7,8). In Gram-positive bacteria such as M. lysodeikticus the peptidoglycan may account for as much as 95% of the wall, the remaining 5% being accounted for by the associated teichuronic acid composed of N-acetylmannosaminuronic acid and glucose (9). Walls of Staphylococcus aureus and Bacillus subtilis possess high contents of teichoic acids and correspondingly lower contents of peptidoglycan (40-50%). In addition to teichoic acids as wall associated polymers, other bacterial species contain polysaccharide components and the mycobacteria contain an array of complex lipids and peptidoglycolipids. Some of the features of the walls of Gram-positive bacteria are summarized in Table 2. In Gram-negative bacteria the peptidoglycan accounts for a much smaller fraction of the total envelope although its chemical structure is remarkably similar throughout these organisms (11).

Biological Properties of Wall Components

The walls of Gram-positive bacteria and the envelopes of

Table 2. Chemical Components of Cell Walls of Selected Gram-positive Bacteria (modified from Rogers et al.,ref. 10)

Bacillus licheniformis	Peptidoglycan, teichoic acid, teichuronic acid
Bacillus subtilis	Peptidoglycan, teichoic acid, teichuronic acid
Micrococcus lysodeikticus	Peptidoglycan, teichuronic acid
Lactobacillus spp.	Peptidoglycan, teichoic acids, polysaccharides
Staphylococcus aureus	Peptidoglycan, teichoic acid, teichuronic acid
Mycobacteria spp	Peptidoglycan, peptidoglycolipids, trehalose dimycolate, acylglucoses (ref. 3)

Gram-negative possess a variety of macromolecular components exhibiting a remarkable array of biological activities and responses in many animal species. The pioneering work of Lederer and his colleagues leading to the elucidation of the minimum structure of peptidoglycan units (muramyldipeptides) responsible for adjuvant effects is a remarkable achievement. Moreover, as shown by Krause and Schleifer and their colleagues peptidoglycan induces formation of antibody populations of limited heterogeneity. In addition to immunostimulatory and pyrogenic effects of peptidoglycans, Krueger *et al.*(12) have made the fascinating discovery that muramyl peptides have somnogenic effects. It is suggested that receptors in the brain have a high affinity for muramylpeptides. A variety of wall polysaccharides, including those of *M.lysodeikticus* and *Propionibacterium acnes* are powerful immunomodulators and possess anti-tumor properties. Lipopolysaccharides of Gram-negative bacteria and of the Gram-positive *L. monocytogenes* have long been known to possess a great variety of immunological and biological responses in humans and animals.

Walls and Plasma Membranes

Since there can be no bacterial cell wall without the existence of the plasma membrane it is only fair to mention it as the site of some of the terminal events in peptidoglycan biosynthesis and assembly and the site of the penicillin targets, the PBP's. Their specific localization on one or both faces of the plasma membrane remains to be conclusively established.

References

1. Park, J.T., Strominger, J.L.: Science 125, 99-101 (1957).
2. Weidel, W., Primosigh, J.: J. Gen. Microbiol. 18, 513-517 (1958).
3. Barksdale, L., Kim, K.-S.: Bacteriol. Rev. 41, 217-372 (1977).
4. Wexler, H., Oppenheim, J.D.: Infect. Immun. 23, 845-857 (1979).
5. Kim, K.-S., Salton, M.R.J., Lev, M.: Microbios 29, 45-61 (1980).
6. Giesbrecht, P., Wecke, J., Reinicke, B.: Int. Rev. Cytol. 44, 225-317 (1976).
7. Konig, H., Kandler, O.: Arch. Microbiol. 123, 295-299 (1979).
8. Kandler, O.: Bakt. Hyg. I. Abt. Orig. C3, 149-160 (1982).
9. Perkins, H.R.: Biochem. J. 86, 475-483 (1963).
10. Rogers, H.J., Perkins, H.R., Ward, J.B.: Microbial Cell Walls and Membranes, Chapman Hall N.Y., pp. 1-56 (1980)
11. Schleifer, K.H., Kandler, O.: Bacteriol. Rev. 36, 407-477 (1972)
12. Krueger, J.M., Pappenheimer, J.R., Karnovsky, M.L.: Proc. Natl. Acad. Sci. USA 79, 6102-6106 (1972)

PART I

PRIMARY AND THREE DIMENSIONAL STRUCTURE OF MUREIN

PRIMARY STRUCTURES OF MUREIN AND PSEUDOMUREIN

Karl Heinz Schleifer
Lehrstuhl für Mikrobiologie, Technische Universität München,
D-8000 München 2

Otto Kandler
Botanisches Institut der Universität München,
D-8000 München 19

Murein is the main cell wall polymer of most eubacteria. It is a heteropolymer consisting of polysaccharide chains which are cross-linked through short peptides (Fig. 1).

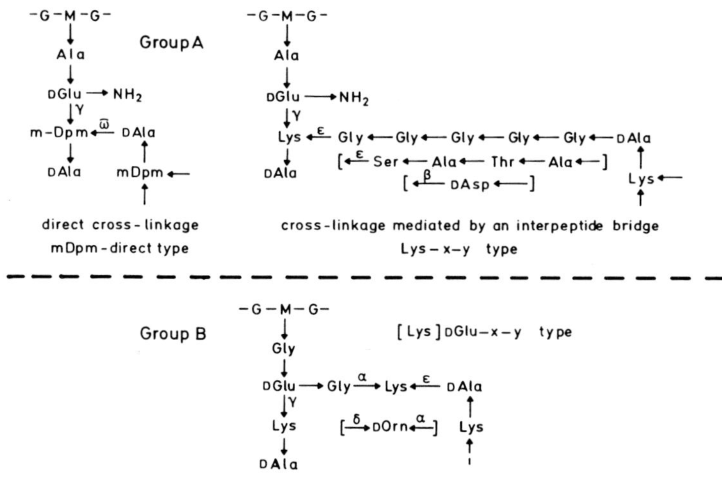

Fig. 1. Primary structures of group A and B mureins. (Designation of murein types according to Schleifer and Kandler 1972).

The Target of Penicillin

The glycan moiety of murein resembles chitin. Like chitin it is made up of alternating ß(1,4)-glycosidically linked units of N-acetylglucosamine. However, in contrast to chitin, each alternate glucosamine residue is substituted by a D-lactic acid ether at C-3. This derivative of glucosamine is called muramic acid. The carboxyl group of muramic acid is substituted by a peptide containing alternate L- and D-amino acids. Adjacent stem peptides may be cross-linked by a peptide bond between the carboxyl group of a D-alanine residue and the distal amino group of a diamino acid residue either directly or via an interpeptide bridge. This gives rise to a huge macromolecule encompassing the entire bacterial cell. The main characteristics of the murein structure can be briefly summarized as follows:

1. The glycan part is rather uniform. It exhibits only few variations, such as acetylation or phosphorylation of the muramyl 6-hydroxyl groups or the occasional absence of N-acyl or peptide substituents. Muramic acid is always of D-gluco configuration. Moreover, the muramic acid residues of some coryneform bacteria and mycobacteria may occur as N-glycolyl derivatives.

2. Depending on the mode of cross-linking, one can distinguish two main groups, named A and B. The cross-linkage of group A mureins extends, as shown in Figure 1, from the distal amino group of the diamino acid in position 3 of one stem peptide to the carboxyl group of D-alanine in position 4 of an adjacent stem peptide. Group B cross-linkage extends between the α-carboxyl group of D-glutamic acid in position 2 of one stem peptide and the D-alanine residue in position 4 of an adjacent peptide.

3. The number of amino acids occurring in various mureins is rather restricted. From 3 to 6 chemically different amino acids can be found in a particular murein. Branched and aromatic amino acids or cysteine, methionine, arginine,

histidine and proline have never been found.

4. The stem peptide always consists of alternating L- and D-amino acids. Even meso-diaminopimelic acid is integrated into this pattern and bound with its L-asymmetric center.
5. Only a restricted variation is found within the stem peptide. The greatest variation could be found within position 3. Up-to-now ten chemically different amino acids have been detected in this position, whereby meso-diaminopimelic acid and L-lysine are the most frequent ones. The α-carboxyl group of D-glutamic acid is usually amidated but in a few rare cases other substituents can also occur. Extensive investigation of the murein structure (Schleifer and Kandler, 1972) of many different bacteria leads to the following classification system (Fig. 2).

Fig. 2: Classification system of the murein

Classification	Abbreviation	Basis for classification
Groups	Roman capital letter (A,B)	Class of cross-linking
Subgroups	Arabic figures (1 - 5)	Kind of interpeptide bridge or lack of it
Variations	Greek letter (α- δ)	Amino acid present in position 3 of the stem peptide
	Prime (')	Replacement of Ala in position 1 of the stem peptide of group A peptidoglycans by Gly
Types	Depending on the different amino acid sequences of the peptide moiety. In particular, the amino acid sequence of the interpeptide bridge.	

A more detailed description of the different subgroups and variations of mureins can be found in the recent review by Schleifer and Stackebrandt (in press).

Until the middle of the seventies it was thought that murein is one of the key characters defining the Kingdom Prokaryota. There were only a few prokaryotes known such as mycoplasma or halobacteria which did not contain murein. However, in recent years the number of non-murein containing bacteria increased rapidly following the detection of the so-called archaebacteria. Comparative analysis of 16S rRNA demonstrated that archaebacteria are no more related to other bacteria (eubacteria) than they are to eukaryotic cells (Woese and Fox, 1977). One of the characteristic phenotypic features of the archaebacteria is the lack of murein in their cell walls. A survey of the various cell wall structures of archaebacteria is given in the Table.

Table: Cell wall and cell envelope structures in archaebacteria

Organism	Rigid Sacculus	Protein envelope	Polymer
Methanobacteriaceae	+	-	Pseudomurein
Methanosarcina	+	-	Heteropolysaccharide
Other Methanogens	-	+	Protein or Glycoprotein
Halobacterium	-	+	Sulfated Glycoprotein
Halococcus	+	-	Sulfated Heteropolysaccharide
Sulfolobus	-	+	Glycoprotein
Thermoproteus	-	+	Glycoprotein
Thermoplasma	-	-	none

Pseudomurein is found only in Methanobacteriales (Kandler, 1979, 1982). A comparison of the primary structures of murein and pseudomurein is shown in Figure 3.

Fig. 3: Fragments of the primary structures of murein and pseudomurein

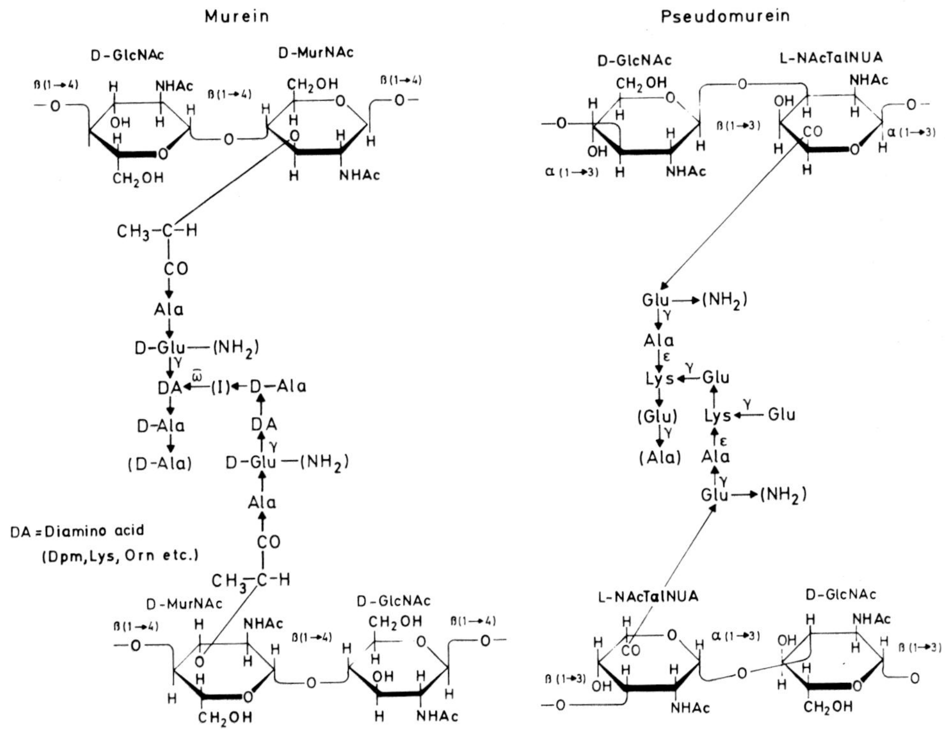

Both are peptidoglycans which are composed of glycan strands cross-linked through short peptides. The glycan strands consist of alternating N-acetyl-D-glucosamine and N-acetyl-L-talosaminuronic acid residues which are ß-1,3 and α-1,3 glycosidically linked, respectively (König et al., in press). The amino acids of the peptide moiety exhibits in contrast to that of the murein exclusively L-configuration. In contrast to typical proteins the distal carboxyl and amino groups of glutamic acid and lysine, respectively, are involved in peptide formations. Pseudomurein with slight modifications could be detected in some species of the order Methanobacteriales (König et al., 1982). Since carbons 3 of D-glucosamine and D-galactosamine possess the same anomeric configuration, it is not astonishing that a mutual replacement is possible. In some strains a mixture of both amino sugars is found. In the pseudomurein of *Mbr. ruminantium* an additional glucosamine is linked to the glycan moiety via a phosphodiester bond.

This strain also contains a pseudomurein in which at least part of the alanine is replaced by threonine. Ornithine is found as an additional component of the pseudomurein of *Mbr. smithii*.

A comparison of the main properties of murein and pseudomurein can be summarized as follows: In the murein the hexosamines possess the D-gluco-configuration, whereas pseudomurein contains hexosamines of the D-gluco- or D-galacto-configuration and an acid derivative of the L-talo-configuration.

The constituents of murein are ß-1,4-linked, those of pseudomurein are ß-1,3 or α-1,3 linked. A replacement of glucosamine by galactosamine in the pseudomurein does not change the structure of a 1,3-glycan. Such a replacement is not possible in the murein since carbons 4 of the two hexosamines exhibit a different anomeric configuration and a replacement of glucosamine by galactosamine would lead to a change of the secondary structure of the glycan strands. Pseudomurein is not attacked by lysozyme since the linkage is different and muramic acid is replaced by L-talosaminuronic acid.
The stem peptides of both polymers consist of a similar set of amino acids but their molar ratios and sequences are quite different. Moreover, murein contains both L- and D-amino acids whereas only L-amino acids are present in pseudomurein.

Cross-linking of the stem peptides proceeds via diamino acids in both polymers.
The resistance against proteases is mainly caused by alternating L- and D-amino acids in the murein or by unusual peptide bonds in pseudomurein. Pseudomurein is resistant to all cell wall-targeted antibiotics, such as ß-lactams, D-cycloserine or vancomycin , which interfere with reactions involving D-amino acid residues. It is, however, sensitive to antibiotics interfering with the lipid cycle of the murein synthesis such as bacitracin or nisin (Hilpert et al., 1981). This indicates that also undecaprenyl-lipid-bound intermediates are involved in the biosynthesis of pseudomurein.

References

1. Hilpert, R., Winter, J., Hammes, W., Kandler, O. Zbl. Bakt. Hyg. I. Abt. Org. C2, 11-20 (1981)
2. Kandler, O.: Naturwissenschaften 6, 95-105 (1979)
3. Kandler, O.: Zbl. Bakt. Hyg. I. Abt. Orig. C3, 144-160 (1982)
4. König, H., Kralik, R., Kandler, O.: Zbl. Bakt. Hyg. I. Abt. Orig. C3, 179-191 (1982)
5. König, H., Kandler, O., Jensen, M., Rietschel, E.Th., Hoppe Seyler (in press)
6. Schleifer, K.H., Kandler, O.: Bacteriol. Rev. 36, 407-477 (1972)
7. Schleifer, K.H., Stackebrandt, E. Ann. Rev. Microbiol. 37 (in press)
8. Woese, C.R., Fox, G.E.: Proc. Natl. Acad. Sci. (Wash.) 74, 5088-5090 (1977)

STRUCTURE OF THE PEPTIDOGLYCAN OF THE CYANOBACTERIUM *SYNECHOCYSTIS* PCC 6714

Uwe Johannes Jürgens, Gerhart Drews, Jürgen Weckesser

Institut für Biologie II, Mikrobiologie, Albert-Ludwigs-Universität, Freiburg
D-7800 Freiburg i. Br.

Dieter Naumann

Institut für Kristallographie der Freien Universität Berlin,
D-1000 Berlin 33

Introduction

The rigid layer of *Synechocystis* sp. strain PCC 6714 consists of a polymer containing mannosamine (ManN), galactosamine (GalN), mannose (Man), glucose (Glc) and phosphorus in addition to the constituents of the A1-type of peptidoglycan (2-5). The presence of non-peptidoglycan components which is also known for other cyanobacteria (1) has been hampering structural studies on the Synechocystis peptidoglycan. This article describes experiments on composition, primary and secondary structure of peptidoglycan.

Results

1) *Chemical composition of peptidoglycan.* The rigid layer was obtained from *Synechocystis* sp. strain PCC 67144 by extraction of cell walls with hot sodium dodecyl sulfate (SDS). Non-peptidoglycan amino acids, amounting to 3.4 % of the dry weight of the SDS-insoluble fraction of the cell wall, were completely removed by digestion with

The Target of Penicillin

pronase and trypsin, respectively, including lysine and arginine. The protease-treated fraction contained MurN, GlcN, L-Ala, D-Ala, D-Glu and meso-A_2pm in molar ratios of 0.9:1.5:1.1:0.8:1.0:1.0 (Tab. 1). In addition, ManN, GalN, Man, Glc and phosphorus were found in this fraction. Removal of this additional polymer and of non-peptidoglycan GlcN could be achieved by treatment with hydrofluoric acid in the cold (48 %, $0^{\circ}C$, 48 h). The result suggests that a charged polysaccharide is very likely bound to the peptidoglycan via phosphodiester bonds.

The amino sugars of the peptidoglycan backbone are completely N-acetylated. Only small amounts of O-acetyl groups were detected. There is evidence that the peptide side chains bear at least a partial amidation of carboxy groups, since amide ammonia is easily set free in high amounts during the first hour of acid hydrolysis: molar ratios of D-Glu-meso-A_2pm-NH_3 = 1.0:0.7:1.5 were found. D-Ala and meso-A_2pm were the only C-terminal amino acids found on hydrazinolysis of peptidoglycan. Thus, tripeptides L-Ala-D-Glu-meso-A_2pm and tetrapeptides L-Ala-D-Glu-meso-A_2pm-D-Ala are suggested to be present as structural components of the peptidoglycan of <u>Synechocystis</u>.

Tab. 1: Chemical composition of peptidoglycan fractions

Component	Amt in peptidoglycan fraction (nmol per mg of fraction dry weight)		
	Pronase-digested	Trypsin-digested	Pronase-digested and hydrofluoric acid (48%, 0 C, 48 h) treated
Amino sugars			
MurN	520	514	837
GlcN	890	876	953
ManN	222	225	-[a]
GalN	15	16	-
Amino acids			
D-Ala	504	503	812
L-Ala	669	667	1077
D-Glu	604	598	959
meso-A_2pm	642	632	933
Neutral sugars			
Man	313	289	-
Glc	33	32	-
Acetyl groups			
N-Acetyl	1402	1372	1346
O-Acetyl	12	10	15
Phosphorus	150	139	-

[a] -, None

2) <u>Isolation and characterization of peptides.</u> Peptides were formed by the partial acid hydrolysis (4 M HCl, $100^{o}C$, 30 min) of peptidoglycan and were separated by combined low-voltage thin layer electrophoresis and thin layer chromatography. Eight peptides were isolated from fluorescamine stained cellulose plates and were characterized by amino acid analysis, gas-liquid chromatography (Fig. 1), and dinitrophenylation studies. They were identified as the dipeptides MurN-L-Ala, L-Ala-D-Glu, γ-D-Glu-meso-A_2pm, meso-A_2pm-D-Ala (mediating direct cross-linkage), meso-A_2pm-D-Ala (terminating the tetrapeptide), tripeptides MurN-L-Ala-D-Glu-meso-A_2pm, L-Ala-D-Glu-meso-A_2pm, and a tetrapeptide L-Ala-D-Glu-meso-A_2pm-D-Ala. The cross-linking dipeptide meso-A_2pm-D-Ala was also found to be sensitive upon degradation with a specific endopeptidase (from E. coli, kindly given by Dr. Höltje).
The peptide analyses show that the peptidoglycan of <u>Synechocystis</u> belongs to the A1γ-peptidoglycan type.

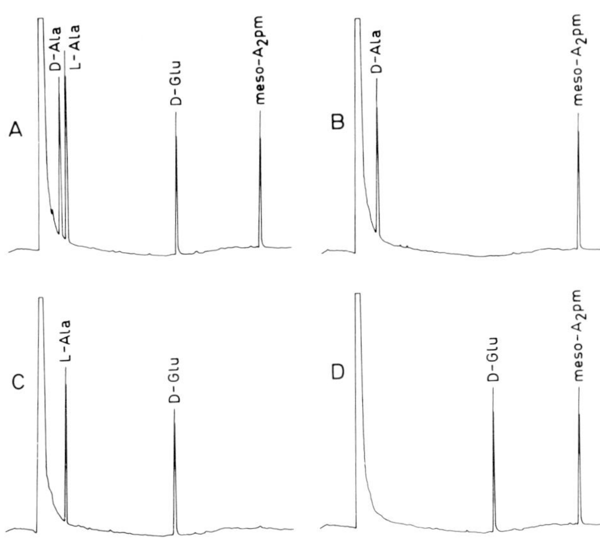

<u>Fig. 1</u>: Separation of amino acid isomers of major peptides by gas-liquid chromatography on a Chirasil-Val glass capillary column as N-Heptafluoro-butyryl-isobutylester derivatives.
(A) L-Ala-D-Glu-meso-A_2pm-D-Ala
(B) meso-A_2pm-D-Ala
(C) L-Ala-D-Glu
(D) D-Glu-meso-A_2pm

3) Degree of cross-linkage. The degree of cross-linkage of the *Synechocystis* peptidoglycan was investigated by two different methods. One is based on 1-fluoro-2,4-dinitrobenzene (FDNB)-treatment (7) of the peptidoglycan which formed mono-2,4-dinitrophenyl (DNP)-diaminopimelate as the only DNP-derivative. About 44 % of the total meso-A_2pm present in the *Synechocystis* peptidoglycan was accessible to the FDNB-reaction. This shows that in 56 % of the peptide side chains the ε-amino groups of meso-A_2pm were blocked as a result of cross-linkage.
Since the amount of C-terminal carboxylic end groups is related to the degree of cross-linkage, the cross-linking index was determined by infrared (IR) spectroscopy (6). The relative absorbance of the group frequency band of carboxylic acid located at about 1730 cm^{-1} in the infrared spectrum (Fig. 2) of the acid form of the *Synechocystis* peptidoglycan was used for the calculation of the degree of cross-linkage. The sugar ring band near 1040 cm^{-1} served as an internal standard (marker for the total amount of subunits present). From the IR-data a cross-linking index of approximately 43 % was calculated. However, this value was not corrected, because the degree of amidation is not yet known.

4) Infrared spectroscopy. The infrared spectrum of the peptidoglycan (acid form) of *Synechocystis* sp. strain PCC 6714 (fig. 2) showed amide A, B, I and II absorption bands centered around 3291 cm^{-1}, 3080 cm^{-1}, 1657 cm^{-1} and 1534 cm^{-1}, respectively. They are very sensitive to conformational alterations in the peptide backbone. Since only one amide I and one amide II band occurs in the infrared spectrum of the peptidoglycan, it is likely that the two -NH and two -C=O groups of the carbohydrate moiety and the five -NH and five -C=O groups of the peptide moiety contribute to the same conformation-dependent amide I (C=O stretching)

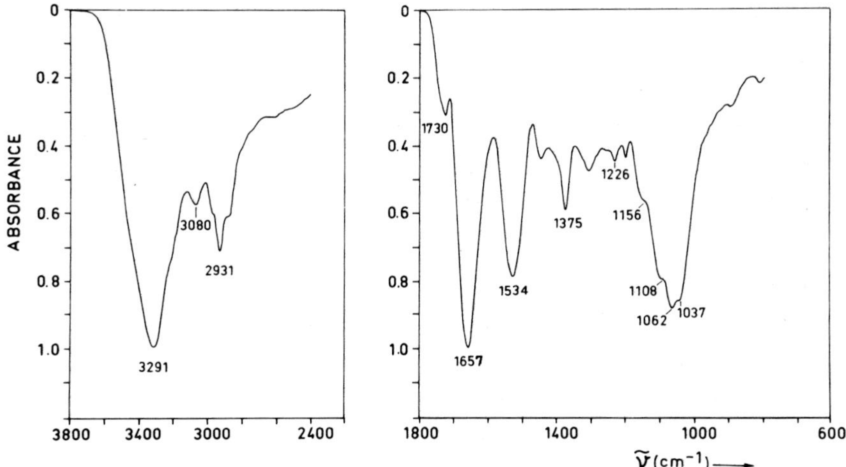

Fig. 2: Infrared spectrum of peptidoglycan recorded on a Fourier-transform/infrared spectrophotometer FT-IR MX-1E.

and amide II (N-H deformation) absorption bands. An absorption band near 1730 cm^{-1} is assigned to hydrogen-bonded >C=O vibrators of carboxylic acid groups (-COOH). The spectral region between 1200 cm^{-1} and 800 cm^{-1} is characterized by several strong absorption bands which are dominated by complex ß(1→4)-linked sugar ring modes. The analysis of band half-widths revealed no crystalline state of order of peptidoglycan.

5) X-Ray diffraction. The exposure of dried powders or dried foils of the Synechocystis rigid layer face-on to X-rays resulted in a Debye-Scherrer-like pattern with Bragg periodicities of about 0.45 nm and 0.94 nm (Fig. 3). Both rings were relatively diffuse. The 0.7 nm-periodicity known for other peptidoglycans, which occured only after cell breakage (6) was not detected in whole cells or in peptidoglycan fractions of Synechocystis. The widths of the interference rings indicate that a portion of the peptidoglycan exhibits a partially ordered although non-crystalline structure. The X-ray diagram of the Synecho-

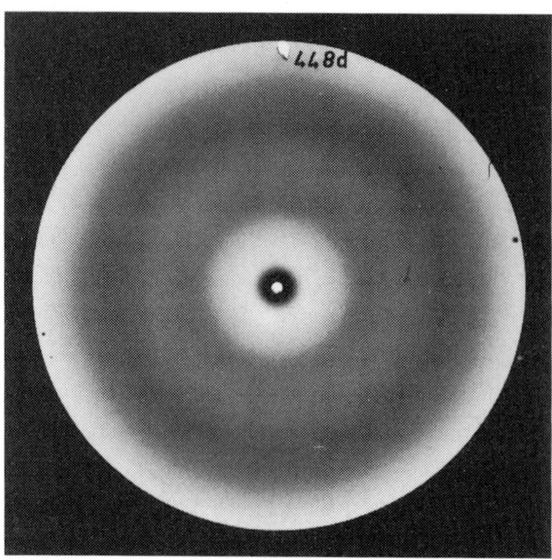

Fig. 3: X-Ray diffraction pattern obtained with Ni-filtered CuK_{α}-radiation in a Kiessig camera from powders or foils of isolated rigid layers from *Synechocystis* sp. strain PCC 6714.

cystis peptidoglycan foil, when viewed from the side (edge-on), showed no meridional reflection (i.e. oriented vertically to the foil plane) corresponding to a periodicity of 4.2 nm as known for *S. aureus* peptidoglycan (6).

References

1. Drews, G., Weckesser, J.: In N.G. Carr and B.A. Whitton (ed.). The biology of cyanobacteria. Blackwell Scientific Publications. Oxford. Great Britain, 333-357 (1982).
2. Jürgens, U.J., Weckesser, J., Drews, G.: Abstr. FEMS Symp. Microbial Envelopes, Saimaanranta, 83 (1980).
3. Jürgens, U.J., Weckesser, J., Drews, G.: Abstr. Int. Symp. Photosynth. Prokaryotes, Bombannes, C 10 (1982).
4. Jürgens, U.J., Warth, R., Weckesser, J.: Zbl. Bakt. Hyg. I. Abt. Orig. C 3, 539 (1982).
5. Jürgens, U.J., Drews, G., Weckesser, J.: J. Bacteriol. 154, in press (1983).
6. Naumann, D., Barnickel, G., Bradaczek, H., Labischinski, H., Giesbrecht, P.: Eur. J. Biochem. 125, 505-515 (1982).
7. Takebe, J.: Biochim. Biophys. Acta 101, 124-126 (1965).

Characterization of Muropeptides Released from Murein of *Caulobacter crescentus* by Endo-N-acetylmuramidase

Zdzislaw Markiewicz
Instytut Mikrobiologii Uniwersytetu Warszawskiego, Nowy Swiat 67, Warsaw, Poland

Bernd Glauner and Uli Schwarz
Max-Planck-Institut für Virusforschung, Spemannstrasse 35, D-7400 Tübingen, F.R.G.

Introduction

The dimorphic bacterium *Caulobacter crescentus* is an attractive prokaryotic model organism for the study of differentiation since it undergoes an obligatory cell cycle in which cell division results in two morphologically distinct cell types, i.e., a swarmer cell and a stalked cell (1). The shape of a bacterial cell is assumed to be maintained by the murein sacculus which completely encloses it and in several cases changes in cell morphology have been shown to be correlated with changes in the activity of murein hydrolases and in murein structure (2, 3, 4). The murein of *C. crescentus* has been described as a typical gram-negative A1γ type (5, 6) though more recent studies indicate a certain modification of this primary structure being involved in morphogenesis of the stalk (7). In the present study we characterized the soluble products released from the murein of *C. crescentus* CB15 by endo-N-acetylmuramidase from *Chalaropsis*. Our findings indicate that the composition of C. crescentus murein visibly differs in some respects from that of other gram-negative bacteria with a similar type of murein - A1γ.

The Target of Penicillin

Results

1) Identification of pentapeptide subunits. A survey for murein hydrolyzing enzymes in *C. crescentus* CB15 revealed the presence of several different activities including a transglycosylase, endopeptidase and amidase. However, no detectable DD-carboxypeptidase and LD-carboxypeptidase activities were found (Markiewicz, unpublished observations). To verify the absence of carboxypeptidase activity in *C. crescentus*, murein sacculi labelled with [^{3}H]diaminopimelic acid, prepared essentially as described (7), were treated with endo-N-acetylmuramidase from *Chalaropsis* and the soluble products were separated by paper chromatography (8). The distribution of the label on the chromatograms closely resembled that for *Escherichia coli*. However, assay for vancomycin binding material according to de Pedro and Schwarz (9) revealed very high pentapeptide content in the three fastest migrating spots. This result was confirmed by HPLC analysis of the digestion products, followed by amino acid analysis of the separated murein subunits, which showed that approximately 50% of all monomeric subunits and 57% of all dimeric muropeptides had pentapeptide side chains. The absence of LD-carboxypeptidase activity was verified by the complete lack of disaccharide tripeptide among the murein subunits. These results clearly indicate that the synthesis of cross-linked murein in C. crescentus does not require carboxypeptidase activity. This is in contrast to Gaffkya *homari* in which the pretreatment of nascent murein with carboxypeptidase is necessary for the occurrence of transpeptidation, or even to *E. coli* where carboxypeptidase has been suggested to have a regulatory function in determining the synthesis of septum-forming or elongation murein (3). The absence of carboxypeptidase in Caulobacter may be related to the cell cycle of this bacterium and may allow for greater plasticity of the murein needed for the transformation

from one morphological type to the other.

2) Substitution of terminal alanine by glycine. Amino acid analysis of C. crescentus murein from bacteria grown in various media revealed the presence of low but reproducible amounts of glycine which were put down to contamination of the isolated sacculi by protein (5, 7). HPLC analysis of the soluble products of murein digestion demonstrated the presence of glycine in approximately half of the pentapeptide carrying subunits. Treatment of the muropeptides with carboxypeptidase from E. coli and HPLC analysis of the reaction products showed that glycine is in position 5 of the pentapeptide in both the disaccharide and bis-disaccharide subunits (to be published).

References

1. Poindexter, J.S.: Microbiol. Rev. 45, 123-179 (1981).
2. Fontana, R., Canepari, P., Satta, G.: J. Bacteriol. 139, 1028-1038 (1979).
3. Markiewicz, Z., Broome-Smith, J.K., Schwarz, U., Spratt, B. G.: Nature 297, 702-704 (1982).
4. Rogers, H.J., Taylor, C.: J. Bacteriol. 135, 1032-1042 (1978).
5. Agabian, N., Unger, B.: J. Bacteriol. 133, 987-994 (1978).
6. Schleifer, K.H., Kandler. O.: Bacteriol. Rev. 36, 407-477 (1972).
7. Poindexter, J.S., Hagenzieker, J.G.: J. Bacteriol. 150, 332-347 (1982).
8. Primosigh, K., Pelzer, H., Maas, D., Weidel, W.: Biochim. Biophys. Acta 46, 68-80 (1961).
9. De Pedro, M.A., Schwarz, U.: FEMS Microbiol. Lett. 9, 215-217 (1980).

THE ANALYSIS OF MUREIN COMPOSITION WITH HIGH-PRESSURE-LIQUID CHROMATOGRAPHY

Bernd Glauner and Uli Schwarz
Max-Planck-Institut für Virusforschung, Abt. Biochemie II
D-7400 Tübingen

Introduction

The metabolism of murein in E. coli seems to involve a host of synthesizing and hydrolyzing enzymes (1) which necessarily would imply multiple recognition sites. This contrasts with the accepted model of a simple murein chemistry (2). Consequently, we reinvestigated murein composition with a new technique, high-pressure-liquid chromatography (HPLC).

Results and Discussion

Sacculi were prepared from E. coli cells as described previously, including digestion with amylase and pronase (3). The purified murein was degraded with muramidase from Chalaropsis (4). Before fractionation, the muropeptides were reduced with sodium borohydride to prevent anomerisation of the muramic acid residues. The fractionation (details see legend to Fig. 1) yielded about 60 well separated peaks as shown in Fig. 1. The composition of the peaks as indicated (Fig. 1) has been determined by amino acid analysis and by partial digestion with murein-hydrolases from E. coli (5,6). The "classical" murein subunits C3, C5, and C6 represent about 70% of the total murein. Some of the other minor components such as X and X' (3), a disaccharide tetrapeptide trimer (7), a disaccharide pentapeptide (8) and a dimer consisting of C6 and X' (9) had been

The Target of Penicillin

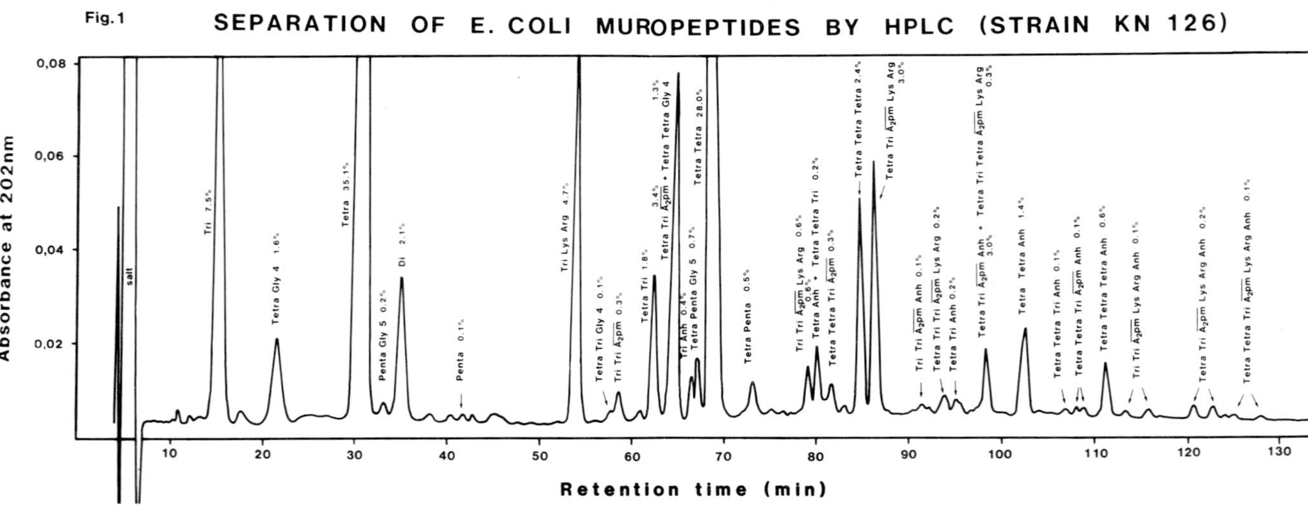

Fig. 1: Sacculi prepared (3) from E. coli KN126 (13) were digested with Chalaropsis muraminidase (4). The muropeptide solution (0.1 ml; 2 mg murein/ml) was reduced with $NaBH_4$ (0.1 ml; 10 mg/ml in 0.5 M borate buffer, pH 9.0; 30' at RT). After adjustment of pH to 4 with H_3PO_4, 50 µl of the solution were fractionated on a reverse phase column (Hibar 250x4, LiChrosorb RP18, 5 µm; Merck, Darmstadt, Germany), equilibrated with 50 mM phosphate buffer, pH 4.69; 0.0005% NaN_3. The eluted compounds (linear gradient of equilibration buffer rising to 15% methanol in the same buffer; 0.5 ml/min; 110 min run time; equipment Waters, Milford, Mass., USA) were detected at 202 nm. The composition of compounds was determined by partial enzymatic hydrolysis and amino acid analysis. Designation of compounds: Di: GlcNAc-MurNAc-L-Ala-D-Glu; Tri: GlCNAc-MurNAc-L-Ala-D-Glu-meso-A_2pm; Tetra: GlcNAc-MurNAc-L-Ala-D-Glu-mesoA_2pm-D-Ala; Penta: GlcNAc-MurNAc-L-Ala-D-Glu-mesoA_2pm-D-Ala-D-Ala; Gly 4(5): glycine in position 4(5) of the peptide chain; Anh: 1-6-anhydro muramic residue; LysArg: Lys-Arg residue at the meso A_2pm from murein-bound lipoprotein; $\overline{A_2pm}$: compound with a mesoA_2pm-mesoA_2pm crossbridge; other oligomers have the normal mesoA_2pm-D-Ala-mesoA_2pm crossbridge. The amounts of the compounds are calculated by integration of the UV-response.

described already, most of the others are novel. Their properties can be summarized as follows: In dimeric and trimeric subunits all possible combinations of different side chains exist. Not only monomeric subunits carry lipoprotein (10), as indicated by residual -lys-arg residues. Dimeric and trimeric subunits as well are linked with lipoprotein, the ratio of lipoprotein on monomers and on dimers is about 2:1. All compounds mentioned also exist in the anhydro-form (3), defining chain ends. That means, that any type of muropeptide can terminate a glycan chain in murein, crosslinked compounds, however, being preferred at both ends (11). On basis of our new data the average chain length is about 35 disaccharide subunits, which is in agreement with previous calculations (9, 11). Finally, a novel type of crossbridge, representing about 1/5 of all crosslinks, was found. Because of its retention time, its amino acid composition and its degradation by E. coli endopeptidase (5) we assume a crossbridge between two diamino pimelic acid residues (meso-2,6 diamino pimelic acid = A_2pm) as it has been found in other organisms (12).

Are the novel compounds accidental or do they fulfill structural and/or regulatory roles in murein assembly and bacterial morphogenesis? Some compounds indeed may be accidental such as subunits with a glycine replacing an alanine residue (Fig. 1). The amount of glycine in murein depends on the medium in which cells grow, which makes an essential function of this subunit quite unlikely. Other compounds, however, play essential roles in murein metabolism. As an example, pulse-labelling experiments _in vivo_ strongly indicate that the novel dimer with the A_2pm-A_2pm crosslinkage is the acceptor for the attachment of lipoprotein to newly synthesized murein. Formation of this very dimer precedes the attachment of lipoprotein (Fig. 2). The kinetics of the two processes match each other. In mutants with impaired attachment of lipoprotein, the amount of the A_2pm-A_2pm dimer is clearly reduced and vice versa (data not shown).

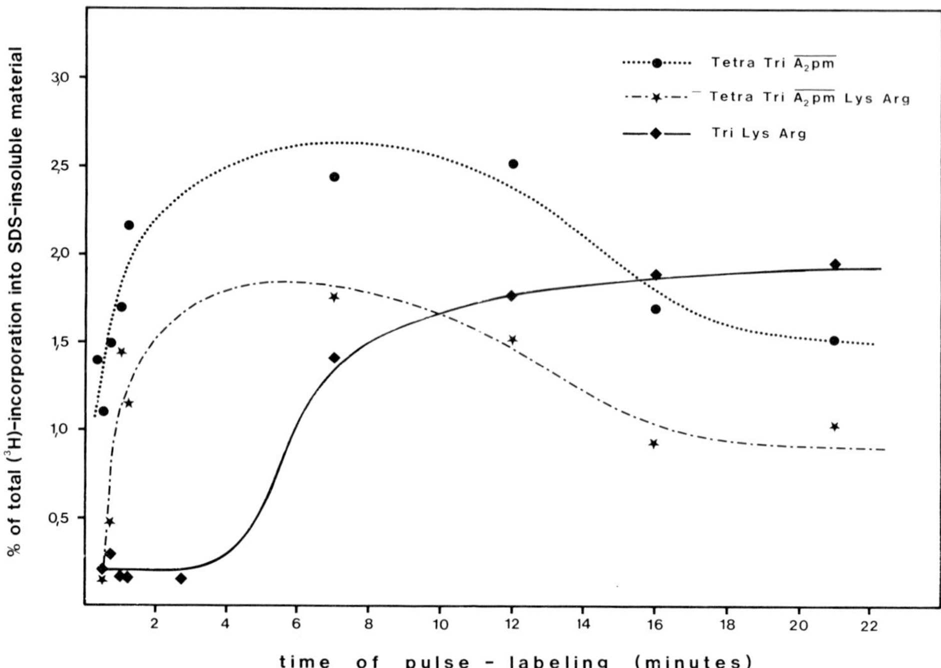

Fig. 2: E. coli W7 (14) was grown at 37°C in supplemented Penassy Medium (Difco, with 20 µg/ml lysine and 2 µg/ml A_2pm); at an OD_{578} of 0.4, 50 ml culture were filtered on a membrane filter (0.45 µM), washed twice with prewarmed, A_2pm-free medium and labelled directly on the filter (up to 60 sec/pulses) with 0.2 ml medium containing ^3H-A_2pm (CEA, Saclay, France; 1 mCi/ml; 50 Ci/mM). For pulses longer than 1 min, cells were resuspended in fresh medium (50 ml; 4 µCi/ml of ^3H-A_2pm), collected by filtration and boiled in 4% SDS (3); analysis of murein composition as described (Fig. 1). Radioactivity was measured in a liquid scintillation counter.

Pulse-labelling experiments revealed another unexpected feature of newly synthesized murein. After very short pulses (20') three murein subunits are found which are rapidly processed. After 60'-pulses their proportional amount in murein becomes negligible (Fig. 3), in aged murein they are undetectable. The composition of these early murein compounds is still unknown.

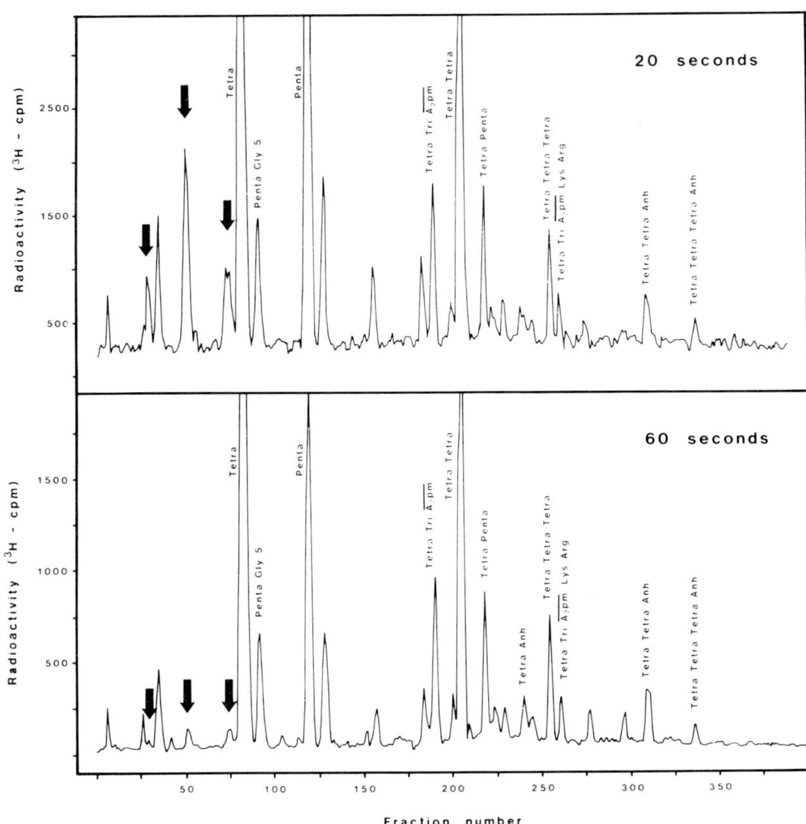

Fig. 3: Murein from E. coli W7 (14) was labelled for 20 and 60 seconds respectively and analyzed by HPLC. Arrows mark "early" compounds with a short life time. For experimental details see legends for Figs. 1 and 2.

Interesting questions arise from the existence of minor murein components. Are they signals recognized by enzymes engaged in murein metabolism? Is part of the regulation of murein biosynthesis and bacterial morphogenesis achieved by substrate modification?

References

1. Höltje, J.-V., Schwarz, U.: In "Molecular Cytology of E. coli, N. Nanninga Ed., Academic Press (London), in press.

2. Weidel, W., Pelzer, H.: Adv. in Enzymol. 26, 193-232 (1964).

3. Höltje, J.-V., Mirelman, D., Sharon, N., Schwarz, U.: J.Bacteriol. 124, 1067-1076 (1975).

4. Hash, J.H., Rothlauf, M.V.: J.Biol.Chem. 242, 5586-5590 (1967).

5. Keck, W., Schwarz, U.: J.Bacteriol. 139, 770-774 (1979).

6. Tamura, T., Imae, Y., Strominger, J.L.: J.Biol.Chem. 251, 414-423 (1976).

7. Gmeiner, J.: J.Bacteriol. 143, 510-512 (1980).

8. De Pedro, M.A., Schwarz, U.: Proc.Natl.Acad.Sci.USA 78, 5856-5860 (1981).

9. Gmeiner, J., Essig, P., Martin, H.: FEBS Lett. 138, 109-111 (1982).

10. Braun, V., Sieglin, U.: Eur.J.Biochem. 13, 336-346 (1970).

11. Schindler, M., Mirelman, D., Schwarz, U.: Eur.J.Biochem. 71, 131-134 (1976).

12. Wietzerbin, J., Das, B.C., Petit, J.-F., Lederer, E., Leyh-Bouille, M., Ghuysen, J.-M.: Biochemistry 13, 3471-3476 (1974).

13. Nagata, T., Horiuchi, T.: Mol.Gen.Genet. 123, 77-88 (1973).

14. Hartmann, R., Höltje, J.-V., Schwarz, U.: Nature (London) 235, 426-429 (1972).

THE ANIONIC NATURE AND SOME PERMEABILITY CHARACTERISTICS OF MUREIN

T.J. Beveridge

Department of Microbiology, College of Biological Science, University of Guelph, Guelph, Ontario, Canada N1G 2W1.

Introduction

With few exceptions all bacteria possess a wall which forms the cell's outer-most boundary and separates its vital constituents from the external milieu. This wall is of fundamental importance since it not only contributes shape and form to the cell, but it must also provide an inanimate boundary through which the cell perceives the surrounding environment and fluctuations therein. Given the stress and strain of the microbial habitat, it is conceivable that this boundary would act as a "microenvironment" which could interact with, buffer, and even modify the stressing influence before it comes in contact with the protoplast. Clearly, the high tensile strength of the wall is able to resist mechanical injury, but recent evidence suggests that it may also have a seiving function (2,13,22-24,26). The wall, to a certain extent, must influence the molecular (22-24) or ionic (2,4,7) form of what gets in and what gets out of cells. But since it takes time to physically alter the chemistry of the framework (i.e., the time for induction → transcription → translation → secretion to occur), a certain degree of accommodation to environmental factors must be built into the pattern. This would help fulfill the economy of design which we expect from all life forms capable of rapid evolution.

It is apparent that the peptidoglycan (PG) which forms the murein sacculus is of fundamental importance to wall integrity; it is the scaffolding upon which the other chemical constituents of the fabric are bonded. There must be mechanisms available to allow expansion (cell elongation) and ingrowth (septation) without effecting apparent imperfections within the strength of its meshwork.

It has been a difficult macromolecule to study. Chemical digests have allowed the identification and molar ratios of its molecular constituents to

The Target of Penicillin

be established (25). Infrared spectroscopy, e^- and X-ray diffraction, and nmr, have presented information concerning the lattice constant and the degree of flexibility within the PG network (1,9,10,15-19). All of these methods have been important for our conceptual view of murein, but we must remember that these are tremendously averaging techniques. Small differences between and within sacculi would not be observed.

The Anionic Nature of Bacterial Walls

Bacterial walls are anionic, and interact with and bind aqueous metallic cations (4-7,14,20). Bacillus walls consist primarily of a PG matrix which is interwoven with linear strands of teichoic and/or teichuronic acids. When B. subtilis 168 walls (54.3% teichoic acid + 44.6% PG) were suspended in 5 mM metal solutions and washed of unbound metal, substantial quantities of metal remained associated with the walls (Table 1). A strict stoichiometric interaction between metallic cation and the available anionic sites within the wall could not account for these amounts (Fig. 1). Presumably, they are the end result of a complex series of chemical reactions initiated by the inaugural binding of cation and are based on the stability of metal in aqueous solution (5-7).

Studies with extracted teichoic polymer indicated that sites were available for metal binding (6), but chemical neutralization of carboxylate groups of PG suggested that this was the primary metal chelating agent (ref.6 and Table 1). Since accurate molar ratios of metal to wall sites was impossible with the salts used in these studies, chloropentaamine osmium III chloride, which is stable in aqueous solution, was used to clarify the issue (5). In this case, the expected stoichiometry between divalent probe and carboxylate groups was observed.

Naively, we felt that these results could be applied to all similar Gram-positive walls. Clearly, this is not the case; our studies on the metal binding capacity of B. licheniformis NCTC 6346 walls (51.7% teichoic acid + 27.4% teichuronic acid + 20.9% PG) presented an entirely different picture (4). These walls did not bind as much metal, nor was PG the principle binding agent (Table 1). In this case, teichuronic and teichoic acids, in concert, developed the major binding capacity of the wall. This points to a fundamental difference between these wall types.

The wall of *Escherichia coli* K-12 contains a layer of PG which is chemically analogous to our bacillus examples (8). This PG can be physically separated to purity and can therefore be compared. In general, the binding capacity of *E. coli* PG is lower than that of *B. subtilis* and *B. licheniformis* (Table 1).

Tenacity of Metal Binding

The binding of metallic cations to walls must be ionic, but it is surprisingly strong. Some of these ions can be replaced by other ions (20); others are more permanently bound (7,14,21). This binding withstands the test of time since walls can be recognized in ancient sediments by their inherent electron scattering ability with electron microscopy (11,12).

We have recently tested the tenacity of metal binding to *B. subtilis* walls by following the biogeochemical events of sediment diagenesis when metal-loaded cells are mixed with a synthetic sediment and incubated under conditions which mimic a low-temperature metamorphic horizon (3). Our initial idea was that the metal would be leached from the wall into the inorganic matrix. In actual fact, the metal remained associated with the cell and cellular material leached metal from the surroundings (Fig. 4). The organic matrix initiated and was responsible for highly distinctive minerals during diagenesis (Fig. 5). Since the PG of the bacteria was responsible for most of the metal input into the system, we assume that these results are a reflection of the tenacity of the PG-metal binding.

Some Permeability Aspects of Murein

We are so familiar with the casual growth and manipulation of bacteria that we sometimes forget the complex series of events that are involved. Most certainly bacterial cells utilize nutrients and grow; they can be plasmolyzed under osmotic stress, and they can be physically stained for microscopic observation. Each of these facts implies physical passage of commodities through the wall. We must distinguish between wall types in a discussion concerning permeability aspects since Gram-positives are emphatically different from the Gram-negative variety. The simplest way to make correlation between the two types is to look at the Gram reaction itself.

The Gram reaction produces a staining response which divides the Eubacteria into two fundamental groups; those that stain Gram-positive and retain the crystal violet/iodide complex (CV/I), and those that are Gram-negative and are decolorized by ethanol treatment. We have determined that the CV/I precipitate is due to the metathetical replacement of the small Cl^- ion of the CV salt with the bulkier I^- of Gram's Iodine solution. We have been able to replace I^- with trichloro (n^2-ethylene) platinum II (TPt^-) to produce CV/TPt which is chemically analogous to the CV/I precipitate. Since it contains Pt we have been able to follow the staining mechanism of *B. subtilis* and *E. coli* by electron microscopy.

From these experiments it has been determined that TPt (a 0.82 x 0.62 nm elipsoid) passes through both Gram-positive and -negative walls and enters the cytoplasm (Fig. 6 and 7). Indeed, crystal violet (a 1.88 x 0.65 "Y"-shaped molecule) also passes through both wall fabrics and interacts with intracellular TPt to form CV/TPt. This complex is dissolved by ethanol and is removed from *E. coli* to leave a colorless cell. Electron microscopy suggests that both outer and plasma membrane were disrupted by the ethanol and that the thin murein sacculus was not a retention barrier for the stain. The implication is that discontinuities ca. 2.0 nm in diameter exist in this murein. *B. subtilis* walls, on the other hand, did provide a barrier to the stain. This, presumably, is a reflection of greater thickness (i.e., greater seiving character) and a condensation of the PG's lattice constant due to the dehydrating quality of the ethanol.

Concluding Remarks

Successful architecture is a proper blend between structure and function. Quite clearly bacterial walls fulfill this mandate; they are ubiquitous among the vast majority of bacteria and have proven the test of time.

I have presented numerous and diverse data in this report. My intention was to offer some perspectives so that we could appreciate the diversity of function which walls must fulfill. Since murein is a major component of walls, it is fitting that a research emphasis be placed on its elucidation. But, we must not be too shallow in our vision; the complex interactions between PG and secondary polymers in Gram-positive walls, and those between PG and outer membrane in Gram-negative walls are, no doubt, all important.

Certainly we do not know all the answers, and often those we answer simply provide more questions.

If nothing else, this report provides more questions.

References

1. Balyuzi, H.H.M., Reaveley, D.A., Burge, R.E.: Nature (London) New Biol. 235, 252-253 (1972).
2. Beveridge, T.J.: Int. Rev. Cytol. 72, 229-317 (1981).
3. Beveridge, T.J., Meloche, J.D., Fyfe, W.S., Murray, R.G.E.: Appl. Environ. Microbiol. 45, March issue (1983).
4. Beveridge, T.J., Forsberg, C.W., Doyle, R.J.: J. Bacteriol. 150, 1438-1448 (1982).
5. Beveridge, T.J., Jack, T.: J. Bacteriol. 149: 1120-1123 (1982).
6. Beveridge, T.J., Murray, R.G.E.: J. Bacteriol. 141, 876-887 (1980).
7. Beveridge, T.J., Murray, R.G.E.: J. Bacteriol. 127, 1502-1518 (1976).
8. Braun, V., Gnirke, H., Henning, U., Rehn, K.: J. Bacteriol. 114, 1264-1270 (1973).
9. Burge, R.E., Fowler, A.G., Reaveley, D.A.: J. Mol. Biol. 117, 927-953 (1977).
10. Burge, R.E., Adams, R. Balyuzi, H.H.M., Reaveley, D.A.: J. Mol. Biol. 117, 955-974 (1977).
11. Degens, E.T., Ittekkot, V.I.: Nature (London) 298, 262-264 (1981).
12. Degens, E.T., Watson, S.W., Remsen, C.C.: Science 168, 1207-1208 (1970)
13. DeRienzo, J.M., Nakamura, K., Inouye, M.: Ann. Rev. Biochem. 47, 481-532 (1978).
14. Doyle, R.J., Matthews, T.H., Streips, U.N.: J. Bacteriol. 143, 471-480 (1980).
15. Formanek, H., Rauscher, R.: Ultramicrosc. 4, 337-342 (1979).
16. Formanek, H., Schleifer, K.H., Seidl, H.P., Lindman, R., Zundel, G.: FEBS Lett. 70, 150-154 (1976).
17. Formanek, H., Formanek, S., Wawra, H.: Eur. J. Biochem. 46, 279-294 (1974).
18. Kelemen, M.V., Rogers, H.J.: Proc. Natl. Acad. Sci. U.S.A. 68, 992-996 (1971).
19. Lapidot, A., Irving, C.S.: Biochem. 18, 704-714 (1979).
20. Marquis, R.E., Mayzel, K., Carstensen, E.L.: Can. J. Microbiol. 22, 975-982 (1976).

21. Matthews, T.H., Doyle, R.J., Streips, U.N.: Curr. Microbiol. 3, 51-53 (1979).
22. Nakae, T.: Biochem. Biophys. Res. Commun. 71, 877-884 (1976a).
23. Nakae, T.: J. Biol. Chem. 251, 2176-2178 (1976b).
24. Nakae, T., Nikaido, H.: J. Biol. Chem. 250, 7359-7365 (1975).
25. Schleifer, K.H., Kandler, O.: Bacteriol. Rev. 36, 407-477 (1972).
26. Stewart, M., Beveridge, T.J.: J. Bacteriol. 142, 302-309 (1980).

Table 1. Metal Binding by Bacterial Walls

Metal	*B. subtilis*			*B. licheniformis*		*E. coli*[a]
	Native[b]	Neutralized[b] COO^-	No[c] Teichoic Acid	Native[a]	No teichoic[c] or teichuronic Acids	Murein
Na	2.697	0	1.497	0.910	0.080	0.290
K	1.944	0	0.782	0.560	0	0.058
Mg	8.226	0.520	7.683	0.400	0.024	0.035
Ca	0.399	0.380	0.012	0.590	0.096	0.038
Mn	0.801	0.732	0.656	0.662	0.004	0.052
FeIII	3.581	2.260	1.720	0.760	0.172	0.100
Ni	0.107	0.024	0.021	0.520	0	0.019
Cu	2.990	0.993	2.488	0.490	0	n.d.
AuIII	0.363	0.214	0.265	0.031	0.012	n.d.

n.d. = not determined

[a] μmol metal per mg dry weight of walls.

[b] Neutralized by the addition of glycine ethyl ester to carbodiimide activated carboxylate groups.

[c] The dry weight of these walls has been adjusted downwards to reflect the loss of mass due to the extraction. All quantities are expressed in μmol.

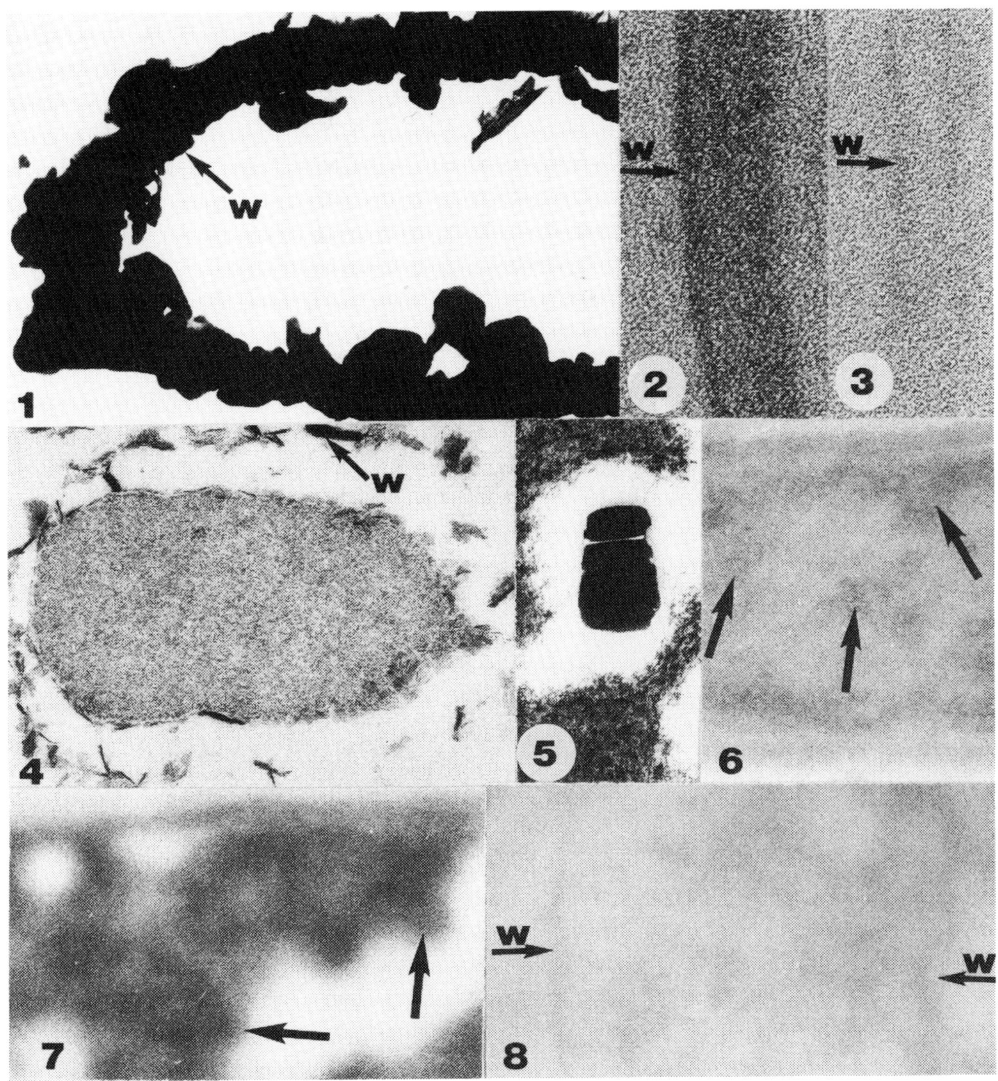

Fig. 1. Unstained B. subtilis wall coated with Au^{o}. 2. Same as 1, but In has been bound. 3. Same as 2, but the COO^{-} groups have been neutralized. Little In has been bound. 4. Diagenic degradation of U-loaded B. subtilis in quartz-magnetite matrix. Platy meta-ankoleite $(K_2(UO_2)(PO_4)2.6H_2O)$ microcrysts are all that remain of the wall. 5. Similar crystal surrounded by the organic matrix. 6. B. subtilis which has been Gram stained using TPt. 7. E. coli, but before ethanol decolorization. 8. E. coli after decolorization. Arrows = CV/TPt precipitate. W = wall.

ELECTROCHEMICAL AND MECHANICAL INTERACTIONS IN BACILLUS SPORE AND VEGETATIVE MUREINS

Robert E. Marquis, Gary R. Bender, Edwin L. Carstensen and Sally Z. Child
Departments of Microbiology and Electrical Engineering, University of Rochester, Rochester, New York, 14642, U. S. A.

Introduction

Knowledge acquired over the past three decades on the chemistry of bacterial mureins or peptidoglycan envelopes is extensive and detailed. Information on their physical properties is less extensive. Views on the biophysical mechanisms of murein functioning *in vivo* are sketchy and confusing. The murein serves as a unimolecular, protective envelope around the protoplast of a bacterial cell and is important in osmotic resistance, ion exchange, antigenicity, toxigenicity, and in the molecular sieving which determines the osmotrophic feeding habits of bacteria. The mechanical functions of mureins seem to be archetypical, especially for gram-positive bacteria. These organisms have a thick, elastic, murein shell and a turgidity associated with internal osmotic pressures as great as 30 atmospheres (3.0 MPa). In the turgid cell, the murein is stretched and under tension. Plasmolysis of the cell results in shrinking not only of the protoplast but also the whole cell (1). Koch et al. (2) have recently proposed roles for murein tension and cell turgor in bacterial wall growth.

Mechanical functions of mureins important for vegetative cells appear to be accentuated in endospores. These biologically extreme cells consist of a central, relatively dehydrated core or protoplast, a thin envelope of vegetative murein, a thick cortex of spore murein, and outer, inverted membrane, protein coat layers, and in many spores, an external, loosely fitting, chemically complex exosporium. In electron micrographs, the cortex appears as a large, electron transparent space, which occupies some 50% of the volume of *Bacillus cereus* terminalis spores (3) or some 40 to 42% of the volume of spores of *Bacillus megaterium* ATCC 19213 (4). These values are close to the values calculated by Beaman et al. (5) on the basis of permeability studies of

The Target of Penicillin

spores of B. megaterium QMB 1551. Moreover, there appears to be a direct relationship between cortex size and heat resistance of spores (6).

Why does the spore elaborate this large murein structure? Calculations by Warth (7) indicate that, to achieve the apparent low water activities in the core required for heat stabilization of enzymes, the turgor pressure of the spore must be some hundreds of atmospheres. Thus, a major need appears to be for an enlarged, elastic, murein cortex to maintain the hydrostatic pressure within the core. The murein of the cortex is admirably designed for this function.

On the basis of the known chemical structure of spore mureins, it is possible roughly to predict physical properties, as done by Rogers (8). Baillie and Murrell (9) have determined the effects of changes in pH and ionic strength (I) on the intrinsic viscosity of soluble fragments of spore mureins of B. cereus terminalis and found that viscosity increased in the pH range from about 2.5 to 5.8 and remained constant in the range from 5.8 to 8.0. However, as Rogers (8) pointed out, the basic need is for experimental studies of cortical mureins in the unhydrolyzed, insoluble state.

Experiments

Our experimental study of mureins has involved two main approaches - one based on assessment of electromechanical volume changes and ion exchange of insoluble murein fragments isolated from mechanically disrupted spores, and another based on dielectric characterization of intact spores, decoated spores and isolated mureins. Clean, refractile spores of B. megaterium ATCC 19213 were prepared as described previously (10). They were decoated by treatment with solutions of 50 mM dithiothreitol and 0.8% (wt/vol) sodium lauryl sulfate at 37°C at pH 10.3 for 90 min. The decoated spores were then disrupted at ice-bath temperature by high-speed grinding with homogenizing glass beads for 30 minutes to 1 hour. The murein fragments were isolated and purified by means of differential centrifugation until a preparation was obtained which was more than 90% solubilized by lysozyme, gave little reactivity with Folin reagent, and contained undetectable levels of organic phosphate. The murein fragments suspended in water had low contrast when viewed under positive phase optics. They retained the general spherical shape of the spores from which they were derived and appeared to have been broken but not into small bits. The preparations contained no refractile or germinated spores.

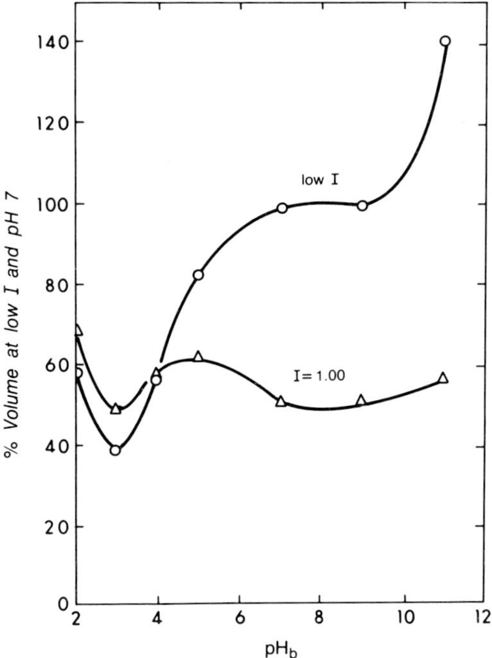

Fig. 1. Volumes of isolated mureins of spores of B. megaterium ATCC 19213 as a function of bulk-phase pH_b and ionic strength (I) in aqueous suspensions. I of 1.00 was established with NaCl; low-I suspensions had only titrating HCl or KOH added.

As shown by the curves presented in Fig. 1, spore mureins swelled and shrank greatly in response to changes in environmental pH and I. At pH 7 and low I, the dextran-impermeable volume of the B. megaterium preparation was 9.5 ml/g, dry weight. For these determinations, we used the thick-suspension or space technique and Dextran T2000 of Pharmacia Corp. which cannot penetrate mureins. The procedure has been described in detail elsewhere (10). This high value of 9.5 ml/g indicates a very open, sparsely cross-linked structure for spore mureins.

When the pH value of the spore murein suspensions in medium of low I was reduced with HCl, the mureins contracted, so that the average volume was only about 40% of the pH-7 value when the suspension pH value was 3. Further acidification with HCl resulted in swelling of the mureins. Titration with KOH to a pH value of 11 resulted in swelling and an increase in volume to some 140% of the pH-7 value. In their most contracted state at a pH value of 3, the murein volume was only about one third of the maximal volume at a pH value of 11. If the I of the suspending medium was increased to 1.00 by addition of NaCl, volume changes were very much moderated, and this result indicates the importance of electrostatic interactions among charged groups in spore mureins in determining expansion and contraction. The volume changes can be interpreted in terms of the known structure of spore mureins (8), as we have interpreted similar changes for vegetative mureins (1). For spore mureins, carboxyl groups which titrate at low pH considerably outnumber amino groups which titrate at high pH. Cation binding in media of low I can be related to net negative charges of spore or vegetative mureins. The spore mureins used here bound an average of 1.0 mequivalents of of exchangeable cations per g, dry weight. They behaved as relatively open, ion exchangers as did vegetative mureins (11). Calcium binding was not unusually avid.

Fig. 2 presents comparative data on the frequency dependence of conductivities and dielectric constants for decoated spores and vegetative cells. Vegetative cells were more highly conducting than decoated spores, even at low frequencies at which conduction is due entirely to movements of ions in enveloping cell wall or cortex structures. The curve for the dielectric constants of vegetative cells shows the typical Maxwell-Wagner or β-dispersion, here with a characteristic frequency of about 6.5 MHz at a dielectric constant of about 200. Data of this sort has been interpreted with previously described models (10) in terms of a highly conducting protoplast surrounded by a highly conducting cell wall with a separating membrane which becomes capacitatively short circuited at frequencies in the neighborhood of the characteristic frequency. The decoated spores had lower conductivities. Moreover, the very low characteristic frequency of less than 1 MHz apparent from the curve for dielectric constants indicates a very poorly conducting core surrounded by a more highly conducting cortex. The cortex has been found (10) to have high conductivity also in intact spores with coats. In contrast, the cortex of intact spores of *Bacillus cereus* has been found (12) to be essentially nonconducting in media of low I. The high conductivity of the cortex of *B. megaterium* spores can be related directly to the large numbers of charged groups in the structure.

Vegetative cells of *B. megaterium* can be made to contract by transferring them from

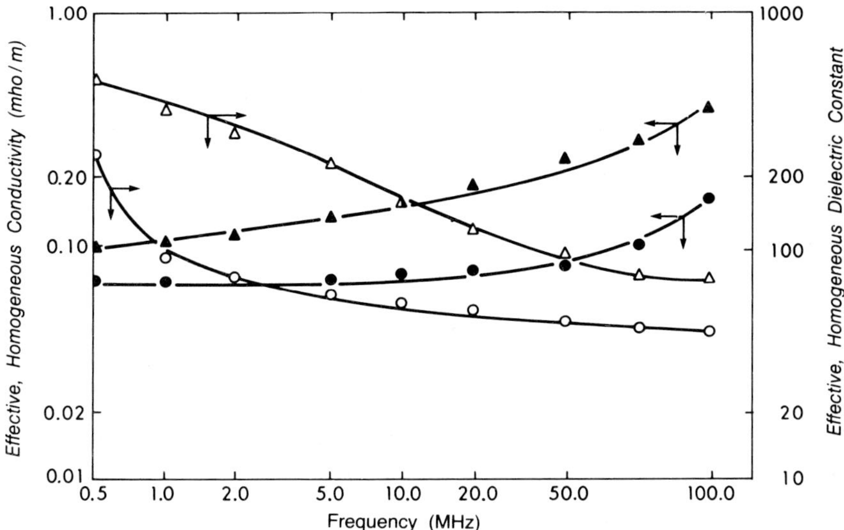

Fig. 2. Effective, homogeneous conductivities and dielectric constants as a function of current frequency for decoated spores (●, ○) and vegetative cells (▲, △) of B. megaterium ATCC 19213 assessed by use of techniques described in detail previously (10). The cells were suspended in deionized water.

dilute to concentrated but non-plasmolyzing NaCl solutions. This contraction has been interpreted (1) in terms of wall contraction. However, decoated spores did not contract in this way. Dextran-impermeable volumes per g, dry weight, for decoated spores in water, 0.05 M or 0.50 M NaCl soutions were, respectively, 2.1, 2.2 and 2.1 ml. Thus, it appears that the high tension in the cortex in vivo reduces salt induced contraction in the intact spore.

Conclusions

1. Insoluble mureins of B. megaterium spores appear to be highly elastic, flexible and well designed for their roles in containing the high hydrostatic pressures of the spore core.
2. The electrochemical properties of spore mureins can be predicted at least roughly from a knowledge of their chemical structure and the known behavior of vegetative mureins.

Acknowledgment

This work was supported by contract DAAF29-80-C-0051 for the U. S. Army Research Office.

References

1. Marquis, R. E.: J. Bacteriol. 95, 775-781 (1968).
2. Koch, A. L., Higgins, M. L., Doyle, R. J.: J. Gen. Microbiol. 128, 927-945 (1982).
3. Warth, A. D.: Adv. Microbial Physiol. 17, 1-45 (1978).
4. Hitchins, A. D., Greene, R. A., Slepecky, R. A.: J. Bacteriol. 110, 392-401 (1972).
5. Beaman, T. C., Greenamyre, J. T., Corner, T. R., Pankratz, H. S., Gerhardt, P.: J. Bacteriol. 150, 870-877 (1982).
6. Murrell, W. G., Warth, A. D.: Spores III, 1-24 (1964).
7. Warth, A. D.: Spore Newsletter 7, 245 (1982).
8. Rogers, H. J.: Spore Research 1976 1, 33-54 (1976).
9. Baillie, E., Murrell, W. G.: Biochem. Biophys. Acta 372, 23-31 (1974).
10. Carstensen, E. L., Marquis, R. E., Child, S. Z., Bender, G. R.: J. Bacteriol. 140, 917-928 (1979).
11. Marquis, R. E., Mayzel, K., Carstensen, E. L.: Can. J. Microbiol. 22, 975-982 (1976).
12. Carstensen, E. L., Marquis, R. E., Gerhardt, P.: J. Bacteriol. 107, 100-113 (1971).

THE STATE OF ORDER OF BACTERIAL PEPTIDOGLYCAN

Harald Labischinski, Gerhard Barnickel, Dieter Naumann
Robert Koch-Institut des Bundesgesundheitsamtes
D-1000 Berlin

Introduction

The process of gaining a closer insight into the principles of three-dimensional architecture of bacterial peptidoglycan is hampered by the existence of controversary opinions about its state of order. Some authors considered the murein-network to exist in a highly ordered chitin-like arrangement, characterized at least partly by extended crystalline regions. Furthermore an amorphous and open structure similar to that of ion exchange resins with practically no ordered elements has been proposed (for review see (1)).
Using comparative X-ray diffraction data, infrared-measurements and stereochemical considerations it will be demonstrated that a non-crystalline, but nevertheless ordered arrangement with well defined structural elements of the murein building blocks has to be taken as the basis for any molecular model of peptidoglycan. The significance of these data for explaining the remarkable flexibility of the cell wall observed will be discussed.

Results and Discussion

Comparative X-ray diffraction pattern were taken on dried foils of murein and chitin essentially as described in (2). The difference in the state of order is immediately apparent (fig.1): For chitin a typical crystalline diffraction pattern characterized by a series of well defined sharp

The Target of Penicillin

reflections was observed. However, in the case of murein only more or less diffuse scattering maxima could be detected, typical for non-crystalline materials.

Correspondent data about a basically different state of order within these two substances were reached by a comparison of their infrared spectra. The bandwidths of the infrared absorption bands of murein were in general high (ca. 70 to 80 cm^{-1}) and no band splitting of the amide I absorption near 1660 cm^{-1} could be observed (3), as found in chitin.

These data may explain the well known huge differences in sensitivity towards lysozyme degradation between chitin and murein samples, although the same chemical linkage is attacked: The sugar chain to be cleaved has to assume a twisted and bended conformation (corresponding roughly to a 2.5 fold screw axis), which in chitin is not observed inside its crystalline regions (5) but can be easily adopted by the non-crystalline murein sugar chains (see contribution of G. Barnickel et al., in this volume). These data about the action of lysozyme stem from the well known X-ray crystallographic studies on the architecture of the active site of lysozyme and the conformation of the substrate bound to it (4). A crystalline state of order of the murein can therefore be excluded. However, there are also several indications against a complete amorphous arrangement:

i) the density of murein was found to be very high, at least in the dry state (2) ii) the infrared absorption data of murein point to a specifically folded peptide arrangement with well defined elements of

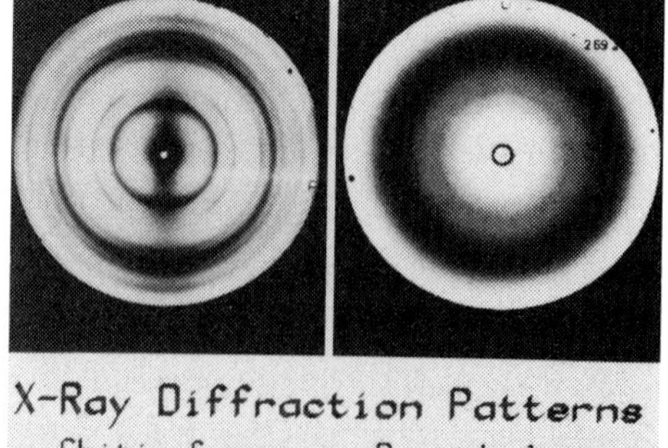

Fig. 1

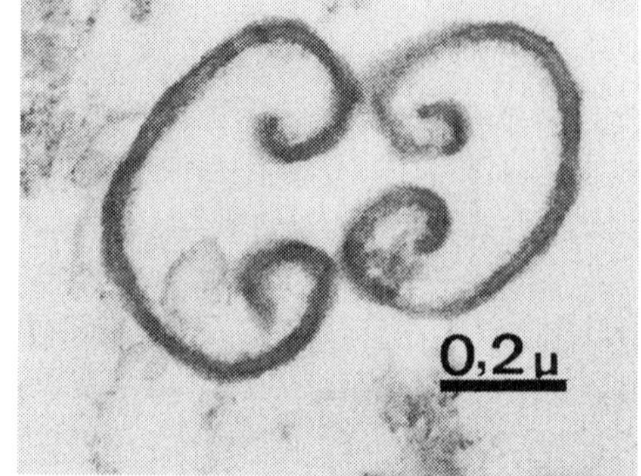

Fig. 2

secondary structure (see contribution of D. Naumann et al., in this volume), and iii) the "rolling in" effect observed in broken wall fragments, which could be observed in EM-preparations of penicillin treated Staphylococcus aureus (fig. 2). This effect may find an interpretation in assuming a different cell wall flexibility at the outer and the inner surface of the wall. Such a flexibility gradient across the cell wall would be obviously inconsistent with an amorphous, completely non-ordered arrangement.

These considerations led us to test in more detail the flexibility of peptidoglycan as a function of osmotic pressure. Although flexibility of bacterial cell walls and changes in cell size after interaction with inorganic salts have been recognized for a long time (1,6,7,8), there is apparently no detailed analysis of the dimensional changes in the murein network as a function of differences in osmotic pressure. Size measurements on Staphylococcus aureus were carried out either by evaluation of photographic exposures obtained by light microscopy with a particle size measuring device (MOP, Kontron) or, alternatively, with a Coulter Counter channelizer.

As depicted in fig. 3, the highly crosslinked staphylococcal cell wall proved to be remarkable flexible; surface variations of nearly 100% could be observed after transfering staphylococci from the logarithmic growth phase to destilled water and then to concentrated sodium chloride or sucrose solutions. Surprisingly, cells from the stationary phase of growth or those having been treated with different bacteriostatic antibiotics showed a drastically reduced flexibility, revealing surface varations of only 35 - 40%. Although we do not yet know the molecular basis for this reduction in flexibility it is tempting to speculate about an invers correlation between flexibility and degree of O-acetylation in staphylococci (see contributions of P. Burghaus et al., and L. Johannsen et al., in this volume).

According to the surface stress theory of microbial morphogenesis (9) in the case of spherical bacteria a plot of the cell diameter versus the reciprocal of the osmolality of the medium should result in a straight line with a slope being proportional to the surface tension of the wall. However, to our surprise, we obtained two straight regions intersecting at the point, where the inner pressure (about 25 atm (9)) equaled the outer pressure (fig. 4). This could simply mean that it is much easier to stretch the murein network than to compress it, starting from its pressureless state.

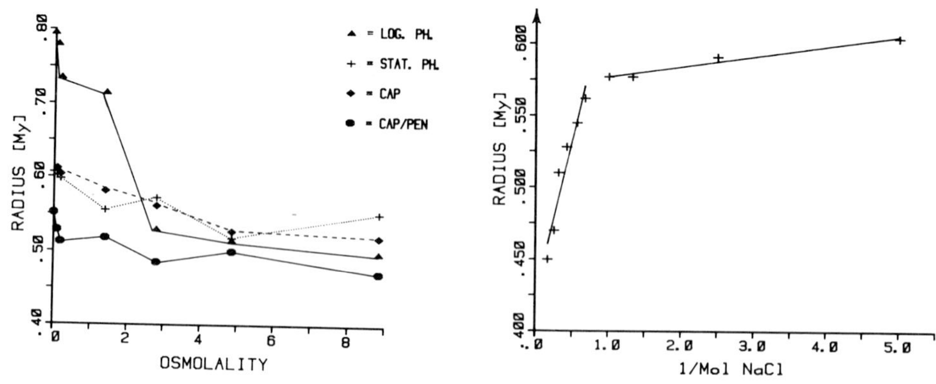

Fig. 3

Radius of S.aureus cells as a function of osmotic pressure

Fig. 4

Radius of log-phase cells of S.aureus versus reciprocal pressure

Again, it is difficult to imagine how this can be managed in case of a completely amorphous arrangement.

Instead, the specifically folded conformation of the peptide part of murein with a nearly ring-shaped backbone arrangement proposed earlier for the non-stressed peptide moiety (10) would be capable of explaining the flexibility of bacterial cell walls since it can be extended in linear dimension by more than 200% (in marked contrast to the sugar chain; fig. 5). These conformational properties of the two constituent parts of murein may also explain why for Bacillus subtilis a marked anisotropy of shrinkage was observed (fig. 6): In spite of a considerable shrinkage (data not shown), the thickness of the bacilli proved to be nearly constant. These data support the view, that the glycan chains in the cylindrical part of the sacculus might be arranged more or less parallel, but not perpendicular to the short axis of the bacteria (7,11).

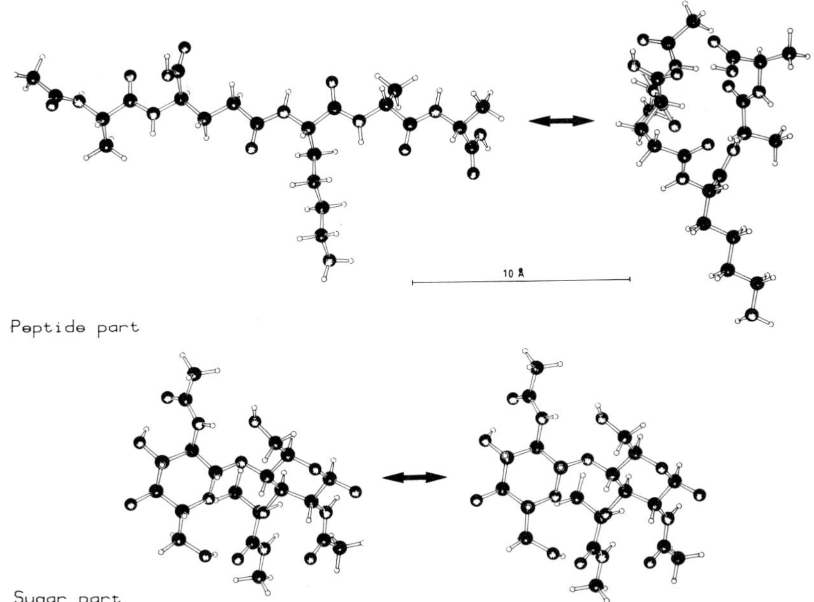

Fig. 5: Comparison between the extended conformations and the minimum energy conformations of the cell wall peptide and disaccharide of murein

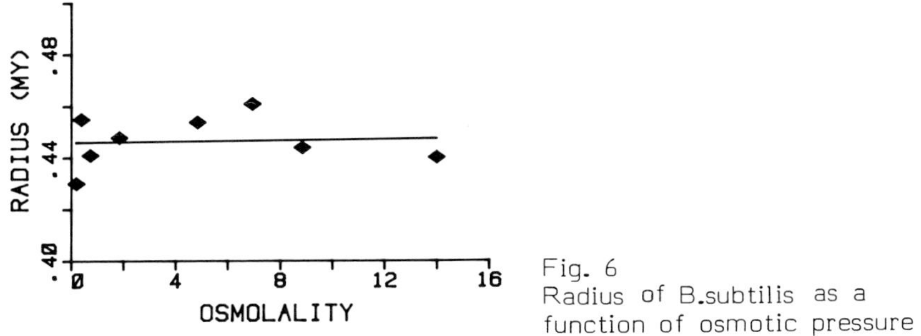

Fig. 6
Radius of B.subtilis as a function of osmotic pressure

In <u>conclusion</u> the data obtained demonstrated: i) It is no longer deniable that in the murein network no crystalline or even microcrystalline chitin-like arrangement exists. ii) An amorphous arrangement of murein is highly unlikely. This deduction is based on the existence of some ordered elements like a certain degree of parallelity in sugar chain packing, sheet like organisation of gram-positive peptidoglycan (2); especially the existence of regular conformational elements in the peptide moiety as well as the flexibility of the murein network are not compatible with an amorphous arrangement of murein.

In order to fulfil its various biological functions in the interplay of rigidity and flexibility the murein has obviously not employed a crystalline organization as an essential principle in which rigidity is determined by crystalline packing forces, allowing no penetration of such small molecules like water. Since there are no covalent bonds between the sugar strands, chitin would become water soluble without crystalline packing just as murein would be after splitting of the covalent peptide bridges by endopeptidase action. Thus for bacterial cell walls, instead of crystallinity apparently the crucial conformational properties of the peptide moiety of murein are used.

Acknowledgements

The skilful assistance of A. Brauer and E. Vorgel is gratefully acknowlegded. This investigation was supported by the Deutsche Forschungsgemeinschaft.

References

1. Beveridge, T.J.: Int. Rev. Cytol. 72, 229-317 (1982).
2. Labischinski, H., Barnickel, G., Bradaczek, H., Giesbrecht, P.: Eur. J. Biochem. 95, 147-155 (1979).
3. Naumann, D., Barnickel, G., Bradaczek, H., Labischinski, H., Giesbrecht, P.: Eur. J. Biochem. 125, 505-515 (1982).
4. Phillips, D.L.: Sci. Amer. 215, 78-90 (1966).
5. Minke, R., Blackwell, J.,: J. Mol. Biol. 120, 167-181 (1978).
6. Marquis, R.E.: J. Bacteriol. 95, 775-781 (1968).
7. Preusser, H.J.: Arch. Microbiol. 68, 150-164 (1969).
8. Isaac, L., Ware, G.C.: J. appl. Bact. 37, 335-339 (1974).
9. Koch, A.L., Higgins, M.L., Doyle, R.: J. Gen. Microbiol. 128, 927-945 (1982).
10. Barnickel, G., Labischinski, H., Bradaczek, H., Giesbrecht, P.: Eur. J. Biochem. 95, 157-165 (1979).
11. Verwer, R.W.H., Nanninga, N., Keck, W., Schwarz, U.: J. Bacteriol. 136, 723-729 (1978).

A THREE DIMENSIONAL MODEL OF THE MUREIN LAYER EXPLAINING ONE STEP OF ITS BIOSYNTHESIS BY SELF-ASSEMBLY.

Helmut Formanek

Botanisches Institut der Universität München, Menzinger Str. 67
D-8000 München 19, Germany

Introduction

Chemical data provide a network structure for murein, where polysaccharide strands are connected with short peptide chains (1). If the periodicity in the chemical composition is also reflected by its sterical arrangement, the three dimensional structure of murein may be described with three parameters. The length and width of its meshes and the thickness of the monolayer of murein.

X-ray diffraction

X-ray diffraction of murein foils of Gram-positive and Gram-negative bacteria taken vertical to the plane of the sacculi showed diffuse Debye-Scherrer rings corresponding to distances of about 4.5 and 10 Å (2). In the direction vertical to the plane of the sacculi of the Gram negative bacterium Spirillum serpens a reflex at 43 Å could be obtained. This periodicity may correspond to the twofold thickness of one murein layer, since only the periodicity of the whole sacculus consisting of two head-to-head superposed layers can be measured.

Electron diffraction

Electron diffraction on murein sacculi adsorbed on hydrophilic

The Target of Penicillin

crystalline films of graphite oxide provide a Debye-Scherrer ring corresponding to a periodicity of 4.5 Å the most intense ring also obtained by X-ray diffraction.

High resolution electron microscopy.

High resolution micrographs taken at 4 K from specifically stained murein of Spirillum serpens adsorbed onto graphite oxide foils where taken with an electron microscope containing a superconducting lens system in the Forschungslaboratorium of Siemens AG München. The micrographs showed a fringe structure with a periodicity of 4.5 Å, which can clearly be seen by light optical diffraction (Fig. 1).

Elementary cell of the murein network.

The 4.5 Å distances of these lines as well as the 4.5 Å Debye-Scherrer rings of both X-ray and electron diffraction may be interpreted as packing periodicities of the peptidoglycan chains (Fig. 2). If the repeating periodicity in the direction of the polysaccharide chain of murein is assumed to be similar with the related ß-1,4 linked polysaccharides, cellulose and chitin, the peptide chain linked to every second carbohydrate residue of murein should repeat with a periodicity of 10.3 Å. This may cause a Debye-Scherrer ring at 9.5 Å if an oblique angled elementary cell is assumed (Fig. 2).

According to these data an elementary cell of the murein network may be constructed with a base of 10.3 x 4.5 $Å^2$ and a thickness of 21.5 Å, into which the smallest repeating unit of the murein lattice fits, the disaccharide tetrapeptide (Fig. 2). Both Fourier Transform and the high density value calculated for this elementary cell can be explained with the experimentally obtained results.

Fourier Transforms of the murein layer

The two dimensional Fourier transform of our murein lattice (Fig. 3) shows a strong 1 0 reflex at 4.5 Å for the packing periodicity of the polysaccharide strands and with 18 % of its intensity the 0 1

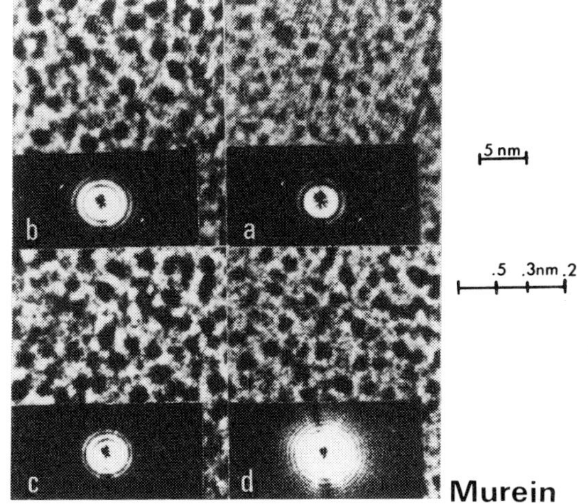

Fig. 1:

High resolution electron micrograph of a murein layer taken with different dose rates (i) and exposure times (t) together with its light optical diffractograms a: i = 0.005 A/cm^2; t = 110 sec. d : i = 0.2 A/cm^2 t = 3 sec.

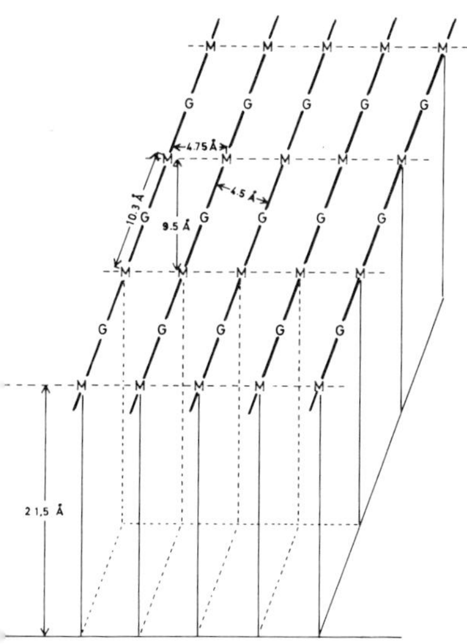

Fig. 2:

Network of murein with elementary cells containing one dissaccharide peptide unit. M = N-acetyl muramic acid, N = N-acetylglucosamine

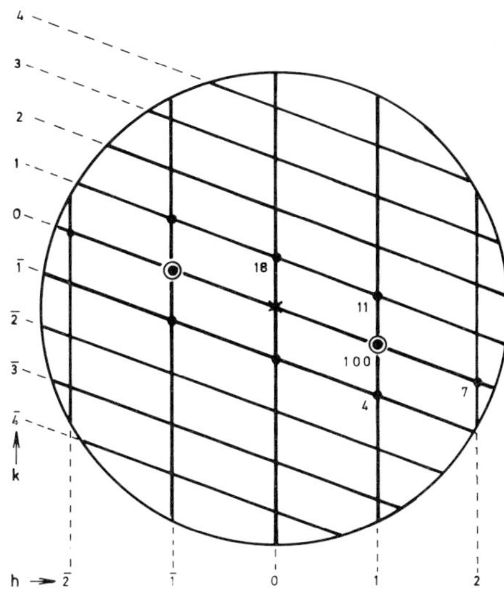

Fig. 3:

Calculated two dimensional Fourier Transform of murein with the dimensions of the elementary cell of Fig. 2. The intensity of the 4.5 Å reflex (10) is taken as 100 %.

reflex at 9.5 Å for the repeating periodicity caused by the peptide chains. In addition there are few other reflexes (02, 11, 1$\bar{1}$) with intensities of about 10 % and lower than the intensity of the 1 0 reflex. A reduction to 1 % of the intensity of this reflex has been observed for the 0 1 reflex, if the peptide chains are not regarded in the calculation. These results are related to the measured and calculated structure factors of chitin (2).

Calculation of Fourier transforms using different packing periodicities of the polysaccharide chains and different arrangements of the linked peptide chains have shown, that the packing periodicities of the polysaccharide chains always provide the dominant reflexes (2).

Sterical requirements of peptidoglycan

Ramachandran type calculations provide two to threefold screw axes in the sterically allowed region of the polysaccharide chain. In case of a not twofold screw axis, the peptide chains of murein point into different direction away from the polysaccharide chain. This can cause a reduction of the dissaccharide peptide periodicity in the direction of the polysaccharide chain from 10.3 Å to 9.5 Å, if a right angled elementary cell is considered for the murein network. On the other hand an obtuse angled elementary cell can also cause a 9.5 Å reflex for a 10.3 Å periodicity and a twofold screw axis (Fig. 2), where all peptides protrude in identical directions and therefore the carbohydrate- and the peptide moiety is separated.

Only flat peptide chains fit into this periodic structure, the pleated sheet (3) and the 2.2_7 helix (2). The hypothetical 2.2_7 helix may explain why the γ-bonds of D-glutamic acid residues are essential for most of the murein structures, while their amidated α-carboxylgroups may perform interchain hydrogen bonds.

Biosynthesis of murein by self-assembly

During the biosynthesis of murein at first an UDP-dissaccharide

pentapeptide unit is performed. If the peptide chain of this unit is in the 2.2 helical conformation an optimum of interchain hydrogen bonds can still be performed. Peptidoglycan strands with twofold screw axes in both polysaccharide and peptide chains are very flat and self-assembly can occur by performing hydrogen bonds between the N-acetylgroups of adjacent peptidoglycan strands. Additional interchain hydrogen bonds may be performed if the α-carboxylgroups of the D-glutamic acid residues are amidated, which is the case in the murein of many kinds of bacteria.
As a consequence of this self-assembly adjacent peptidoglycan strands come in close contact, so that an enzyme can perform crosslinkage.

Comparison between cellulose, chitin and murein

The structures of the related ß-1,4 linked polyglucosides cellulose, chitin and murein can be derived from one another (Fig. 4). Since they all have a sceletal function it is important, that they can be packed with a high density. In case of cellulose and chitin this dense packing is enabled with a twofold screw axis in the polysaccharide chains. The structure of chitin is further stabilized by

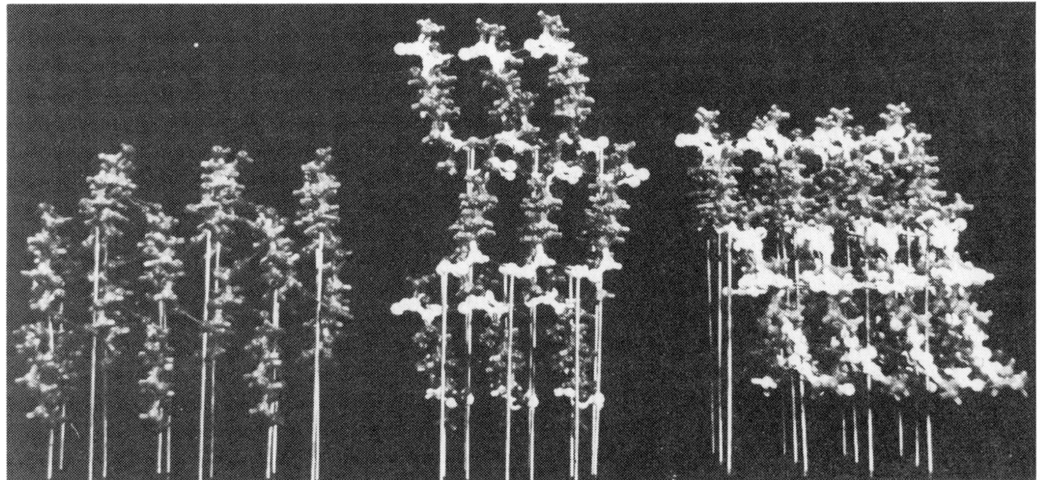

Fig.4: Comparison between the models of cellulose, chitin and murein viewing in the direction of the polysaccharide chains.

interchain hydrogen bonds between the N-acetylgroups of adjacent polysaccharide strands. A similar structure has been assumed as base of our murein model (2).

Digestion of murein by lysozyme

If the murein layer would exhibit an ideal crystalline structure as derived in our model an enzymatic degradation by lysozyme would be impossible, since a hexasaccharide of the polysaccharide chain could not attache to the active site of the enzyme. Digestion of murein may however begin at lattice defects in its micellar, paracrystalline structure. The attachment of the hexasaccharide tripeptide to lysozyme is sterically not hindered by the peptide chains if a separation of the carbohydrate - and peptide moiety is assumed as a result of the twofold screw axes in the peptidoglycan chains.

Conclusions

The model discussed is a simple extension of the two dimensional conception of the murein network to a murein layer. It may be altered or even exchanged by more complicated three dimensional models explaining as many or even more experimental details of structure and function of murein.

References

1. Weidel, W., Pelzer, H.: Bagshaped macromolecules - a new outlook on bacterial cell walls. Advanc. Enzymol. 26, 193-232 (1964)

2. Formanek, H.: Possible models of mureins and their Fourier transforms Z. Naturforsch. 37c, 226-235 (1982)

3. Kelemen, M.V., Rogers, H.J.: Three-dimensional molecular models of bacterial cell wall mucopeptides (peptidoglycans). Proc. Nat. Acad. Sci. USA 68, 922-996 (1971)

COMPUTER AIDED MOLECULAR MODELLING OF THE THREE-DIMENSIONAL STRUCTURE OF BACTERIAL PEPTIDOGLYCAN

Gerhard Barnickel, Dieter Naumann, Hans Bradaczek

Institut für Kristallographie der Freien Universität
D-1000 Berlin

Harald Labischinski und Peter Giesbrecht

Robert-Koch-Institute of the Federal Health Office
D-1000 Berlin

Introduction

Although the primary structure of peptidoglycan is well known (1), no generally accepted model for its molecular architecture exists up to now. There are several reasons for this situation; especially the apparently non crystalline type order of the peptidoglycan (2) has impeded to get detailed information by those experimental methods, sucessfully employed for the determination of single crystals. Furthermore peptidoglycan as an huge macromolecule forms a complex network which must be rigid and flexible at the same time in order to fulfil its biological function.

To get a model of the molecular architecture of murein, a multifacetted approach had to be used, in which the results of different experimental techniques (3,4,5) were combined with theoretical methods of molecular modelling (6,7,8). In this study the method of conformational energy calculations (9) was used to determine the favoured conformations accessible to the polysaccharide and to the peptide moiety, as well as to the repeating unit of the peptidoglycan. Based on these results a model of one strand of murein was constructed, which was capable of explaining many experimental data hitherto obtained for peptidoglycan.

The Target of Penicillin

Results

Peptide moiety:

For the pentapeptide the calculations revealed that in its favoured conformations the peptide adopts a ring-like structure folding back the peptide backbone to the starting point of the chain (6). Responsible for this conformation of the peptide is the gamma bonded glutamic acid and the occurence of alternating D and L amino acid residues.

Sugar moiety:

In the case of the sugar moiety, having a ß 1-4 linkage, two rotations around the glycosidic bonds are possible for each disaccharide unit formed by N-acetyl-glucosamine (=NAG) and N-acetyl-muramic acid (=NAM), defining all mutual orientations of two neighboured sugar residue within the chain. The calculations of possible conformations for a disaccharide without the sidegroups showed, that only a narrow range of different orientations of the sugar ring planes was stereochemically feasible. Among those conformations some were found with coplanar sugar rings but in most cases the sugar rings were twisted to each other to a varying extent.

Considering the influence of the side groups, the most pronounced effect was mediated by the rather bulky lactyl-group of NAM. Introducing this residue into the narrow linkage region of the disaccharide was only possible by rotating one ring relative to the other one. Therefore a total flat arrangement of poly-NAG-NAM as in the case of chitin (poly-NAG-NAG) proved to be impossible (see also 2,10,11).

Using the most favoured conformations of the disaccharide as repeating unit of a polysaccharide strand for further calculations, a lot of favoured conformations were obtained in which the sugar chains proved to be almost extended. These conformations differed only slightly in respect to their shapes.

Repeating unit:

To obtain the most favoured conformations of the repeating unit of peptidoglycan we combined the preferred conformations of NAG-NAM with the ring-like structure of the pentapeptide. As the result of these calculations a most favoured conformation of the disaccharide-pentapeptide was yielded, in which the ring-shaped conformation of the peptide part was maintained (see figure 1). The folding back of the peptide backbone allowed its terminal end to interact strongly with the disaccharide part. The removal of the last D- alanine resulted in a nearly unchanged overall conformation of the repeating unit, although the hydrogen bond pattern between sugar and peptide was altered.

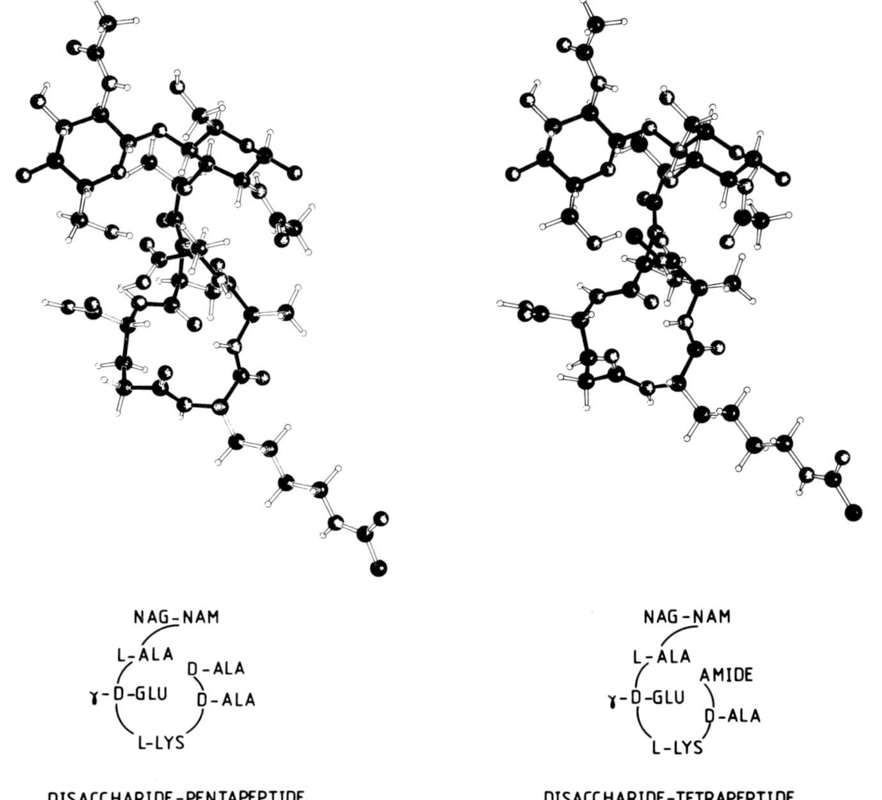

Figure 1: Energetically favoured conformation of the repeating unit of peptidoglycan

Model of a complete peptidoglycan strand:

With the conformation obtained for the repeating unit, a model of a glycan strand of peptiodglycan comprising ten disaccharide units was constructed. The resulting model (Figure 2) should, however, be considered to be still simplified, since due to its assembly process it is somewhat too regular. The glycan part itself revealed a shape containing many pronounced ledges and clefts. Adding the ring-like peptides, some of these clefts were filled, but again the overall shape of the peptidoglycan strand remained virtually unchanged; only the thickness of the strand was increased. In this conformation all sites for crosslinkages proved to be situated on the surface of the molecule. This included the terminal D alanine as well as the starting point of the side chain for the lysine. Moreover, the carboxyl groups of the gamma bonded glutamic acid were also located on the surface of the molecule. Therefore also for bacterial peptidoglycan of the so called B-group (1) a threedimensional crosslinkage pattern can easily be formed.

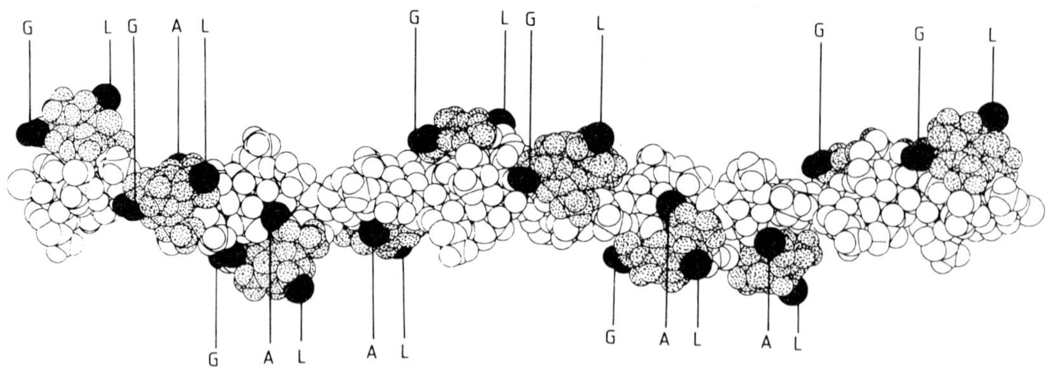

Figure 2: Contours of a space-filling model of a peptidoglycan strand composed of ten repeating units. The sites of possible crosslinkages are denoted with A (last D-alanine), L (starting point for the side-chain of lysine), and G (carboxyl-group of glutamic acid). The peptides are dotted.

Discussion

With these results obtained from modelling of peptidoglycan several experimental observations can now be understood on a molecular level.

Since the polysaccharide chain may adopt a variety of conformations with different degrees of twisting between neighboured sugar residues and, furthermore, because the transitions between the different states of twisting are nearly unhindered, in a dynamic structure the segments of the polysaccharide strand may exhibit distinct conformations. Thus no crystalline regions should exist and a low degree of order would result. Independent of this feature the glycan strands form almost fully extended conformations. Even under the influence of strong mechanical forces an extension of the sugar chain will be restricted. Thus the glycan strands are capable of serving for the overall network of murein as an element of rigidity. On the other hand the possible twisting between the sugar molecules mediates an element of flexibility perpendicular to the strand axis. Concerning the peptide part of the peptidoglycan this structure is folded back to a stable, ring-shaped conformation which, however, can be unwound by mechanical forces and may lead in this way to extended structures. Therefore, drastic increased distances between the glycan strands, cross-linked via the highly flexible side chain of the diamino acid in position 3 of the peptide moiety, can be realized. Therefore peptide and sugar moiety contribute in different ways to rigidity and flexibility of the whole sacculus.

As a main consequence of the twisting of neighboured sugar residues, the peptides protrude to different sides of the strand. Considering only the main directions (within and perpendicular to a murein layer) a simplified model results, which allows for the first time to explain the lower degree of crosslinkage found for gram-negative in comparison to gram-positive bacteria. Since crosslinkage needs the appropiate orientations of the two peptides involved, only those peptides can serve as crosslinking sites which are directed more or less to each other. This condition is only fulfilled for those peptides oriented preferential within the plane of the murein layer.

The peptides located perpendicular to the plane are not suitable for a linkage within the layer, but they may be used for connections to other layers above or below. Therefore in a multilayered gram-positive peptidoglycan much more peptides are available for crosslinking the layers than in gram-negative bacteria.

In conclusion, the new model of peptidoglycan proposed here is consistent with the variations found in the primary structure, and is capable of explaining the dynamic properties of peptidoglycan as an interplay between rigidity and flexibility. Furthermore the higher degree of crosslinkage for gram-positive bacteria compared to gram-negative bacteria could be attributed to sterical features of the peptidoglycan.

References

1. Schleifer, K.H., Kandler, O.: Bacteriol. Rev. 36, 407-477 (1972).
2. Labischinski, H., Barnickel,G., Naumann,D., contribution in this volume
3. Labischinski, H., Barnickel, G., Bradaczek, H., Giesbrecht, P.: Eur .J. Biochem. 95, 147-155 (1979).
4. Labischinski, H., Barnickel, G., Rönspeck, W., Roth, K., Giesbrecht, P.: Zbl. Bakt. suppl. 10, 427-433 (1981).
5. Naumann, D., Barnickel, G., Bradaczek, H., Giesbrecht, P.: Eur. J. Biochem. 125, 505-515 (1982).
6. Barnickel, G., Labischinski, H., Bradaczek, H., Giesbrecht, P.: Eur. J. Biochem. 95, 157-165 (1979).
7. Labischinski, H., Barnickel, G., Leps, B., Bradaczek, H., Giesbrecht, P.: Arch. Microbiol. 127, 195-201 (1980).
8. Yadav, J.S., Barnickel, G., Bradaczek, H., Labischinski, H.: J. Theor. Biol. 95, 167-179 (1982).
9. Momany, F.A., McGuire, R.F., Burgess, A.W., Scheraga, H.A.: J. Phys. Chem. 79, 2361-2381 (1975).
10. Virudachalam, R., Rao, V.S.R.: Biopolymers 18, 571-589 (1979).
11. Burge, R.E., Fowler, A.G., Reaveley, D.A.: J. Mol. Biol. 117, 927-953 (1977).

COMPARISON BETWEEN THE SUGAR CHAIN CONFORMATION IN MUREIN AND PSEUDOMUREIN

Bernd Leps, Gerhard Barnickel, Hans Bradaczek,
Institut für Kristallographie der Freien Universität Berlin
D-1000 Berlin

Harald Labischinski
Robert-Koch-Institut des Bundesgesundheitsamtes
D-1000 Berlin

Introduction

The cell wall of nearly all bacteria contains murein as the most important shape defining and stress bearing component. However some methanobacteria possess another cell wall polymer, which has been called pseudomurein because of its chemical and functional similarity to murein (1). Both murein and pseudomurein consist of polysaccharide strands crosslinked by short oligopeptides; whereas the polysaccharide chains in murein comprise alternating ß-1,4 bonded D-N-acetylglucosamine (=NAG) and D-N-acetylmuramic acid (=NAM), in pseudomurein alternating NAG and L-N-acetyltalosaminuronic acid (=NAT) were found. The linkage between the two sugar residues in pseudomurein, previously presumed to be of the 1,4 type (1,2), was recently shown to be of the 1,3 type (3). Because of the structural similarities it is tempting to assume, that both peptidoglycans might show similar principles in their three dimensional architecture, although pseudomurein cannot be considered to be a simple modification of murein but rather a convergent "invention" in two different phylogenetic lines (2). The method of conformational energy calculation has proved to be very successful in predicting molecular models for biopolymers (see e.g. (4) for review).

The Target of Penicillin

Using this method, a comparison between the structural features of the murein and pseudomurein saccharides is presented.

Methodical remarks

The details of the calculation procedure were as described in (5,6); ring coordinates were taken from (7), side group coordinates from (8) and (for NAT) from a survey of the Crystallographic Data File (9). For NAG, the usual C1 ring conformation was used. Since the configuration of NAT in pseudomurein was recently described (3) to be α-L-N-acetyl-talosaminuronic acid with an equatorially oriented glycosidic oxygen, also for NAT the C1 conformation had to be used (10,11) in spite of the fact, that in this ring conformation the two most bulky substituents both occupy axial positions. Therefore, subsequently the abbreviation NAT will be used to describe this very structure.

Results and discussion

In both, murein and pseudomurein, NAG residues occur as one of the monomer constituents. Comparing the conformations accessible to the other components, NAM and NAT, the most remarkable difference was found in regard to the linkage region between peptide and sugar. Since in NAT the linkage is formed by means of an axial oriented carboxyl group, only one sterically allowed orientation of this substituent was found leading to a rigid connection between the peptide and the sugar part. However, for the corresponding region of NAM, comprised of an equatorial oriented D-lactyl residue, several orientations were possible. (cf. contribution of G. Barnickel et al., in this volume). Typical favoured conformations of the polysaccharide chain of murein and pseudomurein constructed from energetically favoured disaccharide units are shown in Fig. 1 and compared to the

Favoured Conformations for the Sugar Strands in Crystalline Chitin, Murein, and Pseudomurein

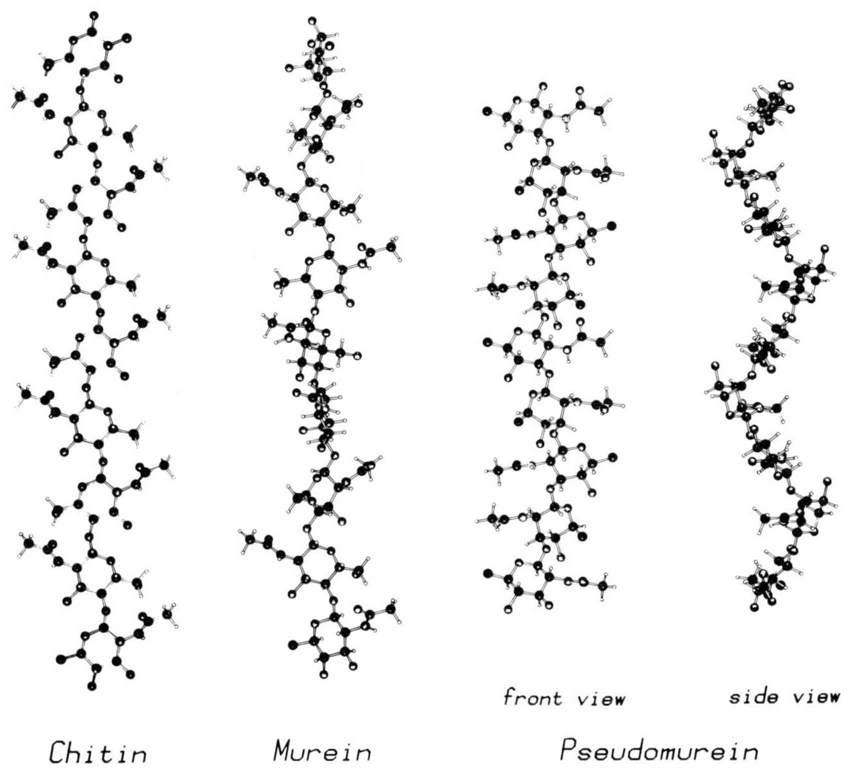

Fig. 1

sugar chain conformation found in crystalline chitin (12). The linkage between the monomers is known to be of the ß-1,4 diequatorial type throughout the whole chain of murein. However, because of the alternating sequence of ß- and α-1,3 linkage types - leading to alternating equatorially and axially linked monomers - the sugar chain of pseudomurein adopted a striking zigzaglike shape.

This conformational feature leads to two distinct consequences:
i) The same number of monomer residues is capable of bridging only a shorter distance in pseudomurein as compared to the murein sugar chain. ii) For pseudomurein the ring substituents of each two consecutive sugar rings point nearly to the same direction in respect to the sugar ring axis, while the subsequent disaccharide is rotated about 180° in respect to the previous unit. However, in the case of murein each monomer residue is twisted against the previous one.
To obtain the range of conformations possible for the sugar chains of both of the peptidoglycans, two parameters usually applied to the regular structures of sugar helices namely N (the number of disaccharide units per helix turn) and H (the length of this unit projected onto the helix axis) can be used conveniently also here. For murein N values ranging between 3.6 and 15.8 with H values of about 1 nm were observed and for pseudomurein two regions for N near the values of 2 and 3 were found, combined with H values ranging from 0.67 to 0.82 nm. Therefore, despite of the length differences mentioned, all sugar chains so far considered favoured extended conformations. However, for both, murein and pseudomurein none of the favoured conformations resembled the structure found in crystalline chitin (with a N value of 1 and a rise per helix turn of 1.03 nm).
Since in murein and pseudomurein each second sugar residue bears one peptide part, the helical parameters described revealed likewise the approximate orientation of the peptides in respect to the sugar chain axis. Therefore, in a simplified description, in pseudomurein two neighboured peptides within one sugar chain would tend to point to opposite sides, but in murein more or less to the same side of sugar strand. Fig. 2 describes the possible relative positions of two subsequent peptides in terms of the allowed angles between these peptides if seen along the sugar axis.

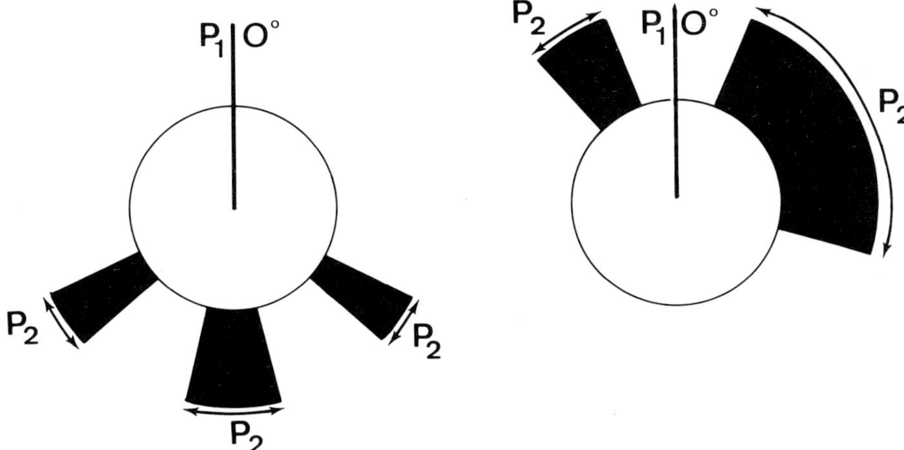

Fig. 2: Schematic graph representing the energetically favoured relative rotations of consecutive peptides along one sugar chain of pseudomurein (left) or murein (right). The circle represents the sugar chain viewed along the sugar chain axis, P1 at zero degree the position of the foremost peptide, P2 the range of favoured relative positions for the consecutive peptide. For both, murein and pseudomurein, P2 was not found at the 0° position as would be expected for a chitin-like arrangement.

At first sight, there seem to exist more different than common features in the polysaccharides of pseudomurein and murein, respectively. Nevertheless, the two cell wall polymers share in common important characteristics of their general architecture. For the sugar chains, the most important common principle seems to be, that relatively straight extended structures are preferred, which enable the sugar moieties to span a considerable distance. Apparently, this is possible even with the relatively short sugar strands found in the bacterial cell wall, from which peptides extrude not only in one direction (as it would result from a chitin-like structure of the glycan part).

Considering our previous results (5, 13), that the peptide part plays an important role for the relative order in both peptidoglycans, our data underline the well known relationship between the similar biological function and the corresponding principles of the three-dimensional architecture of these two biopolymers.

References

1. Kandler, O.: Naturwissenschaften 66, 95-105 (1979).
2. König, H., Kandler, O.: Arch. Microbiol. 121, 271-275 (1979). ibid.: 123, 295-299 (1979).
3. Kandler, O., personal communication
4. Ramachandran, G.N., Sasisekharan, V.: Advan. Protein Chem. 23, 283-438 (1968).
5. Labischinski, H., Barnickel, G., Leps, B., Bradaczek, H., Giesbrecht, P.: Arch. Microbiol. 127, 195-201 (1980).
6. Barnickel, G., Labischinski, H., Bradaczek, H., Giesbrecht, P.: Eur. J. Biochem. 95, 157-165 (1979).
7. Arnott, S., Scott, W.E.: J. Chem. Soc. Perkin Trans. II 3, 324-335 (1972).
8. Momany, F.A., McGuire, R.F., Burgess, A.W., Scheraga, H.A.: J. Phys. Chem. 79, 2361-2381 (1975).
9. Kennard, O., Allen, F.H., Brice, M.D., Hummelink, T.W.A., Motherwell, W.D.S., Rodgers, J.R., Watson, D.G.: Pure and Appl. Chem. 49, 1807-1816 (1977).
10. Tentative Rules for Carbohydrate Nomenclature: Eur. J. Biochem. 21, 455-477 (1971).
11. Conformational Nomenclature for Five- and Six-Membered Ring Forms of Monosaccharides and Their Derivatives: Eur. J. Biochem. 111, 295-298 (1980).
12. Carlström, D.: Biochim. Biophys. Acta 59, 361-364 (1962).
13. Labischinski, H., Barnickel, G., Rönspeck, W., Roth, K., Giesbrecht, P.: Zbl. Bakt. Suppl. 10, 427-433 (1981).

THE CONFORMATIONAL BEHAVIOUR OF THE PEPTIDE PART OF MUREIN - AN INFRARED AND RAMAN SPECTROSCOPIC STUDY

Dieter Naumann[x], Erwin Fischer, Wolfgang Rönspeck, Gerhard Barnickel, Harald Labischinski[x], Hans Bradaczek

Institut für Kristallographie der Freien Universität Berlin, D-1000 Berlin 33;

[x]Robert-Koch-Institut des Bundesgesundheitsamtes, D-1000 Berlin 65

Introduction

The primary structure of the peptide part of murein exhibits some unusual features such as the occurence of alternating D- and L-amino acids and a γ-bonded D-glutamic acid. Due to the ubiquitous existence of peptidoglycan in bacteria and the continuous exposure of human beings to murein a huge variety of biologically and medically important properties have been reported for this macromolecule, including immunological properties (1).
To contribute to a better understanding of structure, biological function and activity of murein and its fractions, we have performed a vibrational spectroscopic study on a variety of peptidoglycans and some of its most important partial components (2).
Our data may contribute to the following structural problems:

(i) Which is the state of order of the peptide part of murein?

(ii) Do there exist conformational similarities between the various murein peptides?

iii) Which conclusions may be drawn with respect to the actual conformations present in the peptide part of murein?

The Target of Penicillin

Results and Discussion

Infrared and Raman data have been obtained about the following materials, investigated under different conditions (crystalline, amorphous and aqueous solution): Peptidoglycan from Staphylococcus aureus, Micrococcus luteus, Bacillus subtilis and the synthesized peptides D-Glu-γ-L-Lys-D-Ala-D-Ala (tetrapeptide), L-Ala-D-Glu-γ-L-Lys-D-Ala-D-Ala (pentapeptide), MurNAc-L-Ala-D-Glu-γ-L-Lys-D-Ala (muramyl-tetrapeptide).
For comparing and discussing conformational properties of these substances we used the well defined and conformationally sensitive amide bands (2, 3), observed in the ir and Raman spectra respectively. Our data resulted in several important conclusions:

1) As may be deduced directly from band positions and half-widths of "amide I", present in the infrared and Raman spectra of murein, the peptide part of this macromolecule proved to be essentially of the non-crystalline type (fig. 1, 2; table 1). Furthermore regular structures like e.g. α-helices or ß-sheets have to be excluded (2, 3). Moreover, the state of order did not significantly change when passing from the solid state to the aqueous suspension (fig. 3). This environmental change was accompanied by a drastic shift of "amide II" band position to higher wavenumbers, whereas "amide I" is only weakly changed (fig. 3; table 1). In conjunction with other results (e.g. influence of metal ions on amide band positions) these findings can be taken as strong evidence against a completely unordered, "random-like" conformation of the peptide part.
2) The spectral characteristics of the tetra- and pentapeptide in the crystalline, amorphous as well as aqueous solution state were markedly different from those found for various peptidoglycans and the synthesized muramyl-tetrapeptide (fig. 4, 5; table 1).

Survey Spectra of Peptidoglycan from Different Bacterial Sources

most intensive bands have been normalized to an absorbance value of 1

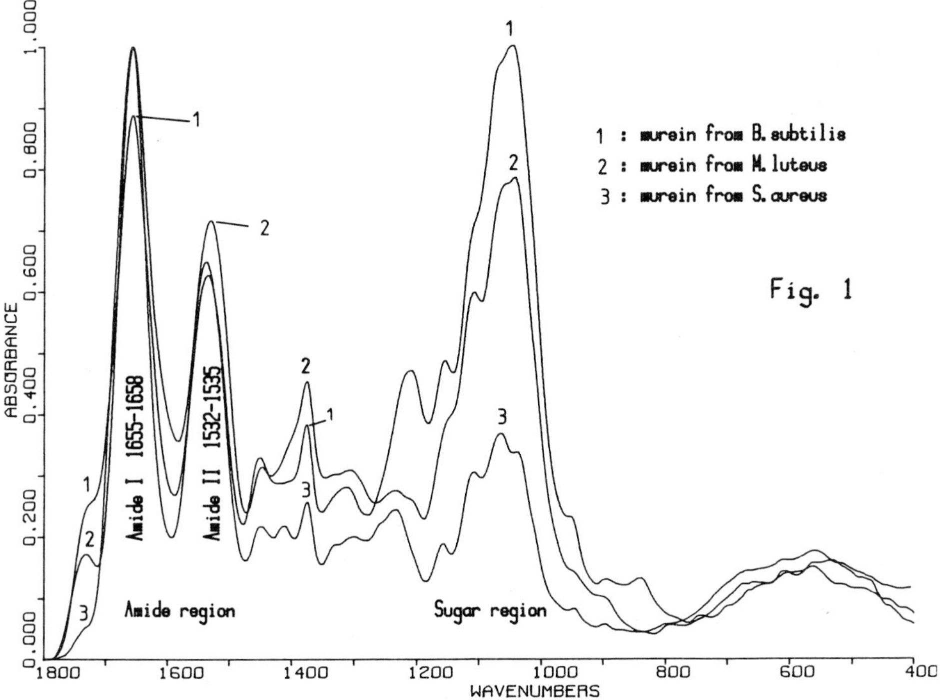

Raman Spektrum of Murein from S. aureus

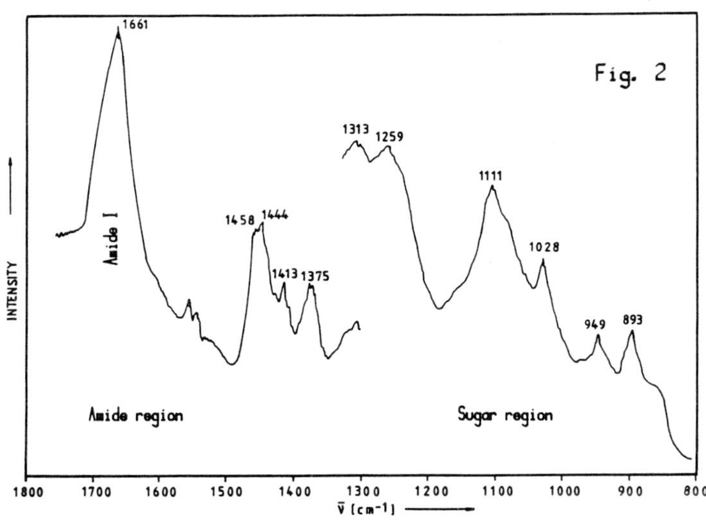

Table 1: Infrared and Raman Data Revealing Conformational Similarity between Murein and Synthesized Muramyl-Tetrapeptide

Position of Amide I Band [cm^{-1}]

Sample	Crystalline Products		Non-Crystalline Products		H_2O Solutions		D_2O Solutions	
	IR	Raman	IR	Raman	IR	Raman	IR	Raman
Murein from S. aureus	--	--	1658	1661	1652	--	1644	--
Muramyl-Tetrapeptide	--	--	1652	--	1646	--	1642	--
Tetrapeptide	1626	1659	1645	1660	1639	1652	1636	1650
Pentapeptide	1630	1661	1644	1662	1639	1655	1635	1653

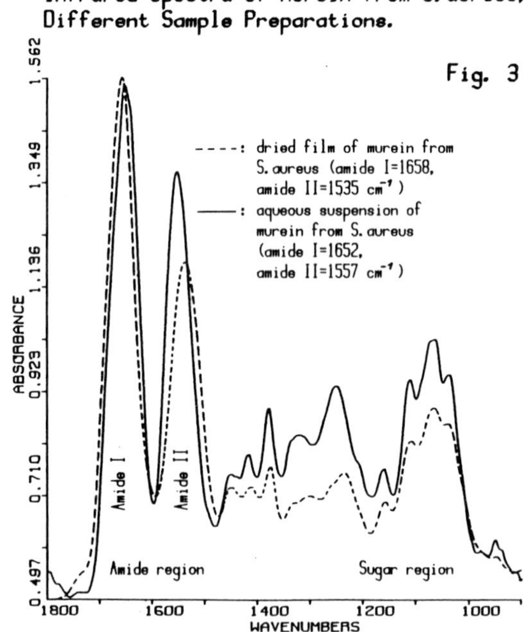

3) In contrast to these findings the corresponding features of murein and the synthesized muramyl-tetrapeptide were indeed quite similar (fig. 5, table 1).

In summary: Our results point to specific, non-covalent bonding interactions between the peptide and the sugar part of native murein which may lead to arrangements comparable to the so-called reversed-turns (3). And indeed, strongly solvated either elongated or "open-turn" like peptide structures should display "amide I" values near 1640 cm^{-1} (4) as actually found for the soluted tetra- and pentapeptide, but not for the different types of murein. Similar conclusions have been drawn independently on the basis of theoretical energy calculations (5,6).

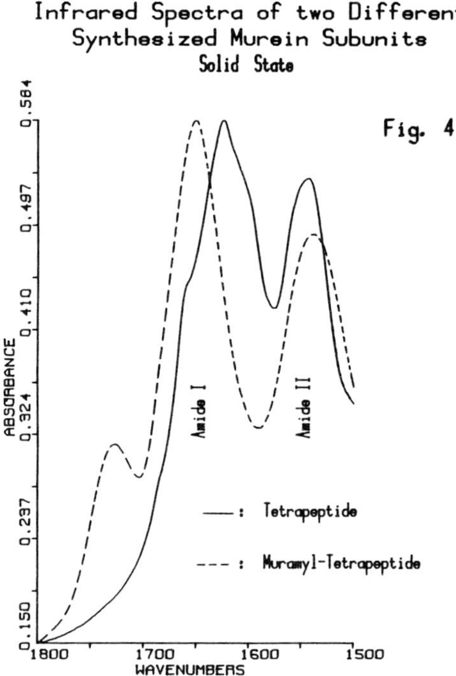

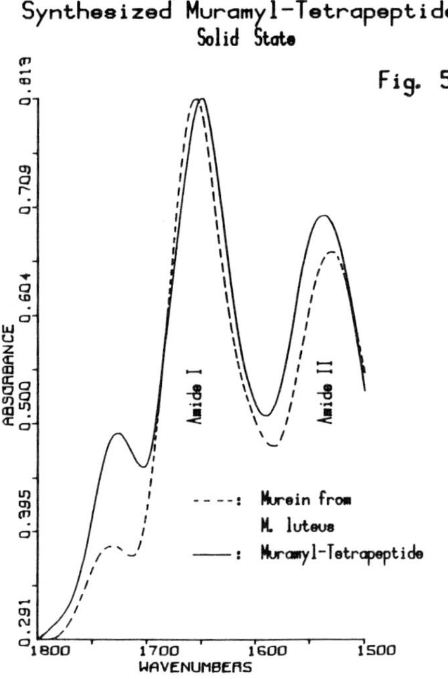

Experimental Data

The ir spectra have been recorded on a FT-IR spectrometer MX-1E from Nicolet, resolution was set to 2 cm^{-1}, wavenumber accuracy was better than ±0.01 cm^{-1}. The Raman spectra were recorded on a Cary Model 82, spectral band width 2-4 cm^{-1}; laser lines (5145 Å, 4880 Å, 100-300 mW) were produced by a Spectra Physics Ar^{+}-Laser.
Murein materials, murein films, solid peptides, aqueous solutions and suspensions were isolated, prepared, synthesized and measured as described elsewhere (2, 7).

References

1. Lederer, E.: Clin. Immun. Newsletter 3, 83-88 (1982).
2. Naumann, D., Barnickel, G., Bradaczek, H., Labischinski, H., Giesbrecht, P.: Eur. J. Biochem. 125, 505-515 (1982).
3. Krimm, S., Bandekar, J.: Biopolymers 19, 1-29 (1980).
4. de Lozé, C., Baron, M.-H., Fillaux, F.: J. Chim. Phys. 75(6), 631-649 (1975).
5. Barnickel, G., Labischinski, H., Bradaczek, H., Giesbrecht, P.: Eur. J. Biochem. 95, 157-165 (1979).
6. Barnickel,G., Naumann, D., Bradaczek, H., Labischinski, H., Giesbrecht, P.: Contribution in this volume.
7. Labischinski, H., Barnickel, G., Rönspeck, W.,Roth, K., Giesbrecht, P.: Zentralbl. Bakteriol. Suppl. 10, 427-433 (1981), in Staphylococci and Staphylococcal Infections (Jeljaszewicz, J., ed.) Gustav Fischer Verlag, Stuttgart.

DYNAMIC ASPECTS OF BACTERIAL CELL WALLS PROBED BY *IN VIVO* ^{15}N NMR

Comparative study of *S. faecalis* and mutant LYT-14 and shape changes in *B. subtilis* mutants

Aviva Lapidot and Livia Inbar
Isotope Department, The Weizmann Institute of Science, Rehovot, Israel

Introduction

Due to the lack of physical information on bacterial cell walls, few attempts have been made to explain the mechanical functions of the wall in terms of the dynamic properties of cell wall polymers.
Nuclear Magnetic Resonance spectroscopy is a powerful tool for probing the conformation, noncovalent bonding interactions and molecular motions of polymeric molecules. However the complexity of the ^{1}H and ^{13}C NMR spectra of intact cells has prevented the technique from being applied to analysis of cellular structural components under physiological conditions, while ^{15}N NMR has been largely ignored because of its low sensitivity and its low natural abundance. The considerable efforts devoted to ^{15}N NMR are certainly justified by the very interesting features of this resonance. The ^{15}N signals have narrow natural widths and dipolar broadening is smaller for an N-H fragment than for a C-H bond. This property, along with the fact that the number of different types of nitrogen functional groups is small, explains the frequent simplicity of ^{15}N spectra. The great range displayed by ^{15}N chemical shifts, as compared to proton and even to ^{13}C NMR, makes ^{15}N NMR especially attractive for the study of biological molecules. The observation of ^{15}N NMR signals in the intact cell depends on the resonance having a relatively narrow linewidth and a large negative nuclear Overhauser enhancement. Normally, we assume that when a proton decoupled resonance is not observed, it has been broadened beyond detection by a large ^{15}N-^{1}H dipole interaction or nulled by a partial nuclear Overhauser effect, both of which result from a low degree of motional freedom of the nitrogenous groups.

The Target of Penicillin

We have shown that the most complex of biosystems are possible subjects for ^{15}N NMR studies by ^{15}N isotopic labelling. Over the last several years we have applied stable isotopic labelling where natural-abundance experiments were precluded or when specific site labelling gave unique information. We have reported that ^{15}N NMR can be used as a probe for structural and dynamic properties of native cellular components (1). The technique has been proven particularly suitable for studying cell wall polymers in intact bacterial cells under physiological conditions (2). It has been used to study the dynamic properties of cell wall envelope components (3), the contribution of noncovalent bonding interaction to molecular motions of peptidoglycan (4-6), the physical interactions between antibiotics and cell wall components (7) and the nature of the molecular motions of the bridge peptide of the peptidoglycan (8). The technique has been further used to study the correlation between the arrangement of cell wall polymers and morphological features in different regions of the bacterial cell wall; the correlation between the shape of defective mutant bacteria with changes in packing of cell wall polymers; the role of autolysins in the packing of cell wall polymers and the action of various antibiotics on cell wall polymers that lead to changes in cell wall organization (9).

The objectives of the research were to provide insights into the formulation of a comprehensive molecular theory of bacterial cell wall shape and function.

Results

1) Comparative *in vivo* ^{15}N NMR study of *Streptococcus faecalis* (ATCC 9790) and its autolytic deficient mutant LYT-14. The action of cell wall autolysins has been revealed by G.D. Shockman and coworkers, who have isolated several autolytic deficient mutants of *S. faecalis* (10) (*S. faecalis* LYT-14 mutant was kindly given to us). The effects of inactivation of cell wall autolysins by $NaDodSO_4$ in comparison with non-treated and with $NaDodSO_4$ treated autolytic deficient mutant were investigated (Fig. 1A,D,F) by ^{15}N NMR spectroscopy. The resonances observed in the parent strain and the LYT-14 mutant were identical.

But the observation of a new resonance at 253.3 ppm which corresponds to acetamido groups of MurNAc (β 1-4) GlcNAc in the NaDodSO$_4$ untreated LYT-14 cells, indicate that the residual amount of cellular autolytic activity in this defective mutant partially induced autolytic cleavage of MurNAc-L-Ala peptide bond and introduced some degree of mobility into the rigid glycan structure.

Exposure of exponential culture of S. faecalis to mitomycin C allowed wall elongation to continue at the expense of septation, resulting in elongated cells (11). In contrast, the protein synthesis inhibited by chloramphenicol resulted in inhibition of cell separation and in the capacity of cells to autolyze leading to wall thickening. These phenomena were demonstrated by ^{15}N NMR (Fig. 1, B,C,E).

^{15}N NMR was used to probe the changes in the mobility of cell wall polymers that occur under conditions leading to disruption of cross links or bridge regions of the peptide in the peptidoglycan. The effect of phosphate concentration on cell wall organization of S. faecalis w.t. and LYT-14 has been studied (Fig. 2). S. faecalis is an ideal system for making definitive spectral assignments, since the specificity of its amino acids requirements were completely characterized. When S. faecalis w.t. and LYT-14 were grown in media in which only $(NH_4)_2SO_4$ was replaced by $^{15}NH_4Cl$, the specifically labeled ^{15}N cells, cell walls and cell wall lysozyme digests enabled us to monitor ^{15}N NMR signals which originated only from amidated polysaccharides and peptidoglycan Gln and Asn amide residues (to be published). ^{15}N NMR spectra of cell wall lysozyme digest have been found to provide a fingerprint of the primary structure and conformation of peptide chains, and the relative amounts of N-acetyl-hexosamines present in the sample (Fig. 3). Quantitative analysis of lysozyme digest spectra by computer simulation can reveal differences in chemical composition of cell wall digests and is probably more sensitive than conventional biochemical methods (2).

2) In vivo ^{15}N NMR studies of shape changes in Bacillus subtilis temperature sensitive mutants. Temperature - sensitive rod mutants of B. subtilis (12) obtained from H.J. Rogers were investigated using ^{15}N NMR spectroscopy to search for changes in cell wall organization during

morphological changes (Fig. 4). The effects of autolysis and antibiotics on cell wall organization of the parent strain in comparison to the mutant strains, have been studied as well. Our ^{15}N NMR observation support the contention that a decrease in negative-charged wall polymers leads to a change from rod to spheres. We have also investigated the reversion of B. subtilis mutants from deformed cocci to normal rods by in vivo ^{15}N NMR. The differences in the spectra observed reflected changes in the mobility along the peptidoglycan chain (to be published).

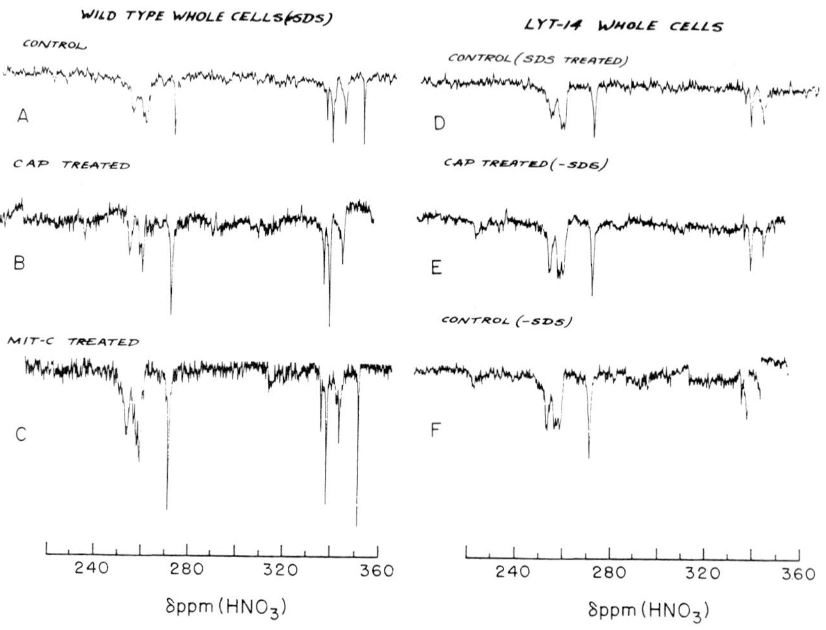

Fig. 1. The proton-decoupled 9.12MHz ^{15}N NMR spectra at 25°C of ^{15}N enriched cells of S. faecalis w.t. (A,B,C) (treated with $NaDodSO_4$ upon completion of growth), and its autolytic deficient mutant LYT-14 (D,E,F). spectral conditions were as in Ref. 2. For assignments of resonances, see refs. 2 and 5.

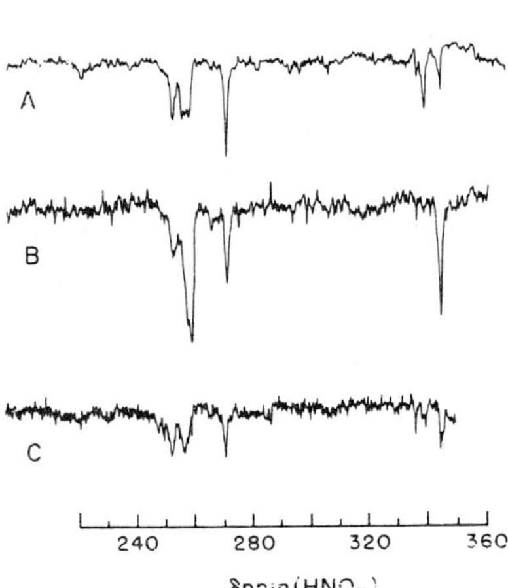

Fig. 2. Effect of phosphate concentration on cell wall polymers organization in S. faecalis LYT-14, grown at pH=7 in phosphate limited medium (0.07M) and in normal phosphate medium (0.3M). The differences in the two growth conditions affected the amount of cross bridging. Free Asn resonances at 338.4 ppm appeared only in the spectra of cells grown at 0.07M phosphate medium (A) whereas free N_{ε}-Lys resonance (343.5 ppm) is significantly increased in cells grown on 0.3M phosphate.

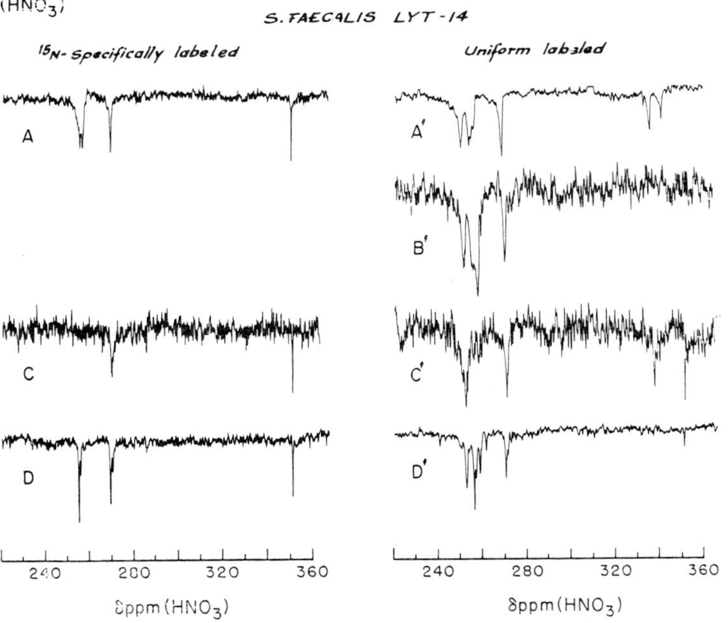

Fig. 3. The proton-decoupled ^{15}N NMR spectra (at 25°C) of cells (A,A'), cell walls (B'), TCA treated cell wall (C,C') and cell wall lysozyme digest (D,D') of S. faecalis LYT-14 grown on limited phosphate medium (0.07M), in which only $(NH_4)_2SO_4$ was replaced by (1g/L) ^{15}N NH_4Cl (enriched to >90%), (A,C,D). Uniformly labeled cells were obtained when cells were grown on a medium in which the synthetic amino acid mixture was replaced with ^{15}N-labeled amino acid mixture (1 g/L) enriched to >90% (A',B',C',D').

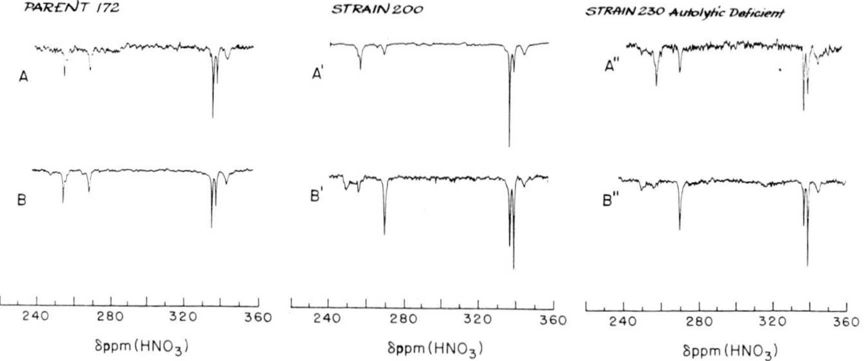

Fig. 4. The proton decoupled ^{15}N NMR spectra of intact cells of B. subtilis parent strain and its mutants rod A 200 and 230 at 32°C (A,A',A") and 42°C (B,B',B"). Cells were grown in a media in which $(NH_4)_2SO_4$ was replaced with ^{15}N NH_4Cl (1 g/L) (enriched to >90%).

Acknowledgements. This study was supported by a grant from BSF to A.L.

References

1. Lapidot, A., Irving, C.S.: Proc. Natl. Acad. Sci. U.S.A. 74, 1988-1992 (1977).

2. Lapidot, A., Irving, C.S.: Biochemistry 18, 704-714 (1979).

3. Irving, C.S., Lapidot, A.: Biochim. Biophys. Acta 470, 251-257 (1977).

4. Lapidot, A., Irving, C.S.: Pepd. Proc. Ann. Pepd. Symp. 5th, 419-422 (1977).

5. Lapidot, A., Irving, C.S.: Jerusalem Symp. NMR Spectrosc. Mol. Biol. 11th, 439-459 (1978).

6. Irving, C.S., Lapidot, A.: Stable Isotopes, Proc. Int. Conf. 3rd, 307-316 (1978).

7. Irving, C.S., Lapidot, A.: Antimicrob. Agents. Chemother. 14, 695-703 (1978).

8. Lapidot, A., Irving, C.S.: Biochemistry 18, 1788-1790 (1979).

9. Inbar, L., Lapidot, A.: Xth Int. Conf. Magnetic Resonance in Biol. Syst. 209 (1982).

10. Cornett, J.B., Redman, B.E., Shockman, G.D.: J. Bact 133, 631-640 (1978).

11. Hinks, R.P., Daneo-Moore, L., Shockman, G.D.: J. Bact. 134, 1074-1080 (1978).

12. Rogers, H.J., Taylor, C.: J. Bact. 135, 1033-1042 (1979).

A NOVEL AND RAPID METHOD FOR IDENTIFYING IMPORTANT CELL WALL PARAMETERS: FOURIER-TRANSFORM INFRARED SPECTROSCOPY

Dieter Naumann, Harald Labischinski, Gerhard Barnickel, Peter Giesbrecht

Robert-Koch-Institut des Bundesgesundheitsamtes,

Nordufer 20, D-1000 Berlin 65

Introduction

The analysis of biological systems, having such a tremendous complexity like bacterial cell walls requires a multifacetted approach by applying highly sophisticated methods. One of these recently developed techniques is the Fourier-Transform Infrared Spectroscopy. FT-IR means essentially Infrared Spectroscopy which, however, is executed now with revolutionary improved instrumental specifications: Dramatically enhanced sensitivity, wavenumber accuracy and scanning speed. This progress is primarily due to the application of interferometric methods and powerful mini-computers to ir-spectroscopy. The inherent advantages of FT-IR directly lead, in praxis, to new possibilities of research: Structural investigations on transient and reacting systems, multicomponent analysis, detection of minute contaminations, fully computerized identification processes and investigations on really complex biological systems under native conditions by applying absorbance subtraction capabilities of FT-IR (1). Using some representative examples, range of application and prospects of this method for identifying important bacterial cell wall parameters are presented.

Results and Discussion

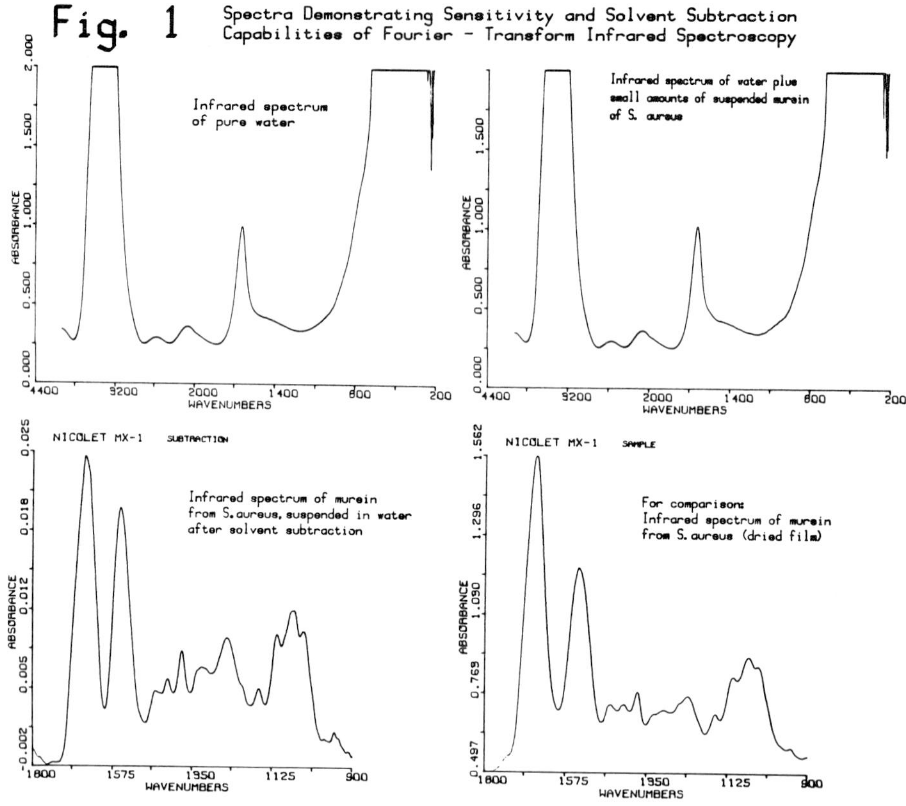

Fig. 1 Spectra Demonstrating Sensitivity and Solvent Subtraction Capabilities of Fourier - Transform Infrared Spectroscopy

In order to check what is now within reach of this FT-IR, we tried to get structural informations from biological systems in aqueous solutions or suspensions, i.e. from systems close to real life conditions. We were especially interested in suspensions of mureins in water (fig. 1): Reference spectrum (water) and sample spectrum (water plus 10^{-2} to 10^{-3} mg of murein) proved to be very similar. Any trial to force informations about native murein out of the sample spectrum <u>alone</u> failed. Only when applying the solvent subtraction capabilities of FT-IR, the spectral characteristics of suspended murein were readily accessible and structural considerations were possible (e.g. the comparison of dried films and aqueous suspensions of murein, investigations on influence of metal ions, enzyme activity etc.).

One of our major goals in applying FT-IR to bacterial cells and cell wall fractions was the detection of possible chemical and structural modifications within the bacterial cell wall as a function of antibiotic treatment, growth time and growth medium - and to correlate corresponding findings to the degradability of cell wall material by lytic enzymes under phagocytic conditions. As the degree of O-acetylation on C-6 position of muramic acid is one possible regulatoric factor controlling degradability of cell walls by lysozyme (2), we recently have developed a special "film technique" for a simple and quantitative ir spectroscopic determination of O-acetyl-groups within cell wall materials (3). To our surprise, applying the same technique to intact bacterial cells yielded reasonable results, too, thus avoiding time consuming and drastic procedure during isolation process of cell walls. It could furthermore be shown that alterations in the O-acetyl content of cell walls may be detected from aqueous suspensions as well (fig. 2). The tiny peak near 1740 cm^{-1} of log. phase cell walls, represented the quantity of approximately 100 nmol O-acetyl groups per mg cell wall in an absolute amount of only 10^{-2} to 10^{-3} mg of material present in the suspension. Upon mild alkali treatment (fig. 2) this band, obviously, was removed. Such a treatment led to some further alterations within the wall material which can, however, better be interpreted by considering the subtraction of both spectra (fig. 3): O-acetyl groups were removed by alkali treatment as the O-acetyl peak near 1740 cm^{-1} went to the negative; acetate ions had gone into solution indicated by a band near 1400 cm^{-1} (symmetrical stretching of $-COO^{-}$) which went to the positive; a band near 1260 cm^{-1} went to the negative indicating that phosphate groups of covalently bound teichoic acid apparently were affected by mild alkali treatment, too.

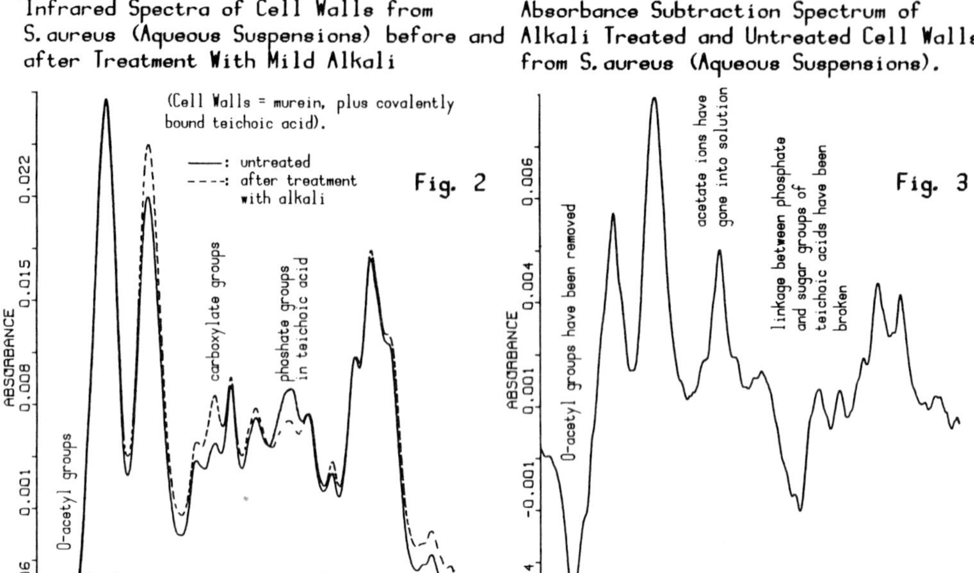

This example clearly demonstrates that minute alterations within cell walls are readily accessible by FT-IR even from aqueous suspensions; this procedure may be performed in a rapid and straightforward manner, yielding a great bulk of structural and quantifiable informations simultaneously. Thus, (as could already be demonstrated for the quick ir-spectroscopic determination of cross-linking indices (4)) FT-IR may be considered indeed as a powerful method to detect sensitively important cell wall parameters. Moreover, even intact, living bacterial cells may now readily be measured to study biochemical processes like wall lysis, autolysis of cells, cells under different osmotic pressures, action of antibiotics etc.. Fig. 4 presents for the first time a high quality ir spectrum of living log. phase cells from Staphylococci.

Most recently we could demonstrate that FT-IR also proved to be convenient for the differentiation and identification of whole bacterial cells. Indeed, a variety of different bacteria

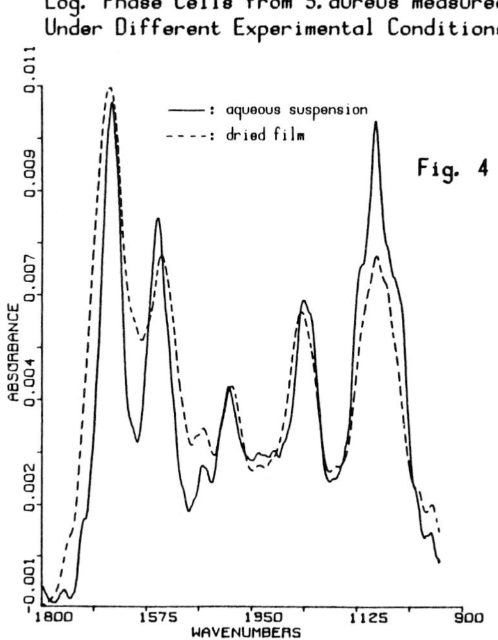

exhibited reproducible "finger-prints" throughout the mid infrared spectral region (data not shown).

Meanwhile, even more complicated identifications could be performed, too: Streptococci of different groups A and B, measured under identical conditions, exhibited rather similar "finger-prints" (fig. 6). Execution of digital subtraction of both spectra (fig. 7) revealed distinct <u>and</u> identifiable differences between both species: Mayor deviations predominantly occur in the so-called "sugar region" of the spectrum indicating the well-known fact that different carbohydrates are present in the cell wall of both groups of Streptococci.

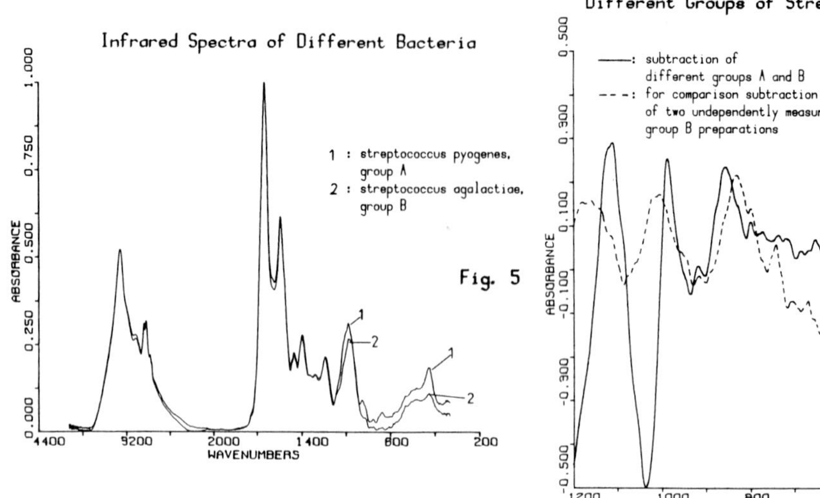

Conclusion

Application of FT-IR to complex biological and biomedical systems has barely scratched the surface. Here it could be examplarily shown that this new technique is indeed suitable to detect rapidly, sensitively and even under "in vivo" conditions important cell wall parameters. Once the power of this technique has been transmitted through scientific community of biologists, biochemists and biomedicals many more applications will be found in the near future.

References

1. Ferraro, J.R., Basile, J.L.: Fourier Transform Infrared Spectroscopy, Application to Chemical Systems, Vol.1 and 2, Academic Press, New York, San Francisco, London, 1978.
2. Johannsen, L., Labischinski, H., Burghaus, P., Giesbrecht, P.: Contribution in this volume.
3. Burghaus, P., Johannsen, L., Naumann, D.,Labischinski, H., Bradaczek, H., Giesbrecht, P.: Contribution in this volume.
4. Naumann, D., Barnickel, G., Labischinski, H., Giesbrecht, P.: Eur. J. Biochem. 125, 505-515 (1982).

CALORIMETRIC STUDIES ON THE BINDING OF VANCOMYCIN - LIKE ANTIBIOTICS TO PEPTIDOGLYCAN

José Laynez

Thermochemistry Laboratory, Chemical Centre,
University of Lund, S - 22007 Lund, Sweden

Alfredo Rodriguez-Tébar, Vicente Arán, David Vázquez

Centro de Biologia Molecular CSIC, UAM
Madrid - 34, Spain

Introduction

It is essential to elucidate the thermodynamic parameters governing the interactions between biological molecules in order to understand the real nature of such interactions. Some binding parameters for quantitative determination of macromolecular interactions, such as equilibrium constants and free energy changes, can be determined indirectly by a variety of methods. However, these studies can be carried out by direct measurement of the heat involved in the process of interaction either to know the parameters of the interaction or simply as an analytical tool when dealing with rather complex systems. Indeed microcalorimetric studies have been widely developed in biochemical research due to the recent availability of new calorimetric devices that allow the measurement of the very low levels of heat involved even when small amounts of materials are used (1,2). We present here data obtained by calorimetric studies on the binding of the antibiotics vancomycin, ristocetin A and teichomycin A_2 to (a) analogs of the peptide moiety of peptidoglycan and (b) peptidoglycan.

The Target of Penicillin

Methods

N-Ac-D-Ala (Sigma Chem. Co., St. Louis, Mi. USA) and N-Ac-D-Ala-D-Ala, that we have synthesized as previously described (3), were used as analogs of the peptide moiety of peptidoglycan. Peptidoglycan from E. coli MC6 (wild type) and JE 5684 (a carboxypeptidase deficient mutant) were obtained as described (4). The antibiotics vancomycin (Ely Lilly, Indiana, USA), ristocetin A (Lundbeck, Copenhagen, Denmark) and teichomycin A_2 (Lepetit, Milano, Italy) were kindly given to us.

Calorimetric measurements were carried out by using a LKB titration assembly (2107-350) attached to a standard LKB batch microcalorimeter (2). Binding experiments were performed by stepwise injection of 10 to 15 µl of either N-Ac-D-Ala or N-Ac-D-Ala-D-Ala solutions into the reaction vessel containing a 4 ml solution of the required antibiotic at a concentration of 0.5 mg/ml. In the binding experiments with peptidoglycan preparations the antibiotics were injected with a syringe into the calorimeter vessel containing a 4 ml suspension of peptidoglycan. The heat of dilution of the peptides and of the antibiotics were cancelled by injection of the same amount of solution in the reference vessel. All solutions were made up in 0.1 M sodium phosphate buffer, pH 7.0 and the calorimetric experiments were performed at 25°C using the same buffer.

Results and Discussion

In order to elucidate the parameters for the interaction of teichomycin A_2 with either N-Ac-D-Ala or peptidoglycan, the corresponding microcalorimetric measurements were worked out.

The values obtained in the titration curves (Figures 1a and 1c) were analyzed using a graphical method (2) which allows the simultaneous calculation of ΔH and K values (Figures 1b and 1d). There is an 1 : 1 interaction of teichomycin A_2 with N-Ac-D-Ala since all curves cut in a single point (Figure 1b). On the other hand the ΔH and K parameters for peptidoglycan interaction could not be obtained since the data obtained in the titration curve do not fit in the 1 : 1 graphical analysis because there is not a single point intersection (Figure 1d). Similar measurements were performed with (a) teichomycin A_2 and N-Ac-D-Ala-D-Ala, (b) ristocetin A and either N-Ac-D-Ala or N-Ac-D-Ala-D-Ala, and (c) vancomycin and either N-Ac-D-Ala or N-Ac-

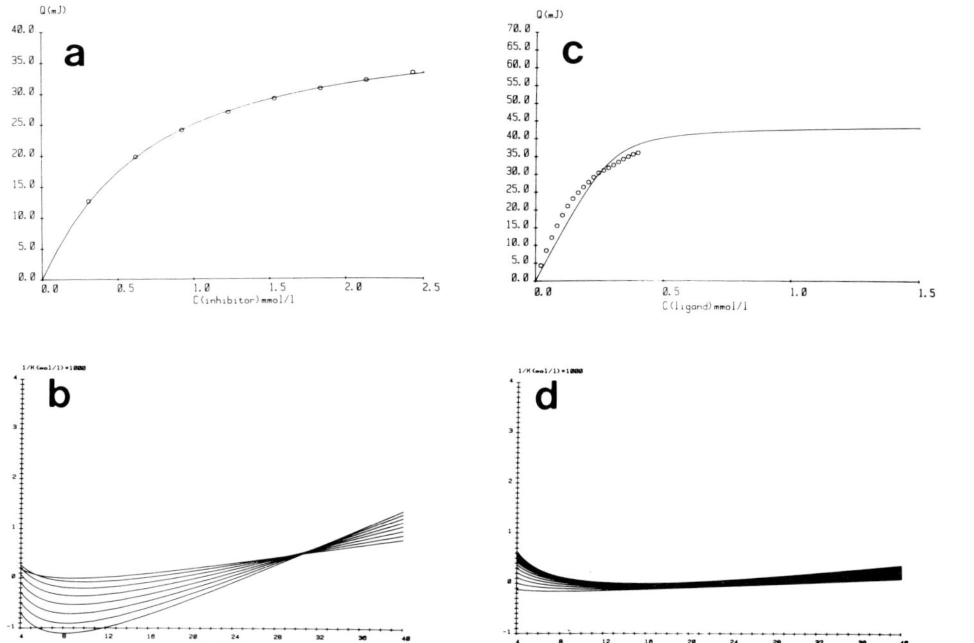

Figure 1. Calorimetric data of teichomycin A_2 interactions. Titration curves: (a) N-Ac-D-Ala; (c) peptidoglycan. Graphical analysis to estimate the ΔH and K values: (b) N-Ac-D-Ala; (d) peptidoglycan. For (a) and (c): (o—o) experimental points; solid line was calculated as explained previously (ref. 2).

D-Ala-D-Ala. The results obtained are summarized in Table 1 and show that N-Ac-D-Ala-D-Ala has a higher affinity than N-Ac-D-Ala for any of the three antibiotics studied, in agreement with data reported by NMR studies with those substrates and vancomycin (5). The values of K with both substrates, N-Ac-D-Ala-D-Ala and N-Ac-D-Ala are in good agreement with those previously reported by spectrophotometric studies using either vancomycin or ristocetin A (3,6). The process is enthalpically driven in all cases while entropic values are negative, possibly as a result of the formation of hydrogen bonds between both ligands (7). Interaction of D-Ala-D-Ala with any of the three antibiotics was not detected in agreement with spectrophotometric studies on the interaction of D-Ala-D-Ala with vancomycin (3).

Table 1

Thermodynamic parameters of binding of vancomycin - like antibiotics to analogs of the peptide moiety of peptidoglycan (a)

Interaction of:	$-\Delta H^o$	K	$-\Delta G^o$	$-\Delta S^o$
N-Ac-D-Ala with:	kJ mol^{-1}	L mol^{-1}	kJ mol^{-1}	$JK^{-1}mol^{-1}$
Vancomycin	32.05	296	14.11	60.17
Ristocetin A	28.35	1110	17.38	36.79
Teichomycin A_2	31.24	1846	18.64	42.26
N-Ac-D-Ala-D-Ala with:				
Vancomycin	34.28	$4.6x10^4$	26.60	25.76
Ristocetin A	30.68	$1.4x10^5$	29.45	4.12
Teichomycin A_2	39.11	$2.0x10^5$	30.28	29.61

(a) mean values of four experiments

Table 2

Apparent dissociation constants (μM) of the vancomycin - like antibiotics for E.coli peptidoglycan

Antibiotic	E.coli strain	
	MC 6	JE 5684
Vancomycin	357	86
Ristocetin A	139	86
Teichomycin A_2	238	77

The ΔH and K parameters for the interaction of the antibiotics with peptidoglycan could not be obtained in any case, as indicated above when teichomycin A_2 was used (Figures 1c and 1d). The data obtained revealed a complex interaction. This is not surprising since E.coli sacculi appear to contain more than one kind of binding sites for vancomycin-like antibiotics.However, apparent dissociation constants for the association of the antibiotics to peptidoglycan can be obtained from reciprocal plots (Table 2). These apparent dissociation constants have only a phenomenological meaning, but can be useful for semiquantitative comparison of the affinities of the antibiotics for peptidoglycan species obtained from the wild type of E.coli and the carboxypeptidase-deficient mutant. In general, affinities of the three antibiotics are higher for peptidoglycan from the E.coli mutant which should be expected to have a higher amount of D-Ala-D-Ala residues exposed to the external milieu. Unlike affinities for substrate analogs, the affinities of the three antibiotics studies for peptidoglycan are roughly similar. This findings might indicate that steric hindrance plays a relevant role in the binding process.

References

1. Spink, C., Wadsö, I.: In: "Methods of Biochemical Analysis" Vol. 23, 1-160 (Glick, D. ed.), John Wiley and Sons, New York 1976.

2. Chen, A., Wadsö, I.: J. Biochem. Biophys. Methods 6, 307-316 (1982).

3. Nieto, M., Perkins, H.R.: Biochem. J. 123, 789-803 (1971).

4. Primosigh, J., Pelzer, H., Maass, D., Weidel, W.: Biochim. Biophys. Acta 46, 68-80 (1961).

5. Brown, J.P., Terenius, L., Freeney, J., Burgen, A.S.V.: Mol. Pharmacol. 11, 126-132 (1975).

6. Perkins, H.R., Nieto, M.: In: "Proceedings Symposium on Molecular Mechanisms of Action on Protein Synthesis and Membranes" (Granada, 1971), 363 (Muñoz, E., García-Ferrándiz, F., Vázquez, D., eds.) Elsevier, Amsterdam (1972).

7. Kalman, J.R., Williams, D.H.: J. Am. Chem. Soc. 102, 906-912 (1980).

PART II

MODELS FOR THE GROWTH OF THE MUREIN SACCULUS

THE SHAPES OF GRAM-NEGATIVE ORGANISMS: VARIABLE-T MODELS

Arthur L. Koch
Department of Biology, Indiana University
Bloomington, IN 47405

Introduction

Gram-negative microorganisms possess only a very thin murein sacculus to resist the stress caused by the internal hydrostatic pressure. The sacculus is less than one molecular layer thick of unstressed peptidoglycan and therefore must enlarge by the prior insertion and crosslinking of new murein before selective cleavages of stress-bearing murein permit enlargement. Since insertion of new murein occurs all over the surface of *Escherichia coli* including completed poles (1), the internal pressure would tend to force the cells into a spherical shape and prevent both cylindrical elongation and cell division. Of course, Gram-negative bacteria do achieve a variety of shapes and do divide. This paradox can be resolved if we postulate that the details of the biochemical mechanism for wall growth vary in different regions of the surface. Depending on the degree and rate of change in biochemical energetics, it is possible to account for rod shapes and the other more complex shapes of certain Gram-negative organisms. It is also possible to understand the biophysical basis of the constriction and cell division in these organisms that do not have cytoskeletons as do eukaryotic cells. It provides the Gram-negative bacteria with an alternative way to transduce biochemical free energy into the mechanical work that is needed to achieve non-spherical shapes and to divide.

The Target of Penicillin

After considering a number of models (2), only one general model seems to account for the important facts about the morphology of *Escherichia coli*. This model, the variable T model, assumes that T, the analogue of surface tension, varies in different regions of the cell surface and varies in different parts of the cell cycle. This is in contradistinction to the Gram-positive situations (3) where T appears to be constant through the cycle and throughout the area where side wall growth is taking place. In the physics of surfaces, surface tension is defined as the work needed to increase the surface area by a unit amount. The work needed to increase the surface area of a bacterium arises in a quite different way from that of a soap bubble, although pressure-volume work is still transduced to force the wall enlargement process (3). When the wall area of the prokaryote is enlarged by cleaving stress-bearing bonds the surface stress is then transferred to previously linked unstressed units and pulls them into an extended configuration. In this way, pressure-volume work is converted into the work of extending the wall.

Possible circumstances that could cause T to vary in order to create shape changes include variability in: the length of the peptidoglycan oligomers, the coupling of the hydrolysis of the second terminal *D*-alanine to the transpeptidation process, the degree of cross-linking, the incorporation of chemically different peptide chains, and the optional incorporation of additional polyglycine (etc.) extensions to the cross-bridges. There are a number of other possibilities not to be summarized so neatly: the environment may be locally altered (similar to processes well-known in higher organisms) due to specially localized ion currents. Thus, if the proton pump is more active in one region of the cell than another, the pH and zeta potential will be altered in that locality and this could vary the local free energy of the transpeptidation reaction. Secondly, the charge on the newly synthesized wall may be different after it matures. For example, new walls with amino groups neutralizing the charges of carboxyl group of aspartic acid residues may be secreted and bound to old wall. Subsequently the amino groups may be cleaved, increasing the net negative charge. This would serve as a way to delay expansion until after the cleavage of the peptidoglycan bridges.

One of the most attractive proposals for Gram-negative rod growth and division is that on the sides and at old poles, single glycan chains are inserted, but that at the constriction site rows of chains formed by first linking a group of new oligomeric glycan chains are inserted into the stress-bearing wall in the same way as single oligomers are inserted elsewhere; e.g., the newly introduced material bridging adjacent regions of the stress-bearing wall. The insertion of a prelinked array of oligopeptidoglycan chains instead of single oligopeptidoglycan chains has almost the equivalent effect of pre-building a septum as done by the Gram-positive organisms. In both cases when cleavages of external wall are made, larger increments of external surface result than result from the insertion of a single oligopeptidoglycan into the external wall.

Diffuse Wall Growth

An equation for numerically computing cylindrically symmetric membrane shapes has been derived (2, 4). Now we need a version that can take into account the possibility that T is a variable. It can be shown that taking T as a variable leads to no change in the mathematics presented in references (2) and (4). The previous equations can be validly applied if T is a function of radius (or time) or anything but a function of distance in the direction of the cylindrical axis. I have since found a better way to formulate the equations for numerical analysis. The revised program calculates a new slope, S, from the old slope, S_0, at an adjacent point where the radius is $\underline{r}_0$ according to:

$$1 + S^2 = [1 + S_0^2]\left[\frac{1 + \Delta S_0/\underline{r}_0}{T_0/T + P\Delta S\ (1+S_0^2)^{1/2}/T}\right]^2,$$

where Δ is the increment interval of the axial distance, $\underline{z}$. This equation can be rearranged to give:

$$S = \sqrt{S_0^2 + [1+S_0^2]\left[\left(\frac{1 + \Delta S_0/\underline{r}_0)}{1 + P\Delta S_0(1+S_0^2)^{1/2}/T_0}\right)^2 - 1\right] + \left(\frac{T-T_0}{T}\right)^2}.$$

If S_0 is zero, then the system is metastable and cylindrical extension will continue, however a slight perturbation is enough to lead to a division or a blow out. For the case where the initial inward slope was only 10^{-6} degrees and T is two fold or more times smaller than the value that give stable cylindrical growth then division ensues. If the initial slope had been 1°, then the curves would be infinitesmally changed. If the initial angle had been outward by 10^{-6} degrees, then the curves would be shifted infinitesimally towards longer poles and the diameter would increase infinitesimally before it decreased. Consequently, special mechanisms to start the inward growth need not be postulated. Moreover, it can be concluded from these simulations that the theory can explain constriction and division by proposing a several-fold decrease in T. Thus, a small alteration in the chemical details of wall formation is sufficient to lead to division.

The Requirements for Constrictive Division

T must decrease at least 2-fold for growth to lead to division if there are only very small fluctuations giving rise to the deviation from metastable elongation. If however, some special mechanism exists that gives the wall a momentary inward growth, then even a smaller change in T will suffice. For example, if T decreases 1.75-fold, above an inward angle of 30° the constriction develops into a V-shape and ultimately leads to division with a rounded pole tip.

The magnitude of the discontinuous shift needed in T for the computer to simulation the electron microscope morphology of dividing *E. coli* is about 3 fold. Examination of many electron photomicrographs show that the poles as initially formed are flatter and then progressively approach a hemispherical shape and become slightly pointed (3).

Figure 1 shows the shape that the nascent division site of a cell would have if a progressive reduction in T occurred in the central region. The Other calculated pole shapes shows that a very gradual decrease in T leads to an elongated pole which a more rapid decrease leads to a flatter one. Surprisingly, a very rapid rate of change in T give a blunter pole than a discontinuous change in T. This is because it quickly converts a chance fluctuation into strong inward growth. This implies that the temporal gradation of switching from one biochemical mechanism to another may have more profound effect on the pole shape than the ultimate extent in the change in T.

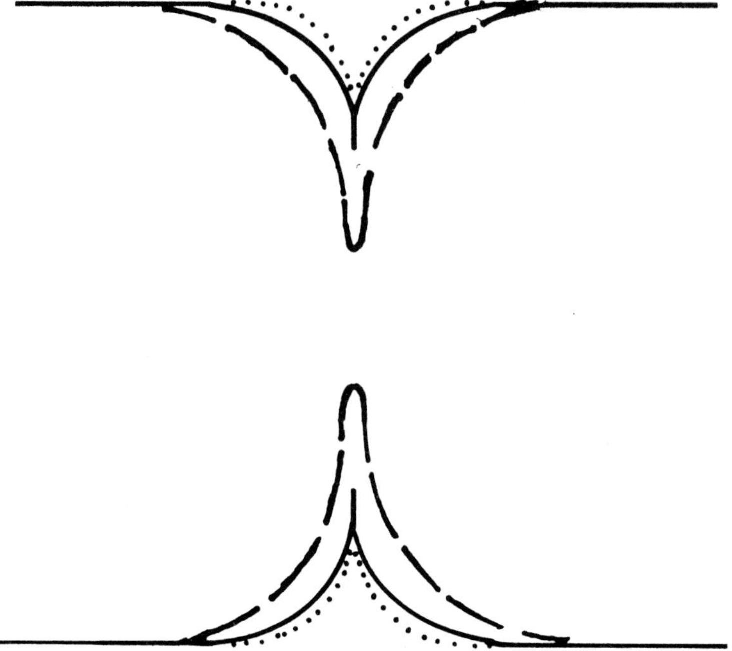

Fig. 1. Calculated shape of a deep constricted site. All curves are for a 3-fold decrease in T from the level that gives cylindrical extension. The dashed curves shows the most gradual shift in T; the solid curve represents a more abrupt change to the 3-fold decrease; the dotted line the most abrupt. If the change is discontinuous the trend reverses and the solid curve results. All calculations are for an inward slope of 10^{-6}.

References

1. Koch, A.L., Verwer, R.W.H., and Nanninga, N.: J. Gen. Micro. 128, 2893-2898 (1982).
2. Koch, A.L.: J. Gen. Micro. 128, 2527-2540 (1982).
3. Koch, A.L.: Adv. Micro. Physiol. 24, 301-367 (1983).
4. Koch, A.L., Higgins, M.L., and Doyle, R.J.: J. Gen. Micro. 128, 927-945 (1982).

MUREIN RING STRUCTURE AND BIOSYNTHESIS IN GRAM-NEGATIVE BACTERIA: SOME TENTATIVE CONTEMPLATIONS

Helmut Pelzer
Thomae Research Laboratories, D-7950 Biberach/Riss

Introduction

Considering the structure of bacterial cell wall murein there exists a gap of knowledge on the orientation of the glycan chains within the sacculus. Obviously, the mechanism of sacculus growth crucially depends on the direction in which the peptidoglycan has to polymerize, in rod shaped bacteria even in such different processes as cylinder elongation and septum formation.
In trying to solve this problem, the present author contemplated a murein model which is compatible with the known biochemical and biological data. For that as the simplest possible and most elaborated murein, the sacculus of E. coli was chosen as a prototype.

Twenty years ago, in the early stage of murein research, Weidel and Pelzer (1) suggested that the glycan of E. coli murein may be arranged longitudinally to the rod length. However, because of difficulties in creating a plausible biochemical mechanism of sacculus growth, this concept was given up in favour of a ring structure hypothesis (2,3). In its detailed form (4) this ring model is consistent with most data on E. coli cell wall growth known as yet.

The ring structure model of rod shaped monolayered murein sacculi

In this ring model the glycan chains of the peptidoglycan are arranged circularly around the cell cylinder like the loops round a cask. They are held together by the octapeptides consisting of L-ala, D-glu, m-dap, and

D-ala (fig. 1). A particular glycan chain must not necessarily cover the whole diameter of the sacculus forming a ring closed by itself. Since the polysaccharides are rather resistant to elongation, the diameter of the cylinder is largely constant during the growth circle of the cell.

The pole caps of the sacculus are also constructed of peptide-bridged polyglycan rings but, in contrast to the cylinder, of successively smaller diameters (fig. 1).

According to muropeptide analyses of E. coli murein, half of its peptides are cross-linked, the rest consists of tetra- and tripeptides, the function of which will be discussed below.

Murein degrading effects of autolysin(s) as deduced from the ring model

When murein synthesis is suddenly stopped, e.g. by a ß-lactam antibiotic, murein degradation evoked by autolysins takes place presumably at "growing points" or "growing zones" (5). Autolysins responsible for that event do not seem to have muramidase but rather muramylamidase or endopeptidase activity. Corresponding to the ring model they split the sacculus preferentially between two neighboured glycan rings. That occurs mainly in areas where the murein is hypersensitive to those enzymes. The morphological consequences are well known (6, 7).

Biochemical mechanism of sacculus growth and septation

The sacculus is a body closed on all sides. For its growth, newly synthesized peptidoglycan must be inserted between preexisting (old) murein. The simplest and most perceptual mechanism seems to be an enzymatic opening of chemical bonds in the sacculus, followed by an immediate filling of the nascent holes with new peptidoglycan (1). The murein hydrolase necessary for this process is, in the longitudinal model, a muramidase or glycosyltransferase (1, 8) or in the ring model an endopeptidase (7). Looked at closely, the longitudinal model can hardly work in this way (fig. 2), whereas the ring model is absolutely predestinated for such a mechanism (fig. 3).

The function of autolysins for sacculus growth is controversial since autolysin deficient mutants have been isolated (9). If these bacteria indeed are lacking a murein hydrolase, an alternative mechanism of insertion must be considered. For that a so far hypothetical "crosslink transferase" has been proposed (4, 10). This enzyme using as the one substrate tetrapeptide moieties of a polyglycan would not be able to join new cross-links (fig. 4) without breaking old ones. Hence, in any case, sacculus growth requires the transpeptidase in addition. Both these enzymes in combination could produce cylinder extension via insertion of new rings or ring fragments.

Because of its obvious simplicity, septum formation is perhaps the strongest evidence for the ring model. It can be started merely by blocking the endopeptidase or crosslink transferase at the growing zone of the cylinder. The transpeptidase then adhere to the inside surface of the cylinder, new glycan rings being followed by further rings of successively smaller diameters. Obviously no peptidoglycan insertion is necessary during this process (fig. 5).

Since incomplete septa do not have to withstand an inside-out difference of osmotic pressure, the rings can be packed most tightly to each other during synthesis. After septum murein has been finally closed, a special autolysinat the cylinder site between the septa is switched on enabling the daughter cells to separate. Owing to the flexibility of the interpeptide angles (fig. 6) the septum murein can now be osmotically pressed outward to form a pole cap. This sudden capping from flat septa would not work in the longitudinal model. In that case the glycan chains have to run radially from the cylinder to the center of the septum. Irrespective of the geometric difficulties at lodging radial glycans into the septum, their resistance to elongation forbids them to stretch as it would be necessary for cap forming.

Conclusions

In rod shaped bacteria containing the E. coli type of murein, three different glycan chain orientations are conceivable: longitudinal (a), circular (b), and diagonal disorganized (c)

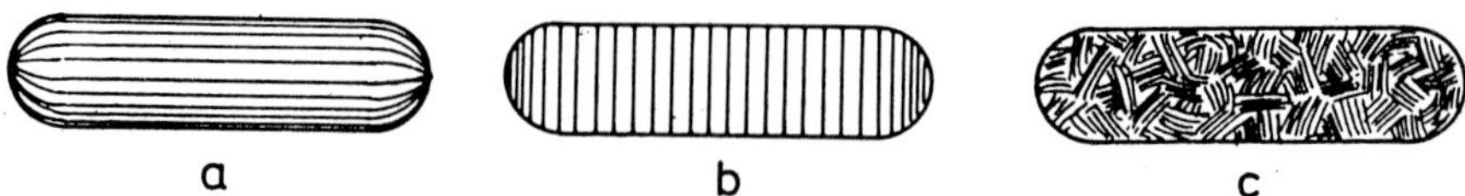

From a theoretical point of view the circular orientation reveals itself as the most plausible one. With this ring model of murein structure all biochemical processes involved in sacculus elongation and septation as well as daughter cell separation can be described without additionnal assumptions. Furthermore, the ring model is compatible with most biological data such as the mode of action of murein synthesis inhibitors.

Certainly, the ring model as discussed here is not easily applicable to Gram-positive bacteria, due to the rather complicated threedimensional structure of their murein.

References

1) Weidel, W., Pelzer, H.: Adv. Enzymol. 26, 193-232 (1964)

2) Pelzer, H.: Abh. Dtsch. Akad. Wsch. Berlin, Klasse Med., Nr. 6, 199-204 (1966)

3) Pelzer, H, in Th. Wieland, G. Pfleiderer, eds: Molekular-Biologie, Umschau-Verlag 147-161 (1967)

4) Pelzer, H.: Habilitationsschrift, Universität Ulm, (1975)

5) Ryter, A., Hirota, Y., Schwarz, U.: J. Mol. Biol. 78, 185-195 (1973)

6) Hahn, F.E., Ciak, J.: Sience 125, 119 (1957)

7) Verwer, R.W.H., Beachey, E.H., Keck, W., Stoub, A.M., Poldermans, J.E.: J. Bacteriol. 141, 327-332 (1980)

8) Hartmann, R., Höltje, J.-V., Schwarz, U.: Nature 235, 426-429 (1972)

9) Kitano, K., Tomasz, A.: J. Bacteriol. 140, 955-963 (1979)

10) Fiedler, F., Glaser, L.: Bioch. Bioph. Acta 300, 467-485 (1973)

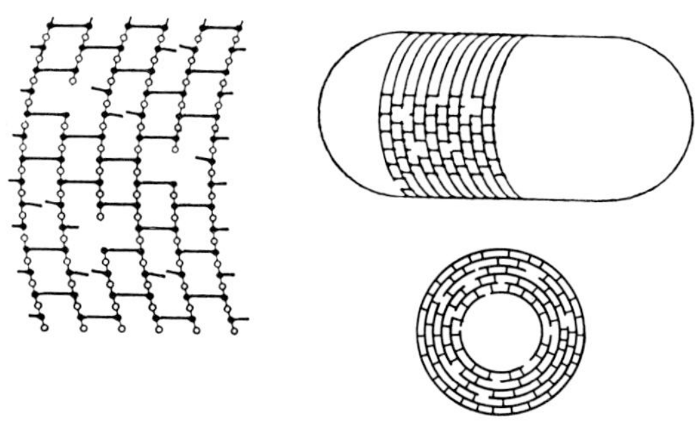

Fig. 1: Schematic representation of the ring structure model of murein sacculi

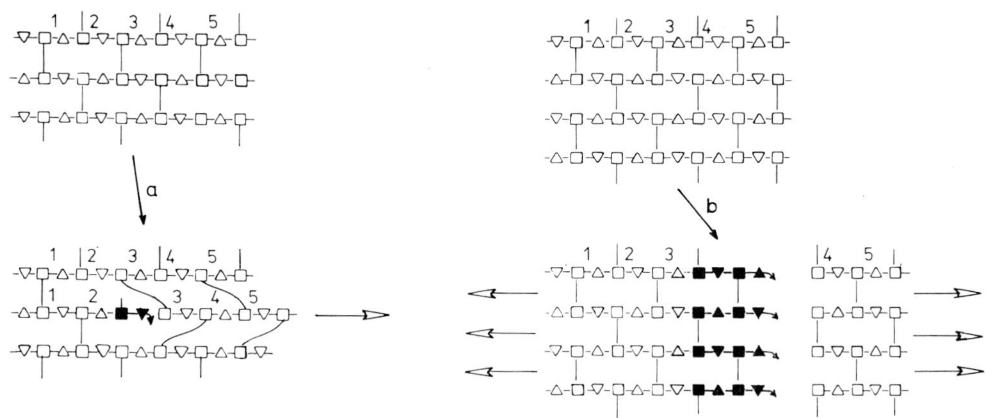

Fig. 2: Problems with the longitudinal model, if insertion of newly synthesized peptidoglycan for cylinder elongation is considered.

a) Isolated insertion of a single polymerizing glycan

b) Simultaneous insertion of several neighboured polymerizing glycans

▽, GlcNAc; □, MANAc; ○, peptides; filled symbols, newly synthesized peptidoglycan;
↘, direction of glycan polymerization.

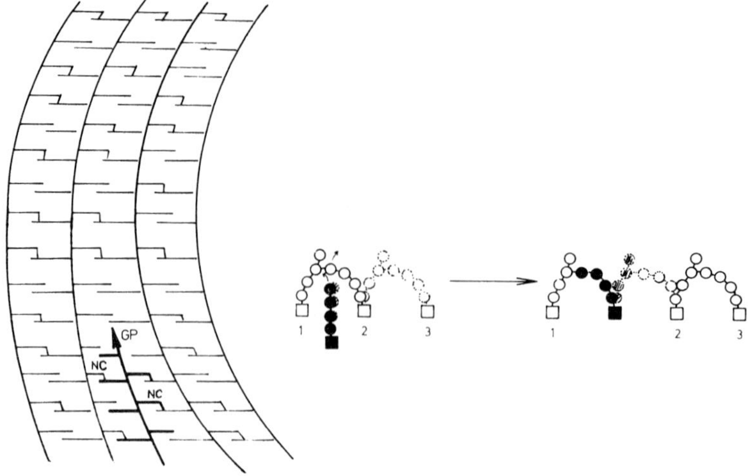

Fig. 3: Cylinder elongation in the ring model
□ and ■ , MANAc; ○ and ● , amino acids;
(, polyclycan; —, tetrapeptides; ┐┴, octapeptides;
GP, glycan growing point; NC, new crosslinks

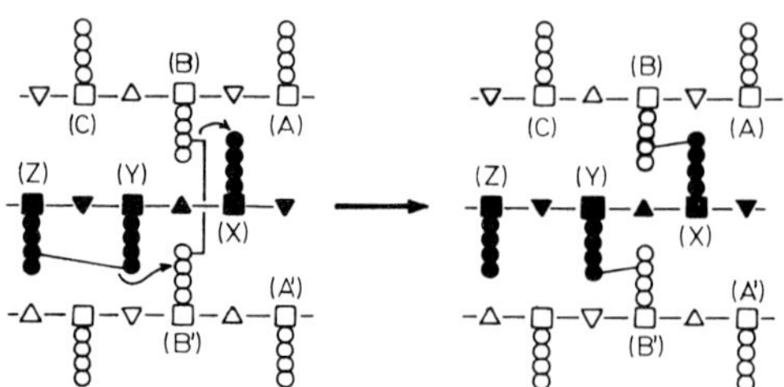

Fig. 4: Mechanism of insertion by the hypothetical cross-link transferase
Z-Y (C4 subunit) ⟶ B'-Y (C3 subunit)
B-B' (C3 subunit) ⟶ B-X (C3 subunit

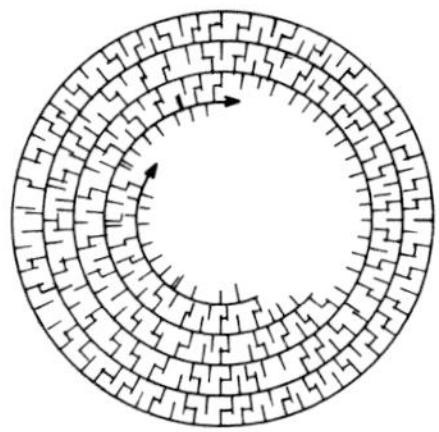

Fig. 5: Septum formation by adherence of successively smaller peptidoglycan rings via transpeptidation

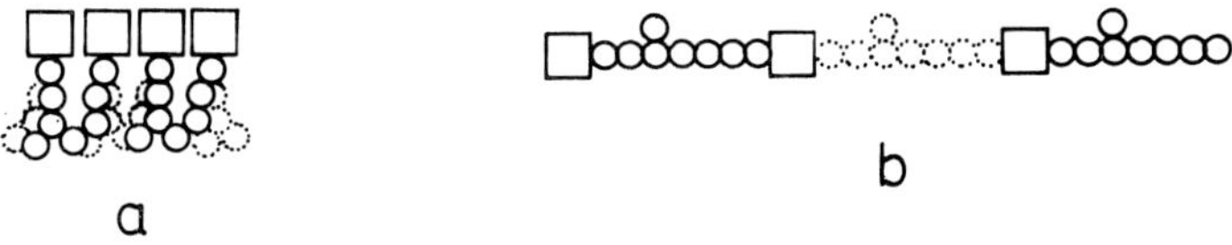

Fig. 6: Flexibility of interpeptide angles

a) tightly packed

b) extremely stretched

□, MANAc; ○, amino acids

REGULATION OF THE INITIATION AND COMPLETION OF ENVELOPE GROWTH SITES IN *STREPTOCOCCUS FAECIUM* (ATCC 9790)

Michael L. Higgins, Carolyn W. Gibson and Lolita Daneo-Moore

Temple University School of Medicine
Philadelphia, Pennsylvania 19140

The envelope of *Streptococcus faecium* is assembled in discrete sites located between adjacent pairs of raised equatorial wall bands, which can easily be observed in electron micrographs. Unlike rod-shaped bacteria, the envelope growth sites of *S. faecium* can be identified unambiguously. Each site produces two polar caps per round of synthesis. Computer reconstruction methods have been developed and applied to carbon-platinum replicas of these cells to estimate the volume and surface areas found separately in sites and polar caps of individual cells (1, 2).

Applying these reconstruction methods to electron micrographs of cells taken from exponential-phase cultures with doubling times ranging between 30 and 110 minutes, it was shown that the size of the polar caps was invariant with respect to growth rate (2). This indicates that the envelope growth occurring in a site is analogous to chromosome replication since a round of synthesis in both cases produces a fixed amount of product which is independent of the growth rate of the culture.

Recently we have been interested in studying the regulation of the initiation of such envelope sites in relation to the cell cycle (3). In rapidly growing cultures, it was observed that most cells initiated new sites before division; however, as the mass doubling time of the culture was increased beyond 60 minutes, an increasing fraction of cells initiated sites after division. The average time that sites were initiated in a cell cycle as a function of growth rate was calculated by introducing the observed frequencies of cells that initiated sites before and after division into a form of the exponential age distribution equation.

The Target of Penicillin

These calculations indicated that for cultures with mass doubling times of less than about 60 minutes, sites are initiated about 10 minutes before division, while in those cultures with mass doubling times greater than 60 minutes, the process became much less precise with a small fraction of the cells initiating sites after division, and causing the average time of initiation to decrease to just a few minutes before division. Moreover, no correlation could be established between the timing of the initiation of envelope sites and either the initiation or completion of rounds of chromosome synthesis. Using an entirely independent technique to study this problem which takes into consideration the exact distribution of cell sizes in a respective population, Arthur Koch has reached similar conclusions (4). Namely, (i) sites are initiated on the average, a few minutes before division, (ii) the timing of these initiations appears to be independent of the timing of chromosome replication, and (iii) the time needed for the assembly of two poles by a single site seems on the average to be a few minutes longer than the mass doubling time of the culture. The implications of these studies are that an envelope growth site may require an amount of time to produce two poles which stretches as a function of the growth rate of the culture.

In the absence of an apparent relationship between the timing of chromosome synthesis and envelope site initiation other regulatory models were investigated. One of the easiest models to check was that sites would be initiated when cells reached a critical size or volume. From the reconstruction of cells obtained from each exponential-phase population analyzed, cells containing small or "birth" sites were selected. The criterion used for selecting such birth sites was that their volume be greater than zero but less than 0.06 μM^3. The assumption was that since the size of these sites was small, the volume of the cells in which they appeared would be close to the cell volume at which the sites initiated. It was found that the cell volume in whole cells which initiated sites after division, and the volume of cell halves which initiated sites before division fell into a fairly narrow range, 0.25 - 0.3 μM^3.

Several of the findings drawn from these studies of exponential-phase cultures were confirmed from the investigation of cells whose chromosome replication was inhibited by the antibiotic Mitomycin C (0.5 ug/ml) (5). Addition of this drug to exponential-phase cultures resulted in rapid inhibition of DNA synthesis. Cell division continued for about 40 minutes, then stopped, but cell mass increased at the pretreatment rate for over 60 minutes. The question was asked how envelope growth sites would accommodate these increases in mass in the absence of normal chromosome replication. These studies indicated that new sites were initiated during the entire 120 minute treatment period, confirming the conclusion from the exponential-phase cell investigations that new sites were initiated independent of chromosome replication.

After the drug inhibited cell division, sites no longer divided (see "-DNA synthesis" in the figure 1). Therefore, it seems that while DNA synthesis was not necessary for sites to appear, these sites could not participate in cell division in the absence of chromosome replication. In addition, when the volume of these sites was studied as a function of treatment time, the treated sites grew until they contained about 0.25 μM^3 of cytoplasm, then stopped. Since 0.25 μM^3 is about the size of the poles produced in untreated cells, this suggested that a site has a finite capacity to enlarge the cell which was independent of whether the site forms two polar caps seen in untreated cells or the undivided sites observed when DNA synthesis was inhibited. Lastly, it was observed that when the cytoplasmic volume contained by the two wall segments joined by a single wall band approaches about 0.25 μM^3, the band splits to form a new site. This is in the same size that was observed in the exponential-phase populations.

Although the model that sites are initiated when cells reach a critical size is attractive, it is our opinion that things are more complicated than this scheme suggests. We feel that the constant cell size at initiation may be an indirect manifestation of many processes involved in cell growth which could be as simple as an increase in cytoplasmic pressure due to sites reaching their limit capacity to enlarge the cell's volume (about 0.25 μM^3) as cells continue to synthesize macromolecules (6), or as complex as a series of enzymes activated in a site specific manner by as yet undissected regulatory mechanisms. The important aspect of this work seems to lie in a system which can quantitate the appearance, topography and growth of envelope growth sites without need to appeal to the indirect measures that must be applied to such problems in rod-shaped bacteria.

With such a system in hand we feel that many of regulatory models for envelope growth site initiation can be tested with a clarity heretofore unattainable.

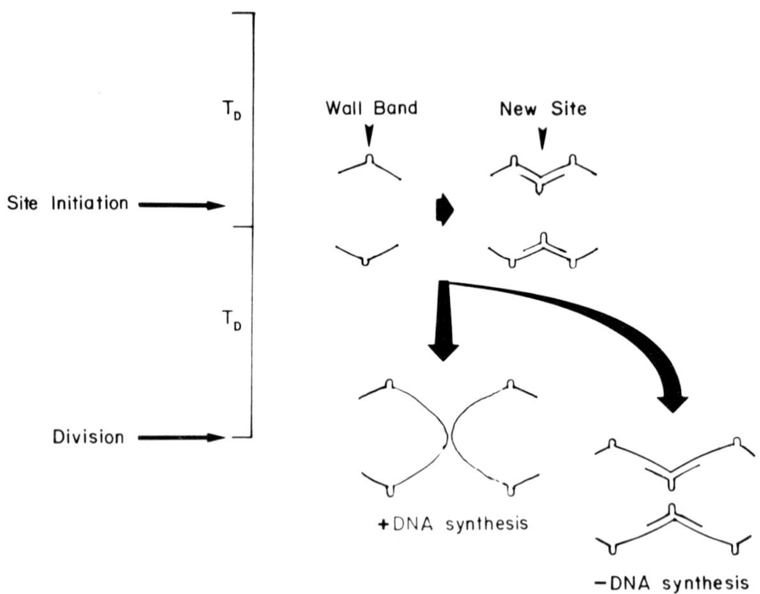

Figure 1. Scheme Regarding the Regulation of Sites of Envelope Growth in *Streptococcus faecium*. On the left, two time intervals each equivalent to a mass doubling time (T_D) of a hypothetical culture are shown. On the average, a site is initiated a few minutes before division, making the time interval from initiation to division slightly longer than T_D. A new site is formed by the assembly of a nascent septum under a wall band and the subsequent splitting of this band to produce two new markers of surface growth. The initiation of a new site occurs when the cytoplasmic volume contained by two wall segments joined by a single band approaches about 0.25 μM^3. While the initiation of new sites appears to be independent of normal chromosome replication, inhibition of DNA synthesis results in sites that do not participate in division. Once initiated, a site can enlarge the volume of the cell by a finite capacity of about 0.25 μM^3 which appears to be unrelated to DNA synthesis or T_D of the culture.

References

1. Higgins, M.L.: J. Bacteriol. 127, 1332-1345 (1976).
2. Edelstein, E.H., Rosenzweig, M.S., Daneo-Moore, L.: J. Bacteriol. 143: 499-505 (1980).
3. Gibson, C.W., Daneo-Moore, L., Higgins, M.L.: J. Bacteriol. (In Press)
4. Koch, A.L., Higgins, M.L.: J. Gen. Microbiol. (submitted).
5. Gibson, C.W., Daneo-Moore, L., Higgins, M.L.: J. Bacteriol. (submitted)
6. Koch, A.L., Higgins, M.L., and Doyle, R.J.: J. Gen. Microbiol. 123:151-161 (1981).

Acknowledgement

This work was supported by Public Health Service Grant AI10971 from the National Institute for Allergy and Infectious Diseases.

A NEW MODEL FOR GROWTH OF THE MUREIN SACCULUS

Lars G. Burman

Department of Clinical Bacteriology, University of Umeå, S-901
85 Umeå Sweden

James T. Park

Department of Molecular Biology and Microbiology, Tufts University,
136 Harrison Avenue, Boston, Massachusetts 02111

We have recently obtained data which allowed us to develop a model to describe the growth of the murein sacculus of Escherichia coli in rather precise terms (1, 2). For purposes of this discussion, we will summarize the significant new data and interpret the date in terms of the model. We will also point out some of the implications of the model and experiments suggested by it.

The revealing new data were obtained by doing an old experiment in a new way. The old experiment is one in which cells are pulse-labeled with (^{14}C)DAP for varying periods of time, samples taken and the relative amount of bisdisaccharide peptide dimer present in the sacculus is determined. The new experiment goes one step further and analyzes the two halves of the dimers separately. One half, the donor, is defined as the half whose pentapeptide was used to drive the transpeptidation reaction which linked it to the amino group of the diaminopimelic acid (DAP) in the other muropeptide (the acceptor). Thus the DAP of the donor muropeptide has a free amino group, whereas the DAP of the acceptor muropeptide does not. The radioactivity in the donor and acceptor muropeptides can therefore be readily determined by isolation of the dimers, dinitrophenylation of the donor's free amino group, acid hydrolysis, and separation of the resulting dinitrophenyl DAP from DAP by chromatography. Our data are not as precise as we would like them to be but the typical result observed in many experiments is as shown in Fig. 1.

The Target of Penicillin

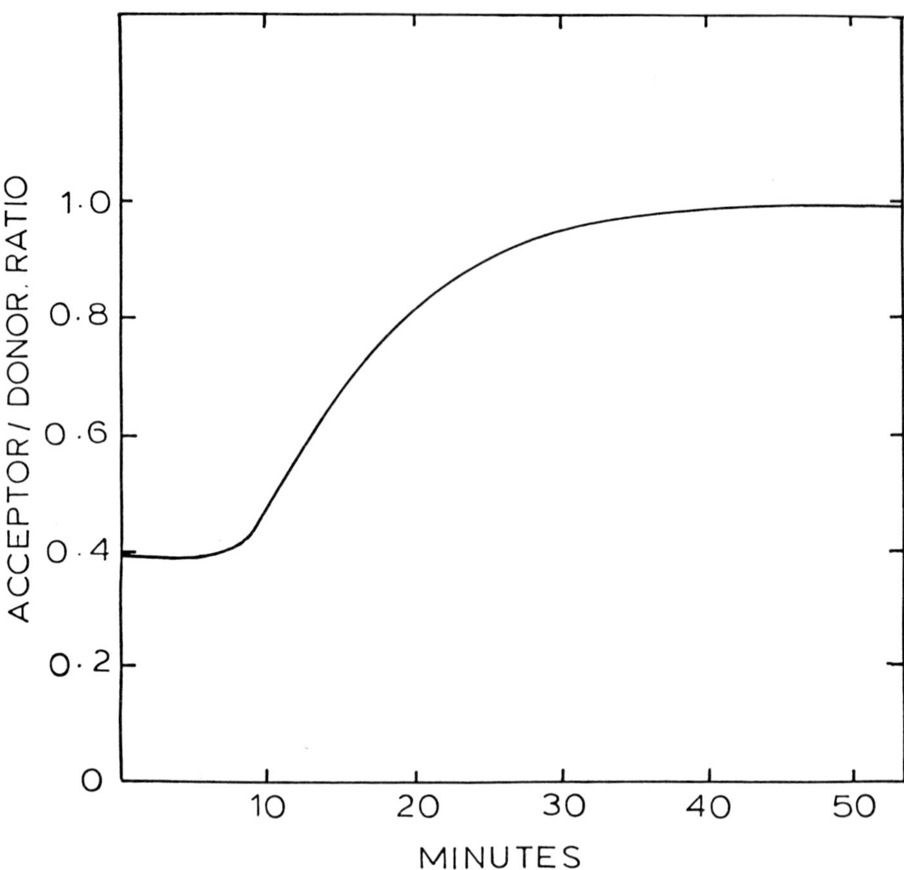

Fig. 1. Typical curve illustrating the changes in the acceptor/donor radioactivity ratio in the dimers from *E. coli* pulse-labeled with (^{14}C)DAP for increasing periods of time.

During the first 8 minutes of labeling with (^{14}C)DAP, the ratio of radioactivity in the donor remains constant. The ratio in different experiments varies between 0.3 and 0.45. During the succeeding 8 minute period, the ratio rises rapidly to a value of 0.5 to 0.7. Thereafter, the acceptor/donor radioactivity ratio continues to rise at a much slower rate and approaches a value of 1.0 after 50 minutes (i.e., one generation). Similar results are obtained with strain W7 (*dap*, *lysA*) prestarved for DAP and with BUG 6 (ts for septation) growing as filaments at 42°. The result with BUG 6 indicates that the curve is determined by the elongation process and is not an artifact of the starvation procedure or the minimal medium employed with W7.

Thus three significant facts emerge from this type of experiment and each provides important insight into the process of elongation. The facts are:

1. The proportion of radioactivity in the acceptors and donors remains constant for about 8 minutes with most of the radioactivity in the donor position.
2. The value of the acceptor/donor radioactivity ratio during this period is around 0.3.
3. During the second 8 minutes of the experiment the ratio rises rapidly

In order to interpret these results, we have made the following assumptions:
The first assumption is that the glycan strands in the murein sacculus are, in fact, arranged perpendicular to the axis of the cell as suggested by the work of Verwer, Nanninga, Keck, and Schwarz (3). The second assumption is that the glycan strands are simultaneously linked to the existing murein while they are being polymerized. This assumption may be questioned because Mett, Bracha, and Mirelman have recently isolated a soluble polymer of murein from E. coli which contains large amounts of pentapeptide and lipoprotein and have suggested that it may be nascent murein (4). The pentapeptide content certainly indicates this is possible but the high lipoprotein content suggest the soluble polymer may be an abnormal product since lipoprotein is not added to newly synthesized murein until some time after incorporation into the sacculus (5,6). More to the point, however, is the fact that the soluble polymer does not behave like a normal precursor since neither they nor we (2) were able to chase this material into the sacculus.

The most persuasive argument that the glycan strands are linked to the sacculus coincident with their formation is the remarkable finding of Matsuhashi and coworkers (7) that two of the proteins believed to be involved in elongation of the sacculus (8), PBP-1A and PBP-1Bs, each have transglycosylase and transpeptidase activity. Surely two-headed enzymes of this type would cross-link the new muropeptide to the sacculus as soon as it was added to the growing glycan chains. The third assumption is that only newly formed strands serve as donors. De Pedro and Schwarz (9) have shown that more than 97% of muropeptides have lost their terminal D-alanine within much less than 3 minutes of their formation and hence can no longer serve as donors.

With these three assumptions in mind, one can interpret the experimental facts as follows:

The Significance of the Constant Acceptor/donor Radioactivity Ratio

The fact that the acceptor/donor radioactivity ratio remains constant for about 8 minutes indicates that the enzyme complexes that add new strands of murein to elongate the sacculus do not come in contact with other new strands during this period. When one considers that over 400,000 muropeptides have been added during this time, there appears to be only one way in which new strands could avoid cross linking to other recently inserted strands. The enzyme complexes involved in elongation must move in one direction around the sacculus and it takes each complex about 8 minutes to synthesize and insert strands that extend the length of the circumference. We believe the enzyme complex must literally travel around the circumference of the cell once in 8 minutes. This is similar to a replicon complex moving along its template DNA at a constant rate. The rate at which the murein elongation complex must move along the strands of its acceptor substrate is about 4 or 5 muropeptides per second.

Complexes that travel around the circumference once in 8 minutes must do so about 6 times in the course of a generation time of 50 minutes in order to double the length of the sacculus. E. coli W7 grown in minimal medium has a unit cell length of about 2200 nm (10). After correction for the length of the poles this gives a length of 1400 nm for the cylindrical portion of the sacculus. If parallel strands are spaced 1.25 nm apart (10), the cylinder would contain 1120 strands of murein. To double the length of the sacculus would therefore require more than 180 enzyme complexes each adding one strand at a time or, alternatively, about 90 complexes inserting two strands at a time. Our reasons for favoring a two strand model are given in the next paragraph.

The Significance of the Initial Value of the Acceptor/donor Radioactivity Ratio

The acceptor/donor radioactivity ratio during the first 8 minutes usually varies between 0.3 and 0.4 depending on the experimental conditions. If strands are added one at a time and the assumption that only new strands serve as donors is correct, the ratio would be 0 for non=septating cells, such as BUG 6 growing at 42°. Since we obtain ratios of greater than 0.3, single strand insertion is ruled out. However, if two strands are added simultaneously by each complex and an equal number of crosslinks is formed between all strands, an acceptor/donor radioactivity ratio of 0.33 would be obtained as can be seen from the diagram in Figure 2A.

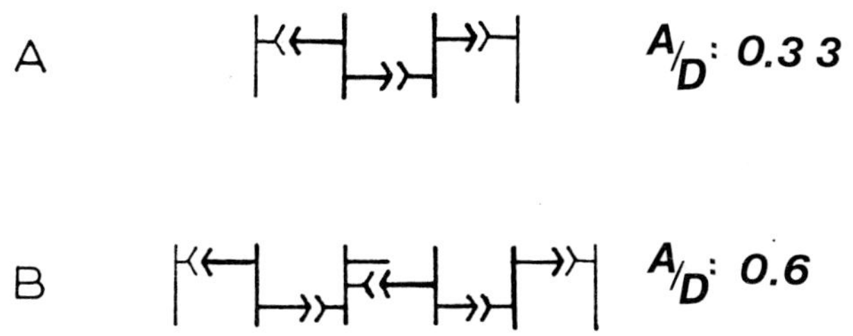

A A/D: 0.3 3

B A/D: 0.6

Fig. 2. Diagram indicating the positions of radioactive acceptor and donor muropeptides after one round in which a pair of radioactive strands are inserted into the sacculus (A) and after two rounds (B). The bold lines represent the new radioactive muropeptides.

Thus insertion of two strands at a time is consistent with the data and a more likely alternative than insertion of more than two strands at a time by each complex. Another reason for proposing a two strand model is that our estimate of the number of enzyme complexes (90 to 100) just about equals the number of molecules of PBP-1A, PBP-1Bs, and PBP-4 present in E. coli according to Spratt (8). We suggest that PBP-1A forms one strand of the pair, PBP-1Bs forms the other, and that PBP-4 is the endopeptidase (11) responsible for separation of existing strands and that these three proteins, together with the D,D carboxypeptidases, PBP-5 and/or PRP-6 (12), work as a unit to insert a new pair of strands and remove the terminal D-alanine from the unutilized pentapeptides. By enzyme complex we simply mean this ordered activity of a series of separate enzymes which must follow one another along a given acceptor-donor pair of strands whether or not they form a tight complex.

We should point out that an equal number of cross-links on both the leading and trailing sides of the pair of strands being inserted is highly unlikely. Various other combinations are possible which would give acceptor/donor radioactivity ratios between 0.25 and 0.4 and further work will be necessary before this refinement of the model can be made. De Pedro and Schwarz (9) have recently shown that newly synthesized murein is loosely crosslinked for a period of 8 to 10 minutes and then "matures" to a more highly crosslinked form gradually during the next generation or two. We have confirmed this result (data not shown). The 8 to 10 minute period during which the dimer content remains low and constant convinces us that the apparent "maturation" is the result of continuous insertion of pairs of strands in the normal manner predicted by the model. However, equal crosslinking between all strands would not lead to increased dimer content after the first round of insertion. Therefore this data indicates that the trailing strand is attached to its neighboring acceptor strand by significantly more crosslinks than is the leading strand of the pair. As each new round of pairs is inserted the more highly crosslinked form accumulates as the endopeptidase selectively attacks the lower crosslinked neighbors to allow the next round to be inserted.

The Significance of the Rapid Rise in the Acceptor/Donor Radioactivity Ratio that Occurs after the Initial 8 Minute Period

By now it should be clear that if it takes 8 minutes for an enzyme complex to travel around the circumference of the sacculus, after 8 minutes there is a good chance it will make contact with the radioactive strands it inserted 8 minutes earlier. The fact that the ratio rises rapidly indicates that these radioactive strands are now being used as acceptors during the second round so that the complex may follow a spiral or corkscrew path always moving in one direction with the same relative positions for the leading and trailing strands in the pair. During the second round, the endopeptidase would be attacking a low crosslinked pair of strands which is rich in radioactive donors so that donor radioactivity is selectively lost from the dimers. One can see how the acceptor/donor radioactivity ratio would change after a second pair of strands is inserted next to the first pair by reference to Figure 2B. Bear in mind that the diagram shown in Figure 2 indicating an equal number of crosslinks between all strands is not representative of the situation at the growth site and is only presented to illustrate why a higher ratio obtains after the second round.

One additional experiment pertinent to the model is a chase experiment. When cells uniformly labeled with (^{14}C)DAP were chased with cold DAP, the acceptor/donor radioactivity ratio rose rapidly for about 8 minutes to a figure 15 or 20 % above the initial value and continued to rise at a much slower rate thereafter (2). As can be seen in Figure 3, this is the increase in ratio that would be expected after each complex added one pair of non-radioactive strands for every six pairs of strands already present. This, of course, is the relative proportions of new and old strands present if the complexes are to double the length of the sacculus by adding new pairs of strands during six trips around the circumference of the cell each taking about eight minutes to complete.

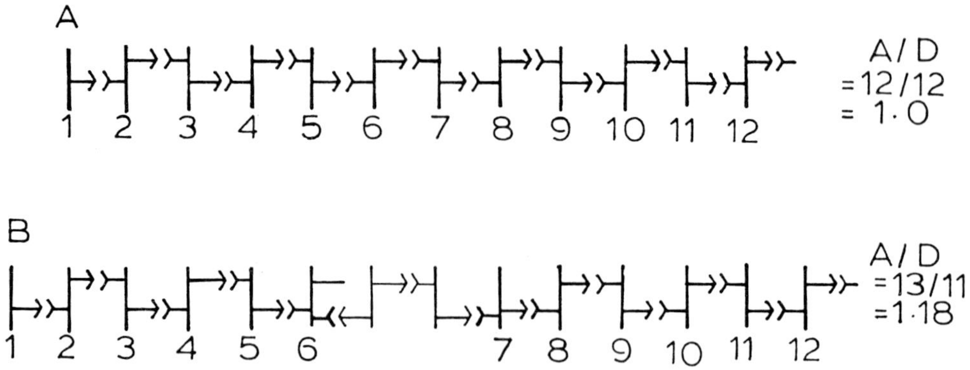

A

1 2 3 4 5 6 7 8 9 10 11 12

A/D =12/12 = 1.0

B

1 2 3 4 5 6 7 8 9 10 11 12

A/D =13/11 =1.18

Fig. 3. Diagram indicating the positions of radioactive acceptor and donor muropeptides in uniformly labeled sacculi (A) and after one round of chase (B). The bold lines represent the radioactive muropeptides.

The model indicates that elongation proceeds in a very orderly fashion. The constant rate of synthesis and insertion of strands may be regarded as a fundamental unit of activity comparable to the fixed rates of synthesis of other important polymers such as DNA, RNA, and protein. Viewed in this way, the model predicts that to change the overall rate of cell elongation, the cell must change the number of functional enzyme complexes per cell. Preliminary experiments suggest that a given enzyme complex completes one round of strand insertion in 8 minutes regardless of the growth rate. We plan to determine whether the number of complexes per cell, as measured by the amount of PBP-1A, PBP-1Bs, or PBP-4 present, varies with the growth rate. The model and the technique of measuring acceptor and donor muropeptides separately open up a number of other avenues to investigation. It should be possible to analyze the response of intact cells to various inhibitors of septation in a more revealing manner. The same applies to learning more about the role of the penicillin-binding proteins by studying individual PBP-defective mutants.

Whether this model applies to the addition of murein to the wall of Gram positive rod-shaped bacteria remains to be determined. Since the orientation of the murein in relation to the enzyme complexes in this membrane is similar in all procaryotes, it would not be surprising if a similar model was followed in all procaryotes that use a multisite mechanism for elongation of the cell.

References

1. Burman, L. G., Park, J. T.: submitted for publication (1983).
2. Burman, L. G., Reichler, J., Park, J. T.: submitted for publication (1983).
3. Verwer, R. W. H., Nanninga, N., Keck, W., Schwarz, U.: J. Bacteriol. 136, 341-352 (1977).
4. Mett, H., Bracha, R., Mirelman, D.: J. Biol. Chem. 255, 9884-9890 (1980).
5. Braun, V., Wolff, H.: J. Bacteriol. 123, 888-897 (1975).
6. Burman, L. G., Park, J. T.: submitted for publication (1983).
7. Matsuhashi, M., Ishino, F., Nakagawa, J., Mitsui, K., Nakajiima-Iijima, S., Tamaki, S.: in β-lactam Antibiotics - Mode of Action, New Developments, and Future Prospects, ed. Salton, M. R. J., Academic Press, New York (1981).
8. Spratt, B. G.: Eur. J. Biochem. 72, 341-352 (1977).
9. De Pedro, M. A., Schwarz, U.: Proc. Natl. Acad. Sci. USA 78, 341-352 (1977).
10. Verwer, R. W. H., Nanninga, N.: J. Bacteriol. 144, 327-336 (1980).
11. Matsuhashi, M., Takagaki, Y., Maruyama, I. N., Tamaki, S., Nishimura, Y., Suzuki, H., Ogino, U., Hirota, Y.: Proc. Natl. Acad. Sci. USA 74, 2976-2979 (1977).
12. Blumberg, P. M., Strominger, J. L.: Bacteriol. Rev. 38, 291-335 (1974).

PEPTIDE CROSSBRIDGES IN THE MUREIN OF *ESCHERICHIA COLI* ARE BROKEN AND REFORMED AS THE BACTERIUM GROWS

E. Wm. Goodell

Natural Science Department
SUNY College of Technology
Utica, N.Y., U.S.A.

Introduction

Growth of a bacterium requires the expansion of its surface, and presumably the intercalation of new murein into the preexisting cell wall or sacculus. A number of different bacteria have been shown to insert murein into more or less discrete growth zones located at the center of the cell or along the long axis of the cell (1-4). Bacteria are unique in possessing covalently crosslinked cell walls; and, therefore, true intercalation of new subunits between preexisting subunits would appear to require hydrolysis of some of the covalent bonds linking the old subunits together.

There are several reasons for supposing that in *E. coli* cleavage of the peptide crossbridges between glycan strands is the reaction which allows intercalation of new murein. First, the glycan strands appear to be wrapped around the circumference of the cell, perpendicular to its long axis (5). As *E. coli* cells grow, they elongate, maintaining an essentially constant diameter (6); a relatively simple way of elongating the sacculus would, therefore, appear to be to cleave peptide crossbridges between old glycan strands and insert new glycan strands between the old strands. Second, two endopeptidases have been found in *E. coli* which, at least *in vitro*, do cleave the peptide crossbridges (7, 8). No murein hydrolases able to cleave the glycosidic bonds in the center

The Target of Penicillin

of glycan strands have been isolated from *E. coli* cells (9). However, in spite of the apparent theoretical necessity for hydrolysis of some covalent bonds in the murein of growing cells, no loss of murein has ever been detected in growing *E. coli* cells (10). This result suggests that either murein hydrolases are not active *in vivo* or that the hydrolysis of bonds is carefully balanced by the formation of bonds between new and preexisting murein.

Results and Discussion

It was decided, therefore, to try to determine whether the crossbridges of *E. coli* were being cleaved and reformed during growth of the bacterium. The method used was based on the fact that the diaminopimelic acid residues in the donor peptide and the acceptor peptide of crossbridges are not identical (Fig. 1). The diaminopimelic acids in the donor peptides have a free amino group which can be removed by nitrous acid after isolation of the murein. The diaminopimelic acid is converted into hydroxyaminopimelic acid (11).
Newly synthesized peptides are pentapeptides and can act as donors when they are inserted into murein; in doing so, they lose their terminal D-alanine (Fig. 1). If these crossbridges are ever cleaved again by the cell's endopeptidases, then the resulting tetrapeptide can only act as an acceptor when it becomes part of a second crossbridge, since it has lost its terminal D-alanine (Fig. 1). If turnover of crossbridges does occur, then ^{3}H-diaminopimelic acid residues incorporated into new murein should slowly be converted from a donor to an acceptor role and should lose their free amino group.

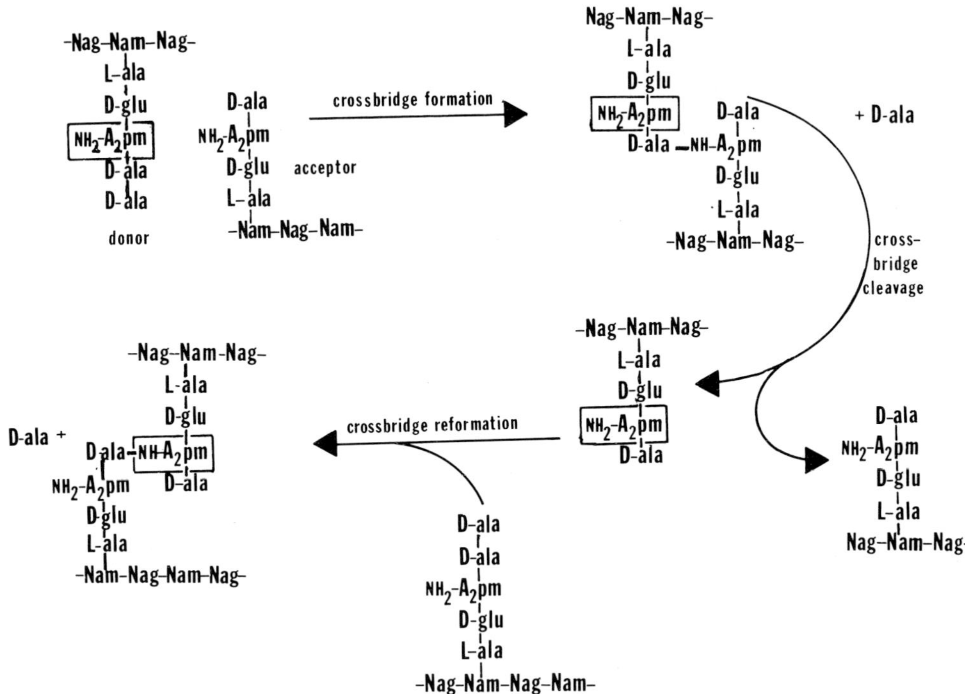

Fig. 1. Proposed turnover of peptide crossbridges in E. coli murein. The participation of a peptide in a crossbridge first as donor and then, after cleavage of the first crossbridge, as an acceptor in a second crossbridge is illustrated. The diaminopimelic acid (A_2pm) residue of this peptide has been enclosed in a rectangle to emphasize that it has a free amino group while the peptide is a donor but that it loses this free amino group when the peptide becomes an acceptor.

E. coli W7 (dap^-, lys^-) was grown in complete medium; the cells were labelled with ^{3}H-diaminopimelic acid for 10 min. and then resuspended in a fresh medium lacking ^{3}H-diaminopimelic acid. Samples from the culture were taken during the pulse and the chase period, and the murein sacculi isolated (13). Free amino groups in the sacculus were removed with nitrous acid (11) and the crossbridges were isolated after lysozyme digestion (12). The amount of ^{3}H-diaminopimelic acid and ^{3}H-hydroxyaminopimelic acid was measured in each sample (11, Fig. 2).

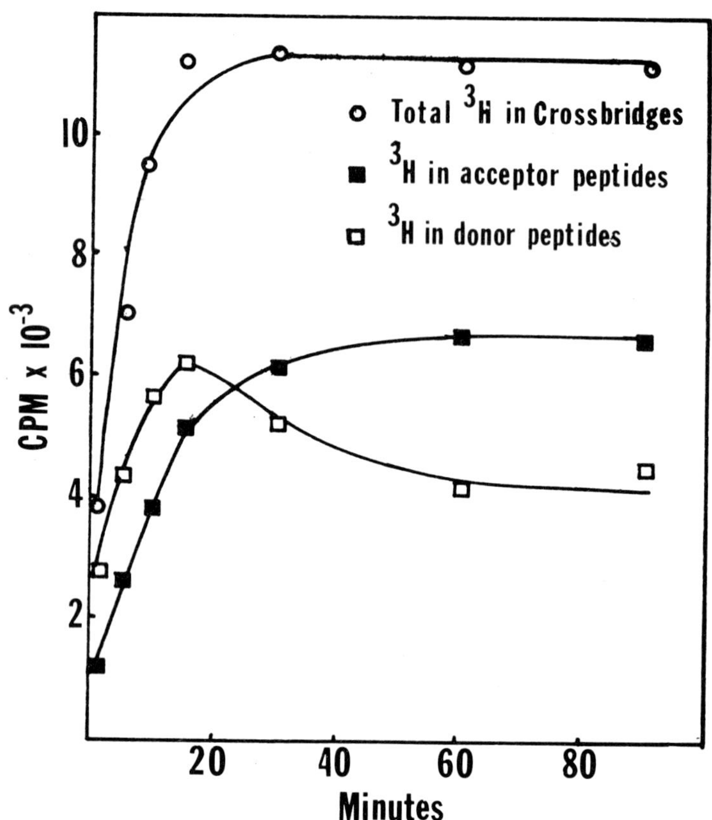

Fig. 2. Loss of label from donor peptides. *E. coli* W7 cells were labelled for 10 min. with ^{3}H-diaminopimelic acid and the label was removed. Samples of bacteria were then removed from the culture after 0, 5, 10, 15, 30, 60, and 90 min. and the murein was isolated. The total amount of label in the murein was measured, and then after chemical analysis, the fraction of label in the crossbridges, the donor peptides and the acceptor peptides was calculated.

Throughout the pulse period (0.5-10 min) the fraction of label in the donor peptide remained constant at ca. 65%. During the chase period (Fig. 2) the percent of label in the donor remained at 65% for the first 15 min. However, this was due to continued incorporation of label into crossbridges (Fig. 2) during this period; as soon as incorporation into crossbridges ceased the percent of label in donor peptides started to drop (Fig. 2). This decrease continued for the next 25 min. (about 1 generation) until roughly 1/3 of the label had been lost from the donors. This result indicates either (i) that crossbridges are undergoing turnover as described in the Introduction or (ii) that previously uncrosslinked ^{3}H-labelled tetrapeptides are being incorporated into new crossbridges as acceptors during this period. However, this latter possibility requires that the amount of label in crossbridges increases over this interval and Fig. 2 shows that this is not the case.

Our finding that only 64% of the peptides were donors during this pulse period suggests that single glycan strands are not being incorporated randomly over the sacculus. If random incorporation were occurring one would expect all new peptides do act as donors since very little pentapeptide exists in the sacculus (13). One can calculate from our results that 5 or 6 glycan strands are being added in short succession or simultaneously to the sacculus.
Label was lost from donor peptides at a steady rate for the first 30 min. after incorporation of label into the crossbridges had ceased; thereafter, loss of label from the donor peptides was much slower (Fig. 2). *E. coli* cells have a growth zone at the center of the cell where most of the insertion of new murein into the sacculus takes place (1). It seems likely from our results that splitting of the peptide crossbridges continues as long as the labelled murein remains in the growth zone and ceases once the murein has left the growth zone and has moved toward the poles of the cell. Finally, though these experiments indicate that cleavage of peptide crossbridges

does occur *in vivo*, they do not indicate which enzyme(s) is responsible. The endopeptidases may not be involved; the murein transpeptidases may cleave old crossbridges as they are forming new crossbridges.

References

1. Ryter, A., Hirota, Y., Schwarz, U.: J.Mol.Biol. 78, 185-195 (1973).
2. Schlaappi, J.-M., Pooley, H.H., Karamata, D.: J.Bacteriol. 149, 329-337 (1982).
3. Higgins, M.L., Shockman, G.D.: J.Bacteriol. 127, 1346-1358 (1976).
4. Cole, R.M.: Bacteriol.Rev. 29, 329-344 (1965).
5. Verwer, R.W.H., Nanninga, N., Keck, W., Schwarz, U.: J. Bacteriol. 136, 723-729 (1978).
6. Grover, N.B., Woldringh, C.L., Zaritsky, A., Rosenberger, R.F.: J.Theoret.Biol. 64, 181-193 (1978).
7. Tamura, T., Imae, Y., Strominger, J.L.: J.Biol.Chem. 251, 414-423 (1976).
8. Keck, W., Schwarz, U.: J.Bacteriol. 139, 770-774 (1979).
9. Mett, H., Keck, W., Funk, A., Schwarz, U.: J.Bacteriol. 144, 45-52 (1980).
10. Rothfield, L., Pearlman-Kothencz, M.: J.Mol.Biol. 44, 477-483 (1969).
11. Fordham, W.D., Gilvarg, C.: J.Biol.Chem. 249, 2478-2482 (1974).
12. Braun, V., Wolff, H.: J.Bacteriol. 123, 888-897 (1975).
13. De Pedro, M.A., Schwarz, U.: Proc.Natl.Acad.Sci.U.S.A. 78, 5856-5860 (1981).

A NOVEL HYPOTHESIS TO EXPLAIN REGULATION OF THE MUREIN SACCULUS SHAPE

Giuseppe Satta, Pietro Canepari and Roberta Fontana.
Istituto di Microbiologia e Virologia dell'Università di Cagliari. Italy.

Introduction

During the last few years we have been studying a pH dependent morphology mutant of Klebsiella pneumoniae (Mir M7) (1, 2). Certain properties of this strain have allowed us to suggest some hypothesis concerning regulation of cell shape in bacterial rods (3, 4, 5). In this study we have performed additional experiments that have confirmed the previous suggestions and have allowed us to propose a more complete hypothesis concerning shape regulation in bacteria.

Material and methods

Wild strains were identified by standard methods. Strains Mir M7 and Mir A12 have been described previously (1). The cell division temperature sensitive (ts) mutants of Escherichia coli BUG6 and AX655 were kindly provided by Dr. J.N.Reeve and by Dr. J.R.Walker. Cell synchronization and trials on the effect of mecillinam on peptidoglycan synthesis were performed and evaluated as described previously (5, 6). Evaluation of peptidoglycan synthesis in toluene permeabilized cells and evaluation of carboxipeptidase and transpeptidase activity were carried out as described by others (7, 9).

The Target of Penicillin

Results

Table 1 shows that inhibition of septum formation in wild coccoid strains and in cocci of morphology mutants causes transition to rod shape in some of both the former and the latter while in others it does not lead to any evident morphology change. Table 2 demonstrates that mecillinam, a specific inhibitor of bacterial lateral wall elongation (10) allows two different cell division ts mutants to divide under non permissive conditions. In synchronous rods of the wild strain K.pneumoniae Mir A12, (parent of Mir M7) the fore mentioned antibiotic caused 50% inhibition of peptidoglycan synthesis when added at the very beginning of the cell cycle. If added 5 minutes (or more) later no effect was observed until the beginning of the following cell cycle when mecillinam caused a similar 50% drop in rate

Tab.1. Effect of septum inhibition on cell shape of various strains.

Wild st.	Shape aft.sept.inhib.	Morph.mut.	Shape aft.sept.inhib.
S.aureus	cocci	SP45	cocci
S.bovis	cocci	CYA^-	cocci
N.perflava	cocci	R3-1	cocci
S.agal.	rods	Mir M7	rods
M.polym.	rods	SP51	rods

Tab.2. Effect of mecillinam on cell division of strains BUG6 and AX655.

Strains (at 42°C)	Mecill.added	Cell numb. ($x10^7$)at 0	45	90 min.
BUG6	yes	3.2	4.4	6
	no	3	3.3	3.2
AX655	yes	2	3.1	3.8
	no	1.8	2	1.8

of peptidoglycan synthesis (not shown). In synchronous rods of Mir M7 (obtained at pH 5.8) the antibiotic inhibited peptidoglycan synthesis when added at any time from the very beginning of cell cycle to 5 min before synchronous division, but had no effect throughout the period of cell number raise (not shown). In toluene permeabilized cells of Mir M7 and Mir A12, cocci of both strains (obtained at pH 7 and by mecillinam treatment respectively) synthesized approximately 50% less peptidoglycan than rods (tab.3). In addition mecillinam inhibited by 50% peptidoglycan synthesis of toluene permeabilized rods of Mir M7 while it had no effect on cocci of the same strain (tab.3). Carboyipeptidase and transpeptidase activity were also analyzed in rods and cocci. The former enzyme showed an approximately similar activity in all cells while the latter was over 70% higher in cocci (obtained at pH 7) than in rods of strain Mir M7. Mecillinam caused an approximate 70% increase of transpeptidase activity associated with transition from rod to coccal shape in Mir M7, but a much lower increase (10%) in Mir A12 (not shown). Cocci of the former strain divided normally while cocci of the latter did not divide. In synchronous rods of both Mir M7 and Mir A12 cells, transpeptidase activity started increasing 10 to 20 min before beginning of cell division, reached a top when cell number

Tab.3. Peptidoglycan synthesis in toluene treated cells and effect of mecillinam.

Strains	Shape	Mecill.	Radiact.incorp.
Mir M7	rods (pH 5.8)	no	750
		yes	390
	cocci (pH 7)	no	380
		yes	375
Mir A12	rods	no	790
		yes	390

began to rise and then decreased for 20 to 30 min. In the first strain the enzyme activity at the height of the rise was 80% higher than at its start while in the other strain it was only 20% higher (not shown).

Discussion

We suggest that in wild bacteria only one or two machineries for the terminal stages of peptidoglycan synthesis exist. In strains where only one machinery is present, cell wall surface is enlarged only during septum formation. These cells are coccoid and their shape is not changed by inhibition of septum formation which, rather, prevents cell surface expansion. In other strains, the probably originally single machinery has evolved to perform two specialized functions which are mutually exclusive and in competition with each other at least at some stage of their activity. Of the two functions one leads to cell wall surface expansion, most often, by causing cell elongation and the other leads to septum formation. In these microorganisms cell morphology depends on the balance between the two competing activities (sites). It will be rod when the site for cell elongation is overcome at regular intervals by that for septum formation, coccobacillar when the intervals during which the site for cell elongation prevails are shorter and coccoid when this site, under ordinary conditions, can not prevail on the other. Bacteria in intermediate stages of evolution of the two specialized functions should exist. The strains that carry two sites but grow as cocci may represent the prefinal stages of this evolutionary process. Other strains in stages prior to this may also be found. In bacteria carrying two competing sites for peptidoglycan assembly cell shape can be altered by all endogenous and exogenous stimuli that either activate or inhibit one or the other of them. Physiological events occurring in the cells also outside the envelope (e.g. DNA replication) may interfere in the balance between the two competing functions and can contribute to shape regulation. Septum inhibition in cocci (carrying the two sites) and in rods should lead to formation of rods and filaments respectively while inhibition of lateral wall in rods should cause formation of cocci. The same effects should be caused by stimulation of the site for lateral wall or of that for septum formation respectively. Similarly, mutations that alter the balance between the two competing functions will lead to formation of cocci when thay cause a constant prevalance of the septal site on that for cell elongation and to formation of filaments under the opposite conditions.

The evolution of two specialized functions for peptidoglycan synthesis involves the development of a much higher organizational complexity and has probably been favoured by some strong pressure giving selective advantages to cells capable of enlarging their envelope and continuing macromolecular synthesis even when not undergoing cell division. Such a complexity is maintained also in the absence of favourable selective pressures. This may likely involve the existance of stringent controls that link cell growth to maintenance of the two coordinate functions. The existance of a control linking formation of new septa to cell elongation has already been proposed (5) and it might well be responsible for preventing cells from losing rod shape under ordinary laboratory growth conditions. A simple control mechanism could be that of the need for an event associated with lateral wall elongation completion for septum formation. At the molecular level evolution of the two specialized functions may, likely, involve evolution of penicillin binding proteins (PBPs) with more specialized functions and development of finer interactions among them. The findings described in this study strongly support these proposals. In particular the observation that synchronous rods of one strain that can grow also under the coccal shape (Mir M7), and one that can not (Mir A12), and which respond in quite a different way to mecillinam, shows that differences in regulation of cell wall elongation in the two strains exist. Such differences could very well be due to loss of the control mechanism by Mir M7. In Mir A12 rods PBP2 activity may be needed to allow the lateral wall site to overcome that for septum formation and initiate cell elongation. After this, event has occurred the control preventing septa until accomplishement of lateral wall elongation may make this PBP activity unnecessary. In Mir M7 where this control, probably due to the high transpeptidase activity is missing PBP2 must rather be constantly active during cell elongation. Since transpeptidase activity was much higher in Mir M7 than in Mir A12 and its activity in the cells rises before cell division it seems likely that for cells to elongate, transpeptidase activity has to be maintained low. This might be the function of PBP2.

References

1) Satta, G., Fontana, R.: J.Gen.Microbiol. 80, 51-63 (1974)
2) Fontana, R., Canepari, P., Satta, G.: J.Bacteriol. 139, 1028-1038 (1979)
3) Satta, G., Fontana, R., Canepari, P., Botta, G.: J.Bacteriol. 137, 727-734 (1979)
4) Satta, G., Canepari, P., Botta, G., Fontana, R.: J.Bacteriol. 142, 43-51 (1980)
5) Satta, G., Botta, G., Canepari, P., Fontana, R.: J.Bacteriol. 148, 10-19 (1981)
6) Mitchison, J.M., Vincent, W.S.: Nature 205, 987-989 (1965)
7) Mirelman, D., Yashouv-Gan, Y., Schwarz, U.: Biochem.J. 15, 1781-1790 (1976)
8) Beck, B.D., Park, J.T.: J.Bacteriol. 126, 1250-1260 (1976)
9) Schrader, W.P., Fan, D.P.: J.Biol.Chem. 249, 4815-4818 (1974)
10) Spratt, B.G.: Proc.Natl.Sci. 72, 2999-3003 (1975)

Supported by contributo n° CT81.018009.04 del CNR to G. Satta.

GENERAL PROPERTIES OF TEMPERATURE-SENSITIVE CELL DIVISION MUTANTS OF STREPTOCOCCUS FAECIUM.

Pietro Canepari, Maria del Mar Lleò
Istituto di Microbiologia dell'Università di Genova
16132 Genova, Italy

Roberta Fontana
Istituto di Microbiologia dell'Università di Padova
35100 Padova, Italy

Giuseppe Satta
Istituto di Microbiologia dell'Università di Cagliari
09100 Cagliari, Italy

Lolita Daneo-Moore, Gerald D. Shockman
Dept of Microbiology and Immunology, Temple University
Philadelphia, PA 19140, USA

Introduction

Cell division regulation is probably more complex in rod-shaped microorganisms, which must coordinate lateral wall elongation with septum formation than in cocci which simply form septa to divide (1, 2). Cell division conditional mutants have proved of great value in the study of this function. However most mutants of this type so far described have been isolated from either Gram-positive or Gram-negative rod-shaped bacteria (3). Strains conditional for cell number increase have also been described in coccoid bacteria dividing in 3 planes (4). In most cases they form completed septa which do not separate as in normal cells. Conditional mutants of coccal strains dividing in 2 planes and unable to form septa would offer a novel approach to the analysis of cell division in Procaryotes. The recent discovery of a relatively simple technique for detecting proteins

The Target of Penicillin

that covalently bind penicillin (PBPs) has provided a further tool for identifying components with key roles in envelope extension and septum formation (5). The importance of PBPs in the regulation of envelope extension and cell division is demonstrated by the finding of mutants either unable to form septa or showing altered cell shape that lack certain PBPs (5). Similar mutants however have not yet been described in streptococci. For these reasons, both mutants conditional for septum formation and mutants that carry altered PBPs of streptococci would be of great value in analyzing regulation of cell division in bacteria. We have recently described over 200 conditional mutants of S. faecium that at 42°C do not increase in cell number and do not form septa (6). In this study we analyze the macromolecular synthesis, the cell division defect and the status of PBPs of such mutants.

Results

1) Cell division and macromolecular synthesis in the ts mutants.
   Based on the behaviour of cell division and macromolecular synthesis at 42°C, over 200 mutants were divided into different groups whose main properties are given in table 1. Two of the five groups did not continue to synthetize proteins after shifting to the non-permissive temperature and the remaining groups synthetized proteins at a rate much reduced compared to the controls grown at the non-permissive temperature. DNA synthesis was inhibited soon after shifting in two groups and continued in all others. In one of these groups, the rate of residual DNA synthesis remained similar to that of the control for one generation; in the others the rate was very low and lasted for a long period. In all strains peptidoglycan synthesis proceeded also at the non-permissive temperature for at least 60-90 min., in some at a faster and in others at a reduced rate.
   When the conditional mutants were shifted back from 42°C to 30°C, both cell division and macromolecular synthesis began again after a lag. Several of the strains divided synchro-

Table 1. Grouping of the ts mutants based on macromolecular synthesis at 42 °C.

group	macromolecular synthesis at 42° C.	% of total strains tested
A	DNA and protein synthesis stop immediately	25
B	DNA and protein synthesis continue for 30 min. at a reduced rate	20
C	DNA and protein synthesis continue for a few min. and then proteolysis occurs	15
D	DNA synthesis continues for a doubling time but protein synthesis stops immediately	30
E	DNA synthesis stops immediately and protein synthesis continues for 30 min. at a reduced rate	10

nously for the first 2 or 3 divisions. In these mutants DNA and protein synthesis started immediately and a few minutes after shifting respectively. In the mutants that at the non-permissive temperature synthetized peptidoglycan at a reduced rate, the synthesis of this polymer started immediately after a shift to the permissive temperature. Peptidoglycan synthesis lagged 15 min. in strains that at 42° C synthetized peptidoglycan at a higher rate than the control.

2) Analysis of the PBPs of the mutants. The PBPs of 193 mutants were analyzed after growth at 30°C or 42°C by binding at the growth temperature. Table 2 summarizes the results obtained and shows that several of the strains carried alterations in PBPs appearance at the non-permissive temperature.

Table 2. *S. faecium* mutants with altered PBPs among 193 strains examined.

PBPs	type of alteration	number of strains
1	lower molecular weight at both temperature	1
2	loss at both temperature	1
3	loss at 42° C	3
5	loss at both temperature	3
	loss at 30° but not at 42° C	3
	overproduced at 42° C	2
other	new PBP with a molecular weight slightly higher than PBP6	1

Some strains showed only minor alterations in intensity and are not shown in Table 2. Of the changes shown in Table 2 each occurred in one strain except for one instance where loss of PBP3 after growth at 42°C was accompanied by overproduction of PBP5. Revertants of the mutants carrying major PBP2 or 3 defects were selected for growth at 42°C. All showed normal PBPs patterns. From one strain lacking PBP3 mutants were isolated as resistant to penicillin (40 µg/ml). These resistant derivatives overproduced PBP5 and were capable of growing at the non-permissive temperature although they still lacked PBP3. As for macromolecular synthesis and growth pattern, mutants lacking PBPs 2 or 3 stopped dividing and synthetizing proteins and DNA a few minutes after the shifting to 42° C. Peptidoglycan synthesis continued for at least 60 min. in every case but at a lower rate than that at 30°C.

Discussion

Cell division mutants have been described as those strains where cell number does not continue to increase while macromolecular synthesis proceeds at a rate comparable to control (7). The temperature-sensitive strains described here do not increase in cell mass (or increase very slowly) when the increase in cell number is inhibited and do not fit this definition. In light of a recent proposal that rod shaped bacteria have at least two sites for peptidoglycan assembly, while coccoid strains have only one (1, 2, 3, 5), it seems plausible that while in rods, when septum formation is inhibited, cell surface can be still expanded by lateral wall extension; in cocci, under similar conditions, surface expansion is completely prevented or drastically limited. Blocking of the extension of the envelope may cause inhibition of macromolecular synthesis due to the lack of space for the new biosynthetic products. All properties of the *S. faecium* ts mutants described here (6) indicate that in many of the strains, the primary defect resides in their incapacity to form septa and not in macromolecular synthesis.
This report is also the first description of cell division mutants of a coccoid bacterium that bear damaged PBPs. The strains showing major alteration and their revertants are of particular interest as tools for studying the possible functions of PBPs in streptococcal growth and division.

Acknowledgments

This work was supported, in part, by NATO grant 288.81 to P.C. and L.D.M.

References

1. Higgins, M.L., Shockman, G.D.: CRC Crit. Rev. Microbiol. 1, 29-72 (1971).

2. Satta, G., Canepari, P., Fontana, R.: This book.

3. Daneo-Moore, L., Shockman, G.D.: In G. Poste and G.L. Nicholson (ed.), Cell Surface Reviews, vol 4., p. 597-715. Elsevier/North Holland Biomedical Press. Amsterdam.

4. Chatterjee, A.N., Wong, W., Young, F.E., Gilpin, R.W.: J. Bacteriol. 125, 961-967 (1976).

5. Spratt, B.G.: Proc. Natl. Acad. Sci. USA 72, 2999-3003 (1975).

6. Canepari, P., Lleò, M.M., Fontana, R., Satta, G., Shockman, G.D., Daneo-Moore, L.: Submitted for publication.

7. Slater, M., Schaechter, M.: Bacteriol. Rev. 38, 199-221 (1974).

PREMATURE DIVISION IN *ESCHERICHIA COLI* IN THE PRESENCE OF TRIS-EDTA

Nanne Nanninga, Tanneke den Blaauwen, Petra M. Nederlof en Piet de Boer
Department of Electron Microscopy and Molecular Cytology, University of Amsterdam, 1018 TV Amsterdam, The Netherlands.

Introduction

In this paper we describe some new observations on the cell division of *Escherichia coli*. It proved to be possible to speed up the completion of cell division so that premature cell division could take place. We wish to discuss the data in the framework of physiological division, physical division and the frequently observed increase in several cellular activities just before division (cf. 13).

Physiological division

This term has been defined as the age at which cytoplasmic events at one end of the cell become independent of cytoplasmic events occurring at the opposite end (4). By measuring survival after T4 phage infection of synchronized *E. coli* Clark (4) could deduce whether a cell reacted as one or two compartments. A two-compartment reaction was seen about 15 min before division (interdivision time 45 min). A comparable compartmentalization was observed by scoring the fraction of plasmolyzed cells in a sucrose-treated cell population as a function of cell length. Prospective daughters (in dividing cells) behaved in the same way as individual non-dividing cells of the same length class. The effect of plasmolysis has been interpreted as a cell dependent cell membrane-cell wall interaction rather than a measure of osmotic pressure (14). Physiological division of the cell envelope would

The Target of Penicillin

thus mean that the envelope has been divided into two compartments which act more or less independently from each other before cell division has occurred. Since the paper of Clark (4) the concept of physiological division has received little attention.

Increased activities before division

With respect to the cell envelope such increase has been found with respect to murein synthesis (10, 12, 15), activity of murein hydrolases (1, 8), phospholipid synthesis (3, 16) and synthesis of some outer membrane proteins (2). Though these activities might be necessary for construction of the division site, it is also conceivable that they are related to the event of physiological division. For the present discussion we will assume the latter possibility.

Physical division

In *E. coli* physical division largely coïncides with cell separation. The question can be asked whether physiological division and physical division can be separated from each other. Early observations of Ward and Glaser (17) on volume increase in synchronized penicillin-treated *E. coli* suggested that a doubling in volume growth rate occurred though cell division was inhibited. Similarly, Donachie et al., (5) observed an increase in length extension rate at a cell age which normally would be followed by cell division 20 min later, after inhibition of cell division through thymine starvation. In the same experiment it could be shown that inhibition of DNA replication (no termination) did not prevent the increase in cell elongation rate.

Premature division

Initially, we found that in Tris-containing growth media division occurred earlier in the presence of EDTA. It proved essential to use synchronized

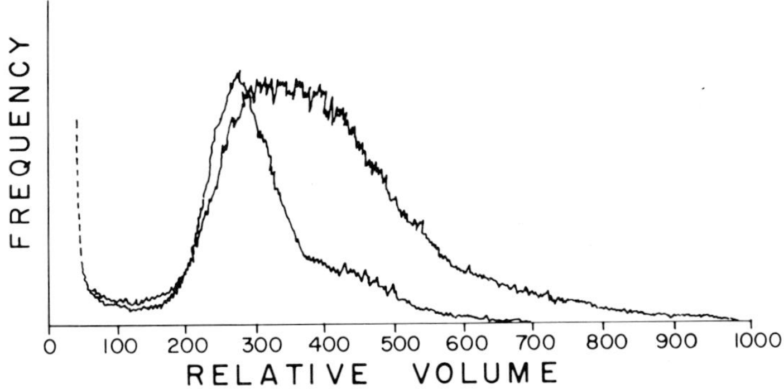

Fig. 1. Centrifugal elutriation. Separation of a fraction of small cells (left narrow peak) from the starting material (right broad peak). For elutriation 400 ml of an exponentially growing culture was concentrated by centrifugation into 4 ml growth medium (cf. Fig. 2). This suspension was directly loaded into the elutriator system at a rotor speed of 5520 rpm and a flow rate of 2.6 ml/min. Two rotor chambers in series were used. Cells were collected at a flow rate of 3.4 ml/min for 20 min in a sterile flask placed on ice.

cultures. Synchronization was achieved after selecting small cells (Fig. 1) by centrifugal elutriation (6, 15) and growing them subsequently. In Fig. 2 an experiment with synchronized _E. coli_ JE 2571 is shown. At the start of synchronous growth EDTA was added at a final concentration of 0.16 mM. As a result division seemed to occur about 15 min earlier. It was necessary to have a combination of Tris and EDTA, rather than EDTA alone. These conditions are thus similar to those necessary for destabilization of the outer membrane (11). EDTA could be replaced by EGTA though the effect was slightly less. A chelator specific for Fe^{2+} (BPDS, (Sigma),bathophenanthroline disulfonate; final conc. 10.8 µg/ml) or Fe^{3+} (Desferal (Ciba) desferrioxamine; final conc. 8 µg/ml)was ineffective. It thus appears likely that interaction of EDTA with Ca^{2+} and Mg^{2+} ions together with association of positively charged Tris ions with the outer membrane is a prerequisite for premature division. Perhaps a local destablization of the outer membrane at the prospective division site is also a requirement for normal cell division to take place. When EDTA was added later during the division cycle at t = 30 min (Fig. 4) or t = 45 min (not shown) premature division was even more clear-cut. Examination of cell number and the fraction of constricted cells during synchronous growth (cf. Fig. 3) suggested that accelerated division

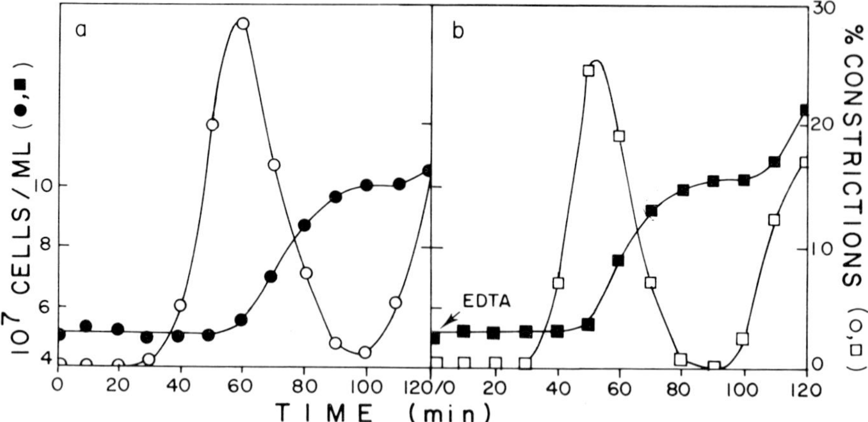

Fig. 2. Premature division. From the elutriator (cf. Fig. 1) 60 ml of a suspension of small cells were collected and divided in two equal portions. These were placed in a shaking waterbath at 37°C to start synchronous growth. To measure cell number (Coulter Counter) 1.5 ml samples were fixed with 0.375 ml 1.2% formaldehyde. From the formaldehyde fixed sample 0.9 ml was taken and fixed with 0.1% OsO_4 (final concentration). Cells were then spread by agar filtration. From the latter the fraction of constricted cells was scored by electron microscopy. (a) Control; (b) EDTA added at 0.16 mM final concentration at t = 0 min. Culture and medium: E. coli JE 2571 (K12, thr^-, leu^-, lac^-, str^R, pil^-, fla^- (obtained from F. de Graaf) was grown with shaking in a waterbath at 37°C. The growth medium contained per liter bidest 10 g Tris Base, 2 g $(NH_4)_2SO4$, 0.2 g $MgSO_4 \cdot 7H_2O$, 0.2 g Cl adjusted to pH 7.2 with 10 N HCl and was supplemented with 0.5 mg $FeSO_4$, 0.15 M $CaCl_2$, 93 mg Na_2HPO_4, 4 mg vit B_I, 40 mg threonine, 40 mg leucine and 0.4% glucose.

is due to a reduction of the T-period (i.e. the period between the onset of constriction and physical division). We also had the impression that constriction started somewhat earlier when EDTA was added at t = 30 min (Fig. 3) It would seem that Tris-EDTA mainly acts on cells committed to division and that commitment as such is less affected.

Returning to the concepts of physiological and physical division one is tempted to suppose that in the presence of Tris-EDTA the two events tend to coïncide. It appears relevant to us to test whether increased activities before division, as briefly mentioned above, also occur at premature division. More specifically, is increased murein synthesis and increased cross-linkage of nascent murein (15) dispensable for division. Answers to this question might help in clarifying which envelope processes are essential or non-essential for cell division.

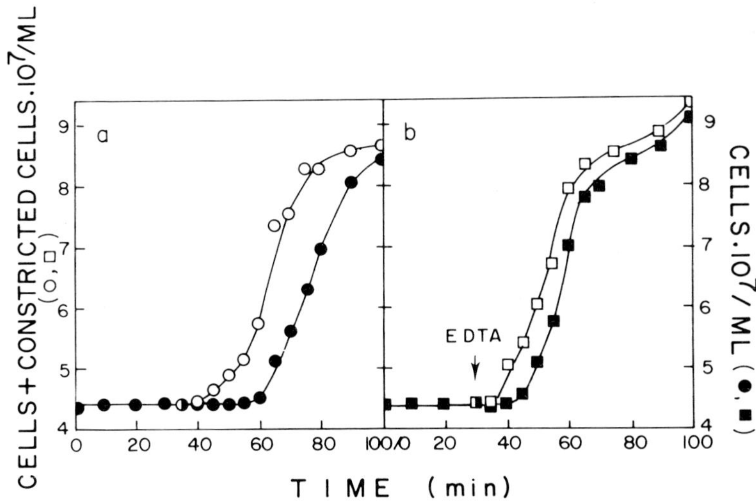

Fig. 3. Effect of EDTA on E. coli JE 2571 (cf. Fig. 2) when added at t = 30 min. a. Control. b. EDTA added at t = 30 (0.16 mM final conc.). (●, ■) increase in cell number as determined with a Coulter Counter. (○, □), increase in cell number + constricted cells. The number of constricted cells was calculated by multiplying cell number and percentages of constricted cells (electron microscopy).

Acknowledgements

We are grateful to E. Pas and J.D. Leutscher for experimental help, to C.E.A. van Wijngaarden for typing the manuscript to J. Woons for preparing the figures and to F. Bienfait for providing the iron chelators.

References

1. Beck, B.D., Park, J.T.: J. Bacteriol. 126, 1250-1260 (1976).
2. Boyd, A., Holland, I.B.: Cell 18, 287-296 (1979).
3. Carty, C.E., Ingram, L.O.: J. Bacteriol. 145, 472-478 (1981).
4. Clark, D.J.: Cold Spring Harbor Symp. Quant. Biol. 33, 823-838 (1968).
5. Donachie, W.D., Begg, K.J., Vicente, M.: Nature (London) 264, 328-333 (1976).
6. Figdor, C.G., Olijhoek, A.J.M., Klencke, S., Nanninga, N., Bont, W.S.: FEMS Microbiol Lett. 10, 349-352 (1981).
7. Hakenbeck, R., Messer, W.: J. Bacteriol. 129, 1234-1238 (1977).
8. Hakenbeck, R., Messer, W.: J. Bacteriol. 129, 1239-1244 (1977).
9. Hartmann, R., Bock-Hennig, S.B., Schwarz, U.: Eur. J. Biochem. 41, 203-208 (1974).
10. Hoffmann, B., Messer, W., Schwarz, U.: J. Supramol. Struct. 1, 29-37 (1972).

11. Leive, L.: Proc. Natl. Acad. Sci. USA 53, 745-750 (1965).
12. Mirelman, D., Yashouv-Gan, Y., Nuchamowitz, Y., Rozenhak, S., Ron, E.Z.: J. Bacteriol. 134, 458-461 (1978).
13. Nanninga, N., Woldringh, C.L, Koppes, L.J.H.: Growth and division of Escherichia coli, p. 225-270. In C. Nicolini (ed.), Cell Growth. Plenum Publishing Corp., New York (1982).
14. Olijhoek, A.J.M., van Eden, C.G., Trueba, F.J., Pas, E., Nanninga, N.: J. Bacteriol. 152, 479-484 (1982).
15. Olijhoek, A.J.M., Klencke, S. Pas, E., Nanninga, N., Schwarz, U.: J. Bacteriol. 152, 1248-1254 (1982).
16. Pierucci, O.: J. Bacteriol. 138, 453-460 (1979).
17. Ward, C.B., Glaser, D.A.: Proc. Natl. Acad. Sci. USA 68, 1061-1064 (1971).

PART III

FUNCTION OF MUREIN HYDROLASES AND CONTROL OF MUREIN DEGRADATION

MUREIN HYDROLASES - ENZYMES IN SEARCH OF A PHYSIOLOGICAL FUNCTION?

Alexander Tomasz

The Rockefeller University
New York, New York

Introduction. The role that cell wall hydrolyzing enzymes may play in the enlargement of the bacterial cell wall murein and in pathological phenomena such as penicillin-induced lysis, has been the subject of much speculation and debate. Until relatively recently, the dominant view has been that participation in wall enlargement and wall degradation are but two different manifestations of the activity of the same hydrolytic enzymes. However, the analysis of murein hydrolase defective mutants and their behavior during penicillin treatment has cast doubt on this. It now seems more likely that these two functions are catalysed by different enzyme systems. My discussion will be restricted to the pneumococcal system. However, the basic conclusions would remain valid if observations made with hydrolase defective mutants of other species had been taken into account.

Autolysins. The discovery of murein hydrolases, enzymes capable of hydrolyzing bonds in the bacterial cell wall, was preceded by the description of the bizarre phenomena of bacterial dissolution or "autolysis" in bacterial cultures exposed to a variety of unfavorable physiological conditions. The involvement of enzymes ("autolysins") in these processes - acting on the major structural component of bacterial cells - the cell wall - was soon recognized. An additional, intriguing feature of these pathological processes was that the unique condition that appeared to invariably provoke autolysis in many bacteria was selective inhibition of wall synthesis.

These observations and the question of why so many species of bacteria would perpetuate such potentially hazardous enzymes (sometimes even referred to in the literature as "suicidases")

The Target of Penicillin

have remained a puzzle until the early 1960's at which time the unique nature of the bacterial cell wall as a giant, cell-sized "macromolecule" was recognized. In a 1964 publication that has provided much stimulus for thinking in the field, Weidel and Pelzer proposed a physiological role for murein hydrolases that could explain both the ubiquity of distribution and the pathological activity of these enzymes (1).

The enzymic balance model. The model proposed by Weidel and Pelzer in 1964 had two main features to it. First, it has put forward a plausible function for hydrolases in wall enlargement. At any given moment of its existence the bacterial cell is surrounded by a giant macromolecule ("sacculus") that has the shape and size of the cell and that is made up of a glycopeptide network containing an uninterrupted system of covalent bonds. Controlled enlargement of such a structure during growth and cell division is unimaginable without the availability of a mechanism for transient breaking of covalent bonds. It was proposed that murein hydrolases could perform this essential physiological function. Specifically, it was suggested that cell wall enlargement required the balanced functioning of hydrolytic and synthetic enzymes, the former producing "nicks" in the glycan backbone of preexisting cell wall. Such "nicks" would then be "re-sealed" by the introduction of a new disaccharide-peptide building block via a synthetic enzyme(s), leading to a net expansion of the wall (Fig. 1A). This notion has led to the second main feature of the model. If wall enlargement involved continuous functioning of hydrolases, in concert with the activity of wall synthetic enzymes, then a selective inhibition of only the latter (for instance by penicillin) would automatically lead to wall degradation (Fig. 1B). Thus, the model could actually predict the phenomenon of bacterial lysis and cell wall degradation that has fascinated so many microbiologists studying the effects of penicillin on bacteria.

G— M— G— M— G— M— G— M— G— M— G— M— G— M— G— M
NICK
G— M— G— M— G— M— G— M G— M— G— M— G— M— G— M
SYNTHESIS
GM
G— M— G— M— G— M— G— M— G— M— G— M— G— M— G— M— G— M
NICK
G— M— G— M— G— M— G— M G — M— G— M— G— M— G— M— G— M
SYNTHESIS
GM
G— M— G— M— G— M— G— M— G — M — G — M— G— M— G— M— G— M— G— M

Fig. 1A

G— M— G— M— G— M— G— M— G— M— G— M— G— M— G— M
NICK (1)
G— M— G— M— G— M— G— M G— M— G— M— G— M— G— M
SYNTHESIS
G— M— G— M— G— M— G
M— G— M— G— M— G— M
G— M— G— M— G— M
G— M— G— M— G— M
G— M— G— M— G
M— G— M— G— M
DEGRADATION (2)

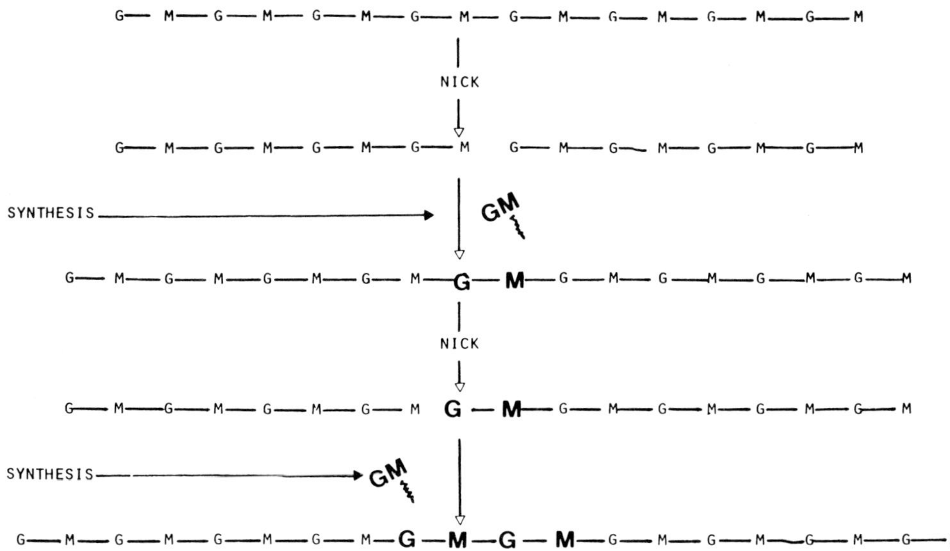

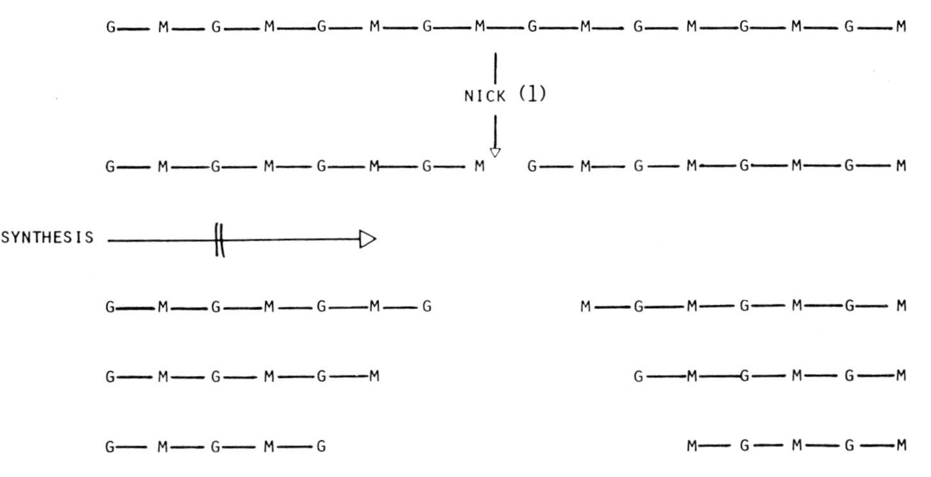

Fig. 1B

Murein hydrolase defective mutants of pneumococci. On the basis of the "enzymic balance" model two main predictions could be made concerning the phenotype of bacterial mutants defective in the activity of a murein hydrolase.

1) Such mutants should not degrade their cell walls and should not lyse during inhibition of cell wall biosynthesis. In short: they should be penicillin tolerant. And
2) Such mutants should exhibit defective cell wall synthesis and cellular growth, under conditions of inhibited hydrolase activity. Such bacteria should only exist as conditional lethal mutants.

By 1970, experimental systems allowing the selective suppression of murein hydrolase activity had been developed in pneumococci and lyt¯ mutants defective in the hydrolase activity (an Nacetyl muramic acid-L-alanine amidase or amidase) had been isolated (2). Characterization of these mutants yielded a surprise: the properties of these bacteria fulfilled the first but not the second prediction of the enzymic balance model. The major revelation of the lyt¯ mutants has been that the loss of 99% of amidase activity did result in a phenotype, namely in a striking and profound alteration of the response of these bacteria to inhibition of wall synthesis. These mutants did not lyse during treatment with penicillin or other cell wall inhibitors; they did not lose radioactive markers from their cell walls and loss of viability occurred with a reduced rate. These properties would cotransform with the defective hydrolase activity during genetic transformation.

On the other hand, the same mutants showed no defect in cell wall synthesis or growth. In other words, none of the lyt¯ isolates were found to be conditional lethal mutants. The conclusion seems inescapable that the cell wall degradation and lysis observable in wild type pneumococci with an inhibited cell wall synthesis must be caused by a hydrolytic enzyme, the activity of which (at least up to 99% of it) is not needed for normal cell wall replication and bacterial growth.

<u>Cell wall synthesis and hydrolase activity</u>. The relationship of amidase activity to cell wall synthesis must be different from that envisioned by the enzymic balance model for an <u>essential</u> hydrolase. The activity in question is clearly non-essential for cell wall enlargement and growth. In addition, the pneumococcal hydrolase is <u>not</u> an endoglycosamidase as originally proposed but an N-acetylmuramic acid-L-alanine amidase. It is difficult to imagine this type of activity integrated, as a continuous function, into a wall synthetic scheme. Therefore, it seems more likely that in normally growing cells the activity of the amidase is suppressed through some kind of (positive or negative) control, which is only upset when wall synthesis is perturbed. If one accepts these arguments, then there is a need for a new expression, a transitive verb to reflect the new situation in which it is no longer obvious why cell wall hydrolysis would be brought about during inhibition of cell wall synthesis. We now must assume that inhibition of cell wall synthesis somehow "triggers" or "provokes"(?) ("unleashes"(?), "de-controls"(?), "induces"(?)) the dormant hydrolytic activity. Our personal semantic preference was the word "triggering". Without implying any specific process, "triggering" refers to any kind of mechanism that releases the control of the murein hydrolase. The nature of this (or these) signal(s) and the events of triggering represents a most important and interesting missing information in the field. A change in the intracellular concentration of some cell wall precursor - a UDP-linked glycopeptide, a murein oligomer or a bactoprenol derivative (3); some alteration of plasma membrane structure or status (as observed, e.g., in the penicillin induced "leakage" phenomena (4); or some alteration of the newly made murein itself (e.g., in degree of peptide crosslinking or secondary modification) - have all been considered as possible "effectors" in the chain of events that must lead from the inhibited cell wall synthetic reaction to the triggering of hydrolase activity and degradation of cell wall material. The recent isolation of pneumococcal mutants that would be only inhibited but not

lysed by penicillin in spite of the fact that they contained normal levels of amidase activity, may be interpreted as supporting evidence for the existence of such a triggering pathway (5).

Enzymes in search of a physiological function? In a sense, these conclusions take us right back to where we started from at the beginning of this brief overview of the field. The question of why cell wall degradation should ensue when wall synthesis is inhibited has re-emerged, still unanswered, accompanied by two new questions. A) What, if any, physiological function do the murein hydrolases that catalyse pathological wall degradation perform in the normally growing cell? B) In view of current information about the biochemistry of cell wall synthesis, is there still a conceptual need for wall hydrolysis or breaking of covalent bonds as an essential part of wall enlargement? Let us follow up both of these questions, starting with the first one.

Physiological role of the amidase. The only physiological anomaly of lyt^- mutants observable (without experimental intervention) is the formation of chains, indicating that the amidase plays a role in the separation of daughter cells at the end of cell division. The involvement of this enzyme in cell separation is demonstrated in a particularly striking manner in lyt^- mutants of pneumococci containing 0.1-1% residual hydrolase activity. These cells grow in short chains of 8-16 bacteria, in contrast to wild type cells which grow mostly as single cells, doublets or very short chains. Addition of high concentrations of amidase inhibitors to the growth medium, such as the Forssman antigen or high concentrations of choline, or growth in ethanolamine-containing medium causes the formation of "infinitely" long chains. This finding suggests that even the low residual activity detectable in the lyt^- mutants is utilized by the cells for a physiological function (cell separation).

<u>Penicillin induced lysis - a premature triggering of cell separation events?</u> It is attractive to speculate that the non-destructive activity (cell separation) may also be "triggered" by a genetically programmed temporary perturbation of wall synthesis at the end of cell cycle, i.e., by signals "mimicked" in (or generated within) cells treated with inhibitors. Lysis of bacteria treated with penicillin or other cell wall inhibitors may then be caused by a premature triggering of these cell separation events, but on the wrong scale, wrong time and at the wrong topographic sites (6). A sketch of this model is outlined in Figure 2. In support of this one may recall that the first morphological effect of penicillin treatment in lyt$^-$ cells is often dechaining by the apparent triggering of the residual (1%) amidase activity. Furthermore, the "infinitely" long chains of lyt$^-$ cells produced upon a more complete inhibition of amidase (e.g., by the addition of choline or ethanolamine) do <u>not</u> show dechaining during exposure to penicillin.

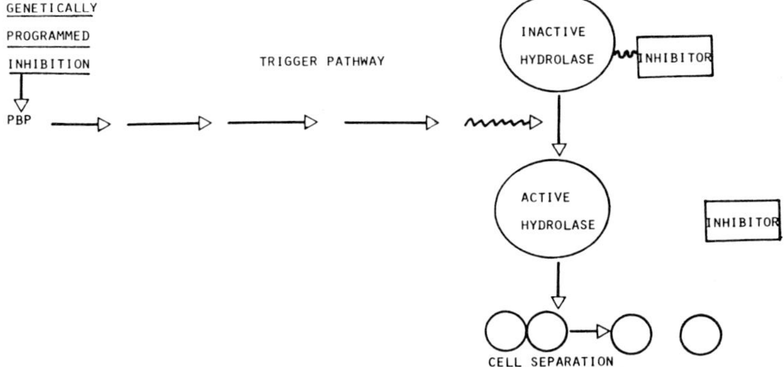

Fig. 2

Separation of daughter cells from one another at the end of cell division is clearly a non-essential function, at least in the test-tube environment. One may argue, however, that such an enzymatic cell separation mechanism has evolved under pressures <u>in vivo</u> in the natural environment of pneumococci, where chain-formation may be a selective disadvantage during encounter with phagocytic cells.

Is there a need for another murein hydrolase in cell wall synthesis? The recognition of the bacterial cell wall as a "sacculus" made of an uninterrupted system of covalent bonds has been one of the powerful *a priori* arguments in favor of a need for covalent bond breakage as an essential aspect of wall expansion. In the original model this was visualized as a "nick" to be introduced into the glycan backbone of murein. The current view of the mode of incorporation of new wall building blocks invokes a different mechanism, namely attachment via a primary transpeptidation. Thus, the actual addition of new elements to the preexisting wall does not require breaking of bonds. Furthermore, recent evidence indicates that, at least in *E. coli*, the orientation of glycan strands is at right angles to the axis of the bacilli and the task of wall expansion involves introduction of new "rings" of glycan in between two preexisting rings and the role is there for a "space-maker" reaction. Whether this has to be a hydrolytic reaction or could also be satisfied by rearrangement of bonds via transpeptidation is not clear yet. The latter type of "space-maker" reaction (as outlined by Pelzer in this vol.) "postpones" the hydrolytic event(s) to the end of cell division, at which time it could be performed by the autolytic murein hydrolase resulting in cell separation. Hydrolytic events (catalyzed by an endopeptidase) and coupled more closely (in time) to the growth and attachment of new murein chains has been envisioned by Burman and Park (see this book).

Bacterial mutants defective in this type of "space-maker" hydrolase would be expected to exhibit major abnormalities in growth and wall synthesis and therefore such cells should only exist as conditional mutants. No such mutant has been described as yet in the literature.

If this type of "space-maker" enzyme(s) exist, it should still be present in the lyt^- mutant which can grow normally. It is not clear what type of catalytic activity one should expect for such an enzyme. "Space-makers" may represent a unique kind of catalytic activity (e.g., endopeptidase) in all bacteria or they may involve the residual hydrolase activities in the lyt$^-$ mutants. In this case, however, different species of bacteria must utilize different types of hydrolases for this function. It is also hard to predict what, if any, pathological effect a continued "space-maker" activity would have in the absence of cell wall synthesis. It should be remembered that all the lyt$^-$ mutants described so far in the literature still lose viability, with a reduced rate, when treated with cell wall inhibitors. The mechanism of this death is not clearly known and may not even be related to cell wall damage. It is also conceivable, however, that this loss of viability is caused by the continued activity of the hypothetical "space-maker" enzymes creating small but irreparable (lethal) nicks in the covalent network of the sacculus.

References

1. Weidel, W., Pelzer, H.: Adv. Enzymol. 26, 193-232 (1964).
2. Tomasz, A., Albino, A., Zanati, E.: Nature 227, 138-140 (1970).
3. Tomasz, A., Waks, S.: Proc. Natl. Acad. Sci. USA 72, 4162-4166 (1975).
4. Tomasz, A.: Rvws. Infect. Dis. 1, 434-467 (1979).
5. Williamson, R., Tomasz, A.: J. Bacteriol. 144, 105-113 (1980).
6. Tomasz, A., Höltje, J.-V.: In Microbiology 1977, ed. J.D. Schlesinger, pp. 209-215 (1977).

THE AUTOLYTIC SYSTEM OF *STREPTOCOCCUS FAECIUM*

Gerald D. Shockman, Takashi Kawamura, John Barrett,
David Dolinger

Department of Microbiology and Immunology
Temple University School of Medicine
Philadelphia, Pennsylvania, USA

Introduction

As a bacterial cell increases in volume and divides, it must maintain the osmotically protective properties of its cell wall while it expands its wall surface area to enclose increased cellular contents. As discussed in this volume by M. L. Higgins, the assembly of new wall surface of streptococci is accompanied by a precisely defined pattern of shape changes.

It has been hypothesized (1,2) that insertion of new murein (peptidoglycan) precursors into a preexisting, protective murein exoskeleton involves the selective and exquisitely regulated action of endogenous murein hydrolases. Activities capable of hydrolyzing bonds in their own mureins have been found in a wide variety of both Gram-positive and Gram-negative species (see 3,4 for recent reviews). Generally, endogenous murein hydrolases have been considered to be synonymous with autolysins. However, it is now clear that not all murein hydrolases are capable of dissolving cells or isolated walls; and therefore, in the strict sense, may not be "autolysins" (see 5 for a review). Non-autolytic murein hydrolases would include activities that hydrolyze only a limited number of bonds essential to the two- or three-dimensional murein network, perhaps not even resulting in the release of soluble products, as well as enzymes that hydrolyze bonds that are not

The Target of Penicillin

essential to maintain the overall structure. Numerous potential functions of murein hydrolases have been proposed (1,2), including roles in cell wall assembly, surface enlargement, morphogenesis and cell division. Clearly, all postulated roles require tightly coordinated and regulated systems.

The Murein Hydrolase System of S. faecium ATCC 9790 - General Considerations

Although many bacterial species contain more than one murein hydrolase, a variety of data indicate that S. faecium contains only a single specificity of autolytic murein hydrolase activity (6-9), a β1-4 N-acetylmuramoylhydrolase (EC 3.2.1.17, muramidase). Over 95% of the autolysin activity recovered from disrupted cells is present in the insoluble cell wall fraction (6), in both a latent, proteinase-activatable form and in an active form, in a ratio of 4:1 to 5:1. Recent experiments (Dolinger and Shockman, unpublished results), using a protein renaturation technique (10), showed the presence of the (denatured) active form of the enzyme in culture supernates. The latent form appears to be a zymogenic precursor of the active form (11). Several types of evidence are consistent with passage of the latent form through the cytoplasmic membrane and its binding to the inner surface of the wall prior to its activation (7,11,12). Both latent and active forms bind to walls with extremely high affinities, requiring high salt concentrations (e.g., 5 M LiCl) or high pH (e.g., 0.01 N NaOH) to remove activity from wall-enzyme complexes or cells. Several types of evidence strongly indicated that, in exponential-phase cells, the autolysin is concentrated in recently assembled wall (7,13) at nascent cross-walls (14).

Evidence exists for a role for autolysin action in the cell cycle, consistent with its participation in surface enlargement and division in a closely regulated fashion (15-19,

L. Daneo-Moore, unpublished data). Cellular autolytic capacity was shown to vary with growth rate and with cell size and age. Notable was a decrease in cellular autolytic capacity about 15 to 25 min before cell division. This potentially suicidal activity appears to be regulated at several levels, including proteinase activation (11) topologically (7,14), and via the inhibitory action of regulatory ligands such as lipoteichoic acids and certain lipids, notably cardiolipin and phosphatidylglycerol (20-22). The roles of these regulatory processes in surface growth and division remain to be established.

Purification and Some Properties of the Autolytic Muramidase of *S. faecium*

Recently, we purified the latent form of the muramidase to near homogeneity (23). Latent activity was extracted from wall-enzyme complexes at high pH, bound to concanavalin A-Sepharose-4B and specifically eluted with α-methyl-D-mannopyranoside. Both inhibition and removal of contaminating proteinases were required to obtain only latent activity which gave one major protein band at M.W. of 130,000 on SDS-PAGE, which also stained with the periodate-Schiff (PAS) reagent. Treatment with trypsin resulted in a progressive decrease in intensity of the band at M.W. 130,000 and the appearance of a major protein and PAS-positive band at M.W. 87,000, along with several minor bands. Elution and renaturation (10) of protein in gel slices showed that the M.W. 130,000 band was latent enzyme and the M.W. 87,000 band was the trypsin-activated form.

The autolysin is truly a glycoenzyme. PAS-positive peptides were obtained. Cold TCA-precipitated enzyme contained neutral hexose, shown to be glucose. β-elimination followed by reduction with $NaBH_4$ converted about half of the glucose to hexitol. Finally, β-elimination and reduction with $[^3H]NaBH_4$

produced ^{3}H-fragments, about one half of which eluted from a gel filtration column coincident with glucose, with the remainder eluting as higher-molecular-weight oligosaccharides. Thus, this enzyme is a glucoenzyme, and perhaps the first proven example of a bacterial glycoenzyme.

The purified muramidase was found to have a somewhat unexpected substrate profile. For example, it failed to dissolve walls of *Micrococcus luteus* and the murein fraction of *S. faecium*. However, after re-N-acetylation, *S. faecium* murein was hydrolyzed (Table 1).

The action of this autolysin on soluble, linear uncrosslinked mureins (s-peptidoglycans), produced by penicillin-inhibited cultures, was studied in some detail (24,25). The use of these substrates, especially the s-peptidoglycan produced by autolysis-defective strains of *S. faecium*, which are about 45 DSP units long and are fully substituted with peptide side chains, permitted kinetic analyses that, heretofore, have not been possible.

In contrast to the action of hen egg-white lysozyme, the *S. faecium* muramidase failed to catalyze transglycosylation. Hydrolysis of the s-peptidoglycan of *S. faecium* resulted in the production of only one product, the disaccharide-peptide structural unit, even after only short periods of hydrolysis. A variety of data, including the results of studies of hydrolysis kinetics and substrate competition, suggested that each enzyme molecule firmly bound to an s-peptidoglycan chain, and hydrolyzed all susceptible bonds before it was released, suggesting a specific enzyme binding site on each s-peptidoglycan chain. Such a site and the direction of hydrolysis of chains was established by using s-peptidoglycans labeled at reducing ends with ^{3}H by reduction of the terminal MurNAc residue with [^{3}H]$NaBH_4$ and labeled at non-reducing ends by the addition of [^{14}]galactose via the action of galactosyltransferase (26).

Table 1. Comparison of properties of the two murein hydrolases of S. faecium[a]

	Con A muramidase	Murein hydrolase-2
Dissolution of:		
S. faecium walls	+	±
M. luteus walls	–[b]	+
S. faecium murein	±	+
S. faecium N-acetylated murein	+	+
Affinity for:		
Con A-Sepharose	High	No
S. faecium walls	High	Low
S. faecium s-peptidoglycan	High	–[b]
M. luteus s-peptidoglycan	High	–[b]
S. faecium-lysozyme lysate-Sepharose[c]	High	Low
M. luteus-lysozyme lysate-Sepharose[d]	No	Yes
Hemoglobin-Sepharose	No	High
Latent, proteinase activatable form	Yes	No
Transglycosylase activity	–[b]	?
Processive exodisaccharidase	Yes	?

[a] Some of the listed properties for murein hydrolase-2 are based on recent, provisional data.

[b] Not detectable.

[c] At pH 7.0 or pH 4.0, low salt concentration.

[d] At pH 7.0, low salt concentration, specificity of binding remains to be determined.

At high enzyme to substrate ratios, ^{14}C-labeled products were released immediately and ^{3}H-labeled products were only released after a 1.5- to 2-min lag. Thus, the muramidase is a processive exodisaccharidase that binds to non-reducing ends of chains and then successively hydrolyzes the bonds between MurNAc and GlcNAc until it reaches the end of that chain. It will be of interest to determine exactly how such an activity may play a role in wall surface enlargement and division.

The Second Murein Hydrolase of *S. faecium*

Recently we discovered and partially purified a second murein hydrolase in *S. faecium* that does not seem to be a glycoprotein. This second enzyme appears to have little ability to dissolve intact walls of *S. faecium*, suggesting that is may not be a true "autolysin," and explaining why it had not been seen earlier. It rapidly dissolves walls of *M. luteus* and the murein fraction of *S. faecium*. The distinct substrate specificities of the two enzymes (Table 1) provide differential assays. The ability of this enzyme to attack the murein fraction of *S. faecium*, suggests that it can hydrolyze at least some bonds in intact walls, although apparently unable to dissolve such walls. This type of subtle and selective "nicking" action could be of considerable importance in post-incorporation modification and "remodeling" of walls.

Although the two activities may have different functions in cells, it seems possible that a defect in one may be partially compensated by the presence of the other. The presence of two murein hydrolases in *S. faecium*, at least in part, explains previous difficulties in obtaining mutants that *completely* lack autolytic activity. Such redundancy of murein hydrolase activity could be a feature common to many bacteria and account for the dearth of truly autolytically negative mutants.

Evidence indicating the presence of two murein transglycosylases in *Escherichia coli* has also been reported (27).

The redundancy of murein hydrolase activities in *S. faecium* and the proposed compensatory activity of one or the other hydrolase is, in part, similar to the presence of multiple penicillin-binding proteins (28) and porin proteins in *E. coli* (29). In the latter case, mutants that lack one, or even two, porin proteins synthesize and incorporate into their membrane a third protein that can function as a porin.

Acknowledgement

This work was supported by U.S. Public Health Service Research Grant AI 05044.

References

1. Daneo-Moore, L., Shockman, G.D.: in Cell Surface Reviews, vol. 4, Elsevier/North Holland, Amsterdam, pp. 597-716 (1977).
2. Rogers, H.J., Perkins, H.R., Ward, J.B.: Microbial Cell Walls and Membranes, Chapman and Hall, London (1980).
3. Weidel, W., Pelzer, H: Adv. Enzymol. 26, 193-232 (1964).
4. Shockman, G.D.: Bacteriol. Revs. 29, 345-358 (1965).
5. Shockman, G.D., Barrett, J.F.: Ann. Revs. Microbiol. 37 (in press).
6. Shockman, G.D., Thompson, J.S., Conover, M.J.: Biochemistry 6, 1054-1065 (1967).
7. Joseph, R., Shockman, G.D.: J. Bacteriol. 127, 1482-1493 (1976).
8. Dezélée, P., Shockman, G.D.: J. Biol. Chem. 250, 6806-6816 (1975).
9. Rosenthal, R.S., Jungkind, D., Daneo-Moore, L., Shockman, G.D.: J. Bacteriol. 124, 398-409 (1975).
10. Hager, D.A., Burgess, R.R.: Anal. Biochem. 109, 76-86 (1980).
11. Pooley, H.M., Shockman, G.D.: J. Bacteriol. 103, 457-466 (1970).

12. Pooley, H.M., Shockman, G.D.: J. Bacteriol. 100, 617-624 (1969).

13. Shockman, G.D., Pooley, H.M., Thompson, J.S.: J. Bacteriol. 94, 1525-1530 (1967).

14. Higgins, M.L., Pooley, H.M., Shockman, G.D.: J. Bacteriol. 103, 504-512 (1970).

15. Shockman, G.D., Daneo-Moore, L., Higgins, M.L.: Ann. N.Y. Acad. Sci. 235, 161-197 (1974).

16. Shockman, G.D., Daneo-Moore, L., McDowell, T.D., Wong, W.: in β-Lactam Antibiotics, Academic Press, New York, pp. 31-66 (1981).

17. Hinks, R.P., Daneo-Moore, L., Shockman, G.D.: J. Bacteriol. 133, 822 (1978).

18. Hinks, R.P., Daneo-Moore, L., Shockman, G.D.: J. Bacteriol. 134, 1188 (1978).

19. Hinks, R.P., Daneo-Moore, L., Shockman, G.D.: J. Bacteriol. 134, 1074 (1978).

20. Cleveland, R.F., Daneo-Moore, L., Wicken, A.J., Shockman, G.D.: J. Bacteriol. 127, 1582-1584 (1976).

21. Cleveland, R.F., Höltje, J.-V., Wicken, A.J., Tomasz, A., Daneo-Moore, L., Shockman, G.D.: Biochem. Biophys. Res. Commun. 67, 1128-1135 (1975).

22. Cleveland, R.F., Wicken, A.J., Daneo-Moore, L., Shockman, G.D.: J. Bacteriol. 126, 192-197 (1976).

23. Kawamura, T., Shockman, G.D.: submitted for publication.

24. Barrett, J.F., Shockman, G.D.: (in preparation).

25. Barrett, J.F., Schramm, V.L., Shockman, G.D.: (in preparation).

26. Schindler, M., Mirelman, D., Schwarz, U.: Eur. J. Biochem. 71, 131-134 (1976).

27. Mett, H., Keck, W., Funk, A., Schwarz, U.: J. Bacteriol. 144, 45-52 (1980).

28. Spratt, B.G.: Phil. Trans. Roy. Soc. London, Ser. B, 289, 273-283 (1980).

29. Nikaido, H.: in Bacterial Outer Membranes: Biogenesis and Function, John Wiley & Sons, New York, pp. 361-407 (1979).

INTERACTION BETWEEN CHOLINE AND THE N-ACETYL-MURAMYL-L-ALANINE-AMIDASE OF *STREPTOCOCCUS PNEUMONIAE*

Thomas Briese and Regine Hakenbeck
Max-Planck Institut für molekulare Genetik, D1000 Berlin 33

Introduction

Lysis in pneumococci is dependent on only one autolysin, an amidase (1). The regulation of this enzyme has been extensively studied by A.Tomasz and his group, and it has been shown that the amidase activity can be influenced by two choline-containing cellular components. The wall teichoic acid (WTA) is necessary for enzyme activity (2); the Forssman-antigen (a lipoteichoic acid, LTA;3) in contrast is able to inhibit the enzyme noncompetitively (4,5), although its teichoic acid part is supposedly very similar to the WTA (6). The choline residues can be replaced *in vivo* by ethanolamine (EA;7). These EA-cells contain amidase-resistant cell walls, an abnormal inactive amidase (E-form) and show a defect in cell separation (8). The E-amidase can be activated *in vitro* by preincubation with choline containing cell walls (substrate) at 4°C; this process is called conversion (9). In an attempt to establish a fast purification procedure, exploiting the known interaction between LTA and amidase, we constructed an LTA-affinity column. This system led to unexpected, surprising results.

Results and Discussion

The gel material used for the LTA-column was prepared by coupling LTA (3) to a Sepharose 4BCl essentially as described in

The Target of Penicillin

Fig.1: Affinity chromatography of amidase with LTA-Sepharose. A 1.1ml LTA-sepharose column was loaded with 30 000U E-enzyme (part.purified by hydroxyl-apatite;5) and washed with 2.5ml of Tris-maleate(TM) buffer (50mM, pH6.8); 1.5ml NaCl in TM; 1ml TM; and 2ml 2.5%Brij in TM. Elution occurred with 5ml of 2%choline plus 2.5% Brij in TM. Fractions were collected (1fr/16min; 1ml/h) and 1µl aliquots tested for enzyme activity using 10µg cell wall in 200µl test volume. For conditions and definition of enzyme unit see (5).

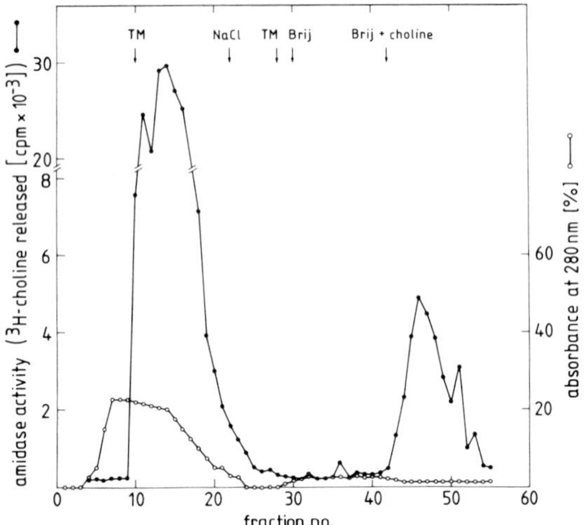

(10). E-amidase was used for the chromatography, and as can be seen in Fig.1, the column was heavily overloaded in order to determine and use its full capacity. After washing with salt and buffer, we tried to elute the bound enzyme with detergent (Brij) in analogy to the described release of LTA-inhibition by detergents (4). Surprisingly, no enzyme could be eluted this way and moreover, the matrix-bound amidase was clearly active in an enzyme test even in the absence of Brij. However, all enzyme activity could be eluted with 2%choline. An explanation for the unexpected behaviour of the amidase on the LTA-column came from an experiment, where the inhibitory activity of LTA was tested after lipase digestion. The treated LTA was no longer able to form micelles and concomitantly, the inhibitory effect was lost. It is quite likely, that micelle formation is also prevented upon coupling of the LTA to sepharose, and that the amidase inhibition by LTA _in vitro_ is due to micelle formation in such a way that the enzyme upon binding to the LTA-choline-residues is trapped in the micelle and thereby access to the high molecular weight substrate is prevented. Another fact discovered by the LTA-chromatography is that the loaded E-amidase was converted to a C-form enzyme upon binding to the LTA-column (table1). That LTA can mediate the conversion was also suggested in experiments outlined in table 2, where the amount and the form of amidase was investigated in cytoplasmic and membrane fractions from either choline- or EA-cells. In EA-cells the expected E-form amidase was found exclusively in the cytoplasm, in choline-cells about half of the enzyme was membrane bound and C-form, whereas in the cytoplasm a large amount of E-form was detected. This is in agreement with the hypothe-

	E-FORM	C-FORM
CRUDE E-AMIDASE	+	
AFTER HYDROXYL-APATITE	+	
IN LTA-SEPHAROSE		+
AFTER " "		+

Table 1: Conversion of amidase during LTA-chromatography. Enzyme activity was tested in the various fractions, with and without preincubation in the presence of substrate at 4°C.

	CHOLINE-CELLS		EA-CELLS	
	ENZYME ACTIVITY (% OF TOTAL)	ENZYME FORM	ENZYME ACTIVITY (% OF TOTAL)	ENZYME FORM
MEMBRANE	58.7	C	0.5	E
CYTOPLASM	40.1	(E)	99.0	E
TOTAL AMOUNT (UNITS)	18 000		16 000	

Table 2: Distribution of amidase activity in fractions from choline- and EA-cells. Membranes and cytoplasm were separated by centrifugation of disrupted cells (48 000 rpm, Ti 50, 3hrs). Aliquots were tested in the presence of 0.2%DOC for total amidase activity, and +/- preincubation for enzyme form.

sis formulated by Höltje and Tomasz (4) that the amidase is synthesized as an inactive E-form and activated to the C-form at a later stage. That the membrane attachment of the amidase is provided via the LTA is supported by the finding that enzyme activity could be strongly stimulated by detergents and completely released by 2% choline (to be published elsewhere). The C-form of the column-eluted amidase was maintained in the presence of choline, but was lost after extensive dialysis of the enzyme. The long time needed for this "back-conversion" (20 days) suggested a very strong binding between choline and the enzyme. Several other effects of choline on the enzyme, depending on the concentration used were observed: i) above 1%, choline alone could induce conversion, this choline-converted enzyme showed the same molecular size as the E-form when analyzed on sucrose gradients and not the high MW-form obtained after conversion with cell walls (5);ii) at 0.06%, the cell wall mediated conversion could be inhibited by 50%; iii) at 0.1% and higher, choline released the enzyme from LTA; iv) above 0.25%, enzyme activity was inhibited in a non-linear way by increasing concentrations of choline. These data suggest, that the binding of the amidase to LTA is weaker then to the substrate-bound WTA, and that the interaction with the WTA needed for enzyme conversion is weaker then the one needed for activity. This could be due to a different amount (or arrangement) of choline residues in the two teichoic acids.
The inhibitory effect of choline can be extended to an *in vivo* situation by adding 2% choline to the growth medium. Under these conditions, lysis of the culture could be prevented even after the onset of lysis, e.g. induced by penicillin (Fig.2a). After prolonged growth in the presence of 2% choline chain formation occurred like in EA-cells (Fig.2b).
The observed interaction between choline and amidase led to an improved purification step, using choline-affinity chromatography. In contrast to LTA-sepharose, some enzyme eluted with NaCl. Still, the residual amidase could be collected only after elution with choline. The procedure resulted in a remarkably high

Fig.2: Inhibition of cellular lysis by choline. a) An exponentially growing culture was divided at an OD=0.06, and the different aliquots treated as follows: 1,untreated control; 2,2% choline only;3-6,penicillin(↓)0.07μg/ml;4-6,2%choline at various times after pen: 4,0min(↓), the same curve was obtained up to 60min after pen (T); 5,105min(Y); 6,140min(Y). b) The picture was taken after 5generation growth in 2%choline-medium.

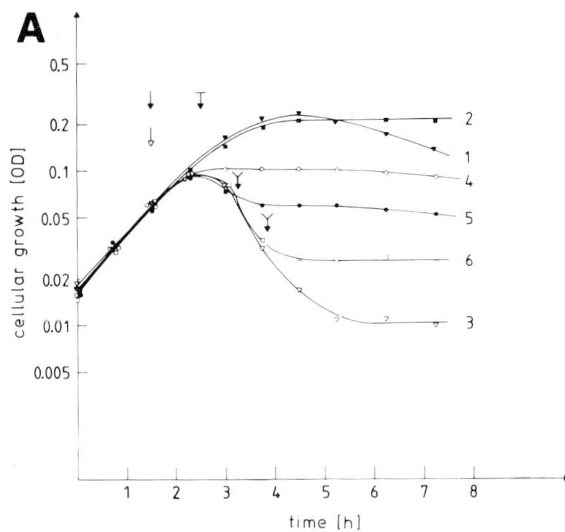

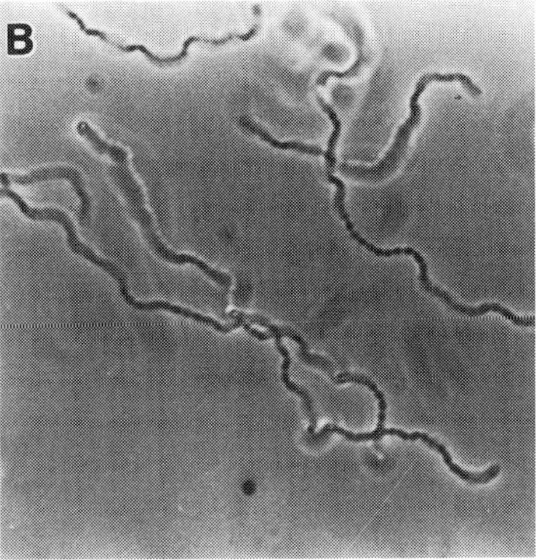

Table 3: Purification of amidase with choline-sepharose

	SPECIFIC ACTIVITY (UNITS/MG)	RECOVERY (%)
CRUDE E-AMIDASE	520	100
HYDROXYLAPATITE	1 290	38
CHOLINE-SEPHAROSE	153 200	24

specific activity of the eluted enzyme combined with a high recovery as shown in table 3.

The results obtained above led to the conclusion that in vivo the LTA might not act as an inhibitor for the autolysin in a classical sense, but could impose a negative effect on the enzyme activity by representing a topological barrier between enzyme and substrate. Such a mechanism would rely strongly on a close vicinity of the LTA-bound choline residues and the lipid moiety of the LTA, mirroring the micelle-trap. Lysis induced by detergents would then not be due to liberation of LTA-bound amidase, but rather to a disturbance of membrane integrity. On the other hand, knowing that the amidase is not a membrane bound protein per se (the E-amidase from EA-cells is a soluble protein;5), the actual step restricting the suicidal activity of the autolysin is more likely the transport through the membrane. It is a well-known fact that choline-starved pneumococci simply stop cellular growth and do not lyse (11), and one might speculate that one of the teichoic acids could be involved in a controlled transport process of the amidase towards the outside of the cell.

References

1. Howard, L.V., Gooder, H.: J. Bacteriol. 117, 796-804 (1974).
2. Höltje, J.-V., Tomasz, A.: J. Biol. Chem. 250, 6072-6076 (1975).
3. Briles, E.B., Tomasz, A.: J. Biol. Chem. 248, 6394-6397 (1973).
4. Höltje, J.-V., Tomasz, A.: Proc. Natl. Acad. Sci. USA 72, 1690-1694 (1975).
5. Höltje, J.-V., Tomasz, A.: J. Biol. Chem. 251, 4199-4207 (1976).
6. Brundish, D.E., Baddiley, J.: Biochem. J. 110, 573-582 (1968).
7. Badger, E.: J. Biol. Chem. 153, 183-191 (1944).
8. Tomasz, A.: Proc. Natl. Acad. Sci. USA 59, 86-93 (1968).
9. Tomasz, A., Westphal, M.: Proc. Natl. Acad. Sci. USA 68, 2627-2630 (1971).
10. March, S.C., Parikh, I., Cuatrecasas, P.: Anal. Biochem. 60, 149-152 (1974).
11. Tomasz, A.: Science 157, 694-697 (1967).

THE INFLUENCE OF LIPOTEICHOIC ACIDS ON THE AUTOLYTIC ACTIVITY OF STAPHYLOCOCCUS AUREUS

Werner Fischer, Hans Uwe Koch
Institut für Physiologische Chemie der Universität Erlangen
D-8520 Erlangen

Lipoteichoic acids (LTAs) are suggested to play a role in the regulation of autolytic enzymes since the inhibitory effect of pneumococcal Forssman antigen on the autolysin of Streptococcus pneumoniae has been discovered (1). Although poorly characterized, the structure of Forssman antigen seems to differ from that of the glycerophosphate-LTAs of other Gram-positive bacteria. Subsequent work revealed a similar effect of glycerophosphate-LTAs, but against the autolysins of those bacteria that have this type of LTA (2).

Dependent on the organism, glycerophosphate LTAs differ in chain length (3) and glycolipid structure (3, 4), but particularly in the substitution of their glycerol residues (5). This diversity prompted us to study the structural requirements of antiautolytic activity (3). Of particular interest was the behaviour of alanine ester substituents because they have been found to considerably modify other biological activities of lipoteichoic acids (6-8).

Test systems

Two test systems were used: A, extracellular autolytic activity of Staphylococcus aureus with cell walls of Micrococcus lysodeikticus as the substrate, and B, spontaneous lysis of S. aureus cells in 0.1 M phosphate buffer of pH 6. Lysis was fol-

The Target of Penicillin

lowed by the decrease in optical density. Both systems gave in principle identical results.

Influence of chain length and lipid moiety

As in previously studied systems (2), unsubstituted LTA was a potent inhibitor of S. aureus autolysin; the concentration effecting 50% inhibition was 1.6×10^{-6} and 8×10^{-7} M in system A and B, respectively.

Differences in chain length between 20 and 37 glycerophosphate residues had no influence. The inhibitory activity, however, drastically decreased when the polymeric chain was replaced by a single glycerophosphate residue. Variabilities of the glycolipid structure, such as the type of linkage between the hexosyl residues, a third fatty acid or a phosphatidyl moiety on the glycolipid, did not alter the activity. But deacylation, as in previously studied systems (2, 9, 10), resulted in a drastic decrease.

Antiautolytic activity therefore depends on a non-specific lipid moiety and a sequence of glycerophosphate residues.

The effect of chain substitution with D-alanine ester and glycosyl residues

A detailed investigation into the relationship between alanine ester content and the antiautolytic effect became possible by two observations: The alanine ester content of S. aureus LTA is modified by the NaCl concentration in the growth medium (Table I), and alanyl LTA can be fractionated by anion exchange chromatography in order of decreasing alanine ester content (11). On this study it became also apparent that, independently of the average, the alanine ester content of S. aureus LTA is kept within a narrow range and alanine-free species could

TABLE I. D-Alanine ester substitution of LTA of S. aureus cells grown at different NaCl concentrations (11)

NaCl (g/l)	AlaGro/Gro	
	Defatted cells	Purified LTA
2	0.78 ± 0.02	0.73 ± 0.03
35	0.57	0.55
75	0.36 ± 0.03	0.32 ± 0.02
100	0.34 ± 0.02	0.30 ± 0.02

not be detected (11). By these procedures a set of lipoteichoic acids with an alanine ester content of 0.2 to 0.7 was prepared and tested as shown in Figure 1. With increasing alanine ester content the dose response gradually decreased and approached

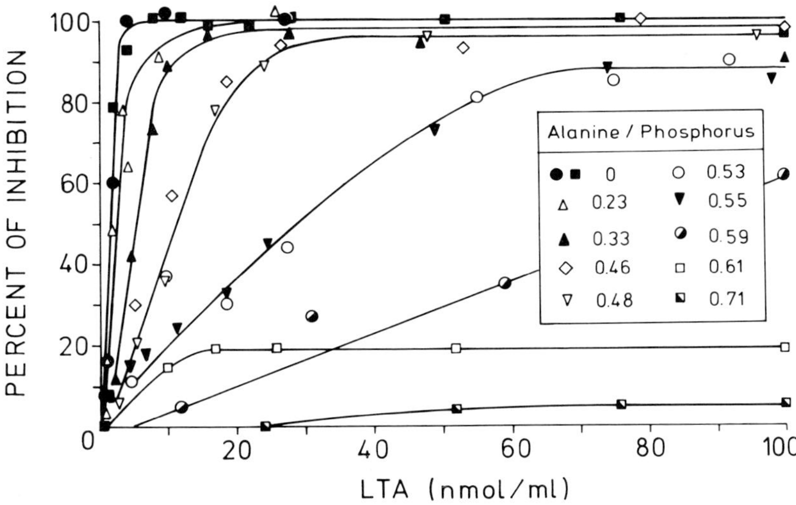

FIGURE 1. Effect of increasing alanine ester content of LTAs on the inhibitory activity against extracellular autolysin of S. aureus (taken from Ref. 3). For percent of inhibition the maximum effect of non-substituted LTA was set at 100%.

zero at a substitution of 0.7. The concentration effecting 50% inhibition increased in a non-linear fashion with increasing alanine ester content (Fig. 2).

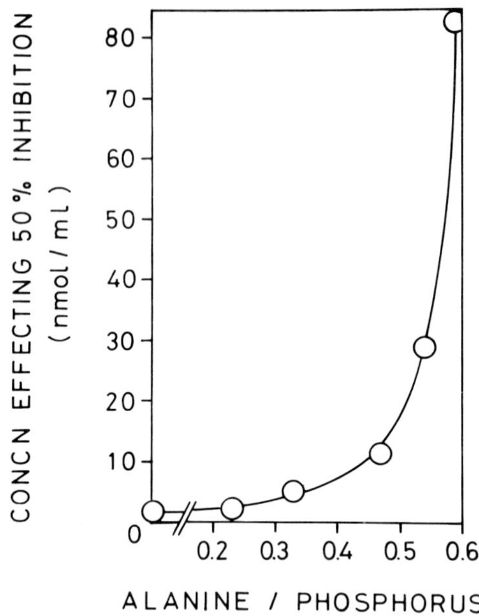

FIGURE 2. Relationship between alanine ester content and inhibitory activity of LTAs against extracellular autolysin of S. aureus (taken from Ref. 3). Concentrations effecting 50% inhibition were taken from Figure 1.

In contrast to alanine ester, glycosyl substituents had virtually no effect (Table II), even if 82% of the glycerophosphate residues were substituted with mono-, di-, and triglucosyl residues. These findings indicate that alanine ester exerts its action through charge compensation rather than by steric hindrance. The antiautolytic activity, therefore, resides in the negative charges of the poly(glycerophosphate) chain.

TABLE II. Inhibitory effect of glycosylated LTAs on extracellular autolytic activity (A) and autolysis of S. aureus (B). The substitution is given as the molar ratio of glycosylglycerol to total glycerol. (Taken from Ref. 3)

Glycosyl Substitution	Concn (μM) effecting 50% inhibition in	
	A	B
None	1.6	0.8
0.33	1.7	-
0.47	2.0	0.8
0.82	-	1.4

This dependence on charge was also seen with lipids (3). The negatively charged cardiolipin and phosphatidyl glycerol were as potent inhibitors as unsubstituted lipoteichoic acid, whereas the zwitterionic phosphatidylcholine and phosphatidylethanolamine were ineffective.

Concluding remarks

The observation, that LTA and cardiolipin loose their antiautolytic activity on deacylation or in the presence of detergent (2, 3, 9, 10) suggests that LTA and cardiolipin interact with the enzyme in the form of micelles and liposomes, respectively. Whether the attachment of the enzyme to these negatively charged particles results in an allosteric inhibition or an exclusion of the active enzyme from peptidoglycan has not yet been established. Depending on which of these alternatives is the correct one, the potential role in vivo would be either to keep the enzyme inactive or to fix the active enzyme in the cell wall-membrane-complex.

The interaction of S. aureus LTA with S. aureus autolysin is modulated by alanine ester in a fashion (Fig. 2) that would be appropriate for a regulation of either the activity or the distribution of the enzyme. Whether or to what extent this potential is used in vivo requires further investigations, including those of the metabolism of alanine ester substituents.

This work was supported by a grant of the Deutsche Forschungsgemeinschaft.

References

1. Höltje, J.V., Tomasz, A.: Proc. Natl. Acad. Sci. USA 72, 1690-1694 (1975).
2. Cleveland, R.F., Höltje, J.V., Wicken, A.J., Tomasz, A., Daneo-Moore, L., Shockman, G.D.: Biochem. Biophys. Res. Commun. 67, 1128-1135 (1975).
3. Fischer, W., Rösel, P., Koch, H.U.: J. Bacteriol. 146, 467-475 (1981).
4. Fischer, W.: Chemistry and Biological Activities of Bacterial Surface Amphiphiles (Shockman, G.D., Wicken, A.J.,eds) pp 209-228, Academic Press, New York 1981.
5. Fischer, W., Koch, H.U.: Chemistry and Biological Activities of Bacterial Surface Amphiphiles (Shockman, G.D., Wicken, A.J., eds) pp 181-194, Academic Press, New York 1981.
6. Heptinstall, S.A., Archibald, R., Baddiley, J.: Nature (London) 225, 519-521 (1970).
7. McCarty, M.: Proc. Natl. Acad. Sci. USA 52, 259-265 (1964)
8. Koch, H.U., Fischer, W., Fiedler, F.: J. Biol. Chem. 257, 9473-9479 (1982).
9. Cleveland, R.F., Daneo-Moore, L., Wicken, A.J., Shockman, G.D.: J. Bacteriol. 127, 1582-1584 (1976).
10. Cleveland, R.F., Wicken, A.J., Daneo-Moore, L., Shockman G.D. : J. Bacteriol. 126, 192-197 (1976).
11. Fischer, W., Rösel,P.: FEBS Lett. 119, 224-226 (1980).

COENZYME A-GLUTATHIONE DISULFIDE: AN ENDOGENOUS INHIBITOR OF THE LYTIC ENZYMES IN ESCHERICHIA COLI

Joachim-Volker Höltje and Ulla Schwarz
Max-Planck-Institut für Virusforschung, Abteilung Biochemie II
D-7400 Tübingen

Introduction

It has been proposed that growth and division of the murein sacculus requires a proper interplay between murein-hydrolizing and murein-synthesizing enzymes (1). Murein hydrolases will provide space for the insertion of additional murein subunits into the pre-existing murein network. Enzymes capable of cleaving specific bonds in the murein are abundant in E. coli (2, 3, 4). However, only the endopeptidases (3) and the transglycosylases (4) accept the intact murein sacculus as a substrate and are capable of lysing the bacterial cell. Thus, these two types of murein hydrolases represent the autolytic system of E. coli. The regulation of the murein hydrolases seems to be one of the crucial factors in controlling growth and division of the bacterium. The penicillin induced lysis of bacteria is only one illustration of this fact. Almost nothing is known about how the murein hydrolases are regulated in E. coli, aside from the observation that they become latent in non-growing cells, avoiding autolysis (5). Nevertheless, hydrolases can be isolated from such cells in a highly active state raising questions about the control mechanism involved. One rapid mechanism to regulate enzyme activity is by regulatory compounds which could act either as inhibitors, activators or modifiers. Such regulators for murein hydrolases have been described as existing in Gram-positive bacteria (6, 7, 8).

The Target of Penicillin

Results

In an attempt to search for regulatory compounds of murein hydrolases in E. coli, cells were extracted with 50% ethanol. The extract was separated into various fractions by Bio Gel P6 gel filtration followed by DEAE-cellulose ion-exchange chromatography. The individual fractions were then tested for their effect on the different murein-hydrolases in *in vitro* assay systems. Two fractions were found to be inhibitory for endopeptidase and one of those also for transglycosylase activity. It turned out that the endopeptidase inhibiting fraction consisted of UDP-muramylpentapeptide, which is one of the substrates for endopeptidase which also exhibits carboxypeptidase activity.

The second inhibitory fraction was further purified by high-pressure-liquid-chromatography to biochemical homogeneity and analysed for its amino acid composition. Besides glycine, cysteine and glutamic acid, two additional modified amino acids were present defining a glutathione derivative. Since the compound shows a UV absorption maximum at 260 nm it was finally recognized that we isolated the mixed disulfide of coenzyme A and glutatione (CoASSG). This substance has previously been described as being present in pro- as well as in eukaryotic cells (9, 10). A comparison of the amino acid analysis of authentic CoASSG with the isolated compound shows an identical amino acid pattern, with taurine and ß-alanine as the two modified amino acids derived from the coenzyme A part of the molecule. An interesting feature of this compound is its capacity to bind iron which can be followed by an increase in the ratio of the absorbance at 250 over 260 nm. The addition of Fe^{3+} to commercially available CoASSG or CoASSG isolated from E. coli increases the spectral ratio from 0.84 to 0.97. Disulfide loaded with iron in this way can be isolated by separation on Sephadex G25.
CoASSG, poor in iron, inhibited both soluble lytic enzymes of

E. coli, the transglycosylase and the endopeptidase, to a degree of 50% at a concentration of about 100 µg/ml. However, iron charged CoASSG dramatically improved in its capacity to specifically inhibit the transglycosylase in comparison with iron free CoASSG (Fig. 1).

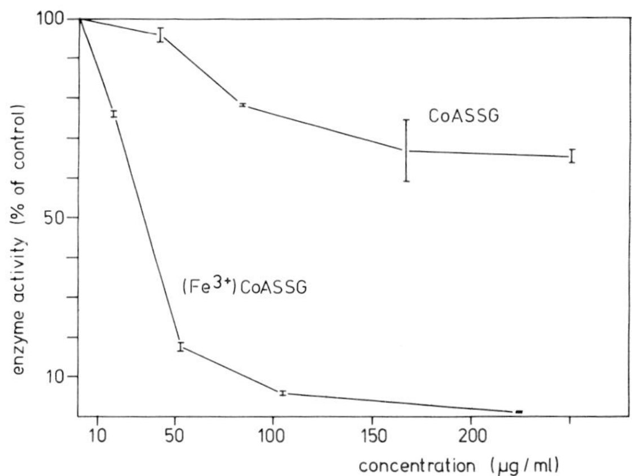

Fig. 1. Inhibition of transglycosylase reaction by CoASSG. CoASSG or Fe-CoASSG was added to the reaction mixture at the indicated concentrations. Tritium-labelled sacculi (5 µg murein; 3×10^3 cpm/µg) were incubated with purified enzyme (2 µg protein) in a total volume of 100 µl buffer (10 mM Tris maleate, 10 mM $MgSO_4$, 1% Triton X-100, pH 5.0) at 37°C for 30 min (4).

50% inhibition is caused by amounts as low as 20 µg per ml. To prove that it is the iron containing CoASSG and not free iron ions that causes the inhibition, the effect of the iron chelator ferrioxamine B was tested (Table 1). Desferrioxamine abolishes the inhibitory effect of Fe^{3+} ions but does not interfere with the inhibition caused by the Fe-CoASSG complex indicating that the observed inhibition is really due to the complex and not to some contamination by free iron ions.

Table 1. Effect of ferrioxamine B on the inhibition of transglycosylase by Fe^{3+} and Fe-CoASSG. Transglycosylase activity was determined as described in Fig. 1.

Addition	Concentration (nM)	% Inhibition of Enzyme	
		-Desferrioxamine	+Desferrioxamine
None	-	0	0.8
(Fe^{3+})CoASSG	0.45	96.5	94.9
Fe^{3+}	0.02	50.0	5.7

A difference in the mechanism of inhibition of the two enzymes is indicated by the fact that dithiothreitol (DTT) neutralises the inhibition of the endopeptidase but not the inhibition of the transglycosylase. This finding could be explained by a stabilisation of the disulfide bond in the iron complexing molecule which seems to be the specific inhibitor for the transglycosylase, whereas the endopeptidase is inhibited by the iron free disulfide compound.

The response of E. coli to CoASSG deficiency was investigated by the use of mutants with defects in the biosynthesis of glutathione (11, 12) which consequently lack CoASSG. It is the typical behaviour of E. coli, but not, for example, B. subtilis, not to lyse when washed, resuspended in phosphate buffer and incubated at 37°C, indicating that the murein hydrolases in E. coli are more efficiently controlled than in B. subtilis. Following this procedure the reaction of two different mutants in the gene gshA coding for the γ-glutamylcysteine synthetase and two different mutants in the gshB gene which codes for the glutathione synthetase, were tested (Fig. 2). Each of the 4 independent CoASSG deficient mutants lysed at a greatly increased rate.

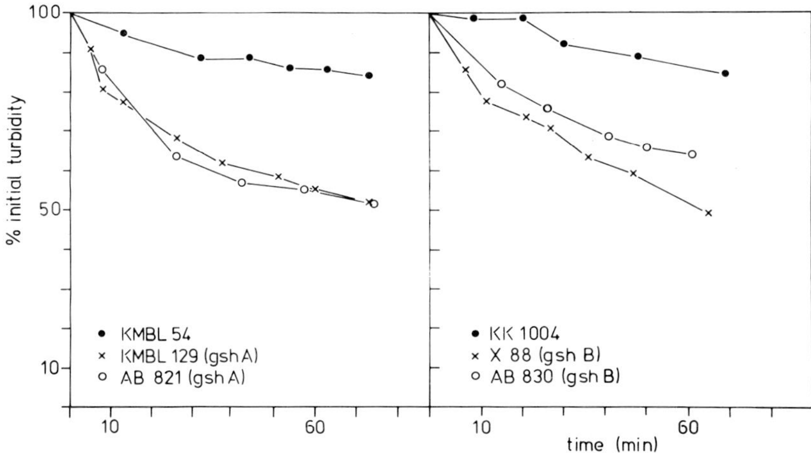

Fig. 2. Lysis response of glutathione deficient mutants of E. coli. Cells were harvested during logarithmic growth (OD_{578} nm 0.4), washed twice with and resuspended in 0.01 M phosphate buffer, pH 6.8, and incubated at 37°C.

Discussion

It is indeed a major problem to demonstrate the true biological role of the isolated CoASSG in regulating the activity of murein-hydrolases in the growing and dividing cell. With this in mind, it is interesting to note that under conditions where, in E. coli, the CoASSG level increases, Gram-positive bacteria tend to lyse as opposed to E. coli, which stays intact. Loewen has shown that the intracellular level of CoASSG varies with different growth conditions (13). For example, the pool of CoASSG increases to about double during oxygen and glucose starvation and more dramatically when growing into stationary phase. In agreement with the postulated role as an endogenous inhibitor for the lytic enzymes, we were able to demonstrate that mutants lacking CoASSG exhibited increased lysis sensitivity compared to wild type cells. It is concluded that CoASSG deficiency results in a loss of an effective control of the autolytic enzymes of the cell. Various nucleotides such as ppGpp or cyclic AMP have either been shown to be or proposed as regulators of some of the cell's activities. CoASSG, a nucleotide derivative, may indeed represent another nucleotide involved in the regulation of an important cellular process, namely, growth and division of the murein sacculus.

Acknowledgement

We wish to thank Uli Schwarz for his encouragement and stimulating discussions throughout these studies.

References

1. Weidel, W., Pelzer, H.: Adv.Enzymol. 26, 193-232 (1964).
2. Pelzer, H.: Z.Naturforsch. 18b, 950-956 (1963).
3. Keck, W., Schwarz, U.: J.Bacteriol. 139, 770-774 (1979).
4. Höltje, J.-V., Mirelman, D., Sharon, N., Schwarz, U.: J.Bacteriol. 124, 1067-1076 (1975).
5. Hartmann, R., Bock-Hennig, S.B., Schwarz, U.: Eur.J.Biochem. 41, 203-208 (1974).
6. Höltje, J.-V., Tomasz, A.: Proc.Natl.Acad.Sci.U.S.A. 72, 1690-1694 (1975).
7. Cleveland, R.F., Höltje, J.-V., Wicken, A.J., Tomasz, A., Daneo-Moore, L., Shockman, G.D.: Biochem.Biophys.Res. Commun. 67, 1128-1135 (1975).
8. Herbold, D.R., Glaser, L.: J.Biol.Chem. 250, 1676-1682 (1975).
9. Ondarza, R.N.: Biochim.Biophys.Acta 107, 112-119 (1965).
10. Loewen, P.C.: Biochem.Biophys.Res.Commun. 70, 1210-1218 (1976).
11. Apontoweil, P., Berends, W.: Biochim.Biophys.Acta 399, 10-22 (1975).
12. Fuchs, J.A., Warner, H.R.: J.Bacteriol. 124, 140-148 (1975).
13. Loewen, P.C.: Can.J.Biochem. 56, 753-759 (1977).

AUTOLYSIS OF *ESCHERICHIA COLI* : INDUCTION AND CONTROL.

Jean van Heijenoort, Mireille Leduc, Yveline van Heijenoort, Rouha Kasra
Institut de Biochimie, Université Paris-Sud, Orsay - France.

Claude Fréhel
Institut Pasteur, Paris - France.

Introduction

E. coli cells generally do not lyse spontaneously at a high rate and some kind of induction is necessary to bring about their autolysis. The conditions of this induction and the properties of the resulting autolyses have not yet been studied to a great extent. Furthermore, it is assumed that one of the main steps in the autolysis of *E. coli* is the degradation of peptidoglycan by specific endogenous hydrolases. Therefore, a closer investigation of the induction and control of the autolytic system of *E. coli* could perhaps help to understand the mechanisms involved in this process and, in particular, the role played by peptidoglycan hydrolases.

Results

1) *Induction of autolysis*. Among the various induction methods examined, a distinction was made between those used for growing cells and those for harvested cells. In the first case, only methods (D-cycloserine, mutants requiring DAP or glucosamine, β-lactams, moenomycin etc...) known for promoting autolysis by interference with peptidoglycan synthesis were considered (1) (2). In fact, the highest rates of autolysis were observed with moenomycin and cephaloridine (Fig. 1A). It is noteworthy that both cephaloridine, at a low concentration, and moenomycin specifically inhibit polymerization steps of peptidoglycan

The Target of Penicillin

synthesis (3) (4). Autolysis of harvested exponential-phase cells of E. coli can be promoted by various shock treatments (Fig. 1B) (2)(5). Shock treatment with 10^{-3}M EDTA is the most efficient.

2) Common properties of E. coli autolyses. Autolyses promoted by these various methods all share certain common properties :

- appreciable rates of autolysis observed only with fast growing cells.
- similar dependency of rates of autolysis on temperature.

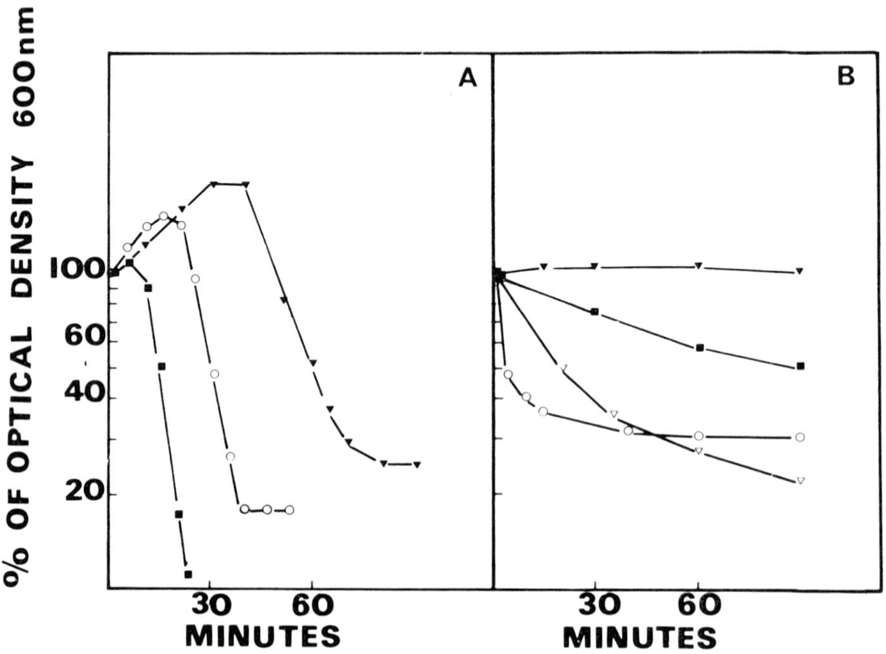

Fig.1: Kinetics of autolysis of E. coli. (A) Effect of moenomycin (10μg/ml) (■) or cephaloridine (5μg/ml) (O) on early exponential-phase cells (10^8/ml) of E. coli K-12 HfrH grown at 37°C in rich medium and (▼) lysis of E. coli W7 (DAP^-, Lys^-) in rich medium after DAP deprivation. (B) Early exponential-phase cells (2 x 10^8/ml) grown at 37°C in rich medium were harvested and autolyses were induced by shock treatment with (▼) 0.05 M sodium chloride (pH 7) ; (■) distilled water and 0.5 M sodium acetate (pH 6.5) ; (O) 10^{-3} M EDTA (pH 6.5) ; (▽) 0.05 M sodium phosphate (pH 7).

- inhibition of autolysis by 10^{-2} M Mg^{++}.
- no effect of uncoupling agents (2,4-dinitrophenol, sodium azide and CCCP), at their usual inhibiting concentrations, on autolysis.

These results suggest that perhaps the same autolysin is functioning in all cases. Recently, with penicillin-tolerant mutants it has been shown that, among the various peptidoglycan hydrolases of E. coli, a N-acetylmuramidase is the most likely primary autolysin (6).

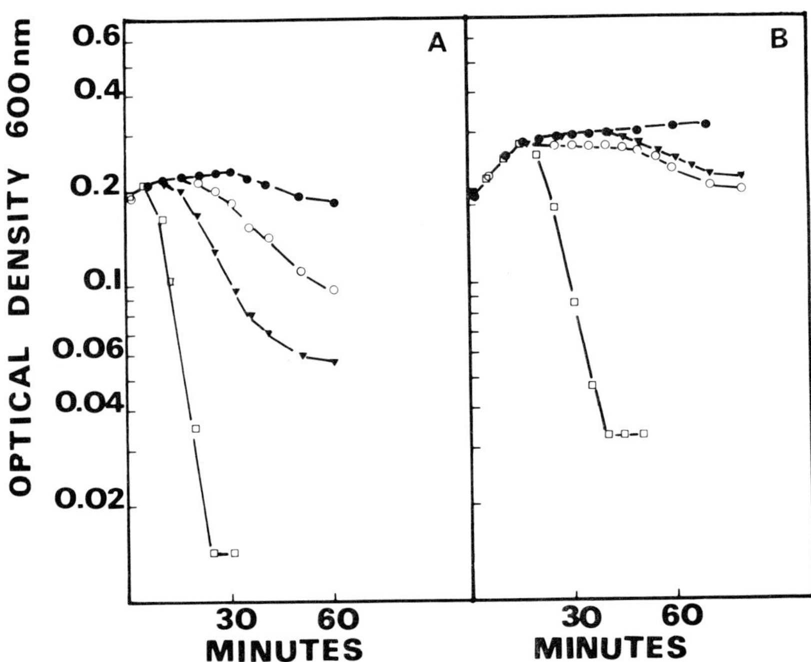

Fig.2: Effect of rifampicin, chloramphenicol, and cerulenin on moenomycin and cephaloridine-induced autolyses of E. coli K 12 HfrH cells growing in rich medium at 37°C. Early exponential-phase cells (10^8 cells per ml) were untreated (□) or treated with either rifampicin (10 μg/ml) for 20 min (○), chloramphenicol (100 μg/ml) for 20 min (●), or cerulenin (100μg/ml) for 30 min (▼) before the addition of the inducing agent, (A) moenomycin (10 μg/ml) or (B) cephaloridine (5μg/ml). It was verified that protein synthesis continued unabated for 40 min after the addition of cerulenin.

3) Differences between the two types of autolyses. However, the two types of autolysis considered differ by certain properties. Inhibition of RNA, protein or fatty acid synthesis greatly reduced the rate and extent of moenomycin or cephaloridine-induced autolysis (Fig. 2), whereas very little or no effect was observed with shock-induced autolysis (2). The highest rates of peptidoglycan degradation were observed during shock-induced autolysis (Table 1). In this case, breakdown paralleled the decrease in optical density. In the autolysis of growing cells peptidoglycan degradation was much more limited and no correlation with decrease of optical density was observed. Electron microscopy studies revealed that during EDTA or moenomycin-induced autolysis peptidoglycan degradation extended over the whole cell surface. However, in shock treated cells a rapid general breakdown into small fragments was observed, whereas with antibiotic treated cells only a slow thinning down of the macromolecule was detected.

Table 1 - Correlation between autolysis and peptidoglycan degradation

Induction method	percentage of optical density decrease at 600 nm.	percentage of peptidoglycan degradation
Water-sodium acetate shock	31	35
EDTA shock	69	66
Sodium phosphate shock	44	45
Moenomycin	73	28
Cephaloridine	82	23
E. coli W7 (DAP^- Lys^-)	58	26

After induction of the autolysis (2) of early exponential-phase cells of E.coli W7 (Lys^-,DAP^-) grown at 37°C in rich medium containing 10^{-4}M ^{3}H DAP (60 mCi/mmol) and incubation for 1 hour at 37°C, the percentage of solubilized peptidoglycan was estimated by the amount of ^{3}H DAP remaining in the trichloroacetic acid-insoluble material.

Discussion

Various explanations, not necessarily excluding one another, can be put forward for the fact that only the most inner part of peptidoglycan appears to be degraded upon autolysis induced by inhibition of its biosynthesis. One possibility could be that the tight anchoring with the outer membrane maintains the upper part of peptidoglycan in a conformation unfavorable for autolysin action. It might also be speculated that active autolysin is only available in very limited amounts in the periplasm or that, autolysin being an inner membrane bound activity, a major part of peptidoglycan remains unaccessible to its action as long as envelope integrity is preserved. Therefore, it is only when the envelope organization is disturbed by shock treatments that the greater part of peptidoglycan becomes amenable to extensive degradation.
Our results are in agreement with previous published data. Recent evidence (7) provides support for a peptidoglycan structure thicker than generally assumed. It has also been observed that lpo mutants in which the outer membrane is less tightly linked to peptidoglycan are more liable to autolysis (8). The distribution of peptidoglycan hydrolases over the whole cell surface (9) and the existence of a barrier observed between these hydrolases and peptidoglycan are also consistent with our results (10). Furthermore, the fact that Mg^{++} inhibits all kinds of autolyses investigated (2), suggest that by its high binding to peptidoglycan, it could maintain this macromolecule in a conformation unfavorable to autolysin attack. Finally, it is presently difficult to speculate on how the arrest of protein or fatty acid synthesis controls autolysin activity.

References

1. Tomasz, A.: Annu. Rev. Microbiol. 33, 113-137 (1979).
2. Leduc, M., Kasra, R., van Heijenoort, J.: J. Bacteriol. 152, 26-34 (1982).
3. Rogers, H.J., Perkins, H.R., Ward, J.B.: Microbial cell walls and membranes. Chapman and Hall Ltd. London (1980).

4. Gale, E.F., Cundliffe, E., Reynolds, P.E., Richmond, M.H. Waring, M.J.: The molecular basis of antibiotic action. John Wiley & Sons Ltd. London (1981).

5. Leduc, M., van Heijenoort, J.: J. Bacteriol. 142, 52-59 (1980).

6. Kitano, K., Williamson, R., Tomasz, A.: FEMS Microbiol. Lett. 7, 133-136 (1980).

7. Hobot, J.A., Carlemalm, E., Kellenberger, E.: Experientia, 37, 1226 (1981).

8. Hirota, Y., Suzuki, H., Nishimura, Y., Yasuda, S.: Proc. Natl. Acad. Sci. U.S.A. 74, 1417-1420 (1977).

9. Goodell, E.W., Schwarz, U.: Eur. J. Biochem. 81, 205-210 (1977).

10. Hartmann, R., Bock-Hennig, S.B., Schwarz, U.: Eur. J. Biochem. 41, 203-208 (1974).

EXPRESSION OF ØX174 LYSIS GENE CLONED INTO DIFFERENT PLASMIDS

Bernhard Henrich, Werner Lubitz, Roland Plapp

FB Biologie, Abt. Mikrobiologie, Universität Kaiserslautern, Postfach 3049, D-6750 Kaiserslautern

Introduction

Infection of *Escherichia coli* with the single-stranded DNA phage ØX174 results in the release of progeny particles by a limited desintegration of the bacterial cell wall (1). The ultimate requirement of phage specific gene*E* product (*gpE*) for this lysis (2) was confirmed by cloning a fragment of ØXRFI DNA, comprising genes*E* and *J*, under transcriptional control of the *lac* promotor giving rise to plasmid pUH12. Induction of the cloned genes provokes lysis of pUH12 transformants (3). Elimination of all ØX174 gene sequences except gene*E* from the cloned DNA fragment revealed that *gpE* alone is sufficient to trigger the lysis event (4). However, the molecular mechanism of this process remains ununderstood, and it has been proposed that *gpE* might interact with the autolytic system of *E.coli* to achieve lysis of the cell (5). The examination of this hypothesis might be facilitated by expression of the ØX174 *E* gene in bacteria with various genetic backgrounds and by isolation of mutants which escape lysis. Such mutants should easily be obtained by use of gene*E*-containing plasmids. However, transformation of different bacterial strains with pUH12 is restricted by the condition that the recipients must contain repressor-overproducing *lac i*-alleles to control transcription of the lysis gene (3). This problem was solved by construction of three combined plasmids which code for different antibiotic resistances and include the ØX174 *E* gene as well as the i^q-allele (6,7) of the *lac* repressor gene. One plasmid of this series was used to characterize some lysis-deficient mutants and to show isogenic complementation of a gene*E* amber mutation.

Results and Discussion

Control of *gpE* production.

Expression of ØX174 *E* gene in pUH12 can be regulated by variation of the concentrations of *lac* repressor and inducer. Therefore, prior to the combination of the lysis gene and the *lac i*-gene in a single plasmid, it was necessary to find out conditions which allow sufficient repression of gene *E* in the absence and maximal gene*E*-expression in the presence of *lac* inducer. The effect of intracellular *lac* repressor concentration on induction of the lysis gene is shown in Fig.1. pUH22 (3), which differs from pUH12 by the *am3*-mutation at the beginning of gene*E* (8), was used as a control. Increasing repression of the *lac* promotor was achieved by use of host strains which carry the repressor-overproducing *i*-alleles i^q and i^{q1} (6,7,9) or lack positive transcriptional control of the *lac* operon.

Strain $F^-z^-\Delta$M15recA(i^+) gave no viable transformants with pUH12. Colonies of strain 71-18(i^q, *suII*) containing pUH12 or pUH22 were relatively small. Growth of both kinds of transformants in liquid media led to spontaneous cell-lysis at an O.D.600 of 0.2 to 0.3 (Fig.1a). Using pUH22, lysis was less dramatical as compared with pUH12 since the *suII*-mutation of the host strain does not support the *wt*-level of gene*E* expression. These observations indicate that neither the i^+-allele nor the i^q-allele of the *lac i*-gene, present in one to two copies per cell, provide sufficient repression of the cloned ØX174 genes.

Under non-induced conditions strains F'$i^{q1}\Delta$M15(i^{q1}) and CAi^{q1} (i^{q1}, cap^-, cya^-), containing pUH12 or pUH22, showed normal cell growth. Addition of IPTG resulted in a very sharp lysis of F'$i^{q1}\Delta$M15/pUH12 cells about 20 min. after induction, and in a delayed lysis of CAi^{q1}/pUH12 cells about 70 min. after induction. No lysis was observed with pUH22 transformants of either strain (Fig.1b,c). The *lac i*-alleles referred to in Fig.1a,b,c were all included in F'factors; another possibility to increase the amount of *lac* repressor is the introduction of a repressor-overproducing multicopy plasmid like pMC7(i^q) (7). As shown in Fig.1d, the time course of lysis obtained with strain DS410/pMC7 containing pUH12 is comparable to that of F'$i^{q1}\Delta$M15/pUH12 transformants. Consequently, the presence of the *lac* i^{q1}-allele on a F'plasmid is sufficient to support normal cell growth of pUH12 transformants as is the *lac* i^q-allele if included in a multicopy

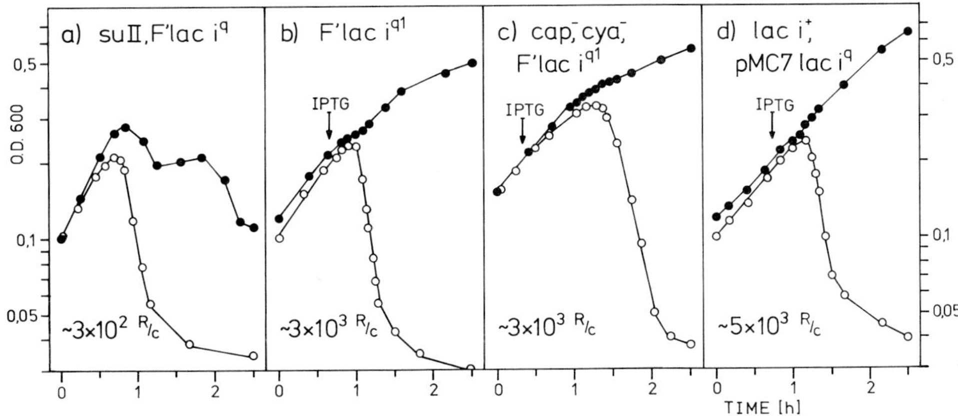

Fig.1: Influence of *lac* repressor concentration on the expression of cloned ØX174 lysis gene. *E.coli* strains a) 71-18 ([*lac pro*]Δ *thi suII, F'lac* $i^q z^- \Delta M15$ *pro*⁺), b) F'i^{q1}ΔM15 ([*lac pro*]Δ *thi ara strA, Ø80 d lac* $z\Delta M15$, *F'lac* $i^{q1} z \Delta M15$ *pro*⁺), c) CAi^{q1} ([*lac pro*]Δ *thi cap cya strA recA, F'lac* i^{q1} *lac*⁺ *pro*⁺), and d) DS410/pMC7 (*minA minB* str^r *lacY xyl mtl thi, pMC7 lac* i^q tc^r), containing plasmid pUH12 (o) or pUH22 (•) were grown at 37°C in LB medium (1% tryptone, 0.5% yeast extract, 0.5% NaCl), supplemented with (a-d) 100 µg/ml ampicillin (ap), and (d) 15 µg/ml tetracycline (tc) respectively. At the times indicated by the arrows the cells were induced with IPTG (isopropyl-ß-D-thiogalactoside) to give a final concentration of 5 mM. Plasmid pMC7 was constructed (7) by crossing the i^q-mutation from an F'factor to pMC1 (10). *R/c*, approximate number of *lac* repressor molecules per cell (6,7,11,12). *O.D.600*, optical density at 600 nm.

plasmid. Besides the amount of *lac* repressor, the concentration of inducer is the second parameter by which gene *E* transcription in pUH12 can be regulated. This correlation was studied by use of two pUH12-containing strains (Fig.2). In both cases the lysis rates increased and simultaneously the time intervals between induction and onset of lysis shortened in the range of 10^{-5} to 10^{-3} M IPTG. At IPTG concentrations above 10^{-2} M no substantial variation of these effects was detectable.

Expression of ØX gene *E* in pUH122, pUH123 and pUH51.

As shown above the successful transformation of bacteria with pUH12 always depends on the preceding or simultaneous introduction of repressor-overproducing *lac i*-alleles. This complication was managed by construction of three plasmids, conferring resistance to ampicillin, tetracycline or to both antibiotics and including both the ØX174 *E* gene and the *lac* i^q-

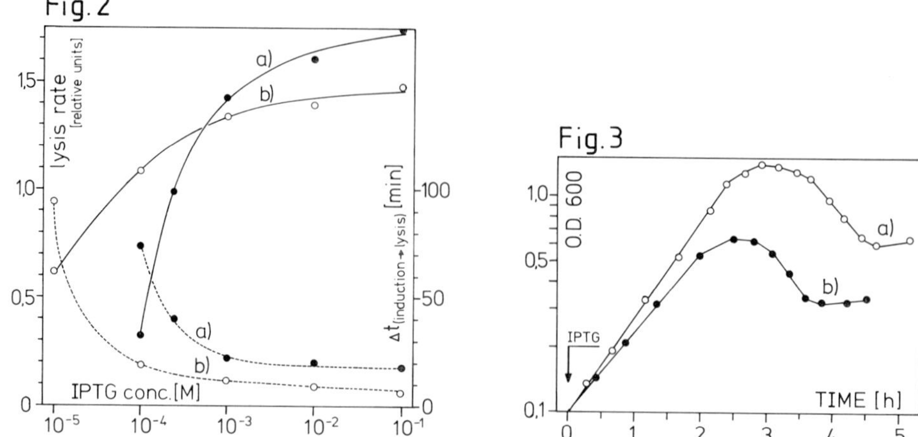

Fig.2: Effect of IPTG concentration on pUH12-mediated cell lysis. Strains a) DS410/pMC7,pUH12 and b) F'iq1 ΔM15/pUH12 were grown in LB medium at 37°C. At an O.D.600 of 0.15 IPTG was added to give final concentrations from 10^{-5} to 10^{-1} M. ├──┤ Lysis rates, slopes of the lysis curves. ├-----┤ Time intervals between induction and onset of lysis.

Fig.3: Induction of the lysis gene cloned in pUH51. Strains a) *E.coli C* PC1363(*pheA*), and b) *E.coli K12* $F^-z^-\Delta$M15recA ([*lac pro*]Δ*thi*, *Ø80 d lac z$^-$* *ΔM15, ara strA recA*), harbouring pUH51, were grown in LB medium with 15 µg/ml tetracycline at 37°C. At an O.D.600 of 0.1 IPTG was added (15 mM final concentration).

allele (Fig.4). This combination seemed to be optimal with respect to the expression of the lysis function (Fig.1). All strains which were transformed with the combined plasmids pUH122, pUH123 and pUH51 showed normal growth in the absence of IPTG and cell-lysis after addition of the inducer (Fig.3).However, with cells carrying these plasmids the production of *gpE*, as expressed by the time interval between induction and lysis-onset and by the lysis rate, is less efficient in comparison with transformants containing genes *E* and *lac i^q* on two separate multicopy plasmids (Fig.1d). Obviously the expression of the lysis gene is affected by the integration of both genes into a single plasmid.

Non-lysing pUH51-clones.

About 6 hours after lysis of induced PC1363/pUH51 cells the optical density of the cultures slowly started to increase. Further transfers of such cultures in liquid media and on agar plates, all containing 15 mM IPTG,

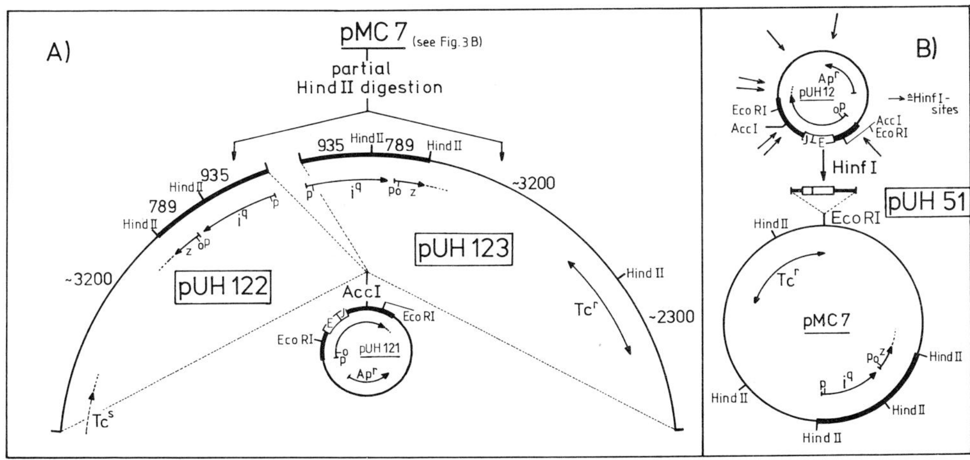

Fig.4: Construction of the plasmids pUH122, pUH123 and pUH51.
A) To construct pUH122 and pUH123 two intermediates had to be prepared: The unique *AccI* site in the polylinker region of pUR222 (13) was destroyed by *AccI* digestion, filling-in with the large fragment of DNA polymerase I and religation. The newly formed plasmid (pUH1) was used to construct pUH121 by inserting the smaller *EcoRI* fragment of pUH11 into the unique *EcoRI* site essentially as described for the construction of pUH12 (3). pUH121 was linearized with *AccI*, and blunt ends were generated by filling-in with DNA polymerase I (large fragment). Of this DNA 0.25 µg were ligated with 0.4 µg of partially *HindII* digested pMC7 DNA in a total volume of 40 µl for 20 h at 4°C. Strain $F^-z^-\Delta M15recA$ was transformed with 20 µl ligation mixture (3) and plated on LB plates with 200 µg/ml ampicillin. As shown by plasmid purification (14) from 40 colonies and analysis on 0.8% agarose gels (3), more than 90% of the clones contained plasmids substantially larger or smaller in size than pUH121. By restriction analysis the smaller plasmids were found to contain deletions, inactivating the lysis gene, whereas among the larger plasmids predominantly pUH122 (5.7 md, ap^r) and pUH123 (7.3 md; ap^r tc^r) were detected. Numbers in Fig.3A give the lengths of pMC7 *HindII* fragments in base pairs (bp). B) To construct pUH51, the plasmid pUH12 was cleaved with *HinfI* and the 893 bp-fragment was purified from a 2% agarose gel (15). Of this fragment 1.3 p moles were mixed with 0.15 pmoles *EcoRI* cut pMC7 DNA. After filling-in the single-stranded ends with DNA polymerase I (large fragment) the DNA fragments were ligated in a total volume of 55 µl for 2 h at 21°C. Strain $F^-z^-\Delta M15recA$ was transformed with 25 µl ligation mixture and plated on LB plates with 15 µg/ml tetracycline. Two out of 60 clones screened for the presence of inserted DNA were larger in size than pMC7 (5.5md). One of them was named pUH51 (6.1md). *E,J*, ØX174 genes. The orientations of the cloned fragments were determined by restriction analysis. Arrows indicate origin and direction of transcription. *p*, promotor; *o*, operator.

allowed the isolation of mutants which did not show any signs of lysis in the presence of *lac* inducer. The plasmid itself seems to be unaffected by this kind of mutation (called L^-) since pUH51, reisolated from such mutants and used to retransform strain PC1363, caused normal cell lysis after addition of IPTG. On the other hand, plasmid-less (cured) L^--mutants (16) did not regain the ability to be lysed by retransformation with pUH51. These observations indicate that the L^--mutation is not located on the plasmid but in the genome of the host. By measurement of ß-galactosidase (17) in L^--clones only 40% of the enzymatic activity found in L^+-clones could be detected. Therefore, taking into account that transcription of cloned ØX174 *E* gene is controlled by the *lac* promotor, the L^--mutation seems to consist rather in a restricted production of *gpE* than in a modification of its cellular targets. This interpretation is consistent with the finding that the *efp* of ØX174 on plasmid-containing and on plasmid-less L^--clones does not differ from the *efp* on L^+-clones (Table 1).

Complementation of a ØX174 gene*E*-mutation.

The lysis deficient ØX174 *E* gene mutant, ØX*am3* (8), was used to infect L^+-clones of PC1363/pUH51. Since lysis of the host bacteria by expression of the cloned lysis gene had to be prevented, IPTG could not be included in plates and media. Nevertheless the *efp* of ØX*am3* on L^+-clones mounted up to 3% of the *efp* on the suppressor strain HF4714 and exceeded the *efp* on

Table 1: Infection of PC1363 transformants with ØX174.

E.coliC strains / phages:	ØX174*wt*	ØX174*am3*
PC1363	1.0	1.0
HF4714	1.2	6.0×10^4
PC1363 /pMC7	0.8	1.5
PC1363 L^+/pUH51	1.1	1.8×10^3
PC1363 L^-/pUH51	0.8	6.0×10^1
PC1363 L^-/ (-)	1.2	1.1

Efficiencies of plating (*efp*) were determined in the presence of 5 mM $CaCl_2$. The table gives relative values, obtained by setting the *efp* on strain PC1363 to one. Strain HF4714 has the genotype *pro arg leu his thy thr* su^+. PC1363 L^-/(-), lysis deficient mutant of PC1363 after removal of pUH51 by treatment with fusaric acid (16).

strain PC1363 by a factor of 2000. Two conclusions can be drawn from this experiment: (I) Even in the absence of IPTG a low-level transcription of gene*E* occurs, which supports the observed complementation effect. (II) Low concentrations of *gpE*, insufficient to achieve lysis in pUH51 transformants, allow effective complementation of the *E am3*-mutation after infection of the bacteria with ØX*am3*. Therefore the development of phage particles within an infected cell in some way seems to amplify the lytic action of *gpE*. This conclusion does not conflict with the finding that *gpE* alone is sufficient to trigger cell lysis (6,7). However, interactions of *gpE* with other phage specific proteins must be expected to cause only quantitative effects in terms of enhancement or delay of *gpE* formation or action.

References

1. Markert, A., Zillig, W.: Virology 25, 88-97 (1965).
2. Hutchison III, C.A., Sinsheimer, R.L.: J. Mol. Biol. 18, 429-447 (1966).
3. Henrich, B., Lubitz, W., Plapp, R.: Mol. Gen. Genet. 185, 493-497 (1982).
4. Young, K.D., Young, R.: J. Virol. 44, 993-1002 (1982).
5. Lubitz, W., Plapp, R.: Curr. Microbiol. 4, 301-304 (1980).
6. Müller-Hill, B., Crapo, L., Gilbert, W.: Proc. Natl. Acad. Sci. USA 59, 1259-1264 (1968).
7. Calos, M.P.: Nature 274, 762-765 (1978).
8. Barrell, B.G., Air, G.M., Hutchison III, C.A.: Nature 264, 34-41 (1976).
9. Calos, M.P., Miller, J.H.: Mol. Gen. Genet. 183, 559-560 (1981).
10. Calos, M.P., Johnsrud, L.: Cell 13, 411-418 (1978).
11. Gilbert, W., Müller-Hill, B. in: Beckwith, J.R., Zipser, D. (eds.) The Lactose Operon, Cold Spring Harbour Laboratory, New York, 1970, pp. 93-109.
12. Müller-Hill, B.: Prog. Biophys. Mol. Biol. 30, 227-252 (1975).
13. Rüther, U., Koenen, M., Otto, K., Müller-Hill, B.: Nucl. Acids Res. 9, 4087-4098 (1981).
14. Birnboim, H.C., Doly, J.: Nucl. Acids Res. 7, 1513-1523 (1979).
15. Henrich, B., Lubitz, W., Fuchs, E.: J. Biochem. Biophys. Methods 6, 149-157 (1982).
16. Maloy, S.R., Nunn, W.D.: J. Bacteriol. 145, 1110-1112 (1981).
17. Miller, J.H.: Experiments in Molecular Genetics, Cold Spring Harbour Laboratory, Cold Spring Harbour, New York 1972.

LOCALISATION OF THE BACTERIOPHAGE ØX174 LYSIS GENE PRODUCT IN THE CELL ENVELOPE OF ESCHERICHIA COLI

U. Bläsi, R. Geisen, W. Lubitz*, B. Henrich and R. Plapp
FB Biologie, Abt. Mikrobiologie, Universität Kaiserslautern, Postfach 3049, D-6750 Kaiserslautern, FRG.

Introduction

ØX174 is a small single-stranded DNA bacteriophage. The DNA of ØX174 has been sequenced (1). All gene products of the ØX genome are either involved in viral DNA replication, capsid maturation or are structural proteins except for the gene E product (gpE) which affects lysis of the host. Infection of E.coli with ØXam3 carrying an amber mutation in gene E leads to the accumulation of large numbers of mature progeny phage particles within the cell, which have to be released by artificial lysis (2). Gene E of ØX174 lies within gene D in a different reading frame and the am3 mutation does not alter the amino acid sequence coded by gene D (3). There are ØX mutants which affect lysis of the infected bacteria carrying mutations in genes other than E. However, by cloning the appropriate ØX-DNA fragment it was shown that gpE is the ultimative requirement (4, 5) and no other phage gene products are needed for cell lysis. The mechanism by which gpE triggers cell lysis is not known. gpE does not seem to have mureinolytic activity, however, as recently proposed (6, 7), it may interact with autolysins of the cell resulting in either activation or uncontrolled action of these enzymes. This would imply that gpE should act at the cytoplasmic membrane. In this communication evidence is presented that gpE, either coded for by ØX in phage infected cells or coded for by a plasmid is incorporated into the cytoplasmic membrane.

*To whom correspondence should be addressed

Results and Discussion

From the nucleotide sequence of ØX174 it has been proposed that gpE is a hydrophobic protein which may have affinity to bacterial membranes (3). In order to detect gpE in ØX infected cells proteins of the cytoplasmic membrane were extracted with Sarkosyl NL 97 from a crude envelope fraction and the remaining outer membrane was solubilized with SDS (8). In Fig.1 the cytoplasmic membrane proteins of ØX174 wt, ØXam3 and uninfected cells are compared (lanes 2-4) by SDS gel electrophoresis. In the 15-25% gradient gel system which gives a good resolution of low molecular weight proteins (9) an additional protein band is visible in the cytoplasmic membrane fraction of the ØX wt infected sample at a molecular weight position of 10 Kd (lane 4 arrow) which is absent in the uninfected control (lane 2) and in the sample from ØXam3 infected cells (lane 3). The position of this additional protein (gpE) in the gel is consistent with its molecular weight of 10,589 as calculated from the nucleotide sequence of the gene.

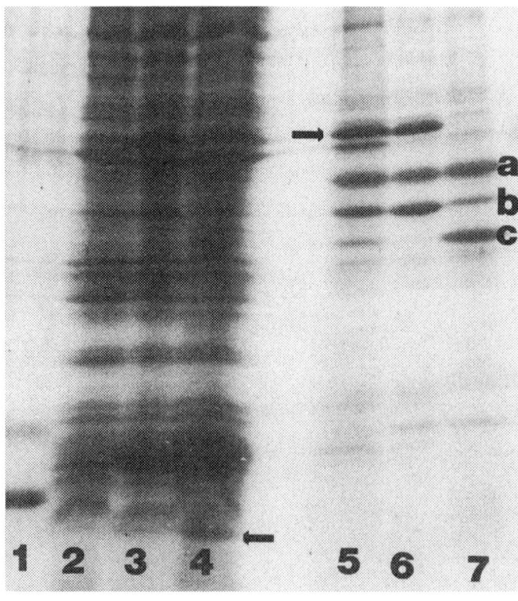

Fig.1: SDS gel electrophoresis of cytoplasmic and outer membrane proteins from E. coli Lane1, molecular weight standarts, 40kd, 18kd, 12kd. Cytoplasmic membrane proteins: lane2, uninfected cells; lane3, ØXam3 infected cells; lane4, ØX wt infected cells (arrow gpE). Outer membrane proteins: lane 5, ØX wt infected cells; lane6, ØX am3 infected cells; lane7, uninfected cells, a: ompF, b: ompC, c: ompA.

The additional protein band visible in the cytoplasmic membrane of ØX wt infected cells is not detectable in the ØXam3 infected sample. The am3 mutation of gene E lies within the 7th amino acid codon leading to a translational stop of gpE at this position (3).

The alterations occurring in the outer membrane of ØX wt and ØXam3 infected cells are shown in Fig.1, lanes 5-7. The most obvious differences between the phage infected samples (lanes 5 and 6) and the control (lane 7) is the appearance of a new major outer membrane protein in the ØX wt and ØXam3 infected sample (arrow) at a position where in the uninfected sample only a faint band is visible. Additionally, the amount of OmpA is greatly reduced and OmpF seems to be augmented in the phage infected samples.

For confirmation of the finding that gpE of ØX174 is located in the cytoplasmic membrane, plasmids carrying the ØX wt gene E (pUH12) or the corresponding ØXam3 fragment (pUH22) (4) were expressed in a mini cell system. The autoradiogram of the plasmids encoded proteins labelled with 3H-leucine is shown in Fig.2.

gpE occurs in the cytoplasmic membrane fraction of mini cells harbouring pUH12 (lane 9 arrow) and is absent in the corresponding fraction of cells harbouring pUH22 (lane 8). However, in contrast to the phage infected samples, small amounts of gpE can also be detected in the outer membrane fraction of mini cells harbouring pUH12 (lane 10). This may be due to its overproduction in cells carrying the multicopy plasmid pUH12. The other lanes, not explained in this context, represent controls, lane 2 shows the background of the mini cell protein synthesis, lane 3 gives the plasmid pUR222 encoded proteins from which plasmids pUH12 and pUH22 are derived.

The separation of the cytoplasmic membrane proteins of mini cells harbouring either pUH12 or pUH22 by two-dimensional gel electrophoresis (10) is given in Fig.3. For identification of gpE in Fig.3a (arrow) its molecular weight and its basic isoelectric point served as indicators. No protein corresponding

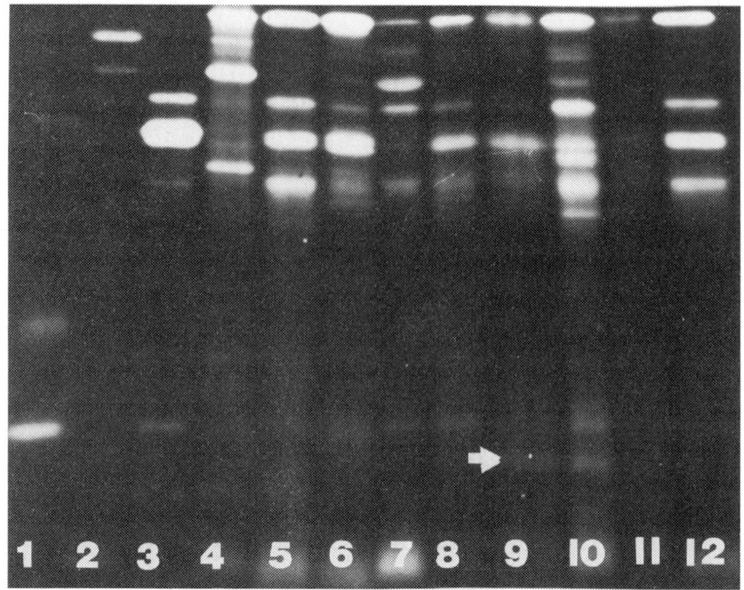

Fig.2: Autoradiogram of 3H-leucine labelled mini cell proteins separated by SDS gel electrophoresis.
DS410 recA is the mini cell strain. Plasmids pUH12 and pUH22 carry the ØXDNA under control of the lac p o region, pMC7 carries the lac repressor gene.
Lane1, molecular weight standarts 30kd, 18kd, 12kd. Lane 2, DS410 background synthesis; lane3, DS410 pUR222; lane4, DS410 pMC7.
DS410 pUH22 pMC7: lane5, total lysate; lane6, cytoplasma fraction; lane7, outer membrane proteins; lane8, cytoplasmic membrane proteins.
DS410 pUH12 pMC7: lane9, cytoplasmic membrane proteins; lane10 outer membrane proteins; lane11, cytoplasma fraction; lane 12, total lysate

to gpE is visible in the cytoplasmic membrane fraction from mini cells harbouring pUH22 (Fig.3b).

From the results presented, it is concluded that ØX gpE is a protein which is found in the cytoplasmic membrane of either ØX174 infected cells or of cells harbouring the cloned ØX geneE on a plasmid.

The location of ØX gpE in the cytoplasmic membrane makes an interaction between the host autolytic system and gpE possible. This mechanism of lysis has been proposed from investigations characterizing the peptidoglycan degrading enzymes in ØX infec-

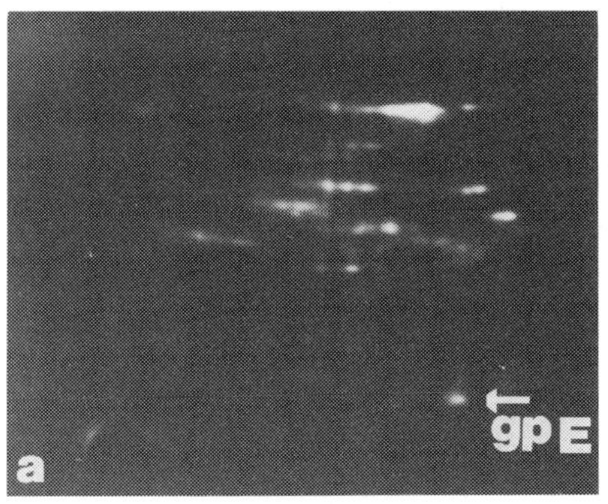

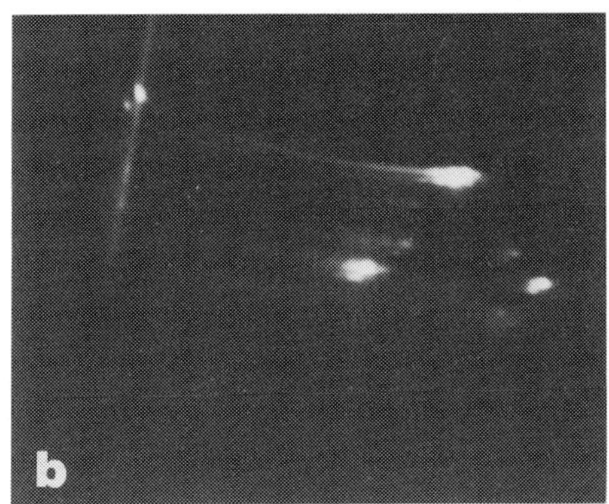

Fig.3: Identification of gpE on two-dimensional gels.

a) Autoradiogram of 3H-leucine labelled cytoplasmic membrane proteins of DS410 pUH12 pMC7, arrow gpE
b) Autoradiogram of 3H-leucine labelled cytoplasmic membrane proteins of DS410 pUH22 pMC7.

Tab.1: Colony forming units ($x10^8$)

OD600nm	E.coli K12	E.coli K12 (pUH12)
0.3	1.5	1.2
0.5	2.4	1.8
0.7	3.8	2.7
0.9	6.1	4.0
1.1	10.0	6.1

Fig.4: Mg^{2+}stabilized E.coliK12(pUH12)

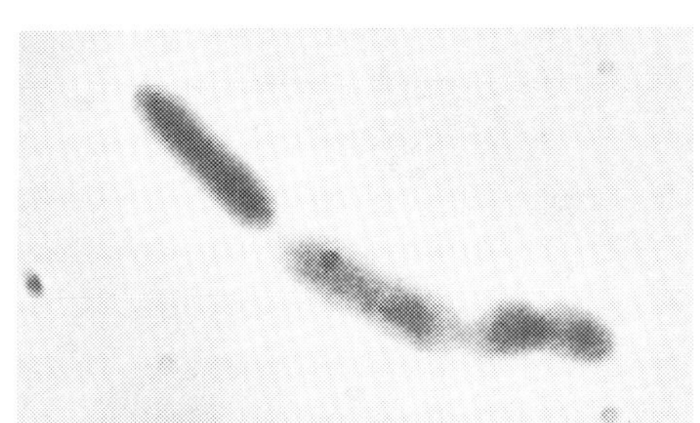

ted cells (6). However, it is not known whether gpE interacts directly with the autolytic enzymes or with regulatory components of the autolytic system.
Autolysis in E. coli can be induced by a variety of procedures, and one of the common features of these treatments is the fact that lytic induction can be antagonized by $MgSO_4$ (11). Addition of $MgSO_4$ (0.2 M final concentration) to E. coli harbouring plasmid pUH12 prevents lysis after induction of the ØX gene E function. Under these conditions cells not only do not lyse but are also able to proliferate (Tab.1). Phase contrast

microscopy reveals, however, that such cells are somewhat elongated and tend not to separate properly (Fig.4). This observation is consistent with the reduced colony forming ability of a culture compared to the uneffected control.
The prevention of gpE induced lysis of E.coli by $MgSO_4$ and the membrane location of gpE support the view that ØX174 caused lysis of E. coli is triggered by interaction of gpE with the autolytic enzyme system.

This work was supported by a grant from the Deutsche Forschungsgemeinschaft (Lu213/4-1).

References

1. Sanger, F., Coulson, A.R., Freidmann, T., Air, G.M., Barrel, B.G., Brown, N.L., Fiddes, J.C., Hutchison, C.A., Slocombe, P.M., Smith, M.: J. Mol. Biol. 125, 225-246 (1978).
2. Hutchison III, C.A., Sinsheimer, R.L.: J. Mol. Biol. 18, 429-447 (1966).
3. Barrell, B.G., Air, G.M., Hutchison III, C.A.: Nature 264, 34-41 (1976).
4. Henrich, B., Lubitz, W., Plapp, R.: Mol. Gen. Genet. 185, 493-497 (1982).
5. Young, K.D., Young, R.: J. Virol. 44, 993-1002 (1982).
6. Lubitz, W., Plapp, R.: Current Microbiol. 4, 301-304 (1980).
7. Lubitz, W., Schmid, R., Plapp, R.: Current Microbiol. 5, 45-50 (1981).
8. Boyd, A., Holland, B.: Cell 18, 287-296 (1979).
9. Lugtenberg, B., Meijers, J., Peters, R., Van Hoek, P., Van Alphen, L.: Febs L. 58, 254-258 (1975).
10. O'Farell, P.Z., Goodman, H.M., O'Farell, P.H.: Cell 12, 1133-1142 (1977).
11. Leduc, M., Van Heijenoort, J.: J. Bacteriol. 142, 52-59 (1980).

INDUCTION OF CELL LYSIS BY MECILLINAM PLUS NOCARDICIN A: ISOLATION AND PROPERTIES OF A TOLERANT MUTANT.

José Berenguer, Miguel A. de Pedro and David Vázquez.
Instituto de Bioquimica de Macromoléculas. C.S.I.C.-U.A.M. Madrid.

Introduction

Mecillinam and nocardicin A are two unusual β-lactam antibiotics (1,2) with a strong cooperative effect against the Gram negative bacterium Escherichia coli. Separately, both mecillinam and nocardicin A, are poorly bacteriolytic against E.coli. However, the simultaneous addition of both drugs to a growing culture of E.coli, at concentrations similar to their respective minimal inhibitory concentrations (MICs, 20 µg/ml for nocardicin A and 0.15 µg/ml for mecillinam against E.coli W7, the strain used in most of our work) induced a drastic lytic response in the cells (3). According to previously reported results (4,5), the induction of cell lysis apparently occurs by a sequential mechanism in which mecillinam acts first, leading to a state of cellular hypersensitivity to nocardicin A which, in turn, triggers cell lysis by a process insensitive to the inhibition of protein synthesis. The sensitizing action of mecillinam depends on its interaction with a native penicillin binding protein 2 (PBP2), and can be prevented by the inhibition of protein synthesis. The mechanism by which nocardicin A triggers cell lysis, in the presence of mecillinam, is still completely unknown. In intact cells, nocardicin A binds to PBP 1a at a concentration similar to its MIC. However, the fact that the PBP 1a deficient mutant SP61 behaves like its parental strain KN126 (6), indicates that the lysis-triggering action of nocardicin A might be unrelated to its interaction with PBP 1a.

The Target of Penicillin

With the intention of identifying the elements involved in the induction of the lytic response of the cells, we recently began working on the isolation and characterization of mutants showing tolerance towards the combination of mecillinam and nocardicin A.

Results

Tolerant mutants were obtained from E.coli W7 (dapA, lysA)(7) by N-methyl-N'-nitro-N-nitrosoguanidine mutagenesis (8) and direct selection of the tolerant phenotype. After mutagenesis, the cells were allowed to recover for 6 hours in Luria broth (LB). Immediately afterwards, mecillinam (10 µg/ml) and nocardicin A (50 µg/ml) were added to the culture. After 2 hours of incubation, the survivors were rescued by centrifugation and resuspended into antibiotic-free medium. After a second, identical, round of selection, single colonies were obtained on LB-agar plates and checked for sensitivity to mecillinam, nocardicin A and for the genetic markers of the parental strain dapA and lysA. 3 out of 400 colonies tested exhibited the selected phenotype and one of them, W7-MNT1, was used for further characterization. Fig 1A shows the comparative behaviour of W7 and W7-MNT1 when mecillinam (10 µg/ml) and nocardicin A (100 µg/ml) were added to exponentially growing cultures. Whereas lysis was readily induced in W7, the tolerant strain W7-MNT1 stopped growing and the optical density of the culture remained constant for at least two generation times. Furthermore, the viability of W7-MNT1 was only slightly affected by this treatment (90% after two hours in the presence of the antibiotics). An unexpected characteristic of W7-MNT1 was the unselected adquisition of tolerance against diaminopimelic acid starvation, as is shown in fig. 1B.
The determination of the MICs of a number of β-lactam antibiotics against W7 and its derivative W7-MNT1, indicated that

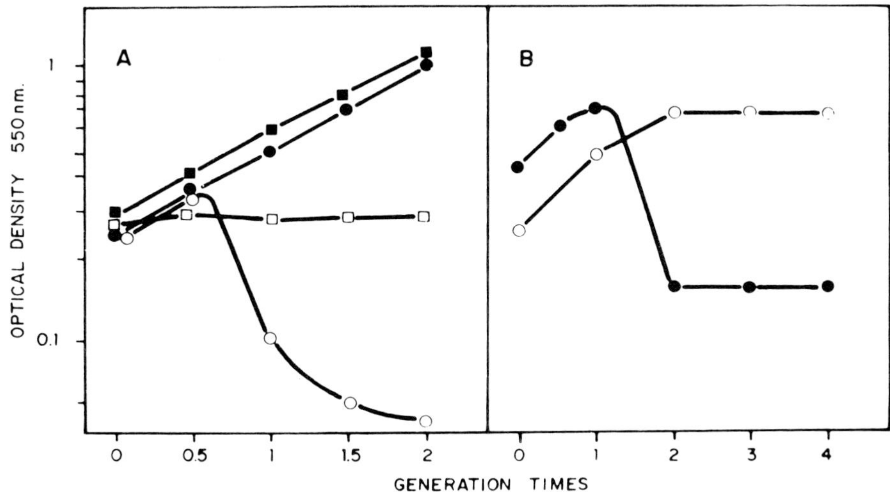

Fig. 1. Tolerance of E.coli W7-MNT1 to mecillinam plus nocardicin A treatment and to diaminopimelic acid starvation.A) Exponentially growing cultures of E.coli W7(●) or W7-MNT1(■) were divided into two subcultures at time 0 min. One subculture of each strain was used as untreated control (●■). Mecillinam (10 µg/ml) and nocardicin A(100 µg/ml) were added to the second subculture at time 0 min(○□). B) Exponentially growing cultures of E.coli W7(●) and of W7-MNT1(○) were starved for diaminopimelic acid at time 0 min by filtration through Millipore HAWP filters and immediately resuspended into diaminopimelic acid-free medium prewarmed at 37 ºC. Growth was monitored by the optical density at 550 nm. Growth medium was LB suplemented with 4 µg/ml of diaminopimelic acid.

both strains were equally sensitive to the β-lactams tested (table 1). Furthermore, in the tolerant strain W7-MNTl, lysis was efficiently induced by other bacteriolytic β-lactams(fig2). The determination of total murein-hydrolase activity in crude preparations from W7 and W7-MNTl, indicated that both strains had a similar murein-hydrolytic capacity. Purified sacculi obtained from both strains were equally degraded by crude murein hydrolase extracts from W7 or W7-MNTl (table 2). In a preliminary analysis of the cell envelope, no significative differences were observed between W7 and W7-MNTl. The two strains had

TABLE 1. Minimal Inhibitory Concentrations of β-lactam antibiotics against E.coli W7 and W7-MNT1.

Antibiotic	Minimal Inhibitory Concentration (μg/ml)	
	W7	W7-MNT1
Ampicillin	1	1
Amoxicillin	3	3
Cefazolin	1	0,5
Cefsulodine	10	5
Ceftizoxime	0,1	0,1
Cephaloridin	1	1
Clavulanic acid	10	10
Mecillinam	0,1	0,5
Nocardicin A	20	25

The minimal inhibitory concentrations were determined by the serial agar dilution technique.

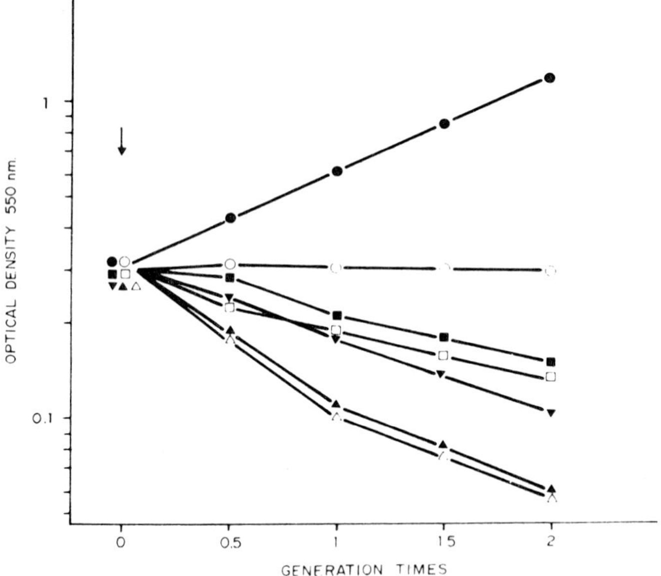

Fig 2. Induction of cell lysis in E.coli W7-MNT1 by β-lactams. A culture of W7-MNT1 growing exponentially in LB medium, was divided into 7 subcultures. At time 0 a different β-lactam was added to each subculture. The optical density at 550 nm of the subcultures was determinated at the indicated times. (●) Control; (○) Plus mecillinam (10 μg/ml) and nocardicin A (100 μg/ml); (■) Cefsulodine (100 μg/ml); (□) Ceftizoxime (30 μg/ml); (▲) Cefazolin (50 μg/ml); (△) Amoxicillin (50 μg/ml); (▼) Cephaloridin (50 μg/ml).

TABLE 2. Digestion of sacculi from E.coli W7 and W7-MNT1 by soluble murein-hydrolases from the same strains.

Enzyme Source	Sacculi source W7	W7-MNT1
W7	26	27
W7-MNT1	21	19

5 ug of ^{3}H-diaminopimelic acid labelled sacculi(specific activity 4x10^{4}cpm per ug of murein) were digested with 80 ug of protein from an enzyme extract obtained as described(9) in a total volume of 200 µl.The reaction was carried out in Tris-ClH 20 mM pH 6.2,20 mM SO_4Mg,0.4% TritonX100 buffer at 37ºC for 60 min.The reaction was stopped by the addition of cetyl-trimethyl ammoniumbromide to a final concentration of 0.5%.After 30 min in an ice-water bath,the samples were centrifuged(10.000xg, 10 min) and the solubilized radioactivity determined by liquid scintillation counting.The values in the table indicate the percentage of the total murein solubilized by the enzyme.

similar electrophoretic patterns for the outer membrane protein and for the PBPs.

Discussion

A mutant,W7-MNT1,tolerant to the induction of cell lysis by the combined action of mecillinam and nocardicin A has been obtained.W7-MNT1 showed normal sensitivity against both drugs,as well as against many other β-lactam tested.Furthermore,W7-MNT1 exhibited a normal bacteriolytic response when treated with other β-lactams capable of inducing cell lysis.This results indicate that the tolerance phenotype of W7-MNT1 is highly specific with respect to the stimulus triggering the bacteriolytic response of the cell.
The adquisition of unselected tolerance to diaminopimelic acid starvation by W7-MNT1 suggests that the mechanisms triggering cell lysis after diaminopimelic acid starvation or mecillinam plus nocardicin A treatment,share some important element,which

does not affect essentially the induction of cell lysis by other agents.

The fact that no significative differences were found between W7 an W7-MNT1 in the murein or murein metabolizing enzymes agrees well with the postulated specificity of tolerance in W7-MNT1. In fact, the alteration of one of these critical elements would very probably affect the lytic response of the cell unespecifically.However, our results from the biochemical analysis of W7-MNT1 are still far from definitive and must be taken with caution. As a whole, the data reported here suggest the existence of more than one specific pathway involved in the induction of cell lysis.

References

1. Lund,F., Tybring,L.:Nature(London)New Biol.236,135-137 (1972)
2. Aoki,H.,Sakai,H.,Koshaka,M.,Konomi,T.,Hosoda,J.,Kuboshy,Y. Igushi,E.,Imanaka,H.:J.Antibiot.29,492-500,(1976)
3. Berenguer,J.,de Pedro,M.A.,Vázquez,D.:Antimicrob.Agents Chemother.21, 195-200, (1982)
4. Berenguer,J.,de Pedro,M.A.,Vázquez,D.:Antimicrob.Agents Chemother. 22, 1070-1072, (1982)
5. Berenguer,J.,de Pedro,M.A.,Vázquez,D.:Eur.J.Biochem.126 155-159, (1982)
6. Spratt,B.G.,Jobamputra,V.,Schwarz,U.:FEBS Lett.79,374-378 (1977)
7. Hartmann,R.,Höltje,J.V.,Schwarz,U.:Nature(London) 235, 426-429, (1972)
8. Adelberg,E.A.,Mandel,M.,Chen,G.C.C.:Biochem.Biophys.Res. Comm.18, 788, (1965)
9. Hakenbeck,R.,Goodell,E.W.,Schwarz,U.:FEBS Lett.40, 261-264, (1974)

INCREASED WALL AUTOLYSIS AND DECREASED PEPTIDOGLYCAN CROSS-LINKING IN METHICILLIN-RESISTANT *STAPHYLOCOCCUS AUREUS* GROWN IN THE PRESENCE OF METHICILLIN

Brian J. Wilkinson, M. Walid Qoronfleh
Microbiology Group, Department of Biological Sciences and Department of Chemistry, Illinois State University, Normal, IL 61761, U.S.A.

Introduction

Wilkinson et al. (13) reported that crude cell walls retaining autolytic activity (CC^W) isolated from methicillin-resistant (MR) *Staphylococcus aureus* strains grown in the presence of methicillin (50 μg/ml) autolysed at about twice the rate of CCW isolated from the same strains grown in the absence of methicillin. Wyke et al. (16) reported that growth of a different MR *S. aureus* strain in the presence of low concentrations of methicillin led to hypo-cross-linked peptidoglycan (PG). This led us to propose the hypothesis that the increased rate of CCW autolysis observed in MR strains grown in the presence of methicillin was due to the presence in the wall of poorly cross-linked PG. Results were obtained that support this hypothesis.

Materials and Methods

1) Strains and growth conditions. MR *S. aureus* strain DU4916, (methicillin minimum inhibitory concentration [MIC] 1600 μg/ml), derived methicillin-susceptible (MS) strain DU4916 (MIC 3.12 μg/ml) and MS strain H (MIC 3.12 μg/ml) were used (13). They were either cultured in PYK broth (13)(DU4916 & DU4916S) or in double strength nutrient broth (4) (H).

The Target of Penicillin

2) Preparation of CCW, purified cell walls (PCW) and PG. This was carried out as described previously (13,15).

3) Autolysin extraction. Exponential phase cells were extracted with 3 M LiCl for 10 min at 4°C as described by Best et al. (1).

4) Assay of autolysin activity. CCW,PCW and PG were suspended in 0.01 M KPO_4 pH 7.0 or autolysin extract to an A_{580nm} of 0.5-0.7 and were incubated at 30°C and the turbidity of the suspension was measured at intervals. Whole cells were washed once in cold H_2O and were resuspended in 0.01 M KPO_4 pH 7.0 to an A_{580nm} of 0.5-0.6, and were incubated with shaking at 30°C, and the turbidity was measured at intervals.

5) Assessment of PG cross-linking. PG was radiolabelled by growing strain DU4916 overnight in PYK medium containing D-[1,6-^{3}H(N)]-glucosamine HCl (32.5 Ci/mmol; 1 μCi/ml) in the presence or absence of methicillin (50 μg/ml). PG was isolated by a modified Park & Hancock method (15). Chromatography of a 4 M HCl, 4 hr, 105°C hydrolysate of [^{3}H]-PG showed that radioactivity was present only in spots corresponding to glucosamine and muramic acid (2). [^{3}H]-PG was digested with Chalaropsis B muramidase (20 units/mg [^{3}H]-PG) in 0.05 M Na acetate buffer pH 4.7 for 4 hr at 37°C. More than 95% of the radioactivity was solubilized by this treatment as revealed by liquid scintillation counting of the supernatant after centrifugation of the digestion mixture (36,000 g for 20 min). The digest was fractionated by High Performance Liquid Chromatography (HPLC) on a TSK-SW 2000 column using 0.01 M KPO_4 pH 6.6 as the eluant (R. S. Rosenthal, unpublished procedure). Column fractions were collected and radioactivity was determined.

Results

1) Autolysis of CCW and lysis of PCW and PG by extracted autolysin. CCW isolated from MR strain DU4916 grown in the presence of methicillin (50 μg/ml, 0.03 MIC) autolysed at about

twice the rate of CCW from cells grown in the absence of the antibiotic (data not shown) thus confirming our earlier observations (13). PCW and PG isolated from strain DU4916 grown in the presence of 50 μg methicillin/ml were significantly more susceptible to extracted autolysin than those isolated from organisms grown in the absence of the drug (Fig. 1). This is good evidence that the increased rate of autolysis of CCW is due to alterations in the substrate (PG) rather than alterations in the enzyme. Concentrations of 1, 10, 50 and 100 μg methicillin/ml caused about the same degree of increased autolysin susceptibility (data not shown). This is similar to the effect of methicillin on PG cross-linking noted by Wyke et al. (16). PG isolated from strain DU4916 grown in the presence of 1000 μg methicillin/ml (0.625 MIC) was extremely susceptible to autolysin.

Wyke et al. (17) reported that growth in the presence of low concentrations of cefoxitin led to a marked reduction in PG cross-linking in MS strain H. Accordingly, the susceptibilities of PCW and PG isolated from strain H grown in the presence of 0.25 μg cefoxitin/ml to extracted autolysin were compared with those isolated from organisms grown in the absence of the drug. PCW and PG isolated from cells grown in the presence of cefoxitin were markedly more susceptible to autolysin than those isolated from cells grown in its absence, providing further support for the idea that increased autolysin susceptibility is due to decreased PG cross-linking.

In similar experiments with strain DU4916S, growth in the presence of methicillin (0.0016-0.16 MIC) lead to PG that showed a decreased autolysin susceptibility.

2) Assessment of PG cross-linking. Peaks in the HPLC profiles of the Chalaropsis B muramidase [^{3}H]-PG digests were tentatively identified as monomers, dimers, trimers and oligomers. A greater proportion of monomers (24.8%) and dimers (21.0%) and a lower proportion of higher oligomers (40.6%) were present in PG isolated from organism grown in the presence of methicillin, indicating decreased cross-linking than in its

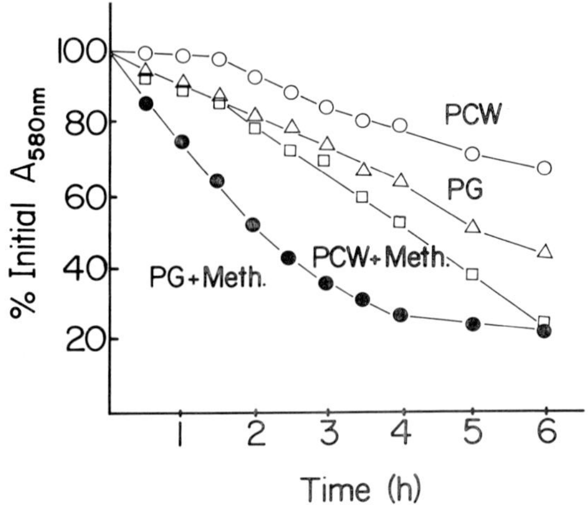

Fig. 1. Lysis of DU4916 PCW and PG by LiCl autolysin extract

absence (monomers, 8.0%; dimers, 12.0%; higher oligomers, 67.3%). These values are similar to those obtained by Wyke et al. (16).

3) Autolysis of whole cells grown in the presence of methicillin. It was of interest to see whether this increased potential for autolysis due to hypo-cross-linked PG was expressed in whole cells grown in the presence of methicillin. Cells grown in the presence or absence of methicillin (50 µg/ml) showed very similar rates of autolysis. Thus, although PG is hypo-cross-linked and the potential for increased autolysis exists, this activity is controlled in some way in whole cells.

Discussion

The results presented here provide good evidence that the increased susceptibility of cell walls isolated from an MR *S. aureus* strain grown in the presence of methicillin to autolysin is due to poorly cross-linked PG. Increased autolysin susceptibility of PCW and PG and decreased PG cross-linking went hand-in-hand. Also, growth of an MS strain in the presence of cefoxitin, which specifically inhibits PBP 4 and leads to poorly cross-linked PG, yielded PCW and PG that showed markedly increased autolysin susceptibilities. We do not know the bonds cleaved by the autolysin(s). Tipper (10) has detected amidase, endo-β-N-acetyl-glucosaminidase and glycine endopeptidase activities in S. aureus.

How do the results presented here fit into the overall picture of the methicillin-resistance phenomenon? Changes have been reported in the PBP affinities for β-lactam antibiotics in MR versus MS strains (3,6,7,16). Changes in PBP 3 (and perhaps PBP 2) in particular seem to be important in methicillin resistance. These PBPs probably represent the primary transpeptidase enzyme responsible for incorporation of PG into the septal region in actively dividing S. aureus cells (9,14,16). It would appear that the presence of hypo-cross-linked PG in MR cells grown in the presence of methicillin is due to inhibition of PBP 4 (4,16). It would be interesting to know whether the enlarged septa seen in MR cells grown in the presence of methicillin represent poorly cross-linked PG. This might suggest that the secondary cross-linking carried out by PBP 4 may function to regulate the proper insertion of PG into the growing cell wall. Although cells grown in the presence of methicillin had hypo-cross-linked PG and thus increased potential for lysis, this was not expressed in whole cells. This indicates that the autolysins are closely regulated in the intact cell.

Growth of MS strain DU4916S in the presence of methicillin did not yield PG with an increased autolysin susceptibility, implying that the PG did not have decreased cross-linking (unpublished observations). In penicillin-susceptible Neisseria gonorrhoeae cross-linking was little affected by penicillin but the degree of O-acetylation of PG was sharply decreased (2,5). S. aureus PG is partially O-acetylated (11) but so far no reports of the effects of β-lactam antibiotics on O-acetylation in this species have appeared.

Most of the studies on the mechanism of methicillin resistance have been carried out with strains that are perhaps atypical in that they do not show the heterogeneity of resistance expressed that is a characteristic feature of many MR strains (8). The biochemical bases of the effects of temperature, pH and salt on methicillin resistance remain to be discovered.

References

1. Best, G. K., Best, N. H., Koval, A. V.: Antimicrob. Agents Chemother. 6, 825-830 (1974).
2. Blundell, J. K., Perkins, H. R.: J. Bacteriol. 147, 633-641 (1981).
3. Brown, D. F. J., Reynolds, P. E.: FEBS Lett. 122, 275-278 (1980).
4. Curtis, N. A. C., Hayes, M. V.: FEMS Microbiol. Lett. 10, 227-229 (1981).
5. Dougherty, T. J.: J. Bacteriol. 153, 429-435 (1983).
6. Hartman, B., Tomasz, A.: Antimicrob. Agents Chemother. 19, 726-735 (1981).
7. Hayes, M. V., Curtis, N. A. C., Wyke, A. W., Ward, J. B.: FEMS Microbiol. Lett. 10, 119-122 (1981).
8. Sabath, L. D.: J. Antimicrob. Chemother. 3 (Suppl. C), 47-51 (1977).
9. Smith, P. F., Wilkinson, B. J.: J. Bacteriol. 148, 610-617 (1981).
10. Tipper, D. J.: J. Bacteriol. 97, 837-847 (1969).
11. Tipper, D. J., Ghuysen, J. M., Strominger, J. L.: Biochemistry 4, 468-473 (1965).
12. Tomasz, A.: Ann. Rev. Microbiol. 33, 113-137 (1979).
13. Wilkinson, B. J., Dorian, K. J., Sabath, L. D.: J. Bacteriol. 136, 976-982 (1978).
14. Wilkinson, B. J., Nadakavukaren, M. J.: Antimicrob. Agents Chemother. In press.
15. Wilkinson, B. J., White, P. J.: J. Gen. Microbiol. 79, 195-204 (1973).
16. Wyke, A. W., Ward, J. B., Hayes, M. V.: Eur. J. Biochem. 127, 553-558 (1982).
17. Wyke, A. W., Ward, J. B., Hayes, M. V., Curtis, N.A.C.: Eur. J. Biochem. 119, 389-393 (1981).

MORPHOLOGICAL STUDY OF THERMOSENSITIVE LYSIS MUTANTS OF BACILLUS SUBTILIS

Cyrille Brandt and Dimitri Karamata

Institut de génétique et de biologie microbiennes
Rue César - Roux 19
CH - 1005 Lausanne

Introduction

Isolation and analysis of bacterial mutants affected in cell wall metabolism provides an efficient tool for the study of cell wall growth. Thermosensitive mutants of B.subtilis which lyse at the non-permissive temperature due to interference with wall synthesis have been described (1,2). We report here an analysis of the pattern of cell lysis and other morphological alterations in a collection of thirty three thermosensitive lysis mutants (ts lss) identified among a large collection of ts mutants of B.subtilis 168 obtained by indirect selection (3). Genetic analysis shows that they are distributed in 8 linkage groups designated A to H. Groups A, B, C, D and F are deficient in different steps of UDP-N-acetyl-muramyl-pentapeptide (UDP-NAM-pentapeptide) synthesis, group G is affected in a late stage of peptidoglycan synthesis whereas groups E and H lyse without being specifically affected in peptidoglycan biosynthesis (C. Brandt and D. Karamata, in preparation).

We examine by phase contrast the behaviour of ts lss mutants and in the case of mutants deficient in peptidoglycan precursor synthesis, we assess the effect on the cell lysis of reduced autolytic activity. Implications for cell wall growth of the morphological changes observed are discussed.

Results

Morphological study of ts lss mutants

After 4 to 5 generations of exponential growth at 30C cultures were shifted to 45C at a rather low cell density (Fig.2). Examination of the cell morphology (Fig.1) allowed the ts lss mutants to be distributed into three classes as follows :
Linkage groups A,B,C,D and F. All mutants deficient in the synthesis of the final cytoplasmic peptidoglycan precursor, present two successive phases upon shift to 45C (Fig.2). The first period ends at about 30 min. when the initially rapid fall in turbidity slows down in most strains. In all strains lysed cells begin to appear about 15 min. after the shift and at the end of the first period represent nearly 20% of the population. Under phase contrast these cells appear completely empty, are distributed nearly at random along chains of cells and their envelopes show no distorsions (Fig.1 d, e).
Following the end of the first period lysis continues at a rate and to an extent which are strain dependent and most likely inversely related to the amount of residual wall synthesis at 45C; many lysing cells exhibit a characteristic granulous appearence (Fig.1 f, h). However, in strains lysing most rapidly (e.g. A29), the majority of cells show a lysis which is clearly progressive. Already 20 min. after shift the unlysed cells begin to show characteristic distortions, the extent of which is directly correlated with the amount of the residual wall synthesis. An increasing proportion of cells become gradually and progressively more swollen, appearing very frequently like balloons. The swellings, symmetrical with respect to the long axis of the bacteria, are localized at various positions along this axis (Fig.1 f,g,h). Rather frequently pairs of neighbouring cells show swellings which arise simultaneously and are symmetrical with respect to their division plane (Fig. 1 f,g). In some instances the latter corresponds to a constriction in the swollen area. It is striking that lysed cells are found at different stages of balloon

Fig.1. Phase contrast micrographs

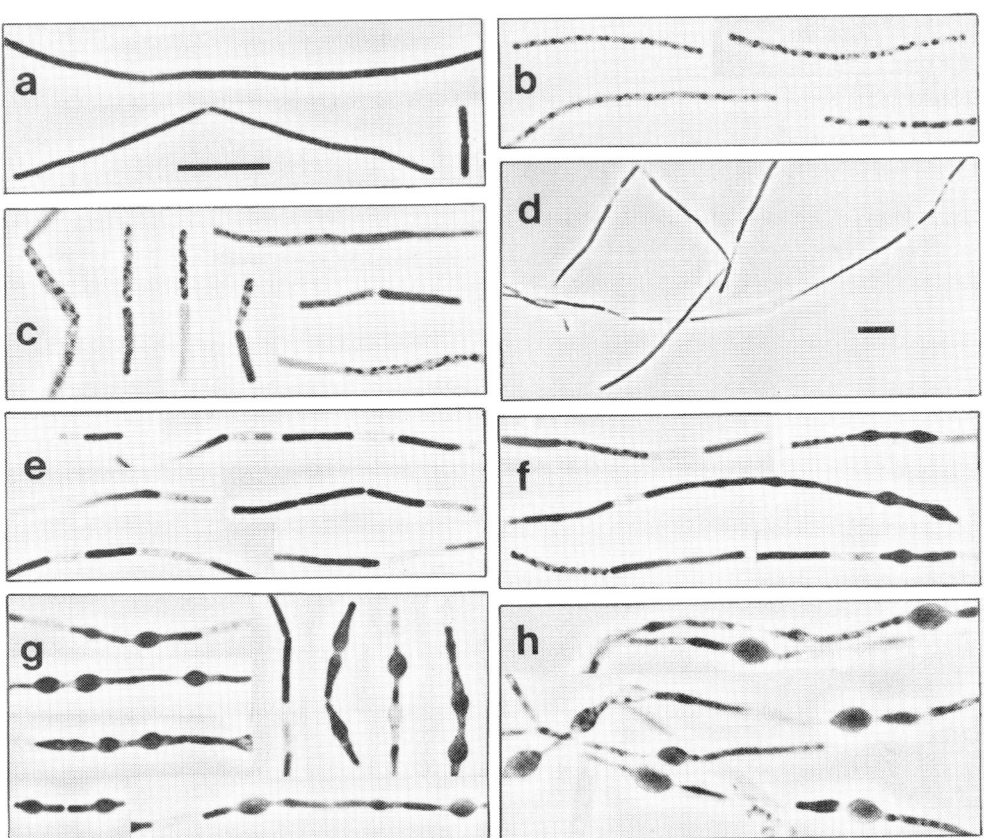

Observations were carried out on mutations back-crossed by congression into strain M25, *ilv-1*, *leu-8*, *lys-31*. Culture conditions are as in Fig.2. a) *ts* D28 and *ts* B23 at 30C. b) *ts* H11, 140 min. at 45C. c) *ts* G20, 80 min. at 45C. d) *ts* D28, 23 min. at 45C. e) *ts* A29, *ts* C32, *ts* D28 and *ts* F16, 18 to 25 min. at 45C. f) *ts* C32, *ts* D28 and *ts* F16, 35 to 40 min. at 45C. g) *ts* C32 and *ts* F16, *lyt*$^+$ and *lyt*$^-$ strains, 40 to 80 min. at 45C. h) *ts* C32 *lyt*$^-$, 120 min. at 45C.Bar is 10 µm.

formation (Fig.1 g,h), suggesting strongly that balloons are due to wall *de novo* synthesized and not to passive deformation by osmotic pressure.

Both the lysis pattern of the first phase and the cell deformations in the second seem to be essentially the same in all mutants belonging to linkage groups A,B,C,D and F.

Linkage group G. One mutant, affected in the final steps of peptidoglycan assembly, shows a small decrease in turbidity beginning more than an hour after shift to 45C. Cells exhibit a small and irregular increase in their diameter together with some minor distortions. They become less and less phase dark suggesting a slow leakage of cell contents. Throughout the lysis period the population appears heterogeneous ; over 50% of cells become granulous (Fig.1 c).

Linkage groups E and H. These strains, which do not have any specific defect in peptidoglycan synthesis, show a drop in turbidity already 15 and 5 minutes, respectively, after the shift. Cells, becoming progressively more granulous (without detectable deformations), present a remarkable homogeneous picture which suggests a simultaneous and slow lysis of the asynchronous population (Fig.1 b).

Characteristics of double mutants ts lss, lyt-1

Lytic negative ts lss double mutants were constructed to assess the role of autolysins in the lysis and development of shape distortions following exhaustion of the pool of cytoplasmic peptidoglycan precursors. The lyt-1 mutation, conferring a very large reduction in autolytic activity, was transferred by congression from strain FJ3 (4) to strains ts A29, ts C32 and ts F16. The lyt^- derivatives of the ts C32 and ts F16 mutants, showing a large residual wall synthesis at 45C, do not differ significantly from the lyt^+ parent either in the lysis pattern (Fig.2) or in the development of swellings. However, in case of ts A29, which has very little residual wall synthesis, the situation is somewhat different. After the initial 30 minutes at 45C, during which no difference between the lyt^+ and the lyt^- derivatives is noticed, turbidity of the former continues to fall while that of the lyt^- derivatives becomes stationary (Fig.2). Balloons, virtually absent in the lyt^+ strain, become visible as in other mutants belonging to groups A,B,C,D and F.

Fig.2. Growth and lysis of strains ts lss, lyt-1 and ts lss, lyt^+.

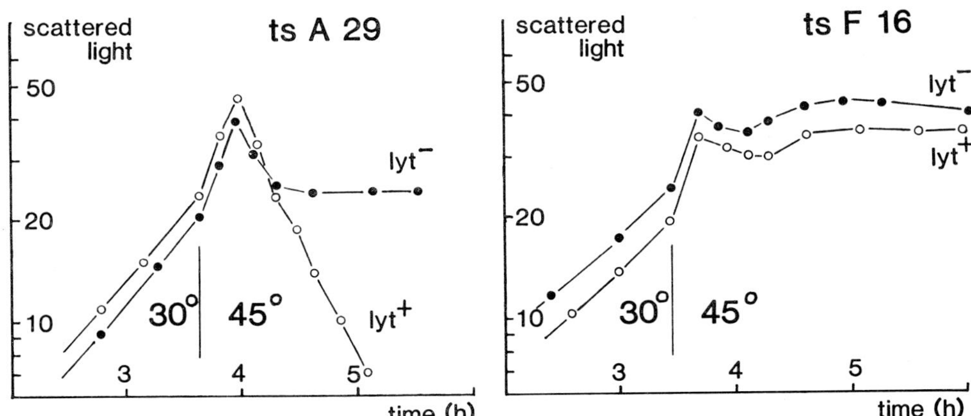

20 ml cultures (strains : see Fig.1 and text) in SA medium (3) were grown with aeration at 30C from a very low initial cell density, the inoculum being already in exponential phase. Cultures were shifted to 45C at an $O.D._{540}$ of about 0,02.

Discussion

Mutants deficient in UDP-NAM-pentapeptide synthesis lyse in two apparently distinct phases at the non-permissive temperature. During the first, lasting about 30 minutes, up to 20% of cells release their contents abruptly without undergoing any obvious change in cell shape. During the second phase cells continue to lyse to an extent dependent on the amount of residual wall synthesis. Provided lysis is not too rapid, a large proportion of cells develop, gradually and asynchronously, characteristic distortions. Empty envelopes often appear granulous and, even when swollen, maintain the shape they had prior to lysis.
We believe that during the first phase the rapid exhaustion of the UDP-NAM-pentapeptide pool results in the lysis of cells in one or several specific stages of the cell cycle (septation, cylindrical wall growth, ...). This lysis is apparently dependent neither on the presence of the major part of the autolytic activity, nor to significant extent on the amount of residual

wall synthesis. It is likely that lysis is due to a rupture of the envelope at specific site(s); a localized autolytic activity might be involved.
During the second phase the lysis mechanism(s) described above appear to be shut off (or slowed down); the following two processes then enter in competition : a degradation of the wall due to autolysins, as revealed by liberation of wall material (unpublished observations), and a wall synthesis, revealed by an important precursor incorporation (unpublished observations). This synthesis seems to be sufficient to fully compensate losses due to wall turn-over in all but the most severely affected mutants, which are unable to maintain their cell integrity except in lytic negative strains. However, besides normal cylindrical wall this secondary incorporation gives rise to swellings and balloons in localized zone within the cell. The distribution of swellings, heterogeneous both within cells (along the cell axis) and among asynchronously dividing cells, usually forming long chains, contrasts with the conspicuously symetrical bulges appearing in pairs of neighbouring cells. The synchrony of such twin cells suggests that the localization of these insertion zones might be correlated to the normal cell cycle.
The lysis phenomena described above, characteristic of all thermosensitive UDP-NAM-pentapeptide deficient mutants so far examined, differ from the lysis patterns of ts lss mutants belonging to groups G, E and H, suggesting that other mechanisms of cell lysis can exist.

References

1. Buxton, R.S.: J.Gen.Microbiol 105, 175-185 (1978)
2. Buxton, R.S., Ward, J.B.: J.Gen.Microbiol 120, 283-293 (1980)
3. Karamata, D., Gross, J.D.: Molec.Gen.Genet 108, 277-287 (1970)
4. Fein, J.E., Rogers, H.J.: J.Bacteriol 127, 1427-1442 (1976)

CHARACTERISTICS OF TOLERANT GROUP A STREPTOCOCCI.

Laurent Gutmann, Danielle Billot-Klein, and Russell Williamson
Laboratoire de Microbiologie,
Institut Biomédical des Cordeliers, Paris, France.

Introduction

The phenomenon of tolerance to β-lactam antibiotics might be important in different clinical situations, particularly infections with some strains of *Staphylococcus aureus* (1), *Streptococcus bovis* (2), enterococci (3), and groups B or C streptococci (4,5). Different mechanisms have been proposed as a basis for tolerance, and the major suggestion is that the lytic and/or bactericidal effects of benzylpenicillin are suppressed in tolerant strains because of a reduction in autolytic activity (6). In some cases the tolerant response has been produced by growth of normally autolytic bacteria at low pH where the autolysin is less active (7), or by the isolation of potentially-autolytic pneumococci in which the triggering of the autolytic system was defective (9). In the case of group A streptococci (*S.pyogenes*), benzylpenicillin is extremely bactericidal, but associated lysis has not been observed (10) and this has been referred to as non-lytic death (6,8). However, tolerant mutants of group A streptococci have been isolated after *in vitro* selection with benzylpenicillin (11) and a clinical tolerant strain has recently been isolated from a patient with group A streptococcal endocarditis after therapy with ampicillin. The characteristics of these mutants have been investigated further, and a pH-dependant tolerance of the wild-type organisms has been demonstrated.

The Target of Penicillin

Results

1) Characteristics of isolated penicillin-tolerant group A streptococci.

The penicillin-tolerant mutants were tested for cross-tolerance to a variety of inhibitors acting at different stages of peptidoglycan synthesis (table 1). The minimum inhibitory concentrations (MICs) of each antibiotic were identical for the tolerant and wild-type strains. However, the bactericidal effects of the antibiotics were significantly reduced for the tolerant organisms. It is interesting to note that the lowest tolerance ratios were obtained with vancomycin, bacitracin, and cycloserine, which had the highest MICs and the least killing activity with the wild-type strain. In contrast, the highest tolerance ratios were obtained with benzylpenicillin and cefotaxime, which inhibit the terminal stage of peptidoglycan synthesis. Since the affinities of benzylpenicillin for the penicillin-binding proteins (PBPs) of either wild-type or tolerant group A streptococci are the same

Table 1 % SURVIVAL OF GROUP A STREPTOCOCCI

		ANTIBIOTIC[a]				
Strain[b]	Phenotype	Pen G	Ctx	Van	Bac	Cyclo
T4/56	Tol^-	0.0067	0.006	0.41	0.047	1.2
T23	Tol^+	6.7(1000)[c]	6.3(1050)	14.7(36)	0.93(20)	28.7(24)
T57	Tol^+	9.3(1390)	11.3(1880)	41.3(100)	4.1 (87)	16.7(14)
SAn	Tol^+	10.0(1490)	ND	ND	ND	ND
T4/56 (PH6.0)	Tol^+	1.8(270)	0.15(25)	3.1(8)	4.0 (85)	5.3(4)

a) antibiotics were used at 10xMIC.

b) MICs (μg/ml) were identical for each strain. Benzylpenicillin (0.006), Cefotaxime (0.006), Vancomycin (0.5), Bacitracin (0.25), Cycloserine (128).

c) Tolerance ratio as the % survival after 4 hours of a tolerant strain in comparison with the wild type.

(11), it is to be expected that cefotaxime would interact similarly in these strains. Strain SAn which was isolated from a blood culture (*in vivo* selected) was as tolerant as the *in vitro* selected strains (T23, T57) to benzylpenicillin. In each of the previous experiments, the exposure of either wild-type or tolerant strains to the antibiotics at 10xMIC caused complete inhibition of growth within one hour, but without detectable lysis.

Since changes in wall composition have been demonstrated to be associated with the tolerant phenotype in other streptococci (12,13), we investigated the sensitivity of wild-type and tolerant strains of group A streptococci to a lytic enzyme from group C streptococcal bacteriophage (C-phage lysin, 14). As shown in Figure 1, the lysis of washed cells from strains T57 and SAn was significantly less than that of the wild-type bacteria when exposed to the same concentration of enzyme. The lysis of the wild-type organisms was even faster when cells from a culture incubated with benzylpenicillin (10xMIC) for 2h were used ; thus the growth in the antibiotic apparently sensitised the bacteria to the enzyme. In contrast, no such effect was observed with the tolerant strains. The lysis of strain T23 was similar to that of wild-type bacteria, but again this tolerant strain was not sensitised after growth in benzylpenicillin. The amount of enzyme able to be bound by the different bacteria was not directly relative to the sensitivity to lysis, since the wild-type strain and T57 bound about 40% of the added enzyme, whereas strains T23 and SAn were able to bind about 65% of the enzyme.

One characteristic of autolytic-deficient, tolerant strains of *S.pneumoniae* and *Bacillus subtilis* is an apparent decreased permeability of the wall (15). Determination of the amount of protein extractable with LiCl from the same weight (50mg/ml) of freeze-dried group A streptococci revealed that 3.40mg/ml

Fig 1. LYSIS OF THE WILD-TYPE T4/56 AND THE TOLERANT MUTANTS (SAn, T57, T23) BY C-PHAGE LYSIN.

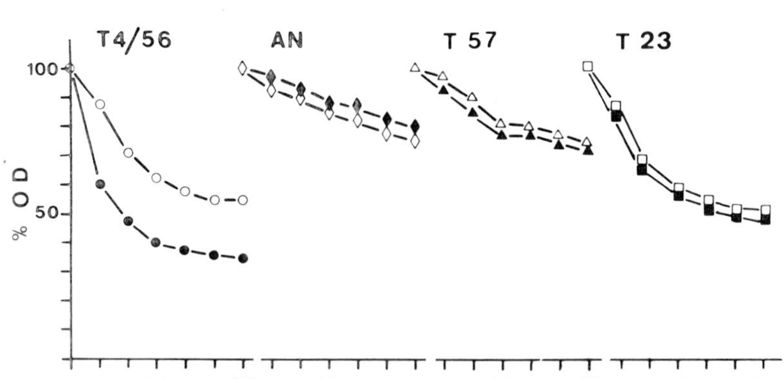

Exponential phase bacteria were incubated with C-phage lysin after growth either without (○, ◇, △, □) or with benzylpenicillin (10xMIC) for 2 hours (●, ◆, ▲, ■).

were obtained from wild-type bacteria, but only 1.56, 1.86 and 2.73mg/ml from the tolerant strains T23, T57 and SAn, respectively.

2) pH-dependant tolerance.

The growth rate of wild-type group A streptococci T4/56 was the same in medium either at an initial pH of 7.4 or 6.0. In addition, the change of pH had no effect on the MICs of the different antibiotics (see legend of table 1) and did not affect the affinity of (^{3}H)-penicillin for the PBPs (data not shown). However, the lethal effects of each antibiotic were significantly reduced at the lower pH (table 1 : see T4/56 pH6). The observed tolerance ratios varied between 4 with cycloserine and 270 for benzylpenicillin. Bacteria grown at pH 6, washed, and resuspended in pH 6 buffer appeared slightly more sensitive to the C-phage lysin than those grown at pH 7.4 In contrast with the apparent sensitisation of wild-type bacteria to the lysin after growth in benzylpenicillin at normal pH, bacteria grown in the antibiotic (10xMIC) at pH 6 for 2h did not become more susceptible to the lytic enzyme.

Discussion

Tolerance to antibiotics in clinical specimens of group A streptococci is a rare occurrence, since only two previous isolations have been reported (16,17). The additional finding of another tolerant stain in an endocarditis infection emphasises the importance of understanding the mechanism(s) responsible for this phenotype. We have used two approaches to this problem ; the selection *in vitro* of tolerant mutants which are comparablc with the *in vivo* selected strain, and the demonstration of a pH-dependant tolerance in the wild-type organism.

Our tolerant strains have many properties in common with a variety of previously described tolerant bacteria. In particular, they are tolerant to antibiotics which inhibit either early or late stages of peptidoglycan synthesis (3,7,8), and they appear to have walls of decreased permeability (15). In addition, some of the strains are less susceptible to lysis by an exogenous lytic enzyme. The lack of increased sensitivity, or sensitisation, after growth of the tolerant strains with benzylpenicillin could suggest that possible changes in wall composition which might be responsible for this effect in the wild-type bacteria do not occur. Tolerance mediated by a change in composition of the wall without loss of autolytic enzyme, has only been reported for ethanolamine-grown pneumococci (13), although a strain of *B.licheniformis* with changes in its content of secondary polymers did become resistant to its autolytic enzymes (18) and therefore could have been tolerant. In most cases tolerance has been shown to be the consequence of a decrease in content (6,8) or an inactivation (pH or protease,7,19) of the endogenous autolytic enzymes. In the case of group A streptococci the basis of non-lytic death (6,8) is not clear. It has been suggested that an autolytic enzyme which had intrinsically low activity or was extremely well regulated could cause no apparent damage to the wall but still be active enough to be responsible for the different events leading to cell death in the presence of antibiotics inhibiting synthesis of peptidoglycan. If such an hypothesis is correct, then one can imagine that in the group A mutants, a deficiency of a possible autolytic enzyme could be the cause of the tolerance and some of the observed changes in the properties of the wall. The hypothesis is supported by the observation of pH-dependant tolerance in the wild-type strain, which has only previously been shown in autolytic bacteria (4,7,20).

This work was supported by a grant from the INSERM to L.Gutmann.

References

1. Sabath, L.D., Wheeler, M., Laverdiere, M., Blazevic, D., Wilkinson, J.B.: Lancet 1, 443-447 (1977).

2. Savitch, C.B., Barry, A.L., Hoeprich, P.D.: Arch. Intern. Med. 138, 931-934 (1978).

3. Krogstad, D., Parquette, A.R.: Antimicrob. Agents Chemother. 17, 965-968 (1980).

4. Kim, K.S., Antony, B.F.: J. Infect. Dis 144, 411-419 (1981).

5. Portnoy, D., Prentis, J., Geoffrey, K.R.: Antimicrob. Agents Chemother. 20, 235-238 (1981).

6. Tomasz, A.: Ann.Rev. Microbiol. 33, 113-137 (1979).

7. Lopez, R., Ronda-Lain, C., Tapia, A., Waks, S.B., Tomasz, A.: Antimicrob. Agents Chemother. 10, 697-706 (1976).

8. Shockman, G.D., Daneo-Moore, L., McDowell, T.D., Wong, W.; β-lactam Antibiotics. M.R.J. Salton & G.D. Shockman, eds. Academic Press, New York (1981).

9. Williamson, R., Tomasz, A.: J. Bacteriol. 144, 105-113 (1980).

10. Horne, D., Tomasz, A.: Antimicrob. Agents Chemother. 11, 888-896 (1977).

11. Gutmann, L., Tomasz, A.: Antimicrob. Agents Chemother. 22, 128-136 (1982).

12. Shungu, D.L., Cornett, J.B., Shockman, G.D.: J. Bacteriol. 142, 741-746 (1980).

13. Tomasz, A., Albino, A., Zanati, E.: Nature (London) 227, 138-140 (1970).

14. Fischetti, V.A., Gotschlich, E.C., Bernheimer, A.W.: J. Exp. Med. 134, 1105-1117 (1971).

15. Williamson, R., Ward, J.B.: J. Gen. Microbiol. 125, 325-334 (1981).

16. Allen, J.L., Sprunt, K.: J. Pediatr. 43, 69-71 (1978).

17. Bayer, A.S., Chow, A.W., Ischida, K., Morrison, J.O., Guze, L.B.: Chemother. 27, 444-451 (1981).

18. Robson, R.L., Baddiley, J.: J. Bacteriol. 129, 1051-1058 (1977).

19. Joliffe, L.D., Doyle, R.J., Streips, U.: Antimicrob. Agents Chemother. 22, 83-89 (1982).

20. Horne, D., Tomasz, A.: Antimicrob. Agents Chemother. 20, 128-135 (1981).

CELL WALL METABOLISM IN LYT-MUTANTS OF BACILLUS SUBTILIS

P.D.Meyer and J.T.M.Wouters

Laboratorium voor Microbiologie, Universiteit van Amsterdam
Nieuwe Achtergracht 127, 1018 WS Amsterdam, The Netherlands

Introduction

The cell wall metabolism of *Bacillus subtilis* provides the organism with a functional cell wall and consists, on the one hand, of synthesis of wall components (peptidoglycan, PG and anionic polymers) and linking these together in an orderly fashion. On the other hand, breakdown of PG takes place (turnover) by wall located autolytic enzymes, propably conferring upon the wall the flexibility necessary for length growth (1). Turnover causes the accumulation in the extracellular fluid of wall material with a composition identical to the one found in the wall. In *B.subtilis* var.*niger* WM it was observed that under certain growth conditions the ratio of PG versus anionic polymers in the supernatant was lower than that found in the cells (2). This was interpreted as being caused by direct secretion of teichoic or teichuronic acid without previous linkage to PG.

To investigate the influence of an alteration in autolytic activity on cell wall metabolism (i.e. on turnover and secretion) *lyt*-mutants of *B.subtilis* var.*niger* WM were isolated and their wall metabolism and composition were determined under various growth conditions.

Materials and methods

1) Organisms and growth conditions. *B.subtilis* var.*niger* WM (WM, lyt^+) was used as the wild type strain. Mutants with a

The Target of Penicillin

reduced lytic activity (*lyt*$^-$) were isolated essentially as described by Fein & Rogers (3) using ethyl methane sulfonate as mutagenic agent. Selection for *lyt*$^-$-strains involved penicillin G treatment, incubation at pH 9.0 (50 mM glycine NaOH) or repeated filtration over a coarse glass sintered filter (G3, see 4). Colonies thus obtained were screened for lytic activity on double layer plates containing Procion Brilliant Red stained walls of WM according to (3). Penicillin G treatment resulted in mutant DM-82, pH 9.0 incubation gave DM-97 and DM-263 was isolated after repeated filtration. DM-20 was a superlytic mutant (*lyt*$^{++}$) obtained during the antibiotic enrichment. All strains were grown in enriched minimal medium (EMM) or minimal medium (MM) as desribed in (5). Chemostat cultivation was under phosphate (P) or magnesium (Mg) limitation (see 6).

2) Turnover measurements. Batch wise grown cells were pulse-labeled with ^{3}H-N-acetylglucosamine and chased as mentioned in (7). Turnover parameters were calculated as desribed in (7) for batch cultures and in (2) for chemostat cultures.

3) Other procedures. Cell walls were isolated and analysed according to (6), while supernatant analysis was carried out as mentioned in (2). Whole cell lysis was measured by following the decrease in optical density at 540 nm of water washed log phase cultures after suspension in 50 mM glycine NaOH (pH 9.0) at 37 ^{0}C.

Results and discussion

Growth rates and lysis rates of the respective strains grown batch wise are compiled in Table 1. The growth rates attained in rich medium showed no great differences, whereas in minimal medium they appeared to be lower in the *lyt*$^-$-strains. These strains grew as long irregular chains of unseparated cells in both media; the chain length seemed less in minimal medium (cf.3). The *lyt*$^{++}$-strain was more motile than WM, whereas the

Table 1 Growth rates and lysis rates of *lyt*-mutants of *B.subtilis* var. *niger*

Strain	*lyt*	EMM		MM	
		$\mu(h^{-1})$	lysis	$\mu(h^{-1})$	lysis
WM	+	1.32	1.00	0.44	1.00
DM-82	–	1.45	0.15	0.29	0.25
DM-97	–	1.45	0.18	0.30	0.28
DM-263	–	1.39	0.24	0.27	0.28
DM-20	++	1.32	9.32	0.45	10.40

EMM, MM: see Materials and methods; μ: specific growth rate
Lysis rates are relative to the lysis of WM

lyt$^-$-strains were almost without flagella.
The rates of lysis of *lyt*$^-$-strains grown in batch culture were about 20 % of that of WM; DM-20 lysed very rapidly after suspension at pH 9.0. Lysis rates of chemostat grown cells of DM-82 and DM-97 were also low (not shown). However, due to the variability encountered in lysis of WM grown under these conditions an easy comparison cannot be made. The *lyt*$^{++}$-strain DM-20 lysed very rapidly after growth in the chemostat, especially under P-limitation.
When the turnover parameters of the various strains were determined in batch cultures in rich medium, we found a lowered rate of turnover in the *lyt*$^-$-strains (see κ and k in Table 2), just as the turnover sensitive fraction of the wall X_B. This latter parameter, however, showed some variability among different experiments with the same organism. DM-20 did not differ

Table 2 Wall turnover parameters in *lyt*-mutants of *B.subtilis* var.*niger* in enriched minimal medium

Strain	*lyt*	$\kappa(h^{-1})$	k(%)	X_B(%)
WM	+	0.24	13	21
DM-82	–	0.097	5	14
DM-97	–	0.13	6	19
DM-263	–	0.17	6	16
DM-20	++	0.28	16	28

For a definition of the turnover parameters, see the text and (7)

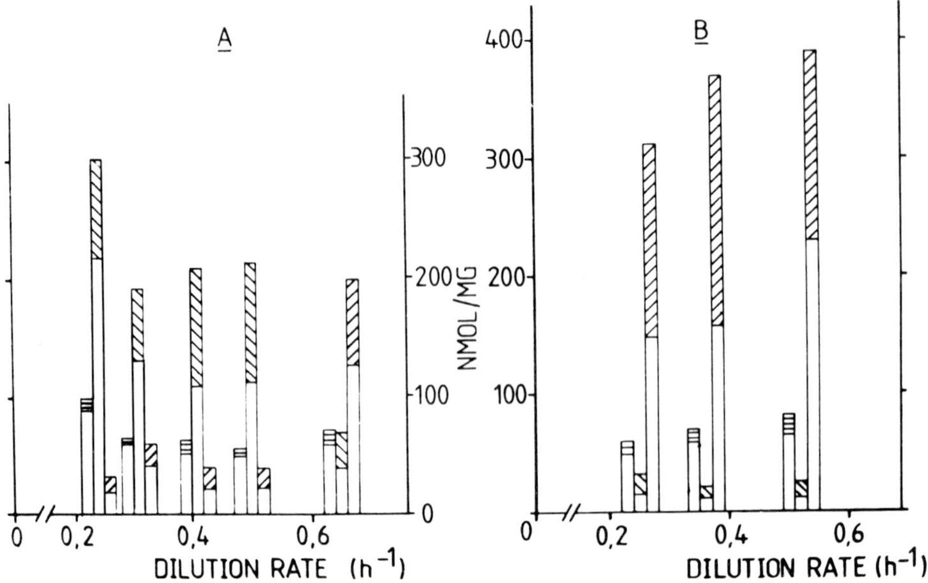

Fig.1 Wall polymers in the cells (open bars) and the supernatant (hatched bars) of DM-82 grown under P-limitation (A) or Mg-limitation (B). PG: left bar, TUA: middle bar, TA: right bar. PG, TUA, TA: see the text.

greatly from the wild type strain, although the turnover parameters showed a tendency of being higher in the *lyt*$^{++}$-strain (see Table 2).

When DM-82 was cultured in the chemostat under P- or Mg-limitation the cell walls showed a composition similar to that of WM, i.e. the wall consisted for about 50 % of PG and 50 % of anionic polymers. The latter were subject to phenotypic variability: teichoic acid (TA) being the main polymer under Mg-limitation and teichuronic acid (TUA) under P-limitation at low dilution rates (Fig.1 and see 6). The cell wall content (determined as the sum of PG, TA and TUA) seemed somewhat lower than in WM grown under these conditions (not shown).

Analysis of the supernatants of these cultures showed that all cell wall constituents were present in the extracellular fluid. From Fig. 1 one can conclude that the ratio of PG versus anionic polymers in the extracellular fluid was always lower than in the walls. This indicates that also in this strain

Table 3 Percentage of peptidoglycan lost per generation time in chemostat cultures of DM-82 at different dilution rates

D(h^{-1})	lim	k(%)	D(h^{-1})	lim	k(%)
0.24	P	5	0.25	Mg	15
0.31	P	6	0.36	Mg	14
0.41	P	15	0.50	Mg	13
0.50	P	8			
0.65	P	16			

D: dilution rate; lim: limitation; k: see the text

secretion of anionic polymers occurs, just as in the wild type strain (see 2).
By comparing the cellular and extracellular amounts of PG one can estimate the percentage of this polymer lost per generation time by turnover (k). The values of this parameter as determined for DM-82 at different growth rates under P- or Mg-limitation are given in Table 3. Turnover rates for PG were low, but comparison with data from WM is hampered by the variability of this parameter (ranging from 10 to 40 %) in the wild type organism (see 2,7). At higher dilution rates, however, peptidoglycan turnover in DM-82 is lower than in WM, since in that strain k is then at least 30 % (2).
Dialysis of the various supernatants obtained from chemostat grown DM-82 resulted in the loss of diaminopimelic acid (DAP) and glucosamine ($GlcNH_2$) containing turnover products (Fig. 2). DAP-containing material was lost for about 60 % by dialysis, whereas $GlcNH_2$-comprising products were lost for about 20 %. Similar experiments performed with supernatants from the wild type strain grown under these conditions showed that the amino acid containing extracellular material was lost for at least 90 % and the amino sugar containing products for about 50 % (2). Apparently, the turnover products from the *lyt*$^-$-strain DM-82 have a higher molecular weight than those from the wild type strain.

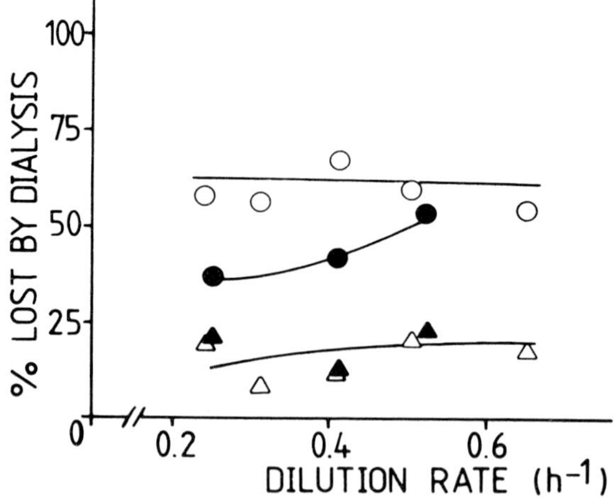

Fig.2 Loss by dialysis of diaminopimelic acid (○,●) and glucosamine (△,▲) containing turnover products in the supernatant of DM-82 grown under P-limitation (open symbols) or Mg-limitation (closed symbols) at different dilution rates.

References

1. Koch, A.L., Mobley, H.L.T., Doyle, R.J., Streips, U.N.: FEMS Microbiol. Lett. *12*, 201-208 (1981)

2. De Boer, W.R., Kruyssen, F.J., Wouters, J.T.M.: J.Bacteriol. *146*, 877-887 (1981)

3. Fein, J.E., Rogers, H.J.: J. Bacteriol. *127*, 1427-1442 (1976)

4. Williamson, R., Ward, J.B.: J. Gen. Microbiol. *125*, 325-334 (1981)

5. De Boer, W.R., Meyer, P.D., Jordens, C.G., Kruyssen, F.J., Wouters, J.T.M.: J. Bacteriol. *149*, 977-984 (1982)

6. Kruyssen, F.J., De Boer, W.R., Wouters, J.T.M.: J. Bacteriol. *144*, 238-246 (1980)

7. De Boer, W.R., Kruyssen, F.J., Wouters, J.T.M.: J. Bacteriol. *145*, 50-60 (1981)

RESTORATION OF PENICILLIN-DAMAGED CELL WALLS BY DE NOVO MUREIN SYNTHESIS AND SUCCESSIVE MUREIN DEGRADATION IN STAPHYLOCOCCI, REVEALING A HITHERTO UNKNOWN MECHANISM OF PENICILLIN ACTION: BLOCKAGE OF AUTOLYTIC WALL PROCESSES BY PENICILLIN.

Peter Giesbrecht, Hiroyuki Morioka*, Dominique Krüger, Thomas Kersten, Jörg Wecke
Robert Koch Institute of the Federal Health Office, D-1000 Berlin 65
*Kyoto Prefectural University of Medicine, Kyoto 206, Japan

Introduction

Bacteriostatic agents may block bacterial growth (1, 3) but many bacteria can regenerate in drug-free media (1). During penicillin treatment, however, the situation is different. It has been reported that penicillin may inhibit growth at low drug concentrations and may cause loss of viability and induce lysis at intermediate drug concentrations while growth may be inhibited at still higher antibiotic concentrations (4). It has been suggested that lysis is caused by blocking cell wall synthesis without blocking autolytic wall processes or simply by an activation of autolytic wall enzymes. The few bacteria surviving "lytic" concentrations of penicillin are regarded mainly as "penicillin insensitive" cells (= "Persister Cells").

Results and Discussion

In this contribution we shall describe some of our main conclusions about the action of penicillin, derived primarily from electron microscopic observations on structural changes in the cell walls of staphylococci exposed to various concentrations of this drug.- Generally, during treatment of staphylococci with penicillin of any concentration, the first effect

The Target of Penicillin

observed by electron microscopy was the cessation of the formation of the so-called splitting system (3) of the cross wall (SP in FIG. 1), which in turn blocked the completion of "normal" cross walls; instead, diffuse wall material was deposited at the tips of the growing cross walls (DW in FIG. 1). However, in such cells the so-called cutting system (3) of the cross wall was frequently found to be active (CU in FIG. 1) although the cross wall was not completed. - After transfer of staphylococci treated with penicillin of very low concentrations (0.01 µg/ml, 3h) into fresh, drug-free medium, a layer of new wall material, the so-called secondary wall, was placed directly beneath the primary wall (A2a in FIG. 4). Only then, the lytic sites of the primary wall induced the disintegration of the primary wall and the exposure of the surface of the secondary wall (A7 in FIG. 4), while in normal staphylococci a continuous replacement of the "primary wall" by newly formed "secondary" wall material took place (wall turnover). Yet, after application of rather high concentrations of this drug (1.0-10.0 µg/ml, 4h) not only growth inhibition took place but many cells were killed without hardly any indication of cell lysis (2). However, in those cells which had survived this drug treatment, complicated steps of restoration could be observed after transfer into fresh medium. In these cells alternating dark and transparent layers which had been identified as layers of autolysins and of new wall material, respectively (3), were placed beneath the primary wall (LW in FIG. 2; A3 in FIG. 4). At this stage, a number of cells lysed by wall rupture in the cross wall region (X in FIG. 2). Eventually, a thicker wall layer, the real secondary wall, was placed beneath the layered wall material (SW in FIG. 3; A4 - A5 in FIG. 4), separated by an additional dark layer (DL in FIG. 3). Only then, the lytic sites beneath the primary wall destroyed all wall structures which had been formed before or during the action of penicillin (A5-A7 in FIG. 4) exposing the entire surface of the secondary wall. These steps of wall restoration resulted in fully intact and infectious staphylococci (A1 in FIG. 4). - Surprisingly, even in those staphylococci which had survived the action of penicillin of "lytic" concentrations (0.1 µg/ml), basically the same sequence of events was observed in drug-free medium as after drug treatment at high concentrations. In this case, however, the multiple layers beneath the primary wall (A3 in FIG. 4) were not the result of the early steps of the restoration processes but could appear already during penicillin treatment. Yet, in both cases,

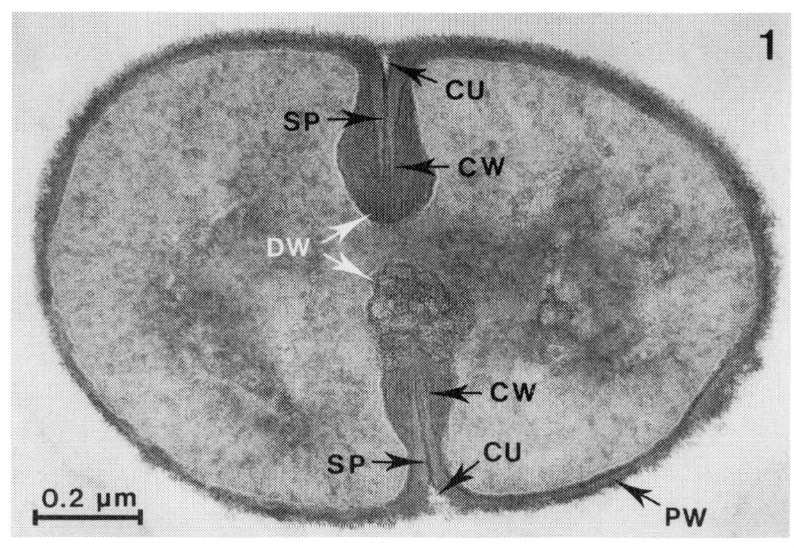

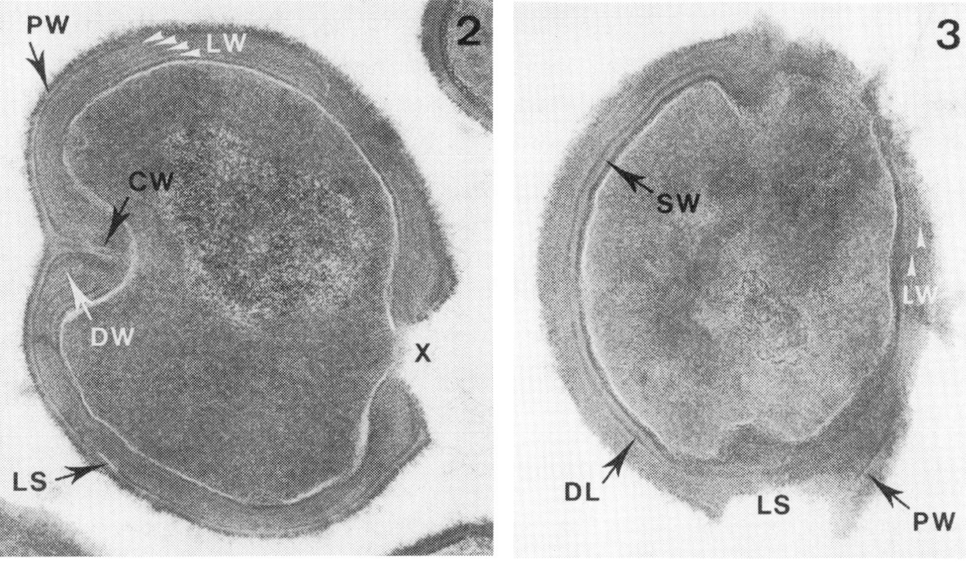

Fig.1: S.aureus during penicillin treatment. Cessation of the formation of the splitting system (SP) and, subsequently, of "normal" cross walls (CW). Deposition of diffuse wall material (=DW) at the intact cross wall. The cutting system (CU) is active although the cross wall is not closed. PW = primary wall. Fig.2: S.aureus during restoration in drug-free medium. Formation of layered wall material (=LW) beneath the primary wall (=PW). CW = cross wall. DW = diffuse wall material. LS = lytic site. X = rupture of the wall. Fig. 3: S.aureus in drug free medium.Formation of the secondary wall (=SW). Disintegration of the primary wall (=PW) and the layered wall material (=LW). DL = dark layer. LS = lytic site.

after transfer into drug-free medium, the subsequent disintegration of the primary wall and the layered wall material proceeded in the same way. The fact that a considerable percentage of penicillin-treated staphylococci were capable of regenerating indicates that re-infection after penicillin therapy might not be based on "penicillin insensitive Persister Cells" but rather on restoration processes of penicillin damaged cells, indicating a novel mechanism of penicillin resistance: "imitated penicillin resistance". Blocking these restoration processes might be an effective method to improve penicillin therapy by eliminating also those cells which had survived the primary attack of penicillin.

The apparent preservation of large parts of the primary wall in staphylococci which had survived the treatment with penicillin of high and of lytic concentrations (A3-A5 in FIG. 4) indicated that penicillin might be capable of blocking autolytic wall processes rather than activating them. First labelling experiments with ^{14}C-acetyl-glucosamine (data not shown) supported this perplexing conclusion: compared to the release of wall material from normal cells the release of wall material during treatment with 10.0 - 0.1µg/ml penicillin was distinctly reduced. These results indicated that, at least in staphylococci, penicillin may act in a highly sophisticated way, apparently different from what has been hitherto discussed.

Based on our results obtained with staphylococci which had SURVIVED the action of penicillin of different concentrations (FIG. 4), we have to conclude: 1.) Penicillin was not capable of activating but of blocking autolytic wall processes, presumably by interfering with the formation of some autolytic wall systems. 2.) Pre-existing wall autolysins, i.e. autolysins of the primary wall formed before or during the very early period of action of penicillin were only blocked reversibly (because after transfer into drug-free medium autolytic processes could be observed at that place). However, wall autolysins newly formed during full action of penicillin - i.e. autolysins of the layered wall material (LW in FIG. 2) and of the splitting system of the cross wall (SP in FIG. 1) - were mostly affected irreversilby (because in drug - free medium hardly any autolytic processes could be observed at that place). 3.) The cutting system of the cross wall (which ruptures the primary wall at the starting point of the cross wall) proved to be rather resistant to the blocking capacity of penicillin (FIG. 1). 4.) After blocking the formation of the splitting system of the cross

wall (SP in FIG. 1) which in turn blocked the formation of intact cross walls, penicillin induced the deposition of diffuse cross wall material.
5.) Lysis of wall structures (A7 in FIG. 4) in staphylococci surviving treatment with penicillin of high concentrations took place mainly in drug-free medium (i.e. much after the action of penicillin). This tremendous wall disintegration after the formation of an intact secondary wall resulted from an apparently uncontrolled re-activation of the blocked, pre-existing autolysins of the primary wall (while after chloramphenicol treatment a rather controlled autolysis occured in more sequential steps of wall degradation (1)).

Based on our results obtained with staphylococci which were KILLED under "lytic" concentrations of penicillin we have to conclude: 1.) Penicillin-induced lysis of staphylococcal cells (i.e. rupture of their protective cell walls and successive lysis of the bacterial cytoplasm) was mainly the result of different resistances of the two autolytic cross wall systems to the capacity of penicillin to block autolytic wall processes: the rather resistant cutting system (CU in FIG. 1) and the irreversibly affected splitting system (SP in FIG. 1) which led to the cessation of normal cross wall formation. But, since the cutting system attempted to perform cell separation by an apparently uncontrolled local lysis of the wall material ("spokes"; (3)), even in cells where the formation of the cross wall was not completed (compare FIG. 2) or where the cells had tried to complete the cross wall by deposition of diffuse and apparently less pressure-resistant cross wall material (FIG. 1), the wall was ruptured at this place and the cell lysed. 2.) An unexpectedly low release of primary wall material was detected during this process of wall rupture. Only later on, presumably after an uncontrolled re-activation of pre-existing autolysins of the primary wall, intense wall disintegration took place (mostly after affection of the cytoplasm). 3.) A shortage of wall material, as has been discussed, was not the very reason for wall rupture since staphylococci treated with penicillin of "lytic" concentrations apparently formed sufficient wall material (FIG. 2). 4.) Penicillin induced killing of staphylococci by rupture of cell walls must be an event completely different from penicillin-induced killing without bacteriolysis (see also (2,4)).

References

1. Giesbrecht,P., Wecke,J.: Zbl.Bakt.Suppl.10,455-459 (1981).
2. Wecke,J., Giesbrecht,P.: Zbl.Bakt.Suppl.10,461-467 (1981).
3. Giesbrecht,P., Wecke,J., Reinicke,B.: Int.Rev.Cytol.44, 225-318 (1976).
4. Tomasz,A.: Ann.Rev.Microbiol. 33, 113-137 (1979).

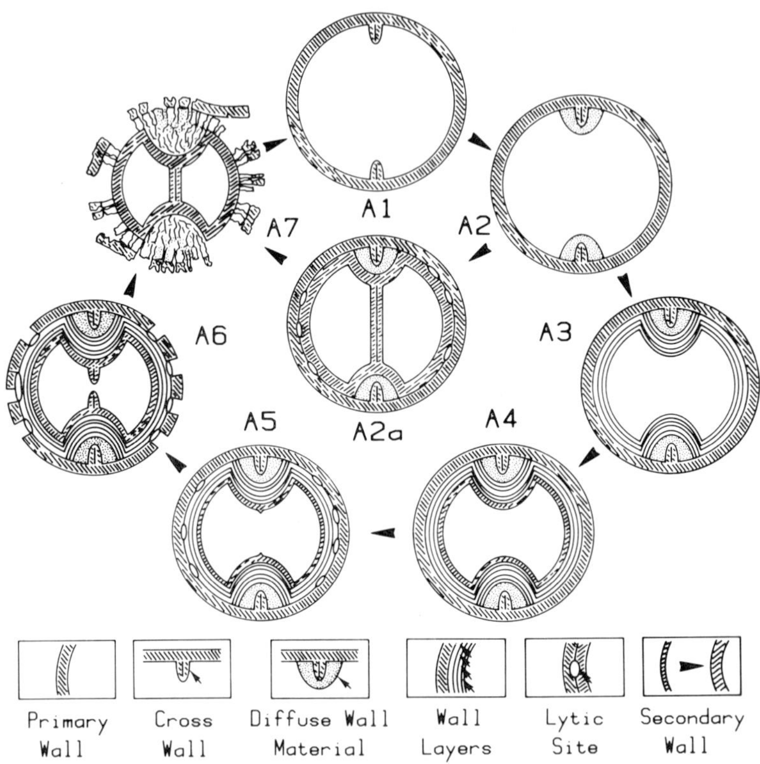

Fig.4: Schematic drawing of the neoformation of wall material beneath the primary wall structures during penicillin treatment in surviving staphylococci (during high doses of penicillin: A2; during intermediate doses: A2-A3; during low doses: A2). After transfer into drug-free medium, restoration processes started by neoformation of a secondary wall (A4-A5) in staphylococci having survived the action of penicillin. Then, by re-activation of the lytic sites of the primary wall which had been blocked by penicillin, the disintegration of all the wall structures, formed before and during the action of the drug, took place (after high and intermediate doses of penicillin: A5-A7; after low doses: A2a-A7). Further explanations appear in the text.

ß-LACTAM ANTIBIOTICS DO NOT TRIGGER THE ACTIVITIES OF AUTOLYSINS IN ETHER PERMEABILIZED ESCHERICHIA COLI

Angelika Raible, Helmut Pelzer
Thomae Research Laboratories, D-7950 Biberach/Riss

Introduction

In Pneumococcus species the action of ß-lactam antibiotics has been demonstrated to include a trigger of autolysin(s) which is responsible for the ultimate cell lysis (1,2). Although this trigger mechanism could later be extented to Escherichia coli (3) it does not seem to be absolutely clarified as yet in so far as inhibition of murein synthesis in general has similar effects in these organisms (4). The trigger hypothesis does not agree with a prior theory (5) claiming that ß-lactams may act solely by inhibition of murein synthesizing enzymes, leaving the murein hydrolases uneffected. According to this "enzyme inbalance model", in growing bacteria murein hydrolases (autolysins) should enable the sacculus to extend by insertion of newly synthesized peptidoglycan. Obviously, in this case no further stimulation or "trigger" of murein hydrolases is essential for a ß-lactam induced lysis of the growing cells.

In the present investigation we were able to show that at least in ether permeabilized cells of E. coli there is no evidence for a special trigger of murein hydrolases by ß-lactam antibiotics even at concentrations which strongly inhibit the de novo synthesis of cross-linked murein.

Methods

Ether treated bacteria (ETB) from growing E. coli ATCC 9637 were prepared as described by Mirelman et al. (6) and used for murein synthesis according to Maass and Pelzer (7). Inhibition of the synthesis of ^{14}C-GlcNAc labelled SDS insoluble material (SIM) as a function of antibiotic concentrations had been determined previously. The IC_{50} were as follows: benzylpenicillin 0.9 µg/ml, cephalexin 0.2 mg/ml, flavomycin 0.02 ug/ml, vancomycin 30 µg/ml.

Results

At optimal substrate concentrations synthesis of SDS insoluble material (SIM) was rapid and nearly constant during 60 min of incubation (Fig. 1A). Using suboptimal substrate concentrations, however, SIM synthesis could be made to cease after approximately 30 min of incubation.

If benzylpenicillin (50 µg/ml) or cephalexin (1.0 mg/ml) was added 30 min before complete cessation of SIM synthesis, further synthesis was stopped immediately followed by a degradation of preformed labelled SIM to about 50 % of the amount present at antibiotic addition (Fig. 1B).

The same degree of inhibition and degradation was obtained when the non ß-lactam antibiotics flavomycin (5 ug/ml) or vancomycin were used (Fig. 1C). Chloramphenicol (20 µg/ml) did not show any effect in this system. Likewise no SIM degradation was obtained if benzylpenicillin (50 µg/ml) was added 60 min after onset of SIM synthesis cessation. Moreover, the vancomycin (100 ug/ml) induced degradation was not further increased by addition of benzylpenicillin (50ug/ml).

The absence of a special trigger mechanism was further demonstrated in the following experiment:
Instead of treatment with an antibiotic the ETB suspension was centrifuged at 10000 x g for 2 min and subsequently resuspended and incubated in the same volume of UDP-GlcNAc-free medium. In this case murein synthesis was discontinued only by the deficiency of one substrate. Even in this case, newly synthesized SIM was degraded by 40 - 50 % (Fig. 2). Also this degradation was not accelerated by simultaneous addition of benzylpenicillin (50 µg/ml).

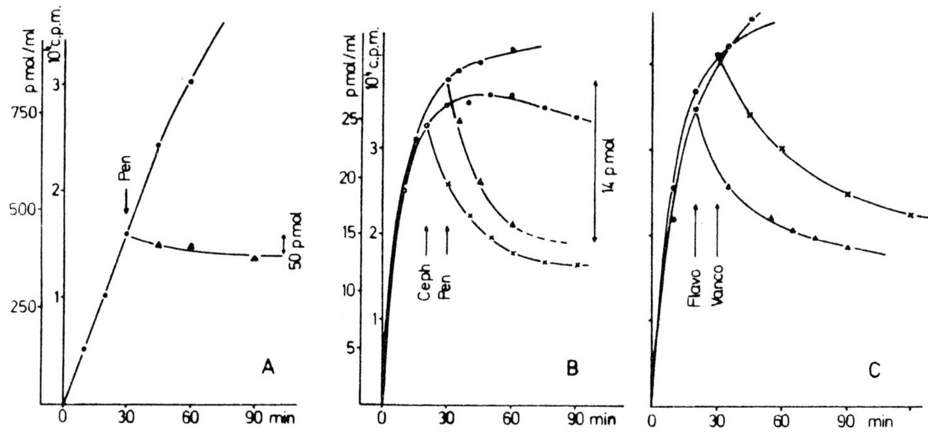

Fig. 1: Synthesis of ^{14}C-GlcNAc labelled SDS insoluble material (SIM) in ETB of E. coli ATCC 9637. Effects of benzylpenicillin (Pen, 50 µg/ml), cephalexin (Ceph, 1000 ug/ml), vancomycin (vanc, 100 µg/ml) and flavomycin (flav, 5 µg/ml).
A) optimal substrate concentrations (4.5 µM UDP-GlcNAc)
B) and C) suboptimal substrate concentrations (0.1 µM UDP-GlcNAc)

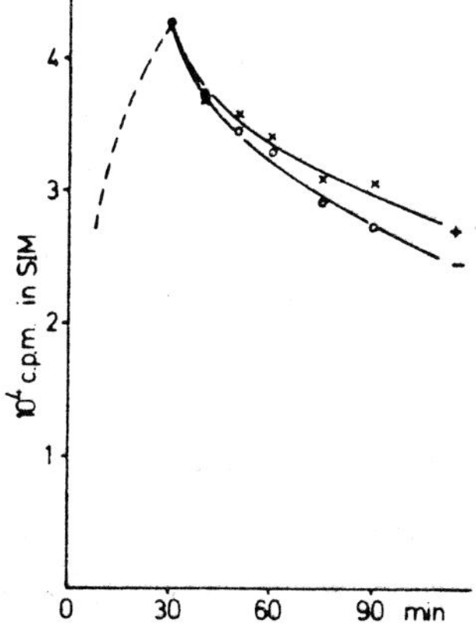

Fig. 2: Degradation of newly synthesized SIM by withdrawel of one crucial substrate for murein synthesis, UDP-GlcNAc, from the medium. Effects with (+) and without (-) benzylpenicillin (50 µg/ml).

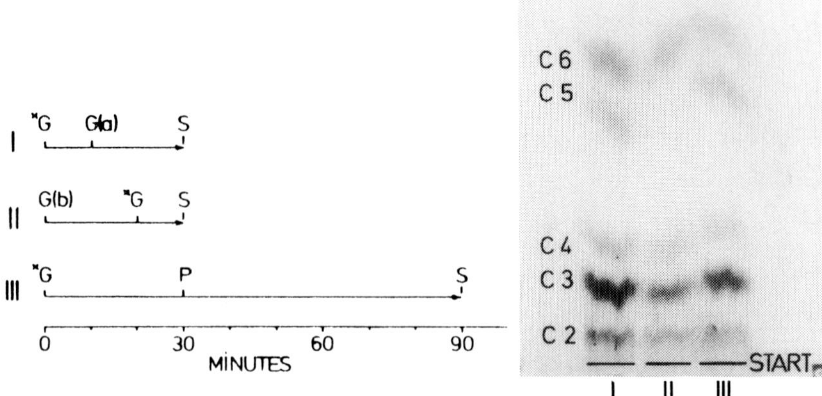

Fig. 3: Elucidation of the murein structure which is susceptible to the autolysins. Incubation regimen and paper chromatograms (autoradiography) of the lysozym digests of SDS insoluble material.

I: Radiolabel in previously synthesized ("old") murein, without antibiotic

II: Radiolabel in very newly synthesized murein, without antibiotic

III: Radiolabel in both murein fractions, 50 % degradation caused by addition of penicillin, 50 µg/ml

ETB (9 mg protein/ml) and UDP-MurNAc-pentapeptide (0.5 mM) present from point zero

*G = ^{14}C-UDP-GlcNAc, 0.1 uM, 270 mCi/mmol

G(a) = UDP-GlcNAc, 9.7 mM

G(b) = UDP-GlcNAc, 4.0 uM

P = Benzylpenicillin 50 ug/ml

S = Na-dodecylsulfate, 4 %, 3 min 100° C

C2-C6 = muropeptides on the autoradiogram of radioactivity aliquots from I - III

Descending paper chromatography in n-butanol/acetic acid/water 4:1:5

Paper chromatography of lysozyme digests of the ^{14}C-SIM before and after antibiotic action showed differences in the C3 through C6 subunit pattern (Fig. 3). Obviously the newly synthesized SIM contained nearly no C5 whereas this subunit was present in previously synthesized ("old") murein as well as in that part which could not be degraded further by the autolysin(s). These results indicate that the autolysins probably hydrolyse only the very newly synthesized part of the murein which might represent the growing area of the sacculus (Fig. 4).

Discussion

In contrast to living bacteria (3), ether permeabilized E. coli produced an immediate onset of murein hydrolysis after the addition of antibiotic or after starvation from an essential substrate. Therefore, a trigger which depends on de novo protein synthesis (3) can be excluded for ETB. Furthermore, identical murein degrading effects were observed with ETB when benzylpenicillin or cephalexin was used (Fig. 1). From that too, principal differences seem to exist between living cells and ETB.

Cephalexin preferentially binds to PBP 3 which is responsible for septum formation (8). Consequently, according to Fig. 4 of the present communication only the newly synthesized murein of septa should become sensitive to autolysin(s) by addition of this antibiotic. In the ETB experiments, degradation of septum murein should be indistinguishable from that of the cylinder murein*). In living cells by contrast, filamentation of the organisms but no lysis can be expected if septum formation is blocked and the corresponding murein is degraded.

The special structure of "old" murein as shown in the present paper may be responsible for its insensitivity to degradation by the autolysin(s). This is in accordance with the proposal that the autolysin(s) cannot sufficiently attach to this particular part of the peptidoglycan (3).

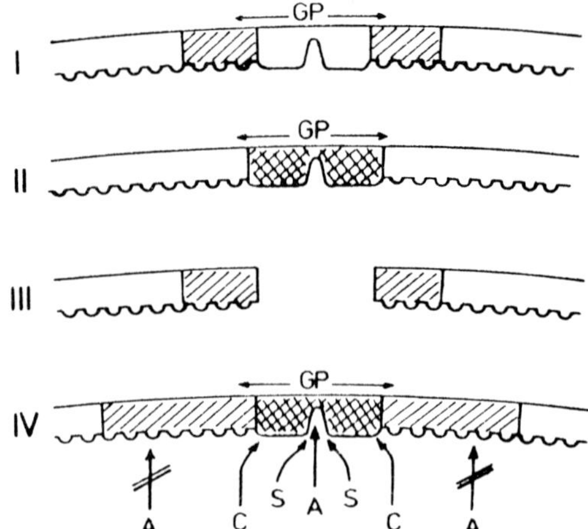

Fig. 4: Schematic representation of the processes shown in Fig. 3. IV gives a comprehensive view of the enzyme actions deduced from the results of the experiments I - III.
A, autolysin(s); C, D-alanylcarboxypeptidase; S, murein synthetases; GP, growing point

Radioactivity in previously synthesized murein, labelled monomers: C5 and C6,

 Radioactivity in newly synthesized murein, labelled monomers: only C6

References

1. Tomasz, A., Waks, S.: Proc. Nat. Acad. Sci. USA 72, 4162 - 4166 (1975)

2. Tomasz, A.: Ann. Rev. Microbiol. 33, 113 - 137 (1979)

3. Kitano, K., Tomasz, A.: Antimicr. Ag. Chemotherapy 16, 838 - 848 (1979)

4. Goodell, W., Tomasz, A.: J. Bacteriol. 144, 1009-1016 (1980)

5. Weidel, W., Pelzer, H.: Adv. Enzymol. 26, 193 - 232 (1964)

6. Mirelman, D., Yashouv-Gan, Y., Schwarz, U.: Biochemistry 15, 1781 - 1790 (1976)

7. Maass, D., Pelzer, H.: Arch. Microbiol. 130, 301 - 306 (1981)

O-ACETYLATION AND HYDROLASES OF GONOCOCCAL PEPTIDOGLYCAN

Harold R. Perkins, Stephen J. Chapman, J. Keith Blundell
Department of Microbiology, University of Liverpool, Liverpool L69 3BX, England.

Introduction

The peptidoglycan of Neisseria gonorrhoeae is known to contain O-acetyl groups, presumably on the C6 position of the muramic acid residues, and to be partially lysozyme resistant in consequence (1), although in at least one strain, RD5, the degree of O-acetylation is much lower and the peptidoglycan is lysozyme-soluble (2). Low concentrations of penicillin have little effect on the overall degree of cross-linking of gonococcal peptidoglycan but cause a sharp decrease in the degree of O-acetylation of the SDS-insoluble fraction, particularly of the uncross-linked components found within it (3). Subsequent work (4) suggested that the cross-linking of newly-synthesized peptidoglycan may also be affected. It has been reported that the peptidoglycan of gonococci undergoes considerable turnover, particularly under the influence of β-lactam antibiotics (5, 6, 7) and we have considered this process from three points of view (i) the ability of ether-treated cells of N.gonorrhoeae to degrade glycosidically-linked and peptidically-linked fragments of peptidoglycan, with or without O-acetylation (8) (ii) the attack upon the artificial substrate p-nitrophenyl-β-N-acetylglucosaminide by sonicated N.gonorrhoeae (9) (iii) the fate of labelled peptidoglycan in growing cultures during autolysis brought about by penicillin.

Results

1) Hydrolysis of peptidoglycan fragments by ether-treated N.gonorrhoeae. The substrates used were the peptidoglycan fragments obtained from Escherichia coli labelled with diamino-[^{14}C]-pimelic acid by the use

of lysozyme. They consisted of monomer fractions (GlcNAc-MurAc-L-Ala-D-isoGlu-meso-A_2pm-D-Ala and the corresponding tripeptide) and dimer fractions. The dimers contained two disaccharide units either peptide-linked or glycosidically-linked or linked by both methods (cyclic dimer) (10). These individual components were purified by paper chromatography before use as substrates, as also were the products of enzyme action. In some experiments similar fragments obtained from N.gonorrhoeae and known to be O-acetylated (3) were used.

Several types of enzyme were found. The endopeptidase that split peptide-linked dimers into monomers had a pH 8 optimum and was sensitive to β-lactam antibiotics, thus resembling the D,D-carboxypeptidase activity formerly described (11). The ID_{50} concentrations were as follows: benzylpenicillin, ampicillin 3×10^{-8}M; methicillin 3×10^{-7}M; mecillinam, 3×10^{-6}M. In all cases there remained some 20% of residual endopeptidase activity, which was not sensitive even to quite high concentrations of all four antibiotics (10^{-4} - 10^{-3}M) and this strongly suggested that N.gonorrhoeae contain a second endopeptidase that is β-lactam resistant, similar to reports for E.coli (14, 15). In this connection it is noteworthy that our work on the progress of peptidoglycan cross-linking in growing gonococci also suggests an early and late transpeptidation process (16), one of which could perhaps be relatively β-lactam resistant.

The fact that we found an L,D-carboxypeptidase activity confirms the indications of earlier studies (12, 13) and this enzyme was not further examined.

The glycosidase activities were of two kinds. The glycosidically-linked dimer was split primarily into two fragments, neither of which proved to be a simple monomer unit. One was MurAc-peptide and the other was trisaccharide peptide with a GlcNAc residue at the reducing end, as indicated by resistance to β-elimination under alkaline conditions (8). The enzyme, with a pH 8.5 optimum, was therefore an endo-β-N-acetylglucosaminidase. Gubish et al. (9) described a similar enzyme from gonococci, but worked at pH 4.5. In our experiments the endo-glycosidase activity at pH 8.5 was completely suppressed by β-lactam antibiotics, the ID_{50} values being similar to those for the endopeptidase. Unlike with the latter enzyme there was no residual activity that resisted high concentrations of the antibiotics.

The second glycosidase was an exo-β-N-acetylglucosaminidase, its presence shown by the fact that non-reducing terminal N-acetylglucosamine residues were cleaved from the various fragments used. The pH optimum of this enzyme was around 7. In subsequent experiments sonicated cells of N.gonorrhoeae strains 1L260 or RD5 were used as a source of enzyme, with p-nitrophenyl-β-N-acetyl-glucosaminide as substrate. In contrast to the results of Gubish et al. (9) there was hardly any activity at pH 4.5, but there was a sharp optimum at pH 7.2 and a possibility of a minor shoulder at pH 8.5. These values were unaffected by changes of buffer.

The effects of O-acetylation upon the endopeptidase and exo-N-acetylglucosaminidase activities of ether-treated cells were also examined. The di-O-acetylated dimer isolated from N.gonorrhoeae (3) yielded initially O-acetylated monomer units, though a direct comparison of the enzyme kinetics of the endopeptidase for this substrate relative to the non-O-acetylated one (Km = 4.4mM) was not made. The O-acetylated monomer units were in turn degraded to release free N-acetylglucosamine, thus indicating that the exo-glycosidase was unaffected by the presence of the O-acetyl substituent. The result also suggested that the O-acetyl group was attached to the muramic acid residue as in many other bacteria although further proof was needed since, in theory at least, O-acetyl esterase activity could also have been present.

To establish the position of the O-acetyl group, O-acetylated and non-O-acetylated monomers were digested with excess commercial purified β-N-acetylglucosaminidase from bovine epididymis. In a parallel experiment the corresponding monomers from Proteus mirabilis in which the O-acetyl group is known to be on C6 of muramic acid (17), were also digested. In each case all the N-acetylglucosamine was set free (detected by tlc) and the other products were either MurAc-peptide or the faster-running

O-Ac-MurAc-peptide, which was identical from both the N.gonorrhoeae and the P.mirabilis samples.

2) The fate of hexosamine-labelled peptidoglycan in growing N.gonorrhoeae 1L260. Cultures were labelled with [^{3}H]-glucosamine as described (3) and 30 min after separation into three flasks, to two of them benzylpenicillin was added at 0.5 µg/ml, a concentration known to lead eventually to loss of labelled peptidoglycan. After a further 60 min excess β-lactamase was added to the control and one penicillin flask, and immediately the [^{3}H]-glucosamine was diluted 42-fold with [^{14}C]-glucosamine. Uptake and survival of label in SDS-insoluble peptidoglycan was followed by differential counting. Figure 1 shows that removal of the penicillin by β-lactamase prevented the large-scale destruction of pre-labelled

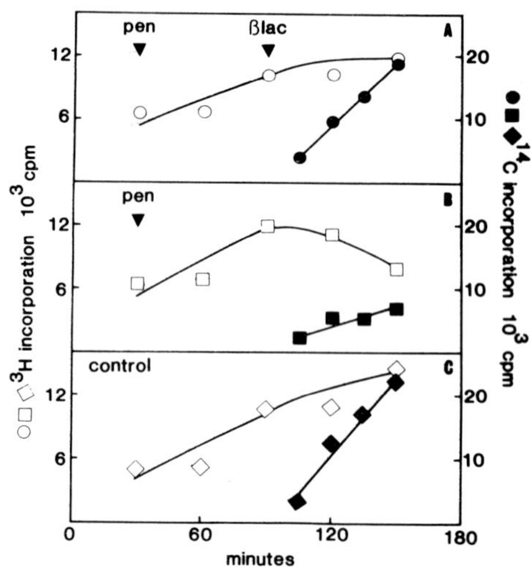

Fig. 1. Labelling of SDS-insoluble peptidoglycan by radioactive glucosamine. [^{3}H]-Glucosamine was present from A_{675} = 0.04 to 0.1. The culture was then divided into three (time 0) and after 30 min benzylpenicillin (0.5 µg/ml) was added to flasks A and B. At time 90 min excess β-lactamase was added to flasks A and C and [^{14}C]-glucosamine (a 42-fold dilution in molar terms) to all. Samples were taken at intervals, treated with boiling 5% SDS for 20 min and collected on GF/C filters for counting.

peptidoglycan that otherwise occurred. The labelling with [^{14}C]-glucosamine showed that rapid uptake was still occurring even in a culture approaching A_{675} = 0.6 and that this rate was maintained in a culture initially treated for 60 min with penicillin, which had then been removed. The latter cells, however, still appeared enlarged and many gross septa were visible (3). In the continuous-penicillin flask peptidoglycan was also being synthesized, though at a lower rate, even when the old label was being lost rapidly and when under the microscope it was mainly cell-debris that could be seen. Even at that stage effects on culture density (A_{675}) were only just beginning to appear. Effects on O-acetylation and cross-linking were also examined. At 90 min in the control and the penicillin flasks the % O-acetylation of the monomer fraction isolated after *Chalaropsis* muramidase B digestion (3) was 52% and 33% respectively whereas in the dimer fraction it remained unaltered at 44%; the %-cross-linking had decreased from 35% to 29%, similar to earlier results for high penicillin concentrations (3). At 150 min the % O-acetylation of ^{3}H-monomer was down to 26% in both penicillin flasks compared with 57% in the control and in the ^{3}H-dimer fraction it was decreased to 35% for penicillin alone and 40% when β-lactamase had been added. New material (^{14}C) had even lower O-acetylation (19%) in the monomer fraction in both penicillin flasks. Cross-linking of the new peptidoglycan was 32% in the control and + β-lactamase flasks but was down to 23% in the penicillin culture, whereas the corresponding values for old material were 35%, 32% and 31%. Thus removal of penicillin had led to a restoration of cross-linking but not of O-acetylation. It is interesting that even in cells grossly affected by penicillin a quite considerable degree of cross-linking was still taking place. The case for a penicillin-resistant transpeptidase as well as an endopeptidase seems very strong.

Our thanks are due to the Medical Research Council, London, for a Grant.

References

1. Blundell, J.K., Smith, G.J., Perkins, H.R.: FEMS Microbiol. Lett. 9, 259-261 (1980).

2. Rosenthal, R.S., Blundell, J.K., Perkins, H.R.: Infect. Immun. 37, 826-829 (1982).

3. Blundell, J.K., Perkins, H.R.: J. Bact. 147, 633-641 (1981).

4. Dougherty, T.J.: J. Bact. 153, 429-435 (1983).

5. Hebeler, B.H., Young, F.E.: J. Bact. 126, 1180-1185 (1976).

6. Goodell, E.W., Fazio, M., Tomasz, A.: Antimicrob. Ag. Chemother. 13, 514-525 (1979).

7. Rosenthal, R.S.: Infect. Immun. 24, 869-878 (1979).

8. Chapman, S.J., Perkins, H.R.: J. Gen. Microbiol. 129, 000-000 (1983). (In press).

9. Gubish, E.R. Jr., Chen, K.C.S., Buchanan, T.M.: J. Bact. 151, 172-176 (1982).

10. Pelzer, H.: Z. Naturforsch. 18b, 950-956 (1963).

11. Davis, R.H., Salton, M.R.J.: Infect. Immun. 12, 1065-1069 (1975).

12. Rosenthal, R.S., Wright, R.M., Sinha, R.K.: Infect. Immun. 28, 867-875 (1980).

13. Sinha, R.K., Rosenthal, R.S.: Infect. Immun. 29, 914-924 (1980).

14. Tomioka, S., Matsuhashi, M.: Biochem. Biophys. Res. Commun. 84, 978-984 (1978).

15. Keck, W., Schwarz, U.: J. Bact. 139, 770-774 (1979).

16. Lear, A.L., Perkins, H.R.: J. Gen. Microbiol. 129, 000-000 (1983). (In press).

17. Fleck, J., Mock, M., Minck, R., Ghuysen, J-M.: Biochim. Biophys. Acta. 233, 489-503 (1971).

ACETYLATION IN DIFFERENT PHASES OF GROWTH OF STAPHYLOCOCCI AND THEIR RELATION TO CELL WALL DEGRADABILITY BY LYSOZYME

Lars Johannsen, Harald Labischinski, Petra Burghaus*, Peter Giesbrecht

Robert Koch-Institut des Bundesgesundheitsamtes
D-1000 Berlin

*Institut für Kristallographie der Freien Universität Berlin
D-1000 Berlin

Introduction

Though the peptidoglycan of bacterial walls is known to be rather constant in regard to its primary structure (1), some variations concerning different substituents have been reported repeatedly. One of the possible variations in the peptidoglycan is the substitution with O-acetyl groups within the muramic acid of this molecule.

In the case of some Gram negative bacteria a decrease in O-acetylation after penicillin treatment could be observed (2,3,4,5). On the other hand we could demonstrate an increase in the O-acetyl content of staphylococcal cell walls grown in the presence of chloramphenicol (=CAP)(6). Together with our finding that such cell walls can be degraded by lysozyme only very slowly (7) this result is of special interest because preserved bacterial cell walls might play an important role in the induction of rheumatic arthritis (7,8,9).

An increase in O-acetyl content of staphylococcal cell walls could also be observed after application of other bacteriostatic drugs (10), suggesting that this increase was mainly due to the inhibition of growth under the influence of the drugs.

The Target of Penicillin

We, therefore, were interested if such an increase in the level of O-acetylation was a special characteristic of the stationary phase of growth or if during growth of bacteria a continuous rise of the O-acetylation occured, leading eventually to the high degree of O-acetylation in the stationary phase. Although it is well known that O-acetylation is one reason for the resistance to lysozyme, it has not yet been shown that a high level of O-acetylation was the very reason for the reduced lysozyme sensitivity of staphylococci treated with chloramphenicol. We, therefore, investigated the kinetics of O-acetylation during growth of Staphylococcus aureus and compared the velocity of degradation of CAP treated staphylococci to untreated log phase cells before and after chemical removal of the O-acetyl groups.

Material and Methods

Staphylococcus aureus SG 511 "Berlin" and "Pelzer" was grown in conical flasks at 30° C at 120 rotations per minute using Brain Heart Infusion Broth. Radioactive labeling and degradation experiments were performed as described previously (11). Cell walls were isolated as described in (6) and O-acetyl groups were determined using gas chromatography essentially following the method described in (12). Removal of O-acetyl groups was accomplished by treatment of the cell walls in 0,05 M NaOH for 4 hours at room temperature.

Results and Discussion

The comparison of the degradability of cell wall material with their content of O-acetyl groups revealed a parallelism between the degree of O-acetylation and the resistance of the cell walls to lysozyme degradation (fig. 1). After 4 hours of lysozyme digestion of cells from different states of growth the cell wall material proved to be dissolved to 100% (log phase), 46% (stat. phase) and to 30% (CAP treated), (B. Reinicke et al., contribution in this volume), with relative O-acetyl values of 1 (log phase), 1.75 (stat. phase), and 2.2 (CAP treated), respectively.

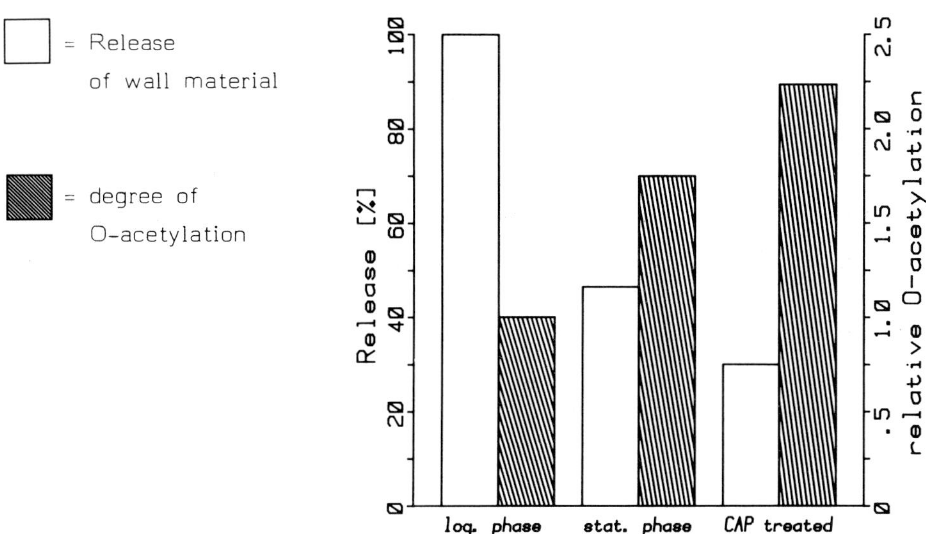

Fig. 1

From the differences in O-acetylation observed one could assume that the increase might have been a continous process occuring during normal growth of staphylococci, the degree of O-acetylation simply being dependent on the duration of stay in the culture. However, there might be also distinct levels in O-acetylation correlated to each of the two phases of growth. In order to differentiate between these possibilities we measured the degree of O-acetylation at different times of growth. To our surprise, the results clearly showed that there were two main levels of O-acetylation during growth: a constant, relative low level of about 120 nmol O-acetyl groups per mg of cell wall in log phase and in stationary phase an again constant but relative high level of about 230 nmol O-acetyl groups per mg of cell wall (fig. 2). Only during transition from log phase to stat. phase the degree of O-acetylation was found to increase continuously. Although it is well known that O-acetylation of peptidoglycan interferes with the action of lysozyme we had to prove that the O-acetyl groups comprised the main factor responsible for the reduced degradability also in our experiments.

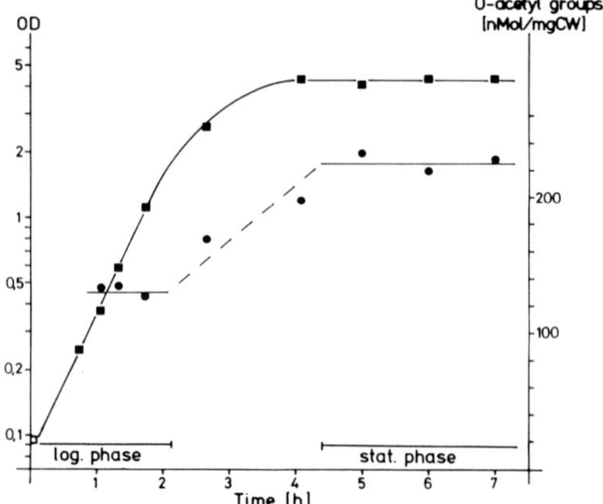

Fig. 2

The degradation rate of CAP treated cell walls was significantly lower than that from log phase of growth (fig. 3). Essential similar results could be obtained with isolated cell walls (data not shown).We, then, removed the O-acetyl groups from the cell walls and compared the slope of degradation curves with that of untreated cells (fig. 3), since this slope is proportional to the rate of degradation. To our surprise, the degradation rate of log phase cells could not be further accelerated by removal of O-acetyl groups.

Our degradation experiments suggested that the main factor responsible for the rise of resistance to lysozyme degradation in walls from staphylococci treated with CAP or harvested from the stationary phase of growth is a high level of O-acetylation. Furthermore, our examination of the cell walls with Fourier transform infrared spectroscopy revealed also some differences in the sugar and phosphorus bands of the spectra, thus pointing to possible variations in teichoic acid content or composition (10).

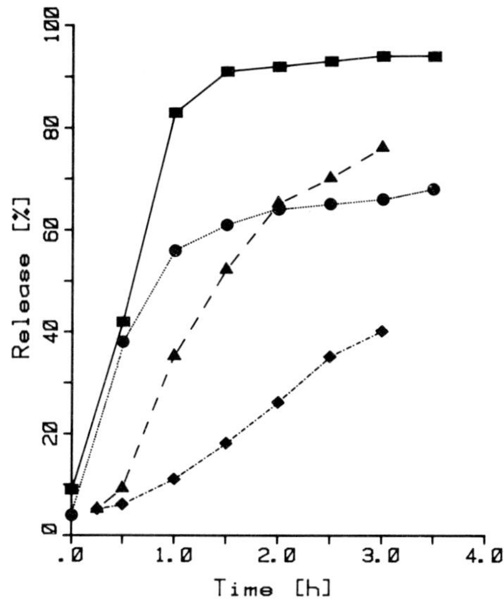

Fig. 3

Lysozyme digestion of CAP treated and untreated cells before and after removal of O-acetyl groups
(-●- = log phase, -■- = log phase, O-acetyl groups removed,
-■- = CAP treated, -▲- = CAP treated, O-acetyl groups removed)

Our data are compatible with the assumption that the increased O-acetyl content of cell walls grown in the presence of bacteriostatic antibiotics might essentially be a result of the growth inhibiting effect of the drugs because a similar effect could be observed in cell wall material from the stationary phase of staphylococci. Taking into account recent findings from other laboratories that penicillin induced a decrease in O-acetylation in certain bacteria (2,3,4,5) one may speculate that O-acetyl groups play an important role in regulating the degradation of cell walls of at least some bacterial species.

It has been concluded (3) that there should be an enzyme outside of the bacteria responsible for attaching O-acetyl groups to the peptidoglycan. Thus the balance of cell wall synthesis and cell wall degradation might well be regulated by different contents of O-acetylation in these bacteria. For gonococci it was proposed that PBP 2 may be a possible candidate for controlling enzymatically the attachment of O-acetyl groups to the peptidoglycan (5). Further investigations are neccessary to elaborate the role of the variation of O-acetyl content in growth regulation, especially under the influence of different antibiotics.

References

1. Schleifer, K.H., Kandler, O.: Bacteriol. Rev. 36, 407-477 (1972).
2. Martin, H.H., Gmeiner, J.: Eur. J. Biochem. 95, 487-495 (1979).
3. Gmeiner, J., Kroll, H.P.: Eur. J. Biochem. 117, 171-177 (1981).
4. Blundell, J.K., Perkins, H.R.: J. Bacteriol. 147, 633-641 (1981).
5. Dougherty, T.J.: J. Bacteriol. 183, 429-435 (1983).
6. Johannsen, L., Labischinski, H., Reinicke, B., Giesbrecht, P.: FEMS Microbiol. Lett. 16, 313-316 (1983).
7. Giesbrecht, P., Wecke, J., Blümel, P., Reinicke, B., Labischinski, H.: in "The Influence of Antibiotics on the Host-Parasite Relationship" (Eickenberg/Hahn/Opferkuch eds.) Springer, Berlin, p. 228-241 (1982).
8. Cromartie, W.J.: in "Arthritis; Models and Mechanisms" (Deicher/Schulz eds.) Springer, Berlin, p. 24-38 (1981).
9. Ginsburg, I., Neemann, N., Lahav, M., Sela, M.N., Quie, P.G.: Zbl. Bakt. Suppl. 10, 851-859 (1981).
10. Burghaus, P., Johannsen, L., Naumann, D., Labischinski, H., Bradaczek, H., Giesbrecht, P.: contribution in this volume.
11. Wecke, J., Lahav, M., Ginsburg, I., Giesbrecht, P.: Arch. Microbiol. 131, 116-123 (1982).
12. Fromme, I., Beilharz, H.: Anal. Biochem. 84, 347-353 (1978).

EFFECT OF PENICILLIN- AND CHLORAMPHENICOL-TREATMENT ON THE MUREIN IN STREPTOCOCCUS PNEUMONIAE: ALTERATION IN STRUCTURE AND SUSCEPTIBILITY TO MUREIN HYDROLASES

Roswitha Nöthling and Regine Hakenbeck
Max-Planck Institut für molekulare Genetik, D1000 Berlin 33

Lars Johannsen
Robert-Koch Institut, D1000 Berlin 65

Introduction

Pneumococci are highly penicillin sensitive organisms with an MIC-value of only 0.007µg/ml for penicillin G. Lysis induced by ß-lactams and other inhibitors of cell wall biosynthesis is dependent on a functional autolytic system (for review see 1). This has well been established for S.pneumoniae by studies of A.Tomasz and colleagues (2,3). Details of the lytic pathway still remain obscure, and only little is known about the lysis inhibiting effect of chloramphenicol. For the pneumococcal system it has been discussed that lysis might be due to the release of an inhibitor of the autolysins(3,4). On the other hand structural changes in the murein (the substrate of autolysins) have been observed after various antibiotic treatments of different species (5,6,7). The question remains whether these alterations could be at least partly responsible for an enhanced activity of these enzymes. Encouraged by initial work carried out in the laboratory of A.Tomasz, we extended this approach to the pneumococcal system. The availability of an autolytic defective mutant cwl (8) gave rise to the possibility of isolating intact cell walls from bacteria that have been treated with penicillin. Under these conditions, pneumococci continue to incorporate precursors into preexisting murein, so that this

The Target of Penicillin

portion of the cell wall can be specifically labelled. In the present communication we will present data concerning changes in the cell wall, isolated from the mutant that had been treated with lysis-inducing (penicillin G;PEN) or lysis-inhibiting (chloramphenicol;CAP) drugs. We investigated the susceptibility of the various cell walls to the pneumococcal amidase and a heterologous murein hydrolase, the M1 muramidase (mutanolysin;9). Furthermore, chemical analysis of cell wall components was carried out.

Results and Discussion

Different cell walls as substrate for murein hydrolases. Exponentially growing cells of the mutant cwl were labelled in the teichoic acid with ^{3}H-choline and at the same time treated with either PEN (0.007μg/ml)or CAP (25μg/ml) for 60min ("new" label). The control remained untreated and was labelled continuously. Cell walls were isolated after shaking the cells with glass beads in a mickle desintegrator and boiling in 2% SDS. Equal amounts of isolated walls were subjected to degradation either by pneumococcal amidase or muramidase (Fig.1a,b). As can be seen, cell walls derived from penicillin treated cells (PEN

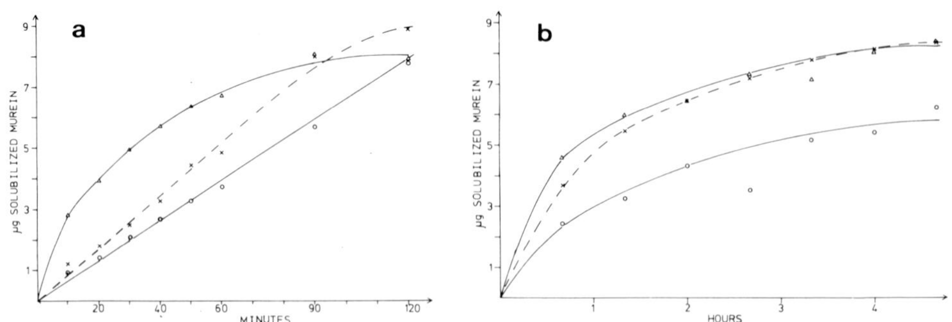

Fig.1: Hydrolysis of cell walls isolated from PEN, CAP and control cells. Walls (10μg and 20 000cpm) were incubated with a) amidase, b) muramidase in 200μl (4) and radioactivity solubilized determined as indicated. The 100% value was obtained after overnight incubation. △—△ PEN ○—○ CAP ×—× control

walls) were attacked by the amidase at a faster rate then "control" walls, and those faster then CAP walls. In contrast, the muramidase acted only poorly on CAP walls, and hardly discriminated between PEN and control walls. When the cell walls were labelled prior to the antibiotic treatment ("old" label) and hydrolyzed with the amidase, the difference between the three types of walls was much smaller suggesting that the portion that was synthesized during PEN treatment was responsible for the effect observed. On the other hand, the slow rate of CAP wall degradation with muramidase was still apparent with old labelled cell walls. Therefore we assume that the CAP induced modification is introduced not during biosynthesis but as a secondary modification into the existing walls. Removal of either the teichoic acid with periodate (10) or O-acetyl groups with 0.01N NaOH (11) had no effect on the muramidase action (data will be presented elsewhere).

Analysis of cell wall components. Table 1 summarizes the amount of murein and teichoic acid components present in the various cell walls. Very little differences between PEN, CAP and control walls could be detected, and determination of O- and N-acetylation also gave very similar results. Therefore the reason for

Table1: Cell wall components in PEN,CAP and control walls. Amino acid and amino sugar analysis was performed as described in (6) using a Biotronik LC6000E amino acid analyzer. For determination of phosphate see (12) and of choline (10). O- and N-acetylation was measured as stated in (13).

	control	PEN	CAP
Glu (molar ratios)	1	1	1
Lys "	0.8	0.8	0.81
Ala "	2.17	2.26	2.15
Ser "	0.27	0.27	0.27
muramic acid "	0.43	0.4	0.43
glucosamine "	0.87	0.87	0.9
choline (μMol/mg)	0.81	0.79	0.83
phosphate (μMol/mg)	1.31	1.41	1.47
N-acetyl groups(nMol/mg)	956	1017	896
O-acetyl groups(nMol/mg)	42	34	38.5

the different substrate properties must reside in other structural properties, such as chain length or distribution of the wall polymers, or the peptide moieties. For analysis of the peptides, cells were labelled with ^{3}H-lysine, cell wall isolated as described above and digested with amidase overnight.

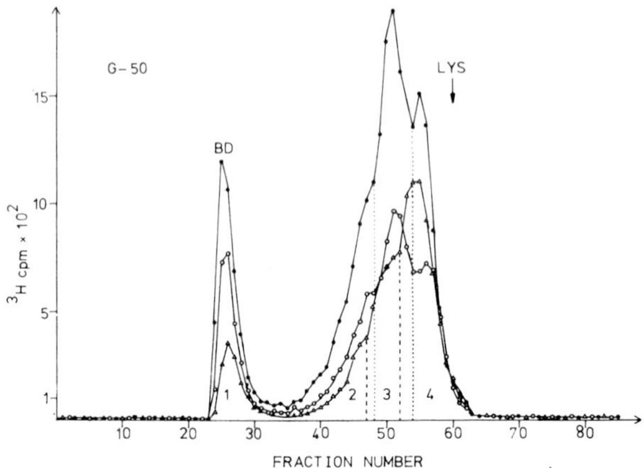

Fig.2: Chromatography of ^{3}H-lysine labelled peptides from PEN, CAP and control walls. 100μg cell walls were hydrolyzed with amidase overnight, centrifuged and the radioactive, soluble peptides chromatographed on a Sephadex G50 column (1.5x32 cm; 50mM Tris-HCl,pH8.3, 0.2M NaCl; flow rate 2ml and 2fractions per h). BD: blue dextran; Lys: lysine. ▵—▵ PEN ○—○ CAP ●—● control

cell walls	incubation time	% ^{3}H-lysine labelled peptides in pool 1	pool 2+3	pool 4
PEN	5 min	22	24	54
"	15 "	19	26	55
"	40 "	19	33	48
"	20 hrs	14	35	51
control	20 "	14	45	41

Table2: Amidase products obtained after different times of hydrolysis. PEN and control walls (^{3}H-lysine labelled) were incubated with amidase for the times indicated. Solubilized peptides were chromatographed on Sephadex G50 and the radioactivity in the different size classes determined (see Fig.2).

The solubilized peptides were chromatographed on a Sephadex G50 column, and their size distribution is shown in Fig.2. Whereas CAP and control walls gave identical elution profiles, peptides derived from PEN walls showed a clear shift towards smaller sized molecules. This is expected, assuming that penicillin inhibits the crosslinking reaction. Surprisingly, these low molecular weight peptides (probably uncrosslinked monomers) appeared as the preferential products after short times of amidase digestion (table2). Since the amidase appears to be an exoenzyme, acting on one murein strand before attacking another (data not shown here) it is conceivable that the non-crosslinked glycan chains that are synthesized in the presence of penicillin contain more attachment sites for the enzyme. The question remains, whether these sites are in reach for the amidase *in vivo*. The release of membrane vesicles that occur early after the addition of penicillin (Hakenbeck et al, submitted for publication; 14) could contribute to the necessary accessibility of the substrate for the autolysin. No evidence could be found for direct activation of the enzyme by penicillin as has been reported for gram negative bacteria (15). Chloramphenicol treated pneumococci synthesize an altered murein as revealed by the reduced action of the muramidase on CAP walls (Fig.1b). This alteration has only little influence on the amidase activity, excluding its responsibility for the CAP induced lysis-inhibiting effect. Again, the situation is different to gram negative bacteria: in E.coli, the cellular autolytic enzymes act clearly much less on murein that has been synthesized during amino acid starvation (which is also a lysis-inhibiting procedure;16). For pneumococci, it is more likely that CAP prevents access to the amidase substrate *in vivo*, possibly by interfering with the enzyme translocation through the membrane.

Acknowledgment: We thank Dr. Yokogawa for the generous gift of M1-muramidase.

References

1. Tomasz, A.: Phil. Trans. R. Soc. Lond. B, 289, 303-308 (1980).
2. Tomasz, A., Albino, A., Zanati, E.: Nature 227, 138-140 (1970).
3. Tomasz, A., Waks, S.: Proc. Natl. Acad. Sci. 72, 4162-4166 (1975).
4. Höltje, J.-V., Tomasz, A.: Proc. Natl. Acad. Sci. 72, 1690-1694 (1975).
5. Goodell, E.W., Fazio, M., Tomasz, A.: Antimicrob. Agents Chemother. 13, 514-526 (1978).
6. Johannsen, L., Labischinski, H., Reinicke, B., Giesbrecht, P.: FEMS Microb. Lett. 16, 313-316 (1983).
7. Martin, H.H., Gmeiner, J.: Eur. J. Biochem. 95, 487-495 (1979).
8. Lacks, S.: J. Bacteriol. 101, 373-383 (1970).
9. Yokogawa, K., Kawata, S., Nishimura, S., Ikeda, Y., Yoshimura, Y.: Antimicrob. Agents Chemother. 6, 156-165 (1974).
10. Tomasz, A.: Science 157, 694-697 (1967).
11. Perkins, H.R.: Biochem. J. 95, 876-882 (1965).
12. Ames, B.A., Dubin, D.T.: J. Biol. Chem. 235, 769-775 (1960).
13. Fromme, I., Beilharz, H.: Anal. Biochem. 84, 347-353 (1978).
14. Horne,D., Hakenbeck, R., Tomasz, A.: J. Bacteriol. 132, 704-717 (1977).
15. Fontana, R., Satta, G., Romanzi, C.A.: Antimicrob. Agents Chemother. 12, 745-747 (1977).
16. Goodell, W., Tomasz, A.: J. Bacteriol. 144, 1009-1016 (1980).

MICROCALORIMETRIC AND ELECTRON MICROSCOPIC INVESTIGATION ON STAPHYLOCOCCI BEFORE AND AFTER TREATMENT WITH PENICILLIN AND CHLORAMPHENICOL AND THEIR COMBINATIONS

Dominique Krüger, Peter Giesbrecht
Robert Koch Institute of the Federal Health Office
D-1000 Berlin-West 65

Introduction

In addition to its manifold application in life science, microcalorimetry has become an interesting method for testing the antibacterial activity of antibiotics (1). The heat produced by bacterial catabolic processes may be recorded before and after treatment with penicillin or chloramphenicol as typical power-time curves which give a direct glimpse into the metabolic situation of the bacterial culture. Our investigations were combined with measurements of the optical density and with electron microscopic methods in order to detect variations of the bacterial cell wall which may be of interest in connection with wall degradability during phagocytosis.

Material and Methods

A flow-through microcalorimeter (LKB 2107, flow rate 40 ml/h) was used. Parallel measurements of turbidity were performed with a flow-through Zeiss PM 6 spectralphotometer. For culture conditions of Staphylococcus aureus, antibiotic treatment, and electron microscopic methods see (2).

Results and Discussion

During growth, normal cells of S. aureus SG 511 generated a characteristic power-time curve (P in Fig. 1) with 3 peaks followed by a steady state period. The period of peak appearances could be related to the logarithmic increase of optical density (OD in Fig. 1).
When adding so-called lytic doses of penicillin(i.e.0.1 µg/ml) to the suspension just after inoculation, heat production and optical density increased rather normally for 90 minutes; then a drastic decrease of the two curves indicated the start of the well-known lytic effect (Fig. 3). Compared to control cells (Fig. 2), electron microscopy revealed lysed cells (Fig. 4) with characteristic alterations at the cross wall (deposition of diffuse wall material) and at the primary wall (deposition of layered wall material; see Giesbrecht et al.,this volume). In contrast to this rather slow reaction of the staphylococci after penicillin treatment, earlier investigations in our laboratory (2, 3) with high doses of penicillin (10 µg /ml) have detected a rather quick metabolic reaction (the heat production stopped already after 20 minutes); as hardly any cell lysis could be detected, only a bacteriostatic (and to a considerable extent also a bactericidal) effect apparently took place after application of high doses of this drug.

On the other hand, after application of chloramphenicol (20 µg/ml) heat production by staphylococci stopped within 4 minutes and the bacteria were not capable of producing peaks in the power-time curve (Fig. 5); furthermore, in this case heat production remained in a steady state after a slight decrease. Nevertheless,optical density increased continuously while no increase of cell number could be detected, indicating that the OD-data were only simulating growth. As revealed by electron microscopy (Fig. 6), the well-known cell wall thickening with chloramphenicol-treated gram-positive bacteria (4) had already taken place, indicating that this wall thickening could be the very reason for the continuous increase in optical density. This chloramphenicol-wall material had a rather low degradability by lysozyme under phagocyte-specific conditions (5).

If, finally, chloramphenicol was combined with penicillin (20 μg/ml chloramphenicol + 0.1 μg/ml penicillin), the power-time curve was virtually of the chloramphenicol-type (Fig.7);however,chloramphenicol-characteristic thickenings of the peripheral wall could not be detected any longer. Only diffuse wall material was deposited at the cross wall region (Fig. 8), resembling the primary effects of penicillin. Hardly any increase in the optical density could be detected (Fig. 7), indicating again that the increase in optical density after chloramphenicol treatment (Fig. 5) must be ascribed essentially to wall thickenings. Cell lysis did not occur during this treatment with drug combinations; this blockage of autolytic wall

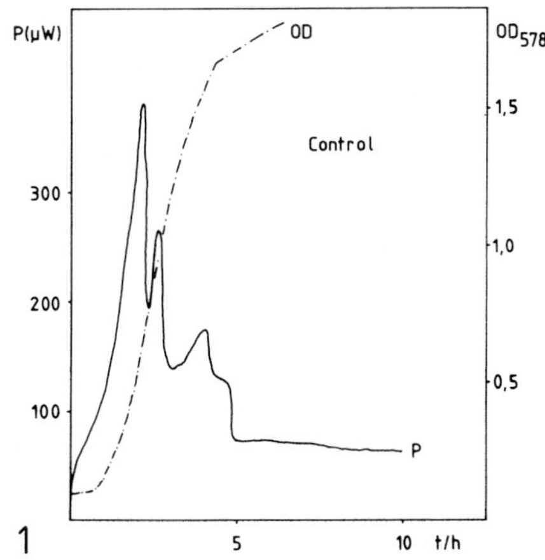

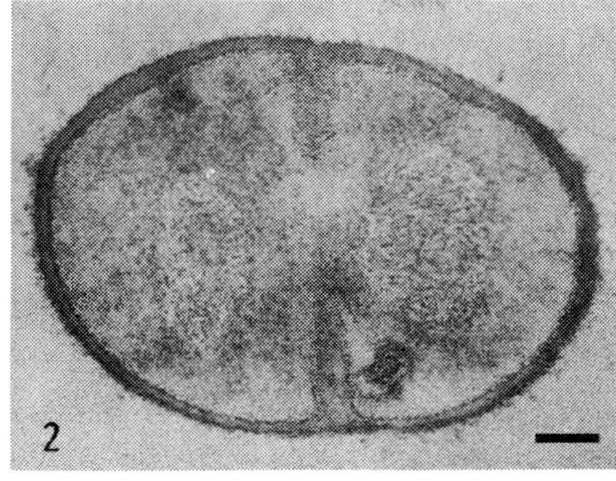

Fig. 1: Typical power-time curve (P), generated by untreated S. aureus SG 511 cells, and optical density (OD) during growth in 2.5 % bacto-peptone.

Fig. 2: Thin section of an untreated staphylococcal cell, bar represents 0.1 μ

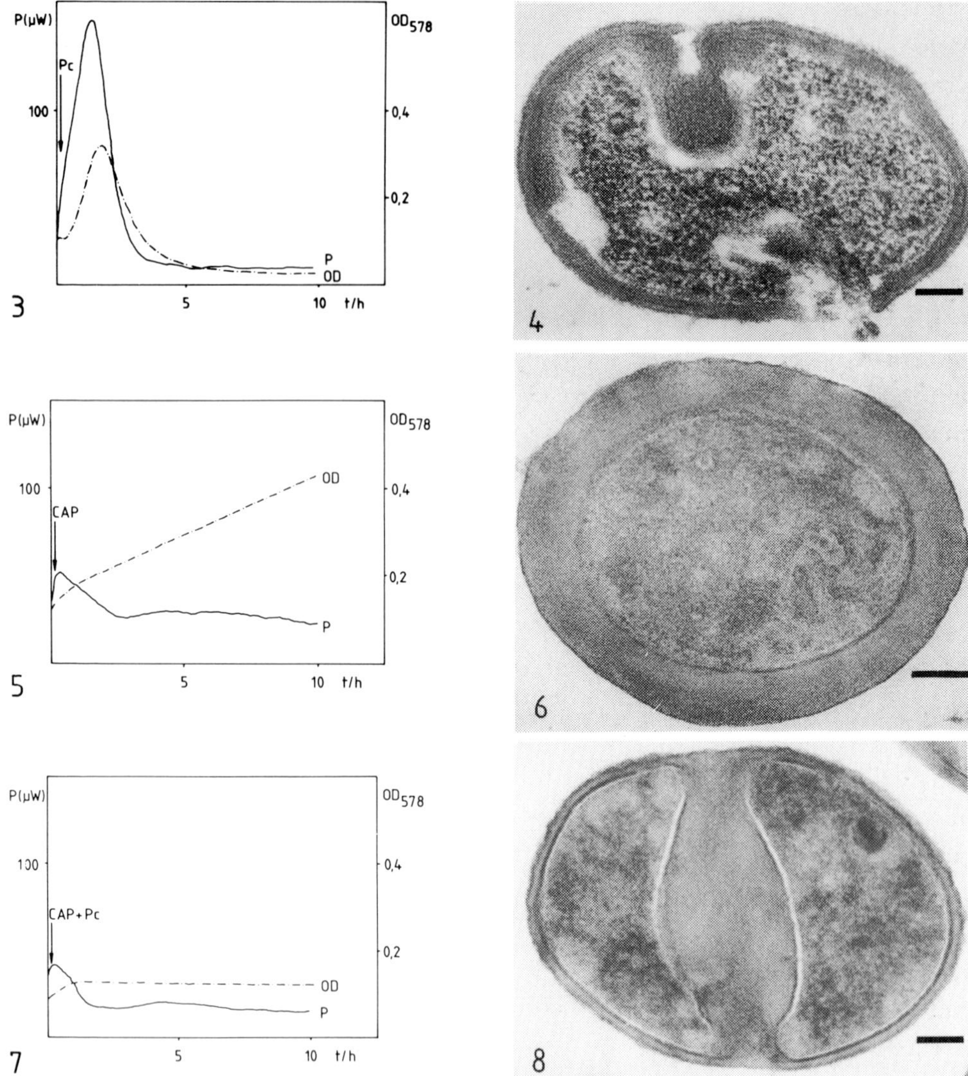

Fig. 3-8: Power-time curves (P), optical densities (OD), and thin sections of staphylococcal cells treated with drugs. Fig.3+4: 6 hours treatment with a lytic dose of penicillin (0.05-0.5 μg/ml ad 10^8 cells/ml). Fig.5+6: 8 hours treatment with 20 μg/ml chloramphenicol. Fig. 7+8: 8 hours treatment with a combination of 20 μg/ml chloramphenicol and 0.1 μg/ml penicillin. The antibiotics were added just after inoculation (arrows in Figs. 3+5+7). Bar represents 0.1 μ

processes may be due to the well-known capacity of chloramphenicol to block autolytic activity (6). Similar alterations of staphylococcal wall structures have been reported to take place after application of a combination of cephalosporin and erythromycin (7).

Our data indicate that the formation of huge amounts of peripheral wall material occurring after application of chloramphenicol can be prevented by a combination of chloramphenicol with penicillin. These data are interesting in view of the fact that the amount of wall material and its reduced degradability during phagocytosis are important factors for the induction of chronic inflammatory processes (arthritis; see (8, 9)).

Summary

Metabolic activity and growth of staphylococci in a complex medium before and after treatment with penicillin or chloramphenicol (and combinations of these drugs) were examined by flow-through microcalorimetry, combined with measurements of the optical density.

Our electron microscopical data indicated that the formation of huge amounts of peripheral wall material beneath the primary wall, normally occurring after application of chloramphenicol, can be prevented by a combination of penicillin and chloramphenicol. Only penicillin-specific depositions of diffuse wall material at the cross wall regions could be detected under these conditions. The medical importance of these findings is discussed.

References

1. Lamprecht,I., Schaarschmidt,B. (eds.):Microcalorimetry in Life Science. Walter de Gruyter, Berlin . New York 1977
2. Wecke,J., Giesbrecht,P.: Zbl.Bakt.Suppl.10,461-467 (1981)
3. Labischinski,H., Barnickel,G., Naumann,D., Wecke,J., Krüger,D., Giesbrecht,P.: Zbl.Bakt.Mikrobiol.Hyg. A I, 251, 455 (1982)
4. Giesbrecht,P.: Zbl.Bakt.I Orig. 187, 452-498 (1962) Giesbrecht,P., Ruska,H.: Klin.Wschr. 46, 575-582 (1968)
5. Johannsen,L., Labischinski,H., Reinicke,B., Giesbrecht,P.: FEMS Microbiol.Lett. 16, 313-316 (1983)
6. Tomasz,A., Waks,S.: Proc.Natl.Acad.Sci.USA 72,4162-4166 (1975)
7. Nishino,T.: Jap.J.Microbiol. 19, 53-63 (1975)
8. Ginsburg,I., Christensen,P., Eliasson,I., Schalen,C.: Acta path.microbiol.scand.Sect. B 90, 161-168 (1982)
9. Giesbrecht,P., Wecke,J., Blümel,P., Reinicke,B., Labischinski,H.: in The Influence of Antibiotics in the Host-Parasite-Relationship (Eickenberg/Hahn/Opferkuch eds.), 228-241, Springer, Berlin . Heidelberg . New York 1982

CORRELATION OF CELL WALL TURNOVER AND AUTOLYTIC ACTIVITY IN FLA$^-$ AND SUPERMOTILE MUTANTS OF BACILLUS SUBTILIS

Harold M. Pooley and Dimitri Karamata

Institut de génétique et de biologie microbiennes
Rue César - Roux 19
CH - 1005 Lausanne

Introduction

The liberation of soluble fragments of the cell wall by growing cultures of many Gram-positive organisms depends upon the action of autolytic enzymes (1). Chain forming mutants containing reduced levels of autolysin have been described and, in a few cases, shown to be associated with reduced cell wall turnover (2,3). To extend and quantify the correlation between autolysin content and wall turnover we examined several lytic reduced strains of *B.subtilis* and a hypermotile mutant. We report a very rapid rate of wall turnover of the latter which is associated with a slower rate of mass increase relative to wild type and discuss the implications for cell surface expansion in *B.subtilis*.

Results

Autolysin activity and cell wall turnover in mutants affected in motility.

The rate of cell wall turnover of continuously labelled cultures of mutants affected in motility was compared with the autolysin activity extracted by 3M LiCl from broken cell preparations of similarly grown cultures.
Lyt-1 and lyt-15 (Table 1), in agreement with a previous report (3), had reduced autolysin contents as did strain TS51, having respectively about 1, 12 and 24% of the wild type content. All

The Target of Penicillin

Table 1. Autolysin content and cell wall turnover

Strain[a]	Relative autolysin[b] %	Wall turnover[c]	Motility[d]
lyt-1	≃ 1	1	-
lyt-15	12	10	-
flaC51	24	20	∓
lyt^+	100	36	+
ifm	190	69	+++

a. The non-motile lytic deficient strains lyt-1 (3), lyt-15 (2), and the ts-flagellaless mutant TSC51 (5) have been described. The supermotily (ifm) mutant was obtained as described (5).
b. Measured at pH8 with SDS inactivated walls of the lyt^+ strain as substrate. Activity (%) relative to lyt^+.
c. Cultures, grown on a glucose/casein hydrolysate medium (6) at 45C were labelled for 5 generations with ^{14}C-N-acetylglucosamine before being filtered and resuspended in unlabelled medium to measure release of radioactivity due to turnover. Rate relative to lyt-1.
d. Measured in semi-solid medium (5) at 45C.

these non motile strains showed a reduced rate of cell wall turnover. The wild type and supermotile strains, containing about 100 and 200 times as much autolysin as lyt-1, have turnover rates 36 and 70 times higher respectively. Over the whole range of strains from lytic deficient to superlytic there is a strong correlation between autolysin content and the rate of cell wall degradation in vivo. It is, however, among the three lytic reduced strains that these parameters show the most remarkable parallel, over a 20 fold range. This strongly suggests that either all or a fixed part of the autolytic activity is actively degrading the cell wall in growing cultures of these strains : the wall bound autolysin providing the rate limiting step in the release of soluble wall fragments. We suggest therefore that reduction in the rate of wall turnover is not only a sign of a reduced cell autolysin content but that the absolute rate of turnover may be a remarkably good indicator of that content over a 20 fold range at least. Evidently, this conclusion is limited to strains having no significant differences in the wall substrate, which is true for the above strains as shown by a similar rate of lysis of autolysin-free wall preparations. In addition, the three lytic reduced strains and other non motile mutants map in at least 5 different loci on the B.subtilis chromosome (in preparation). Therefore, the correlation observed is not confined to mutations in one single gene but is found in mutations at many sites. We suggest that the low turnover, lytic reduced, flagellaless phenotype is associated with mutations in many different fla genes. Fein proposed (7) that autolytic activity might be necessary for flagella insertion into the cell wall, and suggested that the non-motility of lyt-1 and lyt-15 is due to deficiency in autolytic enzymes. We believe, on the contrary, that a block in flagella formation represses production of autolysin.

Cell wall growth and turnover in a superlytic (ifm) strain

While there are several studies of turnover in autolysin reduced strains it seemed interesting to examine a strain showing substantial increases in the content of autolysins and in the rate of wall turnover.

Minimal medium grown cells, with uniformly labelled cell walls (Fig.1), during the first generation after chasing, lost over 70% of the cell wall label to the medium (as compared to 28% for the wild type). This extremely rapid wall turnover, the highest yet reported, has implications for any model of cell wall expansion. It was necessary to confirm that the rapid release of wall fragments was not due to lysing cells. Following a chase of extensively labelled (^{14}C-leucine) cultures of the ifm and wild type strains, a closely similar rate of loss of label to the medium excluded the possibility of significant cell lysis by the ifm strain. The loss of over 70% of old wall means that the majority of the cell surface must be renewed during each generation i.e. between the birth of a cell and the following division not only must wall be synthesised to accomodate the increase in cell surface but the major part of the wall present at the cell's birth must be replaced as well. It may well be that most of remaining label is confined to cell poles which may turn over more slowly than cylindrical wall ; calculations for B.subtilis gro-

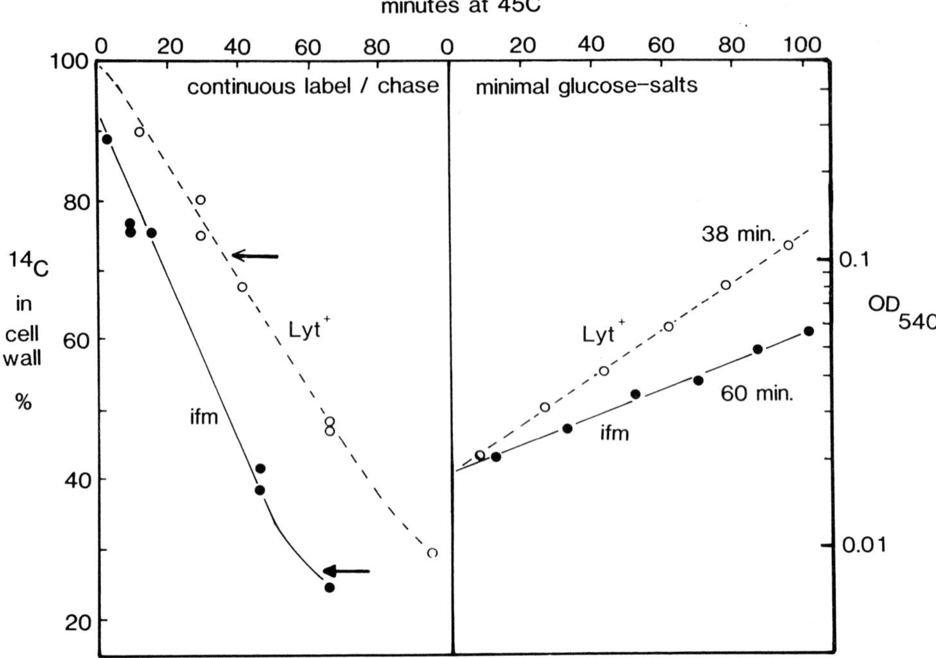

Fig.1. Growth and labelling were as described in Table 1. ^{14}C in the cell wall fraction was calculated by subtracting label incorporated into protein (2). Arrows indicate wall label (%) solubilised after 1 generation. Generation times are shown on the right-hand side.

wing on a similar medium suggest that ca. 26% of the wall is found in poles (8). This would imply that, one generation after chasing, there is no label left in cylindrical wall.

There is no reason to believe that the ifm strain has a wall less thick than that of the wild type, the ratio of cell wall radioactivity to cell mass being the same or higher in the former. Therefore, to maintain the same wall thickness as in the wild type the rate of cell wall synthesis relative to that of overall mass increase must be substantially higher to compensate for loss due to turnover. While the ifm strain has a more rapid wall turnover it's overall growth rate is conspicuously slower than the wild type (Fig.1). Since this is also true for backcrossed ifm transformants it is probable that these two parameters are due to the same mutation(s). We can only speculate on the reasons for rapid cell wall turnover being linked to slower growth. Insertion of new cell wall in B.subtilis is localised at a limited number of sites (6) and if the rate per site in the wild type were already close to the maximum, reduction in the rate of surface expansion in the ifm strain might be the only way of maintaining the same wall thickness as in the wild type. Alternatively, at the wild type rate of surface expansion walls of the mutant might have insufficient rigidity at high rates of autolytic solubilisation. Whatever the explanation, one of the consequences of the high turnover rate is that new wall should reach the outer surface in a shorter time relative to the wild type. There is strong evidence that new wall incorporated near the surface wells up through the thickness of the wall layer (4,9). The time for this displacement corresponds to the lag before a pulse label becomes subject to cell wall turnover (4). We found that the absolute time of the lag before a pulse label starts to turn over was the same in both ifm and wild type strains despite the difference in their growth rates. It follows that in the ifm strain, relative to the lyt^+ strain, there is a smaller increase in surface area by the time that a pulse label reaches the outside of the wall. It is tempting to think that a more rapid inside to outside movement might be achieved at the expense of lateral growth.

It was recently shown (10) in lytic deficient strains that old DNA strands co-segregate with large conserved blocks of old wall implying a physical link between these components. In the ifm strain this association must be able to survive a rather rapid rate of renewal af the cell wall components involved, unless, which appears unlikely, there is no permanent association between the chromosome and the cell surface in this strain.

References

1. Mauck,J., Chan,L., Glaser,L.: J.Biol.Chem 246,1820-1827 (1971)
2. Pooley, H.M. : J. Bacteriol 125, 1127-1138 (1976)
3. Fein, J.E., Rogers, H.J.: J. Bacteriol 127, 1427-1442 (1976)
4. Pooley, H.M.: J. Bacteriol 125, 1139-1147 (1976)
5. Grant,G.F., Simon,M.I.: J. Bacteriol 99, 116-124 (1969)
6. Schlaeppi,J.-M., Pooley,H.M., Karamata, D.: J. Bacteriol 149 329-337 (1982)
7. Fein, J.E.: J. Bacteriol 137, 933-946 (1979)
8. Burdett, I.D.J., Higgins,M.L.: J. Bacteriol 133, 959-971 (1978)
9. Archibald, A.R. : J. Bacteriol 127, 956-960 (1976)
10. Schlaeppi, J.-M., Karamata, D.: J. Bacteriol 152, 1231-1240 (1982)

B. SUBTILIS W23/168 RECOMBINANTS WITH HYBRID CELL WALLS AND PHAGE RESISTANCE PATTERNS.

Michel Monod, Harold Pooley and Dimitri Karamata
Institut de génétique et biologie microbiennes
Rue César - Roux 19
CH - 1005 Lausanne

Introduction

Teichoic acids of cell walls of Gram positive bacteria form part of the receptors of many bacteriophages (1). Bacillus subtilis strains 168 and W23 contain different teichoic acids and show a very different pattern of resistance to a wide range of bacteriophages (2).
The transfer of a bacteriophage receptor from W23 to 168 (2,3) by transformation suggests the intriguing possibility that the teichoic acids of the hybrid transformants might be different from those of the 168 parent. Study of such hybrid strains could provide new insights into the role of the anionic polymers in cell wall growth and metabolism.
Selection for phage resistance phenotypes offers an easy way of obtaining hybrid transformants with resistance patterns intermediary between the parent types.
We report here the isolation, and analysis (genetic and wall chemistry) of several W23/168 hybrid strains. We discuss the implications of these findings on cell wall metabolism and the nature of phage receptors.

Results

1) Isolation of W23/168 phage resistance hybrids. For the isolation of phage resistance hybrids we exploited the fact that W23 is resistant to several bacteriophages to which 168 is sensitive (2).
Competent cells of 168 (ura-26 hisA1 argC4 metC3) - which is sensitive

The Target of Penicillin

to phages Ø29, Ø25, SPP1, PBSY, PBSZ but resistant to PBSX which it harbours - were transformed with saturating concentrations of DNA from a W23 prototroph - sensitive to PBSX, PBSY and Ø25 but resistant to the other phages mentioned above including PBSZ which it harbours. After an 18 hr selection of ura^+ recombinants in liquid medium ca. 10^7 cells were spread onto an LA plate together with either an SPP1 or a Ø25 phage stock (m.o.i. of 10).

SPP1 resistant transformants. Although congression for a reference marker was 7%, SPP1 resistant ($SPP1^r$) colonies appeared at a low frequency (4 x 10^{-4}) they were of two types, 70% being also resistant to Ø29. Among 500 $SPP1^r$ $Ø29^r$ recombinants not a single PBSX sensitive or Ø25 resistant colony was found. Spontaneous $SPP1^r$ mutants occurred at a 10 fold lower frequency and were never $Ø29^r$. Therefore the $SPP1^r$ $Ø29^r$ phenotype is a hybrid resulting from transformation with W23 DNA.

In a back cross, the $SPP1^r$ $Ø29^r$ phenotype was transferred like a single marker with a frequency comparable to that of auxotrophic markers. This phenotype is clearly different from gtaA (non glucosylated teichoic acid), the only known marker with a $Ø29^r$ $Ø25^S$ phenotype, since no linkage could be established between them by transformation. In addition we have isolated several $Ø29^r$ gtaA mutants, all of which were SPP1 sensitive. Thus a new locus determining resistance to SPP1 has been found which appears to show no changes in teichoic acid composition (see below). This type of resistance can be overcome by SPP1 host range mutants which occur spontaneously in SPP1 stocks at a frequency of 10^{-7}.

Ø25 resistant hybrids. We selected for Ø25 resistance since the $PBSX^S$ hybrid, RUB 824, was shown (2) to be resistant to Ø25 in addition to SPP1 and SP02-cl2. Ura^+ transformants (see above) contained 3 types of Ø25 resistant cells ; those resistant to PBSY, PBSX and to PBSZ (I), those resistant to PBSY and PBSX, but sensitive to PBSZ (II), and those resistant to PBSY, but sensitive to PBSX and PBSZ (III).

The frequency of type III was only about 1 in 10^{-4} prototrophs. Transformation with homologous (168) DNA or selection of spontaneous 168 $Ø25^r$ resistant mutants yielded types I and II only, suggesting that type III colonies, whose phenotype is clearly intermediary (Table 1) between the two parent strains, are hybrid transformants and confirming the transfer of the PBSX receptor to the normally resistant 168 reported earlier (3). In back crosses, the PBSX^{s} and Ø25^{r} characters were transferred as one single marker, no separation taking place, suggesting that Ø25 resistance and sensitivity to PBSX are due to the same genetic determinant. While the PBSX^{s} Ø25^{r} phenotype was transferred by congression at a 5 fold lower frequency than several auxotrophic markers, this is still 3 orders of magnitude more frequent than the proportion obtained in the original W23 x 168 cross.

2) Mapping of the hybrid phenotype. PBS1 mediated transduction showed that the PBSY^{s} Ø25^{r} phenotype was cotransducible with hisA (about 30%). The likelihood of a location in the region containing markers involved in the glucosylation of the glycerol phosphate teichoic acid of 168, prompted us to seek for linkage to markers gtaB, gtaA, rodA, and rodC. The PBSX^{s} hybrid was donor in all crosses except in that with rodC. Transformants for an auxotrophic marker were selected and screened for the inheritance of relevant donor markers. While the recombination index between PBSX^{s} and gtaB was about 1, not a single wild type transformant was obtained with all the remaining markers suggesting strongly that DNA from W23 replaced 3 out of 4 markers known to affect teichoic acid synthesis in 168 i.e. gtaA, rodA (50% linked to gtaA by transformation) and rodC (unlinked to rodA by transformation, see 4). Although it is not certain that the whole region between the extreme markers (gtaA and rodC) has been exchanged transfer of several genes from W23 is highly probable.

3) Genetic analysis of RUB 824. DNA from RUB 824 was back crossed into 168 ura-26 hisA1 argC4 metC3. Among 540 his^{+} transformants, 15% were SPP1^{r} Ø29^{r}, 3% were Ø29^{r} Ø25^{r} PBSY^{r} and PBSX^{s}, while 0.5% (3 colonies only) had the parent phenotype. Segregation of two unlinked types of phage resistance took place, corresponding exactly to those which we have isolated by selecting independently for SPP1 resistance and for Ø25 resistance. Their presence in RUB 824 is readily explained by the selection of transformants resistant to both SPP1 and SP02-C12 (resistance to the latter is almost certainly correlated with resistance to Ø25). The requirement for transformants which have inherited two genetically unlinked resistance markers accounts for the low frequency (10^{-7}) of RUB 824 type strains (2).

Table 1. Bacteriophage sensitivity (+) and resistance (-) of parent and hybrid strains of B.subtilis

	Parent strains			Hybrid strains			
				$PBSX^S$	$Ø25^r$	(type III)	$SPP1^r$
Bacteriophage	168	S31	W23	T 1	N 10	RUB 824	13
PBSX	-	+	+	+	+	+	-
PBSY	+	-	+	-	-	-	+
PBSZ	+	+	-	+	+	+	+
Ø29	+	-	-	-	-	-	-
Ø25	+	+	+	-	-	-	+
SPP1	+	-	-	+	+	-	-

A concentrated phage stock was spotted onto a streak of each strain on minimal agar.

Table 2. Teichoic acid components (μM/mg) of the cell walls of parent and hybrid strains of B.subtilis

	Parent strains			Hybrid strains			
				$PBSX^S$	$Ø25^r$	(type III)	$SPP1^r$
Component	168	S31	W23	T 1	N 10	RUB 824	13
Glycerol	0.72	0.16	0.06	0.17	0.08	0.11	0.93
Ribitol	-	0.50	0.56	1.15	0.57	0.66	-
Galactose	-	-	0.10	0.09	0.07	0.06	-
Glucose	0.39	0.42	0.36	0.08	0.07	0.065	0.39
Glc-NH2	0.29	0.29	0.22	0.40	0.21	0.14	0.10
Gal-NH2	0.08	0.086	0.01	0.07	0.07	0.05	0.15
PO_4^{3-}	1.5	not done	1.43	1.26	1.8	1.4	1.55

Cell wall preparations were hydrolysed (3hr at 100°C with 2NHCl under N_2). After neutralisation and treatment with alkaline phosphatase, sugars and amino sugars were reduced with $NaBH_4$, converted to alditol acetates (5) and separated by GC.

4) Chemical analysis of teichoic acids in W23/168 hybrid strains. The transfer of the genetic determinants for the PBSX receptor from W23 to 168 - strains presenting differences in their teichoic acids - suggests that the teichoic acids of the hybrid transformants might be different from those of the 168 recipient.

Chemical analysis of cell wall preparations (Table 2) of $PBSX^S$ hybrids can be summarized as follow :

(i) Ribitol was present, in quantities characteristic of W23.

(ii) Glycerol was present at a 10 fold lower concentration than ribitol, as in W23.

(iii) A substance with a chromatographic mobility identical to that of galactose was found in both W23 and hybrid strains.

(iv) In contrast, galactosamine was found in amounts characteristic of 168 i.e. 5 to 10 times higher than in W23.

(v) The glucose content was strongly reduced as compared to both parent strains.

The wall composition, including glucose content, of an $SPP1^r$ $Ø29^r$ transformant was identical to that of the 168.

Conclusions and Discussion.

We have shown that the spectrum of phage resistance of W23/168 hybrid strains is associated with a similarly hybrid cell wall composition : our results indicate that in the $PBSX^S$ hybrids the ribitol phosphate polymer of *B.subtilis* W23 has replaced the glycerol phosphate polymer of *B.subtilis* 168. This substitution would require the transfer of genes coding for a whole group of enzymes involved in the polymerisation of ribitol phosphate. Genetic analysis shows indeed that the region exchanged includes several genes.

In the whole formed by the different polymers making up the Gram positive cell wall it is at first sight most surprising that replacement of the major teichoic acid by a chemically distinct polymer does not seem to disturb any of the interlocking functions of the envelope i.e. surface expansion, cell division, motility, or cell wall turnover all of which appear normal despite the presence of a "foreign" teichoic acid. The replacement of the glycerol phosphate polymer by the ribitol phosphate one has many implications for our understanding of cell wall synthesis and the roles played by different cell wall polymers. It is interesting that we did not find hybrid strains containing both glycerol and ribitol polymers in comparable amounts. Could this mean that one of the major functions of the cell wall is disturbed by their simultaneous presence or do the genes involved in poly(ribitol phosphate) synthesis and those involved in poly(glycerol phosphate) synthesis form homologous loci on the W23 and 168 chromosomes respectively, so that genetic exchange in this region would automatically lead to the loss of genes essential for the 168 specific teichoic acid synthesis ?

The present results confirm the importance of teichoic acids in bacteriophage receptors. Comparison of resistance patterns of parent and hybrid strains (Table 1) with their teichoic acid content strongly suggests that ribitol phosphate, present in W23, S31 as well as in all PBSX sensitive hybrid strains, forms part of the PBSX receptor. Glaser et al (6) described a PBSX resistant mutant of W23 whose walls contained only the non-glucosylated form of the ribitol phosphate polymer. However the absence of glucosylated teichoic acid is not necessarily accompanied by PBSX resistance since our hybrid W23/168 strains are PBSX sensitive despite an apparently non glucosylated ribitol phosphate.

Comparison of Table 1 and 2 shows that resistance and sensitivity to PBSZ are correlated respectively with the absence and presence of galactosamine, suggesting that a galactosamine containing polymer forms part of the PBSZ receptor.

References

1. Archibald, A.R.: Virus Receptors (Receptors and Recognition, Series B, 7) Ed. L.L. Randall, L. Philipson. Chapman and Hall, London 1980.
2. Yasbin, R.E., Maino, V.C., Young, F.E.: J. Bacteriol. 125: 1120-1126 (1976).
3. Yasbin, R.E., Ledbetter, M.: J. Virol. 25: 703-704 (1978).
4. Karamata, D., Mc Connell, M., Rogers, H.J.: J. Bacteriol. 111: 73-79 (1972).
5. Bulletin 774, Supelco, Inc. Bellafonte, Pa. U.S.A. (1977).
6. Glaser, L., Ionesco, H., Schaeffer, P.: Biochim. Biophys. Acta 124: 415-417 (1966).

PART IV

BIOLOGICAL PROPERTIES AND MEDICAL ASPECTS OF MUREIN

IMMUNOMODULATING PROPERTIES OF NATURAL AND SYNTHETIC MUROPEPTIDES *

Edgar Lederer

Laboratoire de Biochimie, C.N.R.S. 91190 Gif-sur-Yvette,
Institut de Biochimie, Université de Paris-Sud, Centre d'Orsay, 91405
Orsay, France.

Introduction

The biological properties of peptidoglycan derivatives are summarized in table I. The large number of different biological activities could have been ascribed to insufficient purity of some derivatives tested; but most of the biological effects mentioned in this table are also observed with pure, synthetic muramylpeptides.

Table I: Biologic Activities of Cell Walls and Related Compounds *in vivo* (from Kotani 1).

Modulation (mainly potentiation) of antibody-mediated and cell-mediated immune responses.
Stimulation of the reticuloendothelial system.
Increase of natural resistance to microbial infections and to tumor development.
Induction of experimental autoimmune diseases with or without external antigen (AEA or adjuvant arthritis).
Pyrogenicity.
Induction of acute inflammatory reaction and chronic granulomatous lesions in various tissues. Increase in vascular permeability.
Immunogenicities.

* Muropeptides: peptides, containing typical cell wall constituents, such as muramic acid, D-glutamic acid or diaminopimelic acid (DAP). Three categories have immunomodulating properties:
1) muramylpeptides, such as MurNAc-L-ala-D-isogln (MDP)

2) desmuramylpeptides which contain no muramic acid, but DAP instead.

3) desmuramylpeptides containing neither muramic acid nor DAP but lipophilic D-glu oligopeptides such as L-ala-D-isogln-L-ala-glycerol-mycolate.

The Target of Penicillin

MDP and Derivatives

MDP (N-acetyl-muramyl-L-alanyl-D-isoglutamine) 1 , the first synthetic peptidoglycan derivative capable of replacing whole mycobacteria in Freund's complete adjuvant, has been extensively studied since it was described in 1974 (2). MDP is highly active in stimulating antibody production against an antigen injected simultaneously (in the presence of oil, or even in aqueous solution) and produces delayed hypersensitivity against the antigen and stimulates nonspecific resistance. Structure-activity relationships and the mechanisms of action have been reviewed (3-5).

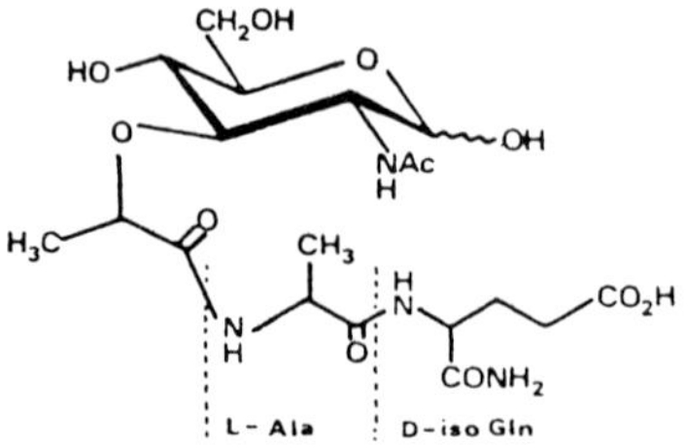

1 MDP

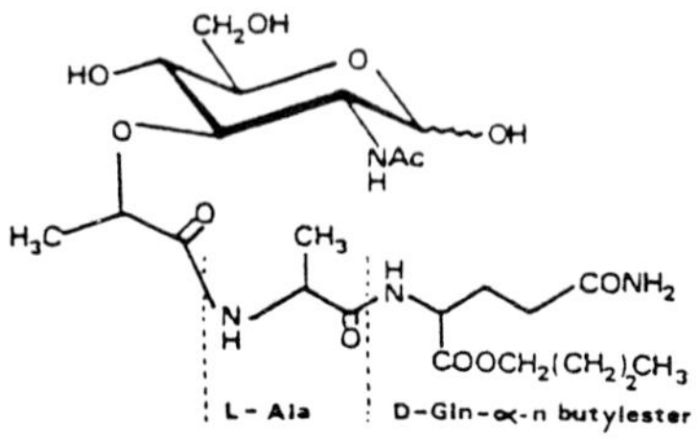

2 Murabutide

Chemical Modifications of MDP

Several structural modifications can increase the biologic activities of MDP. Especially lipophilic derivatives, such as 6-O-acyl esters of MDP (6) or MDP-L-ala-glycerol-mycolate (7) or, MDP-L-ala-phosphatidyl-ethanolamine (8), have been shown to be more active in stimulating non-specific resistance, especially when incorporated in liposomes.

MDP itself is pyrogenic but *murabutide* (N-acetyl-muramyl-L-alanyl-D-glutamine-n-butyl ester) 2 is much less pyrogenic than MDP, but just as active as adjuvant and just as potent for stimulating nonspecific resistance (9). It has successfully passed Phase I and II tests and is now in clinical experimentation.

Desmuramyl-peptides and Derived Peptidolipids

Muramic acid is not essential for immunological activities of murein derivatives.Some active desmuramyl-peptides contain the peptide D-Glu-DAP as essential moiety. The free peptides are usually inactive and need acylation to become immunoadjuvant. Thus, laurylation of a cell wall tetrapeptide from a *Streptomyces* strain gave N^2-[N(lauroyl-L-alanyl)-γ-D-glutamyl]-N^6-glycyl-DD,LL-diamino-2,6,pimelamic acid, which is as adjuvant as MDP and protects mice against *Listeria monocytogenes* (10). An immunoactive peptide FK-156 from a culture of *Streptomyces olivaceogriseus* was found to be D-lactyl-L-alanyl-γ-D-glutamyl(L)meso-diaminopimelyl(L) glycine (11). Even stearoyl-D-glu-meso-DAP is immunostimulant. A highly lipophilic desmuramyl tripeptide: L-ala-D-isogln-L-ala-glycerol-mycolate does not stimulate humoral antibody production, but strongly increases non-specific resistance against bacterial infections (7).

Enhancement of non-specific resistance by MDP

Treatment with MDP or analogs enhances resistance in mice to pathogens, such as *Klebsiella pneumoniae, Pseudomonas aeruginosa, Candida albicans, Salmonella typhimurium, Streptococcus pneumoniae, Trypanosoma cruzi,* and *Toxoplasma gondii* (12). This effect is probably due to the activation of macrophages. It is possible that such an activation requires the presence of other cells of the immune sytem. MDP and its derivatives may prove useful for stimulating immunodepressed patients; it has also been recommended for combined use with antibiotics and for treatment of malnourished individuals (8).

MDP in Tumor Immunology

MDP or its derivatives could find applications as adjuvants, for tumor antigens. In a particular experimental case (the line 10 hepatocarcinoma inoculated in strain 2 guinea pigs), intralesional injection of an emulsion of MDP, or a derivative, with TDM (trehalose-dimycolate or "cord factor") in the presence of 2% paraffin oil, or better, squalane, leads to a complete disappearance of the tumor and metastases and to immunity of the animals against a second challenge with the same tumor line (13,14). Another promising approach has been developed by Fidler (15), who has shown that MDP (or,better, MDP-L-ala-phosphatidyethanolamine) entrapped in specially designed multilamellar liposomes can be used to activate alveolar macrophages, leading to the destruction of pulmonary metastases of a tumor inoculated in mice. For a review see (16).

MDP in Synthetic Vaccines

The extensive studies of Sela et al (17),and Chedid et al (18), have shown that it is possible to obtain strong immune responses with synthetic deca- to octa-decapeptides representing the immunodeterminant moiety of a hormone, a virus, or a bacterial toxin. The latter alone are not sufficiently immunogenic; they can be coupled to a synthetic carrier, such as poly-DL-ala-poly-L-lysine. The resulting larger molecules can then be used as antigen, either by injecting them as such, in the presence of MDP, or by coupling them chemically to MDP. In some cases even the combination of an antigenic decapeptide with MDP-Lys can give strong antibody titers (19). This approach to fully synthetic, pure vaccines has a special appeal for antiviral vaccines.

MDP and Derivatives as Sleep Factors

"Sleep factor" is a trace compound found in human brain and urine that prolongs slow wave sleep, by intracerebroventricular injection into rabbits. From 3,000 litres of human urine, 20μg of pure "sleep factor" were isolated and found to contain muramic acid, alanine, glutamic acid, and diaminopimelic acid (20). Synthetic MDP was then found to be active as well (21). Here again, the D-D isomer and the L-L isomer of MDP are inactive. These studies could lead to orally active MDP derivatives with sleep-promoting activity.

Metabolic Fate of MDP and Derivatives *in vivo*:MDP as vitamins.

Injected in saline, MDP ^{14}C-labelled on the C_1 of the lactyl group of muramic acid, is rapidly eliminated. Thirty minutes after intravenous injection, more than 50% is recovered unchanged in the urine, and 95% in 2 hours (22). MurNAc-L-ala-D-glu-γ-DAP ^{14}C-D-ala-^{14}C-D-ala gives mainly the tetrapeptide D-glu-γ-meso-DAP-^{14}C-D-ala-^{14}C-D-ala. With MurNAc-L-ala-D-glu-γ-^{14}C-meso-DAP, the main product found in the urine is ^{14}C-meso-DAP (23). A uniformly ^{14}C-labelled disaccharide pentapeptide GlcNAc-Mur-NAc-L-ala-D-isogln-γ-meso-DAP-D-ala-D-ala is also rapidly eliminated, the corresponding disaccharide and pentapeptide being found in the urine (24). There are thus several enzymes in mammalian organisms that recognize various muramyl peptides as specific substrates and which might contribute to the regulation of the immune status (25).

A last word about the general importance of peptidoglycan derived muramyl peptides: it is well known that germ-free animals are highly susceptible to infection; their immune system is practically non-efficient and needs stimulation by bacterial products to become fully operative. We are thus led to consider muramyl peptides as vitamins: trace compounds, derived from the food (or the intestinal flora) and indispensable for the normal health (immune status and sleep) of our organism.

References

1. Kotani, S.: Award Lecture, the first Behring Kitasato prize, p.41 (1980).
2. Ellouz, F., Adam, A., Ciorbaru, R., Lederer, E.: Biochem. Biophys. Res. Comm. 59, 1317-1325 (1974).
3. Parant, M.: Springer Seminars Immunopathol. 2, 110-118 (1979).
4. Lederer, E.: J.Med.Chem. 23, 819-825 (1980).
5. Adam, A., Petit, J.F., Lefrancier, P., Lederer, E.: Mol.Cell.Biochem. 41, 27-47 (1981).
6. Azuma, I., Sugimura, K., Yamawaki, M., Uemiya, M., Kusumoto, S., Okada, S., Shiba, T., Yamamura, Y.: Infect. Immun. 20, 600-607 (1978).
7. Parant, M., Audibert, F., Chedid, L., Level, M., Lefrancier, P., Choay, J., Lederer, E.: Infect. Immun. 27, 826-831 (1980).
8. Sackmann, W., Dietrich, F.M.: in Current Chemother. Immunother. Amer. Soc. Microbiol. Washington, 2, 1162-1164 (1982).
9. Chedid, L., Parant, M., Audibert, F., Riveau, G., Parant, F., Lederer, E., Choay, J., Lefrancier, P.: Infect. Immun. 35, 417-424 (1982).
10. Werner, G.H., Floc'h, F., Bouchaudon, J., Zerial, A., Migliore-Samour, D., Jollès, P.: in Current Concepts in Human Immunol. Cancer Immunomod. 645-662 (1982), Elsevier Biomed. Press.

11. Gotoh,T.,Nakahara,F.R.,Nishiura,T.,Hashimoto,M., Kino,T., Kuroda,Y., Okuhara,M., Kohsaka,M., Aoki,H.,Imanaka,H.: J.Antibiotics, 35, 1286-1292 (1982).

12. Fraser-Smith,E.B.,Waters,R.V., Matthews,T.R.: Infect.Immun. 35, 105-110 (1982).

13. McLaughlin,C., Schwartzman,S.M.,Horner,B.L.,Jones,G.H.,Moffat,J.G., Nestor,J.J.,Tegg,D.: Science,208, 415-416 (1980).

14. Yarkoni,E., Lederer,E.,Rapp;H.J.: Infect.Immun.32, 273-276 (1981).

15. Fidler,I.J.,Poste,G.: Springer Seminars Immunopathol. 5, 161-174 (1982)

16. Lederer,E.,Chedid,L.: in Immunological Approaches to Cancer Therapeutics. E.Mihich Ed., Wiley-Inters, J.Wiley & Sons, 107-135 (1982).

17. Langbeheim,H.,Arnon,R.,Sela,M.: Proc.Natl.Acad.Sci.USA 73, 4636-4640 (1971).

18. Audibert,F.,Jolivet,M.,Chedid,L.,Arnon,R., Sela,M.: Natl.Acad.Sci. USA 79, 5042-5046 (1982).

19. Carelli,C.,Audibert,F.,Gaillard,J.,Chedid,L.: Proc.Natl.Acad.Sci. USA, 79, 5392-5395 (1982).

20. Krueger,J.M., Pappenheimer,J.R., Karnovsky,M.L. J.Biol.Chem. 257, 1664-1669 (1982).

21. Krueger,J.M., Pappenheimer,J.R., Karnovsky,M.L.:Proc.Natl.Acad.Sci. USA 79, 6102-6106 (1982).

22. Parant,M.,Parant,F.,Chedid,L.,Yapo,A., Petit,J.F.,Lederer,E.: Int.J. Immunopharmac. 1, 35-41 (1979).

23. Yapo,A.,Petit,J.F.,Lederer,E.,Parant,M.,Parant,F., Chedid,L.: Int.J. Immunopharmac. 4, 143-149 (1982).

24. Valinger,Z., Ladesić,B.,Tomasič,J.: Biochim.Biophys.Acta 701,63-71 (1982).

25. The muramidase lysozyme, or even pronase, or enzymatically digested cell walls given per os, have an immunostimulating effect on guinea pigs. Namba,Y., Hidaka,Y., Taki,K., Morimoto,T.: Infect. Immun. 31, 580-583 (1981).

Acknowledgments: The work of the author was supported by CNRS and, in part by grants from DGRST, INSERM, Fondation de la Recherche Médicale, Ligue Nationale Française contre le Cancer, The Cancer Research Institute, New York, and a contract with ANVAR, Laboratoires Choay, Institut Pasteur and Sanofi. The author thanks Dr.Arlette Adam for critical reading of the manuscript.

THE IMMUNOCHEMISTRY OF PEPTIDOGLYCAN

Peter H. Seidl, Norbert Franken, Karl H. Schleifer
Lehrstuhl Mikrobiologie, Technische Universität München
D-8000 München

Introduction

Peptidoglycan reveals several striking biological properties such as adjuvant activity, endotoxin-like properties, arthritogeneity and immunogenicity. Specific antibodies are produced to peptidoglycan in animals and in man. The immunochemical aspects of peptidoglycan and the biological role of peptidoglycan antibodies have been studied.

Results and Discussion

Antigenic Determinants of Peptidoglycan Characterized in Animal Antisera

Three antigenic determinants of peptidoglycan have been characterized in animal antisera to peptidoglycan:

a) Antigenic properties of the pentapeptide of the peptide subunit L-Ala-D-Glu(L-Lys-D-Ala-D-Ala).

The pentapeptide L-Ala-D-Glu(L-Lys-D-Ala-D-Ala) and not the predominantly occurring tetrapeptide L-Ala-D-Glu(L-Lys-D-Ala) was determined as the antigenic determinant of the peptide subunit. The antibodies are directed to the COOH terminus of the pentapeptide, and D-alanyl-D-alanine is the immunodominant group. The contribution of D-Glu, L-Lys to the antigenic site is of less importance, that of L-Ala is insignificant.

The Target of Penicillin

b) Antigenic properties of the glycan strand
Specific antibodies to the glycan moiety of peptidoglycan were found frequently in animal antisera to Gram-positive cocci. Normally, antibodies are directed to N-acetylglucosamine and not against N-acetylmuramic acid as the determinant sugar. The antigenic site was in the size of a tetrasaccharide. Antisera to Micrococcus luteus, however, contained additional antibodies highly specific to N-acetylmuramic acid and not to N-acetylglucosamine. A free carboxyl group of muramic acid was absolutely necessary for binding of the antibodies. These data are consistent with primary structure of M. luteus peptidoglycan. About 50% of the muramic acid residues in M. luteus peptidoglycan are not substituted by peptide subunits, and exhibit an unsubstituted carboxyl group.

c) Antigenic properties of the interpeptide bridge
The interpeptide bridges of peptidoglycan are also immunogenic. Specific antibodies reacted with the oligoglycine interpeptide bridge of staphylococci and with the oligo-L-alanine interpeptide bridge of streptococci or micrococci, respectively. No cross-reactivity was found at all between the oligo-glycine and oligo-L-alanine interpeptide bridges. Specific antibodies against the interpeptide bridges can therefore be employed for the serological differentiation of bacteria containing different peptidoglycan types.

The antigenic determinants of peptidoglycan in animal antisera were predominantly characterized by studying the quantitative precipitin reaction and its inhibition. Bacterial antisera containing antibodies to peptidoglycan or antisera to synthetic protein-peptide-conjugates (Fig. 1) were used as the antibody source. Highly purified peptidoglycans with different primary structures or peptidyl-proteins carrying peptide substituents with structural similarity to peptidogglycan fragments (compare Fig. 1) were employed as an antigen. For a review of these studies see reference 1.

a) Gly-Gly-Gly-Gly-Gly- -Ahx-
Gly-Gly-Gly-Gly-Gly- -Ahx- Protein

b) Protein $-CH_2CO$-Gly-L-Ala-L-Ala-D-Ala-D-Ala
$-CH_2CO$-Gly-L-Ala-L-Ala-D-Ala-D-Ala

Fig. 1. Scheme of synthetic antigens carrying peptide substituents with similarity a) to the oligoglycine interpeptide bridge of staphylococi; b) to the peptide subunit pentapeptide L-Ala-D-Glu(L-Lys-D-Ala-D-Ala).

Crossreactivity of Peptidoglycan Peptide L-Ala-D-Glu(L-Lys-D-Ala_2) with ß-Lactam Antibiotics

Due to proposed structural analogy between peptidoglycan precursor peptides R-D-Ala_2 and the penicillin molecule (2), cross-reactivity of peptides R-D-Ala_2 and ß-lactam antibiotics was examined. Binding of ^{125}iodolabelled L-Ala-D-Glu(L-Lys-D-Ala_2) to antibodies with specificity for peptides R-D-Ala_2 was not only inhibited by peptides with the C-terminal sequence R-D-Ala-D-Ala (e.g. Ac_2-L-Lys-D-Ala_2; L-Ala-D-Glu(L-Lys-D-Ala_2), but also strongly by ß-lactam antibiotics. Moreover, IgG isolated from antisera to peptides R-D-Ala_2 bound specifically to ^{35}S-penicilline. The data are summarized as follows: Antibodies to the precursor sequence R-D-Ala-D-Ala specifically bound penicillins and cephalosporins (35 investigated). 6-aminopenicillanic acid or 7-aminocephalosporanic acid were characterized as the immunodominant group of the ß-lactam molecules. An intact ß-lactam ring was absolutely necessary. In average, specific binding of cephalosporins was slightly less than that of penicillins. Originally, cross-reactivity between peptidoglycan precursor sequence R-D-Ala_2 and the ß-lactam molecule was demonstrated using animal antisera to peptidoglycan or to synthetic antigens carrying peptides R-D-Ala_2 (compare Fig. 1b). Recently, specific binding of antibodies directed against L-Ala-D-Glu(L-Lys-D-Ala_2) to ß-lactam antibiotics was also demonstrated in human sera. However, due to the complex specificity of antibodies to the peptide subunit in human sera (see below: Antigenic determinants of peptidoglycan in human sera), binding occurs only in cases where antibodies to L-Ala-D-Glu(L-Lys-D-Ala_2) are exclusively or predominantly directed to the C-terminal sequence R-D-Ala-D-Ala. No binding was demonstrated in cases where D-Glu and/or L-Lys contributed significantly to the antigenic site and the contribution of the sequence R-D-Ala-D-Ala to the binding properties of antibodies directed against L-Ala-D-Glu(L-Lys-D-Ala_2) was of minor importance. In any case, serum levels of ß-lactam antibiotics necessary to demonstrate cross-reactivity are reached in therapy only in rare cases.

Secretion of Soluble Peptidoglycans into the Growth Medium by Growing Cells, Demonstrated by a Radioimmunoassay

Soluble peptidoglycans secreted into the medium by growing bacterial cells were quantitatively measured by a radioactive hapten binding inhibition assay. Binding of ^{125}iodo-labelled L-Ala-D-Glu(L-Lys-D-Ala_2) to specific antibodies directed against the sequence R-D-Ala-D-Ala were not only specifically inhibited by peptides R-D-Ala_2, but also strongly by culture filtates. As an antibody source, antisera to the synthetic antigen albumin-$(CH_2CO\text{-Gly-L-}Ala_2\text{-D-}Ala_2)_{39}$ was used (compare Fig. 1). The results are summarized as follows: All Gram-positive organisms investigated, including staphylococci, micrococci, streptococci, corynebacteria, revealing peptides R-D-Ala_2 in their peptidoglycans, secreted soluble peptidoglycans into the medium. In contrast, secretion was not detected in the culture filtrates of Gram-negatives including *E. coli*, *Enterobacter*, *Salmonella*, *Pseudomonas*.

As a negative control, culture filtrates of yeasts or of Gram-positive bacteria lacking peptides R-D-Ala_2 in their peptidoglycan (e.g. B. subtilis) were used. Thus, a very simple and sensitive test system (10 µl of culture filtrate are sufficient) is available to study the secretion of soluble peptidoglycans into the medium. This secretion is of great medical importance, since this material is discussed to be responsible for the biological activities of peptidoglycans including immunogenicity (3).

Specific Detection of Peptide Subunit Pentapeptides and of Interpeptide Bridges in the Cell Walls of Gram-Positive Bacteria by the Indirect Ferritin Technique.

The peptide subunit pentapeptide L-Ala-D-Glu(L-Lys-D-Ala_2), the pentaglycine interpeptide bridge of staphylococci and the L-Ala_{2-3} interpeptide bridge of streptococci were specifically detected in the cell wall of Gram-positive bacteria by the indirect ferritin technique, using monospecific antibodies. Monospecific antibodies to these peptidoglycan structures were obtained by immunizing rabbits with synthetic antigens (compare Fig. 1 and Ref. 1). The specificity of ferritin labelling was proved by immunoelectronmicroscopical controls and by correlation of ferritin labelling with the chemical cell wall structure of bacteria employed.

Biological Role of Antibodies to Peptidoglycan in Human Sera.

Peptidoglycan antibodies are frequent in human sera. In order to evaluate their biological role, specificity of these antibodies, immunoglobulin classes involved in immune response, change of titers during time course in normal human sera, elevation of titers in patients with Gram-positive infections and age dependency of titers is currently studied. The following data were obtained:

In human sera, L-Ala-D-Glu(L-Lys-D-Ala_2) is an antigenic determinant, as described for animal antisera. However, specificity is much more complex in human sera. Like in animal antisera, in some human sera antibodies to L-Ala-D-Glu(L-Lys-D-Ala_2) are exclusively or predominantly directed to C-terminal R-D-Ala_2 as an immunodominant group. In contrast to animal antisera, L-Lys and/or D-Glu significantly contribute to the antigenic site, the sequence R-D-Ala_2 playing a minor role. Moreover, in contrast to animal hyperimmune sera, L-Ala-D-Glu(L-Lys-D-Ala) was characterized as an additional antigenic determinant in human sera. Normally, no cross-reactivity between L-Ala-D-Glu(L-Lys-D-Ala) and L-Ala-D-Glu(L-Lys-D-Ala_2) was observed.

Immunglobulin classes.
Specific IgG and IgM antibodies were characterized in human sera to peptidoglycan peptides R-D-Ala_2. These data were obtained in ELISA-techniques using synthetic antigens (compare Fig. 1b) or by radioimmunological studies of isolated immunglobulins.

Titers to L-Ala-D-Glu(L-Lys-D-Ala_2) in human sera.
Study of titers to L-Ala-D-Glu(L-Lys-D-Ala_2) in normal human sera during time course revealed a remarkable constancy even over a period of 18 months. Titers to L-Ala-D-Glu(L-Lys-D-Ala_2) are clearly age dependent, the levels of the adults are reached at 14 years. Antibody titers to L-Ala-D-Glu(L-Lys-D-Ala_2) are strongly elevated in patients' sera with Gram-positive infections (e.g. pneumonia, streptococcal nephritis, osteomyelitis, lymphadenitis, scarlet fever)

References

1. Schleifer, K.H., Seidl, P.H.: Microbiology, 339 (1977).
2. Tipper, D.J., Strominger, J.L.: PNAS 54, 1133 (1965).
3. Zeiger, A.R., Wong, W., Chatterjee, A.N., Young, F.E., Tuazon C.U.: Infect. Immun. 37, 1112 (1982).

A NEW MUREIN HYDROLASE IN HUMAN SERUM

Stefan Mollner and Volkmar Braun
Mikrobiologie II, Universität Tübingen
D-7400 Tübingen

Introduction

In order to study the relevance of lysozyme for cell lysis and killing of E. coli in human serum, lysozyme was inactivated by antibodies. Lysozyme activity was tested by the release of soluble radioactivity from ^{3}H-Dap-murein sacculi. In spite of antibody inactivation of lysozyme a murein hydrolyzing activity was left in the serum.
This enzyme will be characterized in the following.

Results

Enzyme specifity

The muropeptides C5/C6* and X/X'** are substrates for the hydrolase which means that the specifity is different from lysozyme. The ^{3}H-Dap-labeled products after degradation of C5/C6 and X/X' show identical mobility on paper chromatography (1), suggesting that the cleavage must occur on the 3-O-D-lactyl ether bond of the muramic acid, or in the peptide side chain of C5/C6 and X/X' respectively (Fig. 1a, Fig. 1b).

* (N-acetyl-glucosaminyl-1,4-N-acetyl-muramyl-tri(tetra)-peptide)

** (1,6-anhydromuramyl derivate of C5/C6)

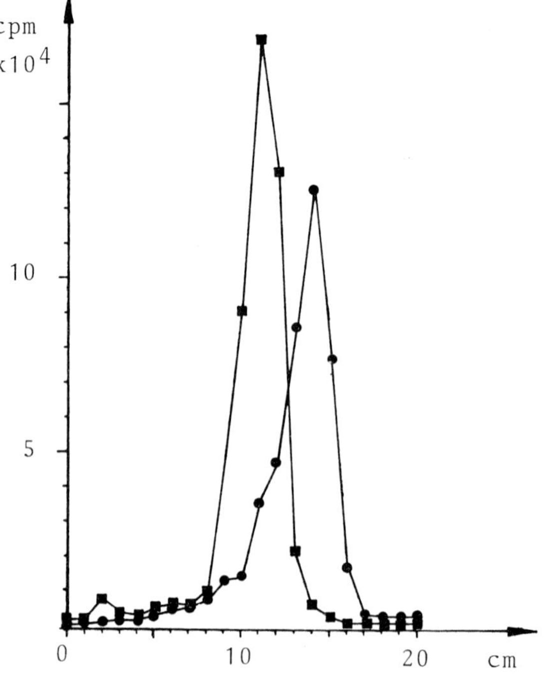

Fig. 1a Paper chromatography of ^{3}H-Dap C5/C6. ●—● C5/C6 in buffer only. ■—■ C5/C6 treated with enzyme.

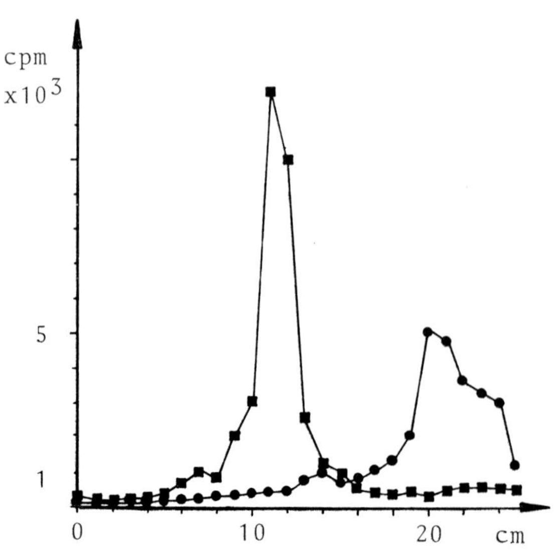

Fig. 1b Paper chromatography of ^{3}H-Dap X/X'. ●—● X/X' in buffer only. ■—■ X/X' treated with enzyme.

Tab. 1 Composition of the products in the various spots on the paper chromatogram of enzyme-cleaved peptidoglycan of E. coli.

Enzyme-released murein fragments of E. coli were separated by paper chromatography and their analysis showed that the enzyme is an amidase which cleaves the N-acetylmuramyl-L-alanine bond (muramic acid was not determined quantitatively).

Mobility (cm)	Glu	Ala	Dap	Glucosamine
		[nMol]		
0	9.10	12.76	5.40	20.10
2.5- 5.0	32.01	49.59	26.16	$<$0.05
9.0-10.5	18.17	18.03	14.52	$<$0.05
10.5-13.0	23.80	32.96	16.01	1.35

After degradation of murein sacculi with the enzyme all of the glucosamine and muramic acid stayed at the origin of the paper chromatogram indicating that the polysaccharide chains were not cleaved. However, alanine, diaminopimelate and glutamic acid were present as peptides in three spots of the chromatogram showing that they have been released from the glycan chains.

Partial purification

Human serum was dialysed against 50 mM Tris/HCl, pH 7.9 and separated on a DEAE - Sephadex ion-exchange column. The amidase-containing fractions were pooled and gel filtration was carried out on Sephadex G 200 SF. Polyacrylamide gel electrophoresis of the G 200 SF fractions showed a 85.000 band (Fig. 2b). The elution position of the activity corresponded with the appearence of this band (Fig. 2a). The molecular weight of the active enzyme (Fig. 3) was the same as the size determined on SDS gels.

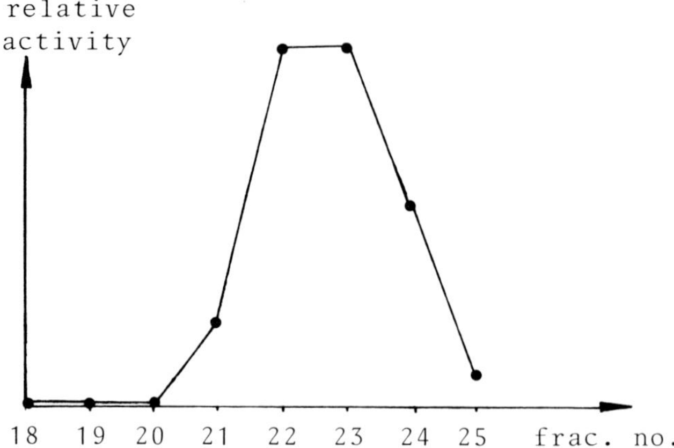

Fig. 2 a Determination of the enzyme activity of fractions from Sephadex G 200 SF. 20 µl of the fractions were incubated with aliquots of ^{3}H-Dap-murein sacculi of E. coli. The relative activity was determined as radioactivity released into the supernatant after centrifugation.

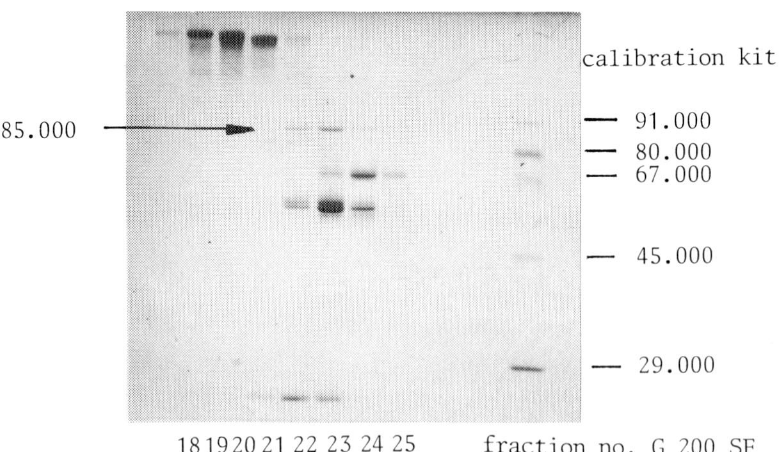

Fig. 2b Polyacrylamide gel electrophoresis (2) of fractions from gel filtration.

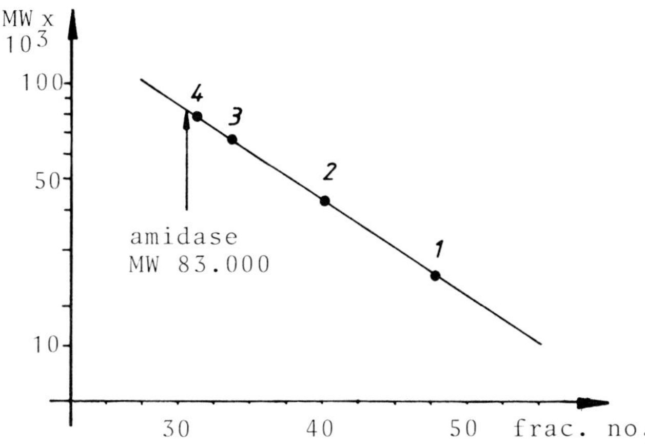

Fig. 3 Determination of the molecular weight by gel filtration on Sephadex G 200 SF. Calibration proteins: 1 chymotrypsinogen, 2 ovalbumin, 3 bovine serum albumin, 4 human transferrin. The calibration proteins and the amidase were separated by cochromatography. The amidase was identified by the release of radioactivity from ^{3}H-Dap-murein sacculi.

Bactericidal effects on E. coli

Figure 4 demonstrates lysis of E. coli ($5x10^8$/ml) treated with isolated amidase under conditions of osmotic shock (3). After a 8 min period of slow lysis the amidase-treated cells showed in contrast to the untreated cells enhanced lysis.

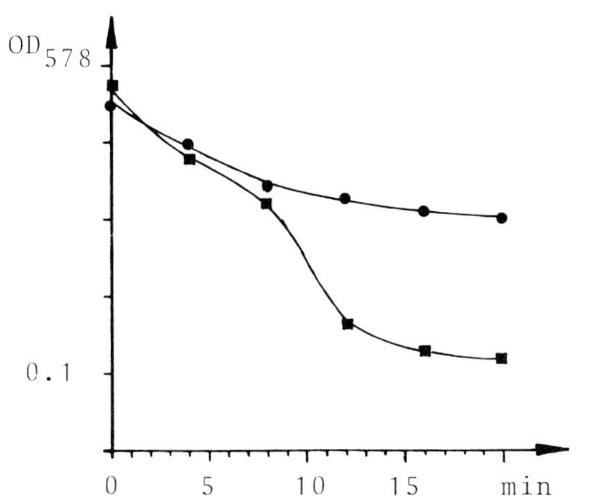

Fig. 4 Lysis kinetics of E. coli after EDTA-treatment and osmotic shock with and without an enriched amidase fraction.
●—● Only osmotic shock.
■—■ Osmotic shock and amidase.

Tab. 2 Killing rates of *E. coli* after osmotic shock with and without amidase treatment. Osmotic shock was carried out by the method described by Nossal (4).

	Killing rate
Osmotic shock and amidase	86%
Only osmotic shock	56%
Only amidase	0%
No osmotic shock, no amidase	0%

Conclusions

-In addition to lysozyme there is a second murein hydrolase in human serum with a molecular weight of about 85.000.
-The amidase hydrolyses the N-acetylmuramyl-alanine bond.
-The amidase completely cleaves the peptide side chains of murein sacculi including those that contain covalently linked lipoprotein.
-The enzyme has only a lytic and a killing effect on *E. coli* when the barrier function of the outer membrane is overcome under conditions of osmotic shock. It may add to the bactericidal activity of serum after complement binding to cells.

References

1 Primosigh, J., Pelzer, H., Maass, D., Weidel, W.: Biochim. Biophys. Acta 46, 58 (1961)

2 Lugtenberg, B., Meijers, J., Peters, R., van der Hoek, P., van Alphen, L.: FEBS Lett. 58, 254-258 (1975)

3 Witholt, B., van Heerikhuizen, H., De Leij, L.: Biochim. Biophys. Acta 443, 534-544 (1976)

4 Nossal, N.G., Heppel, L.A.: J. Biol. Chem. 241, 3055-3062 (1966)

LYSOZYME-RESISTANT *O*-ACETYLATED PEPTIDOGLYCAN OF *NEISSERIA GONORRHOEAE*: STRAIN VARIATION, RESISTANCE TO HUMAN PEPTIDOGLYCAN HYDROLASES, AND PATHOBIOLOGICAL PROPERTIES

Raoul S. Rosenthal*, Scott C. Swim, W. Joseph Folkening
Indiana University School of Medicine

Bruce H. Petersen, Rebecca L. Fouts, and Kalindi Phadke
Eli Lilly Inc.
Indianapolis, Indiana, U.S.A.

Introduction

Previous studies from our laboratory (1-5) argue that *Neisseria gonorrhoeae* represents a unique model organism in which peptidoglycan (PG)-host interactions during natural infection might be particularly direct and extensive. Recently, we have found that purified gonococcal PG fragments, analogous to those likely released *in vivo* by the action of bacterial- or host-derived PG hydrolases, possess diverse biological activities of possible pathological significance. These include intrinsic toxicity, activation of complement, modulation of blastogenesis, and arthritogenicity. The structural bases of these activities have not been completely defined but the latter three appear to depend on large mol. wt. glycosidically-linked fragments. This requirement for intact glycan chains takes on added interest in light of observations (6,7) that, of three genetically-independent sets of gonococci so far examined, two possessed PG that was substituted extensively with *O*-acetyl derivatives, a modification that conferred resistance of the PG to hen-egg white lysozyme (HEW-LZ) but not to Chalaropsis B lysozyme. If *O*-acetylation also mediates resistance of PG to human PG hydrolases, then this pro-

The Target of Penicillin

perty might promote persistence of PG *in vivo* in the large mol. wt. biologically-active forms. Accordingly, objectives of the current study were to (i) examine the susceptibility of extensively-*O*-acetylated PG (*O*-PG) and *O*-acetyl-deficient PG (non-*O*-PG) to degradation by hydrolase activity present in normal human sera (NHS) and by purified human polymorphonuclear lysozyme (PMN-LZ), (ii) determine how wide-spread HEW-LZ-resistant *O*-PG is among gonococci, and (iii) evaluate the role of *O*-acetylation in PG-mediated biological responses.

Experimental

Strain distribution of HEW-LZ resistant O-PG. Glucosamine-labled PG was purified as described (1,7) from 14 strains of *N. gonorrhoeae* that, collectively, represented various sub-sets of gonococcal isolates. To determine HEW-LZ resistance, PG was treated exhaustively with HEW-LZ (7), and the size distribution of HEW-LZ-soluble PG at completion of the reaction was determined by molecular sieve HPLC using a Varian TSK SW2000 column, a method that distinguished low mol. wt. PG somewhat better than traditional gel filtration on Sephadex (3), and did so in 40 min rather than 24 h. The extent of HEW-LZ resistance was expressed as the % of PG that was larger than disaccharide peptide tetramers (including insoluble PG removed by centrifugation). To determine percent *O*-acetylation, insoluble PG was first converted almost totally (ca. 95%) to uncrosslinked monomers by the combined action of Chalaropsis muramidase followed by *E. coli* endopeptidase (prepared by a modification of the method of Keck and Schwarz; 8); subsequent quantitation of radioactivity in *O*-acetylated and non-*O*-acetylated monomers was accomplished by paper chromatography in butanol: acetic acid:water (4:1:5,UP). By this method, % *O*-acetylation is defined in respect to total monomer subunit equivalents, irrespective of the distribution of *O*-acetyl derivatives among the individual crosslinked fragments in native PG.

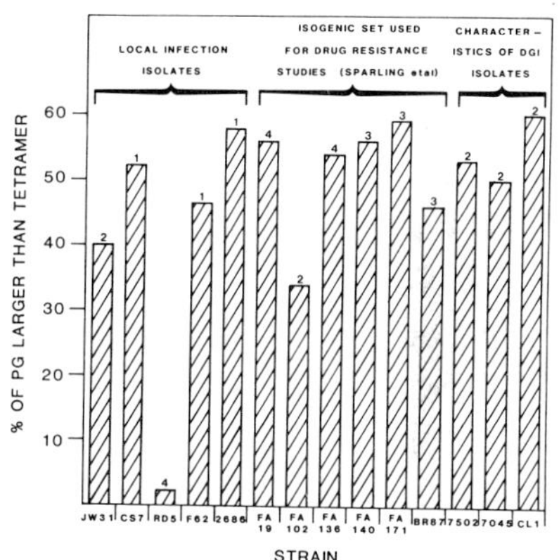

Fig. 1. HEW-LZ resistance of gonococcal PGs. Values are the means of the no. of experiments shown in parenthesis.

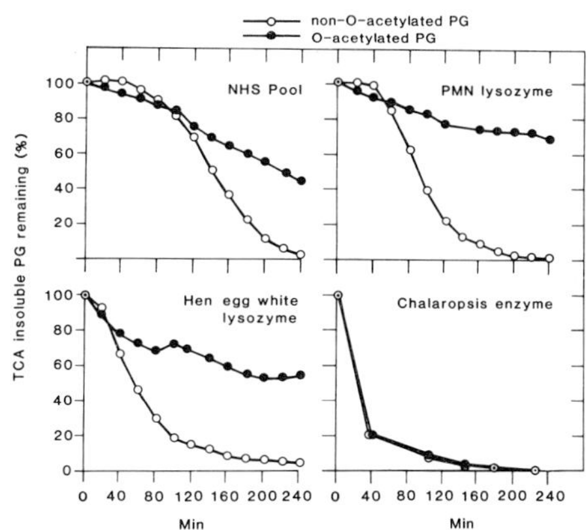

Fig. 2. Hydrolysis of gonococcal PG by PG-degrading enzymes.

The PG of most of the gonococcal strains examined was extensively _O_-acetylated (range, 47-52%) and HEW-LZ resistant (range, 40-60% larger than tetramers; Fig.1). Only the PG of RD_5, the strain with the highest rate of PG turnover among gonococci so far examined (1) and the prototype of gonococci having _O_-acetyl-deficient PG (7), had greatly reduced _O_-acetylation (15%) and exhibited virtually no HEW-LZ resistance (Fig.1). Extent of HEW-LZ resistance and % _O_-acetylation did not appear to correlate specifically with (i) type of colonial variant (p+ _vs_ p- and op _vs_ tr;9), (ii) type of clinical isolate (local _vs_ disseminated), or (iii) with the possible exception of FA102 (a penicillin-resistant strain carrying the _pen_A2 chromosomal marker), the presence of one or more drug resistance loci among an isogenic set of gonococci (10). For any given strain, the % _O_-acetylation in isolated monomers, derived from PG treated with Chalaropsis muramidase and no endopeptidase, and purified by HPLC, was slightly less than that of the % _O_-acetylation in total native PG.

Resistance of O-PG to human PG-degrading enzymes. Twelve µg/ml each of ^{3}H-_O_-PG from strain FA19 and ^{14}C-non-_O_-PG from strain RD_5 were mixed together and treated with one of four sources of PG hydrolases. The initial rate of hydrolysis of O-PG by a pool of four NHS (1:100 dilution) or by PMN-LZ (0.03 U/ml), like HEW-LZ, was 2- to 4-fold less than that of its non-O-PG counterpart in the same tube (Fig.2). Chalaropsis B enzyme, on the other hand, hydrolyzed both PGs at similar rates (Fig.2). O-PG was also relatively resistant to degradation by normal calf and rat serum and by PG hydrolase associated with washed intact human peripheral blood lymphocytes. As evidence that the differential digestion of O-PG and non-O-PG by NHS was not unique to the particular NHS pool used, the mean rate of digestion of non-O-PG by individual NHS (n=31) exceeded by 2.5-fold ($p<.001$) that of O-PG in the same tube. When reactions were allowed to go to completion, all of the non-O-PG was rendered soluble, i.e., present in the supernatant following ultra-centrifugation, by all four sources of PG hydrolase; however, PMN-LZ (like HEW-LZ) could solubilize only about half of the O-PG, while NHS (like Chalaropsis enzyme) solubilized all of the O-PG. This suggested that human serum contains PG hydrolase activity with substrate specificities in addition to those of PMN-LZ.

To characterize PG fragments solubilized by human PG hydrolases, reactions were allowed to go to completion and the soluble PG was fractionated by gel filtration (3). After treatment with PMN-LZ (similar to HEW-LZ), virtually all of the non-O-PG was in the size range of PG monomers (ca. 1000 daltons) through hexamers, but the distribution of soluble O-PG was shifted greatly in favor of high mol. wt. oligomers greater than hexamers. Since complete digestion of glycosidic linkages by Chalaropsis enzyme yields PG fragments no greater in size than tetramers (3,7), these high mol. wt. fragments represent incompletely-degraded glycan chains. Unexpectedly, the final NHS-digestion products of both PGs were uniformly smaller in size than the free disaccharide of PG, suggesting that NHS contained, in addition to the well-known muramidase, an amidase (and probably an additional

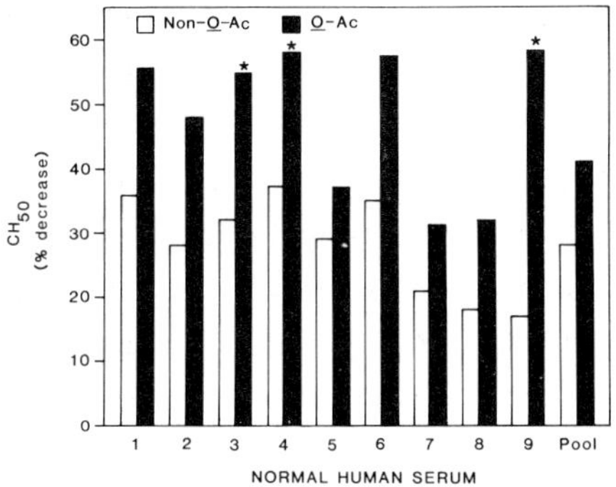

Fig. 3. Consumption of complement in NHS by non-O-acetylated and O-acetylated S-PG. Due to the nature of the assay, starred values are actually equal to or greater than that shown.

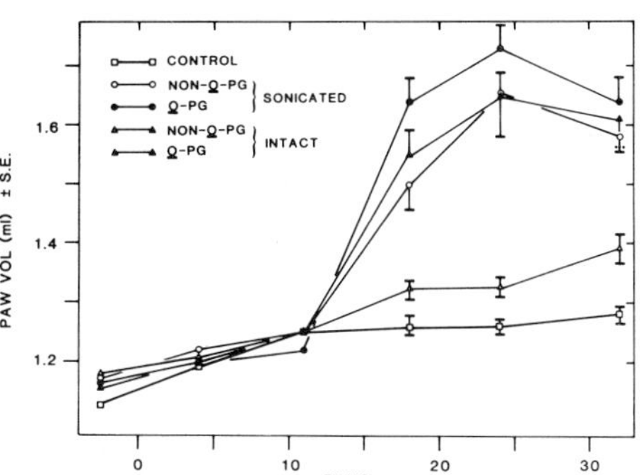

Fig. 4. Arthritogenicity of gonococcal PG in rats.

glycosidase) activity capable of degrading PG. At intermediate times of incubation, i.e., prior to completion, NHS digestion of O-PG yielded fragments of higher mol. wt. than corresponding fragments derived from non-O-PG.

Biological properties of O-PG vs non-O-PG. The biological activities of O-PG and non-O-PG were compared directly using systems in which the extent of the PG-mediated response might be modulated by endogenous PG hydrolase activity. The PGs were used as either intact (insoluble) preparations or as sonicated (S)-PG (soluble, > ca. 3×10^6 daltons). Consumption of human complement by PG (500 µg/ml of serum) was determined using a hemolytic assay as described (5). Arthritis induction in groups of 10 male Lewis rats was determined using a model developed for type II collagen arthritis (11). Paw swelling induced by intra-dermal injection of 200 µg of PG in incomplete Freunds adjuvant (IFA) or of IFA alone (negative control) was determined by mercury displacement. Extensive O-acetylation potentiated both S-PG-mediated consumption of complement in 10 of 10 NHS (Fig.3) and intact-PG-mediated arthritogenicity using the rat model (Fig.4).

Conclusions

1. Extensive _O_-acetylation was found to mediate intrinsic resistance of gonococcal PG to human PG-degrading enzymes, a property that promoted persistence of PG in the form of glycosidically-linked oligomers of the repeating disaccharide sub-unit _in vitro_, and may do so _in vivo_.
2. Although _O_-PG was relatively resistant to both sources of human PG-hydrolase activity employed (serum and PMN-LZ), there were distinct differences in the ultimate extents of degradation and in the sizes of the final products of these two enzyme activities. This was attributed to the presence in serum of glycosidase (in addition to the known muramidase) and amidase activities capable of degrading PG.
3. Extensive _O_-acetylation potentiated gonococcal PG-mediated complement consumption and arthritogenicity.
4. Extensive _O_-acetylation of PG was found to be wide-spread among gonococci and, thus, most strains are a potential source of hydrolase-resistant, biologically active PG.

References

1. Rosenthal, R.S.: Infect. Immun. 24, 869-878 (1979).
2. Sinha, R.K., Rosenthal, R.S.: Infect. Immun. 29, 914-925 (1980).
3. Rosenthal, R.S., Wright, R.M., Sinha, R.K.: Infect. Immun. 28, 867-875 (1980).
4. Sinha, R.K. Rosenthal, R.S.: Antimicrob. Agents Chemother. 20, 98-103 (1981).
5. Petersen, B.H., Rosenthal, R.S.: Infect. Immun. 35, 442-448 (1982).
6. Blundell, J.K., Smith, G.J., Perkins, H.R.: FEMS Microbiology Lett. 9, 259-261 (1980).
7. Rosenthal, R.S., Blundell, J.K., Perkins, H.R.: Infect. Immun. 37, 826-829 (1982.)
8. Keck, W., Schwarz, U.: J. Bacteriol., 139, 770-774(1979).
9. Swanson, J.: Infect. Immun. 19, 320-331 (1978).
10. Guymon, L.F., Walstad, D.L., Sparling, P.F.: J. Bacteriol. 136, 391-401 (1978).
11. Phadke, K., Carroll, J., Nanda, S.: Clin. exp. Immunol. 47, 579-586 (1982).

This work was supported by PHS grant AI-14826 from the NIH.

THE INFLUENCE OF DIFFERENT ANTIBIOTICS ON THE DEGREE OF O-ACETYLATION OF STAPHYLOCOCCAL CELL WALLS

Petra Burghaus*, Lars Johannsen, Dieter Naumann, Harald Labischinski, Hans Bradaczek*, Peter Giesbrecht

Robert Koch-Institut des Bundesgesundheitsamtes
D-1000 Berlin

*Institut für Kristallographie, Freie Universität Berlin
D-1000 Berlin

Introduction

It is well known that the application of chloramphenicol (=CAP) induces thickening of staphylococcal cell walls (1). We investigated such cell walls to examine if their primary structure differs from the normally grown ones. We could show that CAP treated cell walls contained 40 - 160% (fig. 1) more O-acetyl groups as compared to the log phase cell walls while there were only minor changes in their amino acid and amino sugar composition (2). For control cell walls an amount of O-acetyl substituted muramic acid in percent of total muramic acid content of approximately 35% was estimated previously (2). Recently we found that, apparently, an increased O-acetylation was the very reason for the lower degradability of CAP treated staphylococci by lysozyme (contribution of L. Johannsen et al., in this volume).

The decrease in degradability may be of great medical interest because it is known that insufficiently degraded cell walls may be responsible for the induction of rheumatic arthritis (3,4,5). Considering the great importance of antibiotics for curing infectious diseases, we were interested in the question if an increased content of O-acetyl groups in staphylococcal cell walls is the specific result of CAP application or if other drugs may induce similar effects.

The Target of Penicillin

Therefore we tested if also the application of the bacteriostatic agents erythromycin (=ERY) and a trimethoprim sulphamethoxazole combination (=TSC) may lead to an increase of O-acetylation in staphylococci.
The quantitative determination of O-acetyl groups was accomplished by gas chromatography (=GC). These data were supported by Fourier transform infrared spectroscopy (=FT-IR), that proved to be a very promising tool in probing cell wall material of bacteria ((6) and contribution of D. Naumann et al., in this volume).

Materials and Methods

Staphylococcus aureus SG 511 "Berlin" was grown with aeration in 3.7% Brain Heart Infusion broth in 500 ml flask or fermenter at 37° C. Harvesting of untreated cells and application of antibiotics were accomplished during the logarithmic phase of growth; treatment with antibiotics lasted for 16 h (10µg/ml ERY respectively 200 µg/ml TSC, i.e. 100 µg/ml of each of its components). Isolation of cell walls was carried out as described earlier (2). Removal and measurement of O-acetyl groups by GC were accomplished as described in (7).
For quantitative FT-IR measurements equal amounts of isolated and lyphilized cell walls were suspended in water and dried to ir transparent films in vacuum over P_4O_{10} using either reflecting or ir transparent materials for measuring in direct reflection or in transmission respectively. For quantitative determination the peak height of the O-acetyl band near 1737 cm^{-1} was used.

Results and Discussion

Using GC and FT-IR we found that ERY as well as TSC increased the O-acetyl content of Staphylococcus aureus.

Tab. 1
Determination of O-Acetyl Groups by Gas Chromatography

Antibiotic Treatment	Increase of O-Acetyl Groups*
Chloramphenicol 20 µg/ml	40 to 160 %
Erythromycin 10 µg/ml	50 to 90 %
Trimethoprim and Sulphamethoxazole 100 µg/ml each	60 %

*as compared to log phase cell walls

Fig. 1 Expanded Presentation of the O-Acetyl Peak from Spectra of Cell Walls Before and After Treatment with Trimethoprim and Sulphamethoxazole

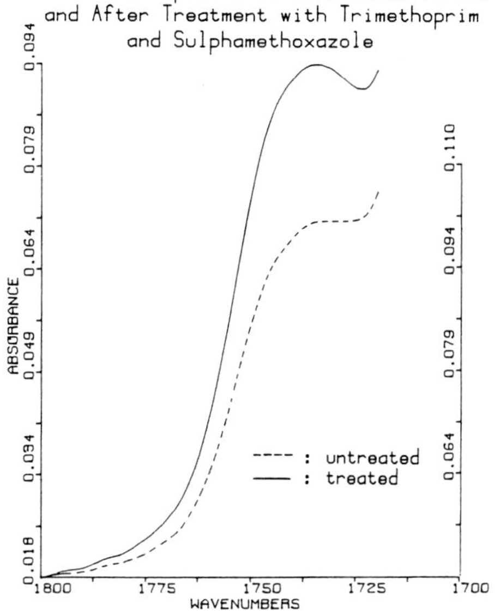

After treatment with ERY in flask cultures we revealed by GC a 50 to 90% higher O-acetylation as compared to untreated cell walls of the logarithmic phase of growth. The exposure to TSC induced an increase of about 60% (tab. 1).

Our GC data have also shown that the O-acetyl content per weight of staphylococcal cell walls seemed to be rather variable in dependence on the actual conditions of growth: There were differences in O-acetylation between fermenter and flask cultures and, for example, the increase in O-acetylation after treatment with TSC was only 20% when the cells had been cultivated in another nutrient broth (2.5% bactopepton containing 0.5% NaCl).

Therefore we have always isolated both, treated and untreated cell walls, under the same conditions to determine the percentage of their difference in O-acetyl content. Irrespective of the diferent O-acetyl level observed the percentage increase in antibiotic treated cell walls was always evident.

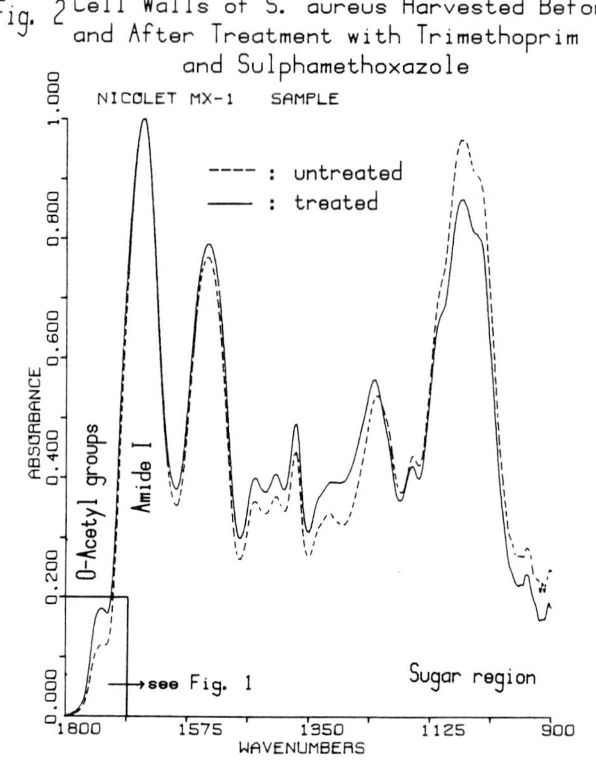

Fig. 2 Cell Walls of S. aureus Harvested Before and After Treatment with Trimethoprim and Sulphamethoxazole

In accordance with the GC data, our ir measurements revealed a higher absorbance of the O-acetyl band near 1737 cm^{-1} for cell walls exposed to ERY (spectra not shown) and TSC.

These results could be obtained by a direct comparison of the peak heights of the O-acetyl band from samples of the same weight using equally scaled absorbance axes (fig. 1) and also after normalizing the spectra to equal absorbance of the so called amide I peak (fig. 2).

However these ir measurements did not only reveal differences in the O-acetyl band of the spectra. For cell wall preparations from the bacteria exposed to ERY or to TSC, a lower absorbance was observed in the sugar region from 1150 to 900 cm^{-1} as compared to untreated samples (fig. 2). These finding may also indicate changes in the teichoic acid content of cell walls after drug treatment.

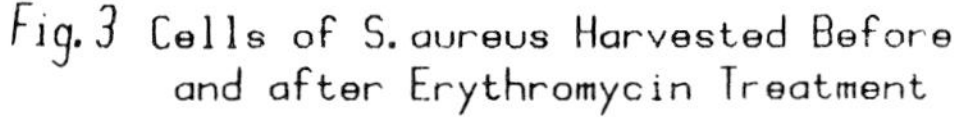

Fig. 3 Cells of S. aureus Harvested Before and after Erythromycin Treatment

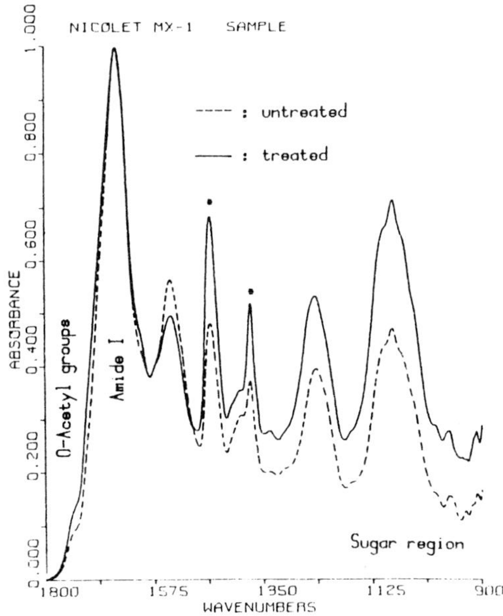

The bands marked (*) were produced by nujol an immersion liquid used to reduce scattering at film surface.

First attempts have shown that it is possible to prepare films not only from isolated cell walls but also from intact cells which were of sufficient quality for quantitative analysis (fig. 3). This is of great interest since the time-wasting isolation of cell walls could now be prevented. Because of the possibly different content of wall enzymes in staphylococci before and after drug treatment, normalizing of spectra to equal absorbance of amide I band was considered to be a first approximation. Further investigations are neccessary to find out a suitable standard to realize an exact quantitative determination (e.g. the relation between peak height of O-acetyl band and sample weight).

In conclusion we could show by gas chromatography and Fourier transform infrared spectroscopy that the increase of O-acetylation of staphylococcal cell walls is not a specific effect of chloramphenicol as similar results could be obtained with erythromycin and a combination of trimethoprim and sulphamethoxazole in spite of the differences in their molecular mode of action. On the other hand, the exposure to the bactericidal antibiotic penicillin was reported to induce a decrease in the degree of O-acetylation in Proteus mirabilis as well as in Neisseria gonorrhoeae (8,9,10,11).
It should be noted, however, that the increase in O-acetylation in the case of TSC application may be due to exhaustion of the nutrient broth, since TSC is well known to have only little influence on S. aureus grown in BHI medium. We observed an increased O-acetyl level in cell walls harvested from the stationary phase of growth as compared to that found in cells from the logarithmic phase (contribution of L. Johannsen et al., in this volume).
Thus a high content of O-acetyl groups may be correlated mainly with a "stationary state" of bacterial growth, either reached during the stationary phase of growth or induced by the inhibition of growth after application of bacteriostatic drugs.

References

1. Giesbrecht, P., Ruska, H.: Klin. Wschr. 46, 575-582 (1968).
2. Johannsen, L., Labischinski, H., Reinicke, B., Giesbrecht, P.: FEMS Microbiol. Lett. 16, 313-316 (1983).
3. Cromartie, W.J.: in "Arthritis; Models and Mechanisms" (H. Deicher and L.C. Schulz eds.), Springer, Berlin p. 24-38 (1981).
4. Ginsburg, I., Neemann, N., Lahav, M., Sela, M.N., Quie, P.G.: Zbl. Bakt. Suppl. 10, 851-859 (1981).
5. Wecke, J., Lahav, M., Ginsburg, I., Giesbrecht, P.: Arch. Microbiol. 131, 116-123 (1982).
6. Naumann, D., Barnickel, G., Bradaczek, H., Labischinski, H., Giesbrecht, P.: Eur. J. Biochem. 125, 505-515 (1982).
7. Fromme, I., Beilharz, R.: Anal. Biochem. 84, 347-353 (1978).
8. Martin, H.H., Gmeiner, J.: Eur. J. Biochem. 95, 487-495 (1979).
9. Gmeiner, J., Kroll, H.P.: Eur. J. Biochem. 117, 171-177 (1981).
10. Blundell, J.K., Perkins, H.P.: J. Bacteriol. 147, 633-641 (1981).
11. Dougherty, T.J.: J. Bacteriol. 153, 429-435 (1983).

CELL WALL DEGRADATION OF STAPHYLOCOCCUS AUREUS BY AUTOLYSINS AND LYSOZYME

Peter Blümel, Bernhard Reinicke, Meir Lahav*, Peter Giesbrecht
Robert Koch Institute of the Federal Health Office, Berlin 65
*Hadassah Medical School of the Hebrew University, Jerusalem

Introduction

There is a large number of informations concerning the relationship between the degradability of the cell walls of certain bacteria like streptococci and staphylococci and the induction of inflammatory processes in the host (1, 2, 3). It has been concluded from in-vitro experiments that lysozyme apparently is not capable of lysing the walls of streptococci and staphylococci indicating that the idea of the central role of lysozyme as the key enzyme in bacterial cell wall degradation might be too simple. Furthermore, it has been shown that the degradation of staphylococcal cell walls can be accelerated by cationic proteins via an activation of bacterial autolytic wall enzymes (1, 4). This lead to the conclusion that the enzymatic action of lysozyme could be accompanied by such activating effects on autolytic wall enzymes, since lysozyme is a cationic protein itself.
It is the aim of this investigation to examine in an in-vitro system the enzymatic specifities being involved in the cell wall degradation of S. aureus, especially in respect of a possible combined action of activated wall autolysins and lysozyme. On the other hand, the combined action of autolysins and ribonuclease (a cationic protein devoid of any muralytic activity) was analysed.

The Target of Penicillin

Material and Methods

Intact cells or isolated cell walls of Staphylococcus aureus SG 511 from the logarithmic growth phase were used in the experiments. The conditions of cultivation, radiolabelling, and of enzymatic release measurements have been published elsewhere (5). N-terminal amino acids were labelled by dansyl chloride (Pierce, Rochfort, USA) (6). The analyses of the DANS-amino acids were performed by high performance liquid chromatography, LX-PD system (Pye-Unicam, U.K.) in combination with a UV-fluorescence detector type 82.00 (Knauer,West Germany). The amounts of reducing sugar were determined after reaction with sodium borohydride (7)(Merck,Darmstadt),by means of an amino acid analyzer LC 6000 (Biotronic, Frankfurt).

Results and Discussion

Isolated radiolabelled cell walls of S. aureus were chosen to study the degrading effects of lysozyme and of ribonuclease without being influenced by the action of the bacterial autolysins. The combined actions of lysozyme and autolysins, on the one hand, and of ribonuclease and autolysins on the other, were followed by adding viable non-radioactive staphylococci as autolysin-doners to the system. This was necessary,because the isolated cell walls had lost their own autolysins by SDS treatment during the isolation procedure. The pH of the degradation mixture was 5.0 according to the pH existing in the phagolysosomes. The supernatants of these mixtures were investigated in respect of the release rates of soluble cell wall material and in respect of bacterial cell wall hydrolases involved. - The kinetics of the cell wall release of our staphylococcal strain was measured in terms of release of the radiolabel to the supernatant (Fig.1). The curves representing the action of autolysins as compared with those reflecting the actions of lysozyme and ribonuclease showed a relatively

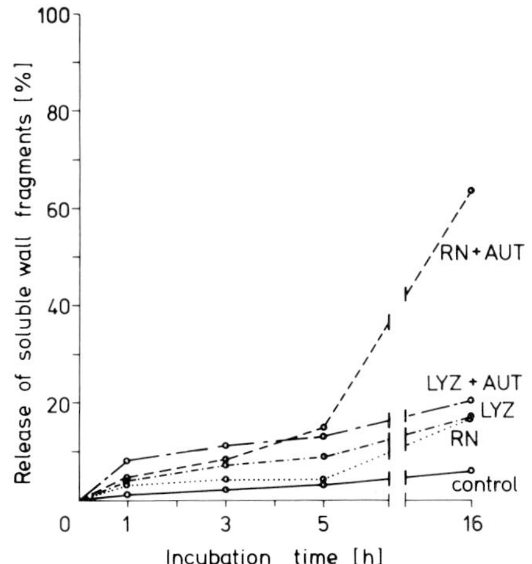

Fig.1: Release of soluble fragments from isolated walls of S.aureus (500 µg/ml) under the influence of autolysins (200 µg/ml vital staphylococci as autolysin donators), lysozyme (10 µg/ml), and ribonuclease (10 µg/ml). AUT = autolysins; LYZ = lysozyme; RN = ribonuclease.

Lytic agents in the incubation mixture	% terminal constituents in soluble wall fragments			
	NGL	MUR	GLY	ALA
Autolysins	9.0	0	13.3	8.8
Lysozyme	4.7	55.6	3.8	4.4
Autolysins + Lysozyme	9.0	50.0	10.7	12.5
Ribonuclease	4.7	0	8.6	7.0
Autolysins + Ribonuclease	16.6	0	29.6	13.4

Table 1: Contents of terminal glucosamine (NGL), terminal muramic acid (MUR), terminal glycine (GLY), and terminal lanine (ALA) in soluble fragments from isolated walls of S.aureus (500 µg/ml) after 16 hours of incubation at pH 5 under the influence of auf autolysins (200 µg/ml vital staphylococci as autolysin donators), lysozyme (10 µg/ml), and ribonuclease (10 µg/ml)

close similarity to each other. Unexpectedly, ribonuclease appeared to express some muralytic activity by itself. This may be considered as an indication for incomplete removal of the wall autolysins. However, whereas the combined action of autolysins and lysozyme showed no significant alterations of the release rate, the simultaneous action of autolysins and ribonuclease showed a dramatic increase of cell wall release under the conditions employed. - Additional informations were obtained from the determination of the sites where the different enzyme systems cleave the peptidoglycan chains. Terminal glucosamine, muramic acid, glycine, and alanine in the cleavage products stand for the action of glucosaminidase, muraminidase, endopeptidase, and amidase respectively (Table 1).

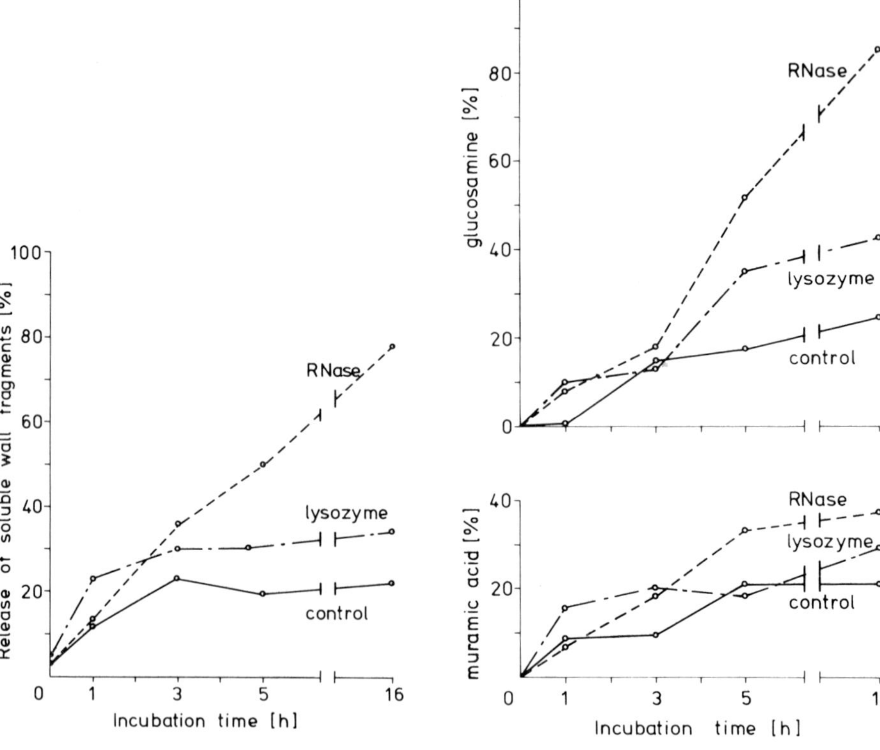

Fig. 2: Release of soluble wall fragments from S.aureus under the influence of 10 µg/ml lysozyme or 10 µg/ml ribonuclease

Fig. 3: Contents of glucosamine and muramic acid in soluble wall fragments released from S.aureus under the influence of 10 µg/ml lysozyme or 10 µg/ml ribonuclease

Our experimental data revealed the following informations: 1) Three autolytic activities were found in our staphylococcal strain confirming the results obtained with other staphylococcal strains. 2) A muraminidase could not be detected. 3) Lysozyme was capable of degrading isolated cell walls of S. aureus, which is contrary to what is presently known. 4) Lysozyme acted significantly as a muraminidase. 5) The combined action of autolysins and lysozyme showed a more synergistic action of the enzymes with a certain activating effect on the amidase. 6) The combined action of wall autolysins and ribonuclease showed an activation of all the three autolytic activities.- In order to investigate whether in these experiments the autolysins, having been excreted from the donor cells, reflected the native autolysin pattern, and whether the integrity of the substrate structure had been preserved, we studied the degradation of viable staphylococci by lysozyme and ribonuclease under the same experimental conditions as mentioned above (Fig. 2, 3). Again, the kinetics of cell wall release showed a particular characteristic for the combined action of autolysins and ribonuclease. Thus the data obtained from the experiments with intact bacteria can be considered to be a valuable proof of accuracy for the results obtained from the experiments with isolated cell walls of staphylococci, as well as for the kinetics of the release of glucosamine and muramic acid.

The data presented confirm that ribonuclease as a cationic compound was capable of activating staphylococcal wall autolysins. With respect to cell wall degradation of bacteria in phagocytes, we have to consider, therefore, that due to the presence of numerous cationic compounds in the lysosomes, the more direct muralytic action of lysozyme can be considerably enhanced via autolysin activation. Future investigations will have to provide an extension of the enzymatic analyses of the effects of other charged compounds that influence the autolysins of staphylococci.

In conclusion:Three autolytic activities were found in Staphylococcus aureus SG 511, being an endopeptidase, an amidase, and a glucosaminidase. A muramidase such as lysozyme apparently is not part of the autolytic system of this bacterium. The combined action of wall autolysins and lysozyme showed that lysozyme acted in our system with isolated cell walls and viable cells of Staphylococcus to a considerable degree as a muralytic enzyme. The capability for activating the autolytic wall enzymes by lysozyme was not as significant as it proved to be for the activation by ribonuclease which showed exclusively an activating potential in regard to the three autolytic wall enzymes.

References

1. Ginsburg,I.: in P.Jacque/J.Dingle (eds.) Lysosomes in Applied Biology and Therapeutics. North Holland Publ. 1979, 327-406
2. Ginsburg,I., Christensen,P., Eliasson,I., Schalen,C.: Acta path.microbiol.scand.Sect.B 90, 161-168 (1982)
3. Giesbrecht,P., Wecke,J., Blümel,P.,Reinicke,B.,Labischinski,H.: in Eickenberg/Hahn/Opferkuch (eds.) The Influence of Antibiotics on the Host-Parasite-Relationship, 228-241 Springer, Berlin . Heidelberg . New York 1982
4. Wecke,J., Lahav,M., Ginsburg,I., Giesbrecht,P.: Arch.Microbiol. 131, 116-123 (1982)
5. Reinicke,B., Blümel,P., Giesbrecht,P.: Arch.Microbiol. in press
6. Wiedmeier,V.T., Porterfield,S.P., Hendrich,C.E.: Physiologist 24, 410-415 (1981)
7. Sarwadeker,J.S., Sloneker,J.H., Jeanes,A.: Anal.Chem. 37, 1602 (1965)

CELL WALL DEGRADATION OF ANTIBIOTIC-TREATED STAPHYLOCOCCI UNDER PHAGOCYTE-SPECIFIC CONDITIONS:

Distinction between Penicillin-Induced Lytic Effects and Wall Alterations Caused by Anticoagulants.

Jörg Wecke[1], Meir Lahav[2], Isaac Ginsburg[2], Elli Kwa[1], Peter Giesbrecht[1]

1) Robert Koch Institute of the Federal Health Office, Nordufer 20, D-1000 Berlin-West 65

2) Hebrew University Hadassah Med. School, Jerusalem

Introduction

In previous studies it could be shown that wall-autolytic enzymes of staphylococci can be activated by the cationic protein ribonuclease.Even the effect of lysozyme on staphylococci must be partly ascribed to its capability of activating pre-existing wall autolysins besides its enzymatic capacity (1). On the other hand,several anionic polyelectrolytes like chondroitine sulfate,DNA or polyanethol sulfonic acid ("Liquoid") can inhibit the spontaneous lysis of staphylococci (2). The mechanism of this inhibition, however, is largely unknown.

Therefore, the purpose of the present paper is to describe the role of the anionic compound liquoid in the cell wall degradation process, especially in the penicillin-induced lysis of staphylococci.

Material and Methods

Bacterial strains: Strains SG 511 and Oxford of Staphylococcus aureus were used which had proved to be especially sensitive

The Target of Penicillin

to antibiotics. For cultivation see (1).
Substances: Penicillin G potassium (Serva), polyanethol sulfonic acid (Sigma) were used.
Electron Microscopy: The methods applied for thin-sectioning have been published earlier (1).

Results and Discussion

1) The influence of an anionic polyelectrolyte on growing staphylococci.

After the addition of liquoid to growing staphylococci no suppression of growth could be detected. The scanning electron microscopical examination of liquoid-treated cells revealed a considerable aggregation of the staphylococci (Fig. 1) in contrast to normal cells. However, scanning electron microscopy did not indicate whether these clusters of staphylococci were formed due to some unspecific aggregation or to some inhibition of cell separation. By use of the thin-sectioning technique it was possible to answer this question. Frequently, two successively built cross walls could be detected in one and the same cell. It was easy to differentiate between the first and the second division plane, because, generally, the first cross wall is thicker from apposition than the second one (Fig. 2). A lot of clustered cells contained even more than two cross walls of different thickness. Such symmetrically "pseudomulticellular bacteria" (3) can be caused by a certain inhibition of the autolytic system, which usually separates the daughter cells along the middle axis of the cross wall (cutting and splitting systems (4)).

2) Suppression of penicillin-induced bacteriolysis by liquoid.

The bacteriolytic effect of penicillin can be observed only within a relatively small dosė range. Under these conditions

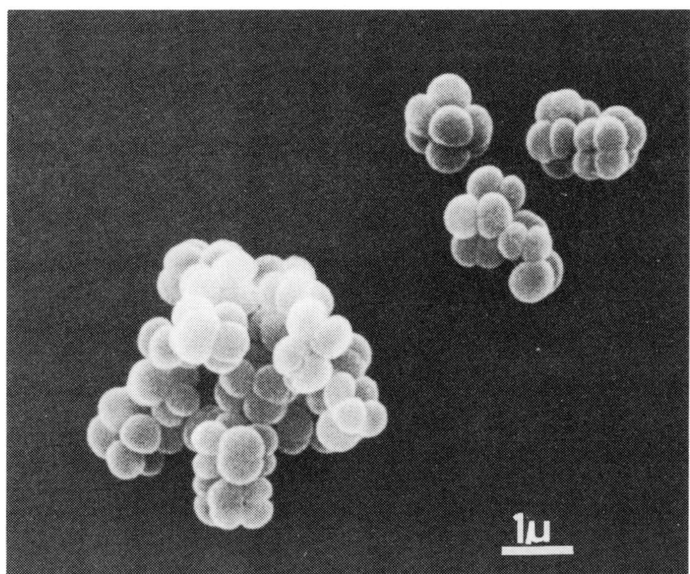

Fig. 1: Scanning electron microscopical view of "aggregated" staphylococci after treatment with 2 mg/ml Liquoid

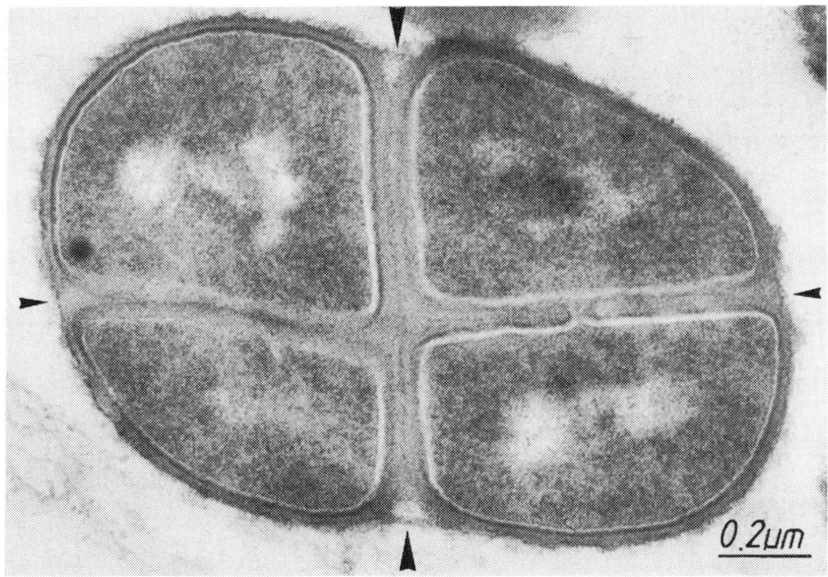

Fig. 2: Thin section of liquoid-treated staphylococci. Pseudomulticellular staphylococcal cell (the first division plane is marked by a large arrow, the second one by a small arrow)

(0.1 µg/ml penicillin) the growth (extinction) curve of a logarithmically growing suspension of staphylococci was rising at first and then declining (Fig.3).However,the decline of the growth curve was retarded, if liquoid was added to the culture. This effect was observed after the addition of liquoid at different times between 1 and 6 hours (Fig.3). This lysis-protection effect could be demonstrated over a range of 5 µg - 10 mg/ml of the anionic polyelectrolyte.

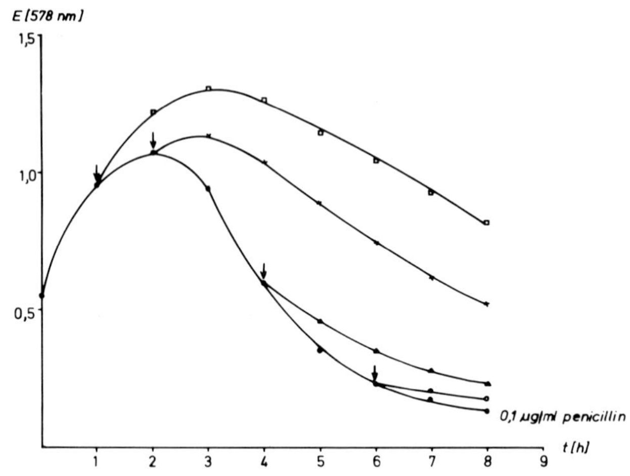

Fig. 3: The suppression of bacteriolysis of penicillin-treated staphylococci after addition of 10 µg/ml Liquoid (↓)

3) The ultrastructural alterations of penicillin-treated staphylococci after addition of an anionic polyelectrolyte.

Protection by liquoid against penicillin-induced bacteriolysis could also be demonstrated in electron micrographs (compare Fig. 4 with Fig. 5). Many of the liquoid-treated cells were intact. In some cases, the cross wall was enormously thickened. While the autolytic systems of the cross wall were inhibited by liquoid, some disturbed morphogenesis of the bacteria continued. In some cases, asymmetrically pseudomulticellular bacteria resulted from this treatment

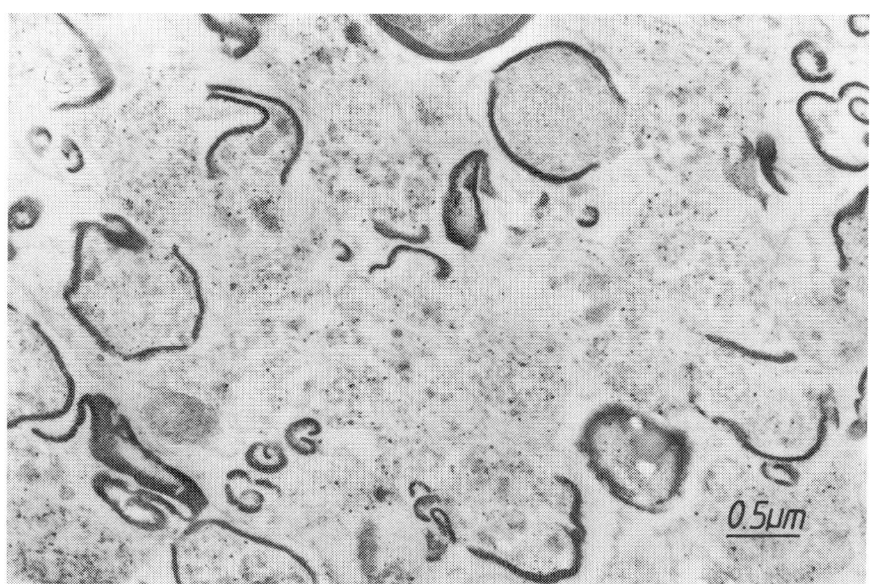

Fig. 4: Thin section of lysed staphylococci after 6 h of penicillin-treatment (0.1 µg/ml)

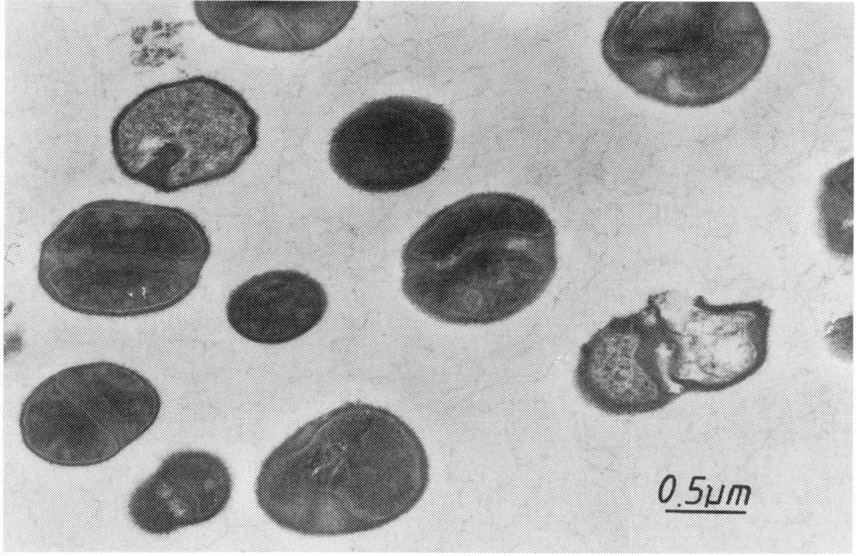

Fig. 5: Thin section of staphylococci after a combined treatment with penicillin and Liquoid.
Only a few bacteria are lysed

due to the formation of a new cross wall in only one of the two daughter cells.

Summary

Electron microscopical data from the modification of the autolytic wall systems of staphylococci by an anticoagulant polyanethol sulfonic acid ("Liquoid") were presented. This anionic polyelectrolyte caused an inhibition of cross wall separation which resulted in the formation of "pseudomulticellular" bacteria. A considerable degree of suppression of the autolytic wall enzymes normally inducing cell separation was discussed to be the reason for that. Such a suppression of the autolytic wall enzymes by liquoid could also be demonstrated in penicillin-treated staphylococci. Extinction curves as well as electron micrographs proved that the penicillin-induced bacteriolysis was mostly inhibited, while the penicillin-specific morphogenetic reactions continued.

References

1. Wecke,J., Lahav,M., Ginsburg,I., Giesbrecht,P.: Arch. Microbiol. 131, 116-123 (1982)
2. Ginsburg,I., Christensen,P., Eliasson,I., Schalen,C: Acta path.microbiol.scand.Sect.B 90, 161-168 (1982)
3. Giesbrecht,P., Wecke,J.: Zbl.Bakt.Suppl.10, 455-459 (1981)
4. Giesbrecht,P., Wecke,J., Reinicke,B: Int.Rev.Cytol. 44, 225-317 (1976)

INDUCED AUTOLYTIC WALL PROCESSES IN HEAT-INACTIVATED STAPHYLOCOCCUS AUREUS

Meir Lahav, Isaac Ginsburg
Hebrew University Hadassah Medical School, Jerusalem, Israel

Thomas Kersten, Jörg Wecke, Peter Giesbrecht
Robert Koch Institute of the Federal Health Office
Nordufer 20, D-1000 Berlin 65

Introduction

In previous studies it could be shown that autolytic wall enzymes of bacteria can be activated by some cationic proteins (1). Recently we obtained results from chemical and end group determination of the cleavage products from peptidoglycan confirming our data that even lysozyme acted not only as a muralytic enzyme but also as a cationic protein (2). In order to elucidate the mechanisms of the activation of autolytic wall processes by cationic proteins and of the direct respectively indirect muralytic actions of lysozyme we performed investigations on heated cells.

Material and Methods

Radioactively labelled cell walls of log-phase cells of Staphylococcus aureus strain "Oxford" obtained from the collection of the Hadassah Medical School, Jerusalem, were used. Three different types of cells have been investigated:

a) normal cells,
b) cells, heat-inactivated for 1 hour at 60°C
c) cells, heat-inactivated for 1 hour at 100°C.

The Target of Penicillin

Conditions of growth and labelling of bacterial cell wall as well as the determination of the release of cell wall material have been described formerly (1). In our experiments, suspensions of bacteria ($OD^{1\ cm}_{578\ nm}$ = 1) were incubated for defined times at 37°C in a shaker in 0.1 M acetate buffer at pH 5.0 in the presence of 0.5 mg/ml ribonuclease-A or egg-white-lysozyme purchased from Sigma Chemical Co., St. Louis, USA. Bacteria incubated in buffer served as controls.

Results and Discussion

The different time patterns of release of cell wall material from labelled S. aureus in the presence of 0.5 mg/ml ribonuclease or lysozyme and from control cells are shown in Figs. 1,2, 3.

In the controls it could be seen that heat-treatment of staphylococci led to a decrease in the release of cell wall material: After 20 hours, normal cells released 57 % while 60°C heat-inactivated cells released only 7 % and 100°C heat-inactivated cells only 3 %.

Cells heat-inactivated at 60°C showed subsequently in the presence of ribonuclease or lysozyme a considerable higher release of wall material than the controls (Fig. 2).

However, heat-inactivating by 100°C revealed no significant differences in the release of wall material between the control cells and bacteria which were incubated after heating with ribonuclease or with lysozyme (Fig. 3).

In order to analyse whether the target structure of the autolytic system was changed by heating, we added intact unlabelled staphylococci (= normal cells which are known to secrete parts of their wall autolysins into the growth medium) to the heat-inactivated bacteria (Fig. 4: 60°C; Fig. 5: 100°C).

After a supplementary addition of ribonuclease respectively lysozyme to the mixture of normal cells and heat-inactivated bacteria an enhanced release of wall material compared to

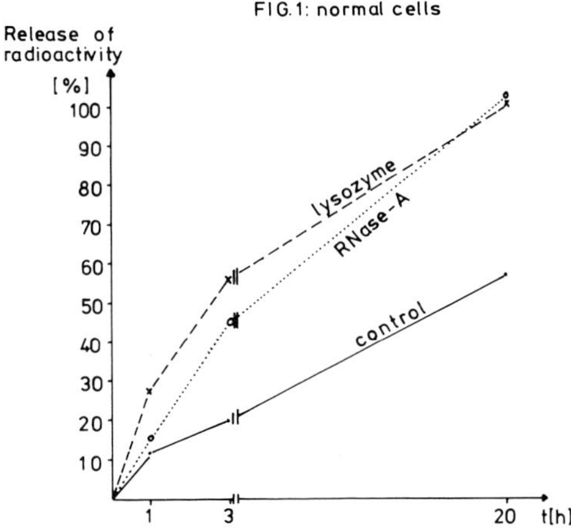

Fig. 1. Release of cell wall material from normal cells of Staphylococcus aureus strain "Oxford". Cells were incubated at 37°C for different times in 0.1M acetate buffer pH 5.0 or in the presence of 0.5 mg/ml lysozyme respectively ribonuclease.

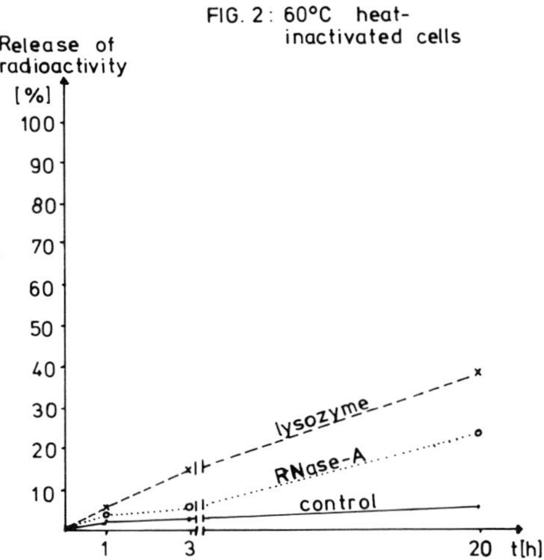

Fig.2. Release of cell wall material from 60°C heat-inactivated cells of Staphylococcus aureus strain "Oxford"

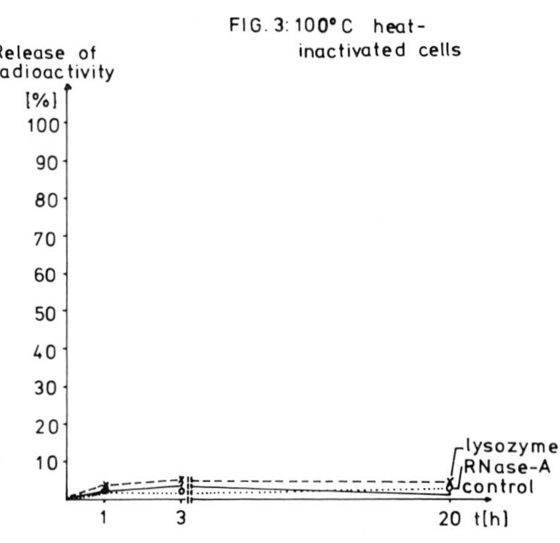

Fig. 3. Release of cell wall material from 100°C heat-inactivated cells of Staphylococcus aureus strain "Oxford".

that of normal cells could be observed (Fig. 1), indicating an enhanced secretion of wall autolysins from the added normal cells.

These results could be confirmed by electron microscopical studies, as after addition of lysozyme a degradation of the walls took place (compare Fig. 6 with Fig. 7).Virtually the same results could be demonstrated after addition of ribonuclease.

On the basis of these results we may assume that in staphylococci inactivated by 60°C, apparently the autolysins were only partially destroyed while after applying 100°C, autolytic wall activity was no longer detectable. However, the sensitivity of the heated walls to external autolysins was in no case diminished. Yet the target structure for the muralytic capacity of lysozyme within the cell wall obviously was altered by heating at 100°C as lysozyme exerted no detectable

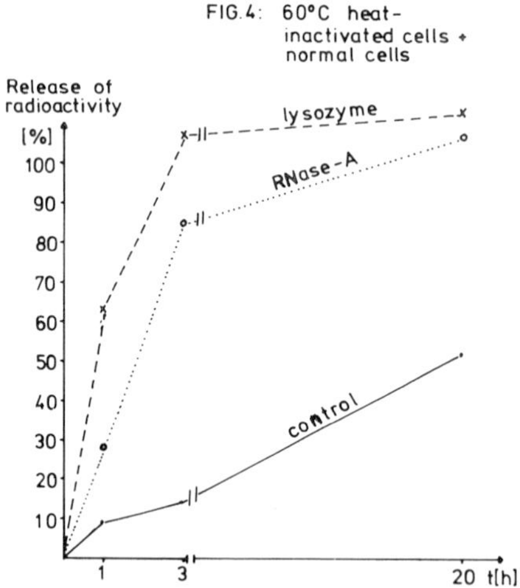

Fig.4. Release of cell wall material from 60° heat-inactivated cells of Staphylococcus aureus strain "Oxford" by lysozyme and ribonuclease in the presence of normal cells.

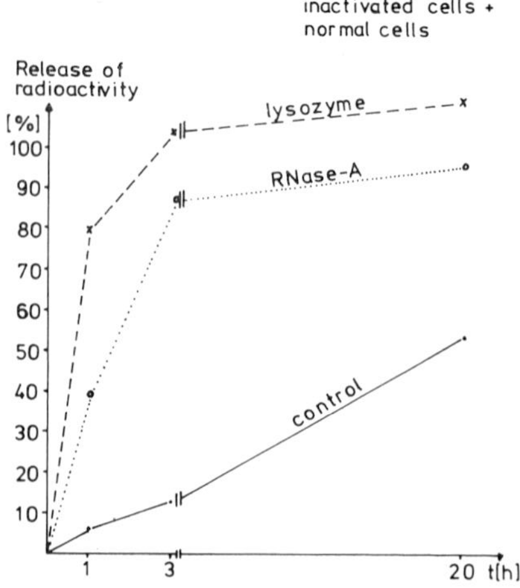

Fig.5. Release of cell wall material from 100°C heat-inactivated cells of Staphylococcus aureus strain "Oxford" by lysozyme and ribonuclease in the presence of normal cells.

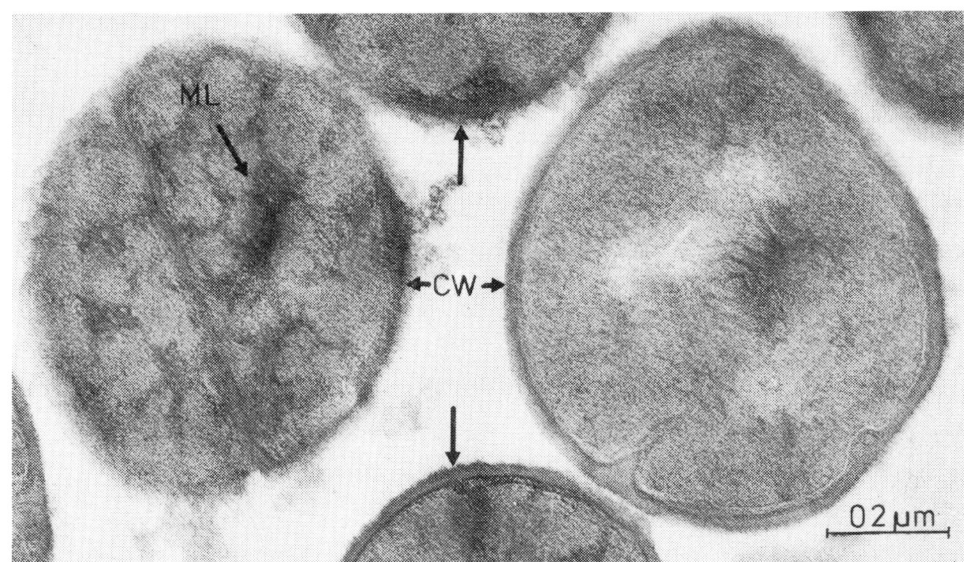

Fig. 6. Staphylococcus aureus: Mixture of boiled (100°C) cells and normal cells after 3 hours incubation. (ML = melted lipids; CW = cell wall) The cell walls are mostly intact.

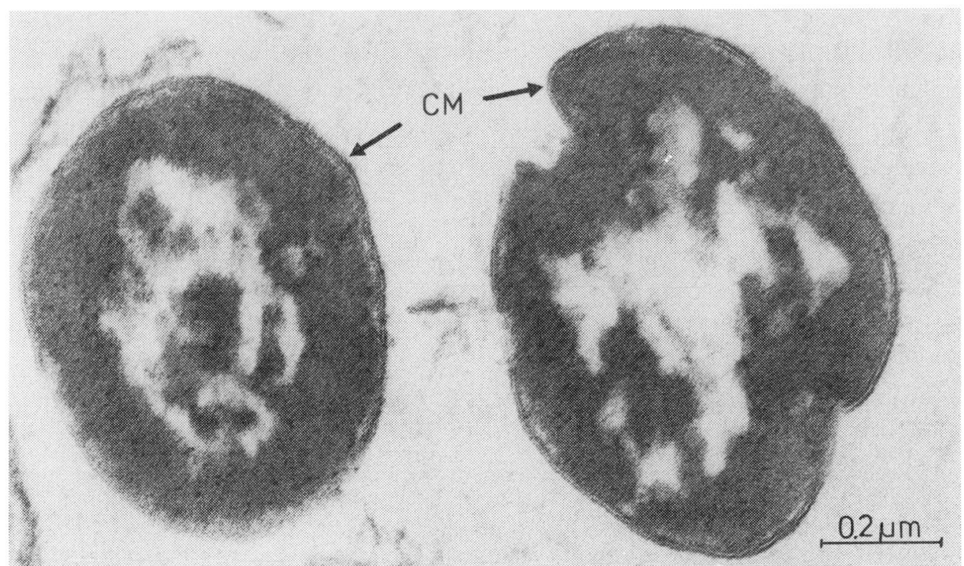

Fig. 7. Staphylococcus aureus: Mixture of boiled (100°C) cells and normal cells after 3 hours incubation together with 1 mg/ml lysozyme. (CM = cytoplasmic membrane) The cell walls of most of the bacteria have been disintegrated.

muralytic activity on cell walls after pre-treatment with 100°C (Fig. 3). With regard to recent findings (2) we must, therefore, conclude that the target structure of cell wall for lysozyme as a muralytic enzyme was affected by heat pre-treatment. The observed accelerations of the release of labelled wall material by the combined action of external autolysins and lysozyme or ribonuclease (compare Fig. 1 with Fig. 4 or Fig. 5) may be due to a "loosening" of the cell walls during heat-treatment.

Summary

Autolytic wall enzymes of staphylococci could be completely inhibited by heat-treatment of bacteria at 100°C, while only a partial inhibition occured at 60°C. When normal cells, which are known to secrete parts of their wall autolysins into the growth medium, were added to heat-inactivated cells in the presence of lysozyme or ribonuclease, increased rates of wall degradation could be observed indicating that the sensitivity of the heated walls to external autolysins had not been reduced. However, the target structure for lysozyme at which lysozyme is known to act as a muralytic enzyme, apparently had been affected within the cell walls by a heat-treatment at 100°C.

References

1. Wecke,J., Lahav,M., Ginsburg,I., Giesbrecht,P.: Arch. Microbiol. 131, 116-123 (1982)
2. Blümel,P., Reinicke,B., Lahav,M., Giesbrecht,P.: This volume.

CATIONIC POLYELECTROLYTES ACTIVATE AUTOLYTIC WALL ENZYMES IN STAPHYLOCOCCUS AUREUS: Modulation by anionic polyelectrolytes in relation to the survival of bacterial constituents in tissues.

Isaac Ginsburg and Meir Lahav
Department of Oral Biology, Hebrew University-Hadassah Faculty of Dental Medicine, Jerusalem 91010, Israel.

Introduction

Although a wealth of knowledge exists today on the biochemical pathways of biosynthesis, turnover and autolysis of bacterial cell wall components *in vitro* (1, 2), surprisingly very little is actually known about the mechanisms of biodegradation of microbial constituents *in vivo*. One should differentiate between bactericidal and bacteriolytic processes induced by leukocytes since killed, but non-degraded, microbial cells may persist within macrophages to trigger chronic inflammation (3, 4). The present communication further supports our contention (5, 6) that the degradation of microbial cell wall components by leukocytes may be due to activation, by leukocytic cationic proteins, of autolytic wall enzymes rather than to the direct cleavage of the cells by lysosomal hydrolases. The modulation of bacteriolysis by anionic polyelectrolytes will be described and discussed in relation to the pathogenesis of chronic inflammation and sequelae.

Results

1) *Effect of polycationic agents on wall lysis*: *Staph. aureus* labeled with ^{14}C-N-acetylglucosamine (NAGA) was harvested from the log. phase, washed in saline and exposed for 15 hrs at 37° in 0.1M acetate buffer pH 5.0 to a variety of cationic polyelectrolytes, and the percentage of solubilized radioactivity was measured. Fig. 1 shows that all the

The Target of Penicillin

cationic agents, except lysozyme, induced a bell-shaped curve of wall lysis. Thus, while at low concentrations (10^{-9} - 10^{-7}M) the agents induced bacteriolysis, this process was blocked by higher concentrations of the polycationic agents (10^{-6} - 10^{-5}M), suggesting that the activation of the autolytic wall enzymes is modulated by these cationic agents. The data also suggest that being highly cationic, these agents might have acted by interacting with the protoplast membrane as suggested (7), and that membrane damage may be linked with the activation of autolytic wall enzymes (6). We further studied this phenomenon by subjecting M. lysodeikticus (M.L.) to the effect of human lysozyme after the cells had been precoated with cationic polyelectrolytes. Bacteriolysis was also drastically inhibited by polyarginine, polylysine, polyornithin and by protamine, when employed at 10^{-6}M. Although the mechanisms of inhibition of lysis are not fully known, it is possible that all these agents stabilized the bacterial protoplasts following the degradation of the peptidoglycan, as suggested (8).

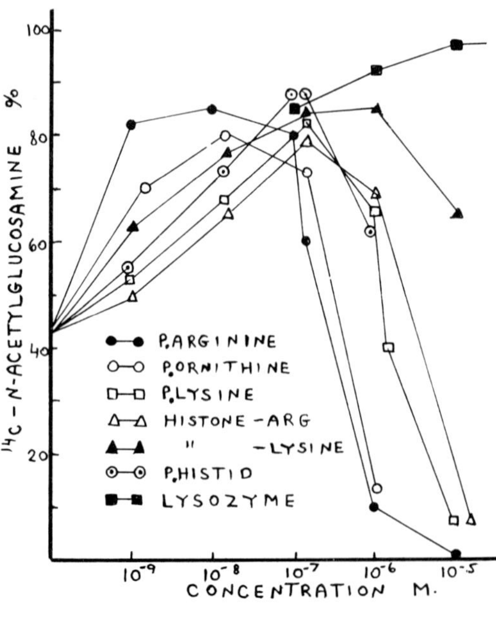

Fig. 1:
Effect of polycationic agents on cell wall lysis. Note that the spontaneous wall lysis can be inhibited or activated.

We also found that liquoid (polyanethole sulfonic acid, a synthetic anticoagulant) at very low concentration also markedly inhibited the lysis of M.L. by lysozyme as well as the degradation of the cell walls of Staph. aureus following treatment either with buffer alone (spontaneous lysis) or with a variety of cationic and membrane-damaging agents (5). This anionic agent may be thus used as a probe to study activation of autolysins.

2) Mechanisms of cell wall lysis: To further eludicate the mechanisms of wall lysis we preheated ^{14}C-NAGA-staphylococci to 100°C (to destroy their autolytic wall enzymes) and employed them as targets for LYZ or to other cationic agents in the absence or presence of added intact unlabeled staphylococci (see 6).

REACTION MIXTURE		% WALL LYSIS
^{14}C-STAPHYLOCOCCI*	+ BUFFER	30
"	+ LYSOZYME (LYZ) 50 µg/mL**	87
^{14}C-STAPHYLOCOCCI-100°C	+ LYZ	15
"	+ STAPHYLOCOCCI	40
"	+ STAPHYLOCOCCI + LYZ	90
"	+ STAPHYLOCOCCI – LIQUOID*** + LYZ	15
"	+ " – DEX-SO_3*** + LYZ	25
^{14}C-STAPH-C.WALLS****		2
"	+ LYZ	5
"	+ STAPHYLOCOCCI	35
"	+ " + LYZ	95

Table 1. Lysis of heat-treated staphylococci by lysozyme
*) labeled with NAGA;
**) grown with 1 mg/ml of the polyanion;
***) cell walls cleaned off cytoplasmic constituents.

Table 1 shows that whereas intact labeled staphylococci were lysed by LYZ (or by RNase--not shown), heat-treated cells were refractory to lysis but could be lysed if intact unlabeled cells were added to the system (also see 9). No lysis occurred, however, if liquoid-grown cells were used instead of regular cells. The Table also shows that staphylococcal cell walls presumably devoid of autolysins can be lysed by LYZ upon addition of intact staphylococci which act as donors of autolysins. Since liquoid proved a potent inhibitor of cell wall lysis, we also investigated bacteriolysis by exposing staphylococci for short periods either to LYZ or to liquoid followed by the removal of unbound agents. Fig. 2 (part B) shows that the pretreatment of bacteria with liquoid for 15 min was sufficient to block both the spontaneous and the LYZ-induced lysis. Similarly, the pretreatment of the cells with LYZ followed by its removal by washing and the subsequent addition of liquoid also blocked lysis (part B), suggesting that liquoid was capable of either preventing the access of LYZ to its target on the cell membrane or that it inactivated the autolysins which had been triggered by LYZ action (10). Since a similar pattern of inhibition was also obtained with Dextran-SO_3 and with other anionic substances, likely to be present in tissues, it is plausible that such anionic agents may also modulate cell wall lysis *in vivo*.

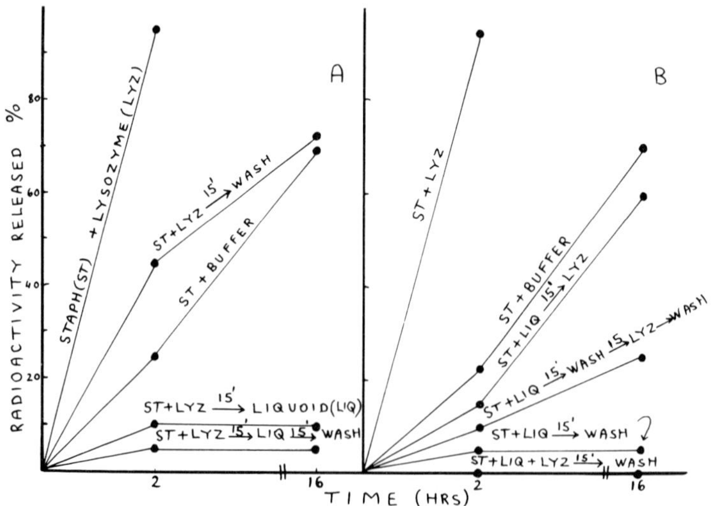

Fig. 2. Effects of liquoid on lysis of *Staph. aureus* by lysozyme. The bacteria were either pretreated with lysozyme or with liquoid, washed and then treated again by either of these agents.

Note that liquoid totally shuts down cell wall lysis.

3) Effect of anionic agents on lysostaphin lysis: To establish whether liquoid and other anionic agents also inactivated another lytic system we tested lysis of staphylococci by lysostaphin. We also found that both liquoid and Dextran-SO_3 markedly inhibited lysostaphin action when tested at pH 5.0 but not at pH 7.4. On the other hand, neither heparin nor DNA, hyaluronate or poly-glutamate had any effect at either pH value employed. Lysostaphin lysis could also be inhibited by poly-L-arginine.

4) Degradation of staphylococci *in vivo*: The ability of liquoid to block autolytic wall enzyme *in vitro* also suggested that liquoid-grown staphylococci would perhaps also resist degradation by leukocytes *in vivo*. Preliminary results have indeed shown that whereas ordinary

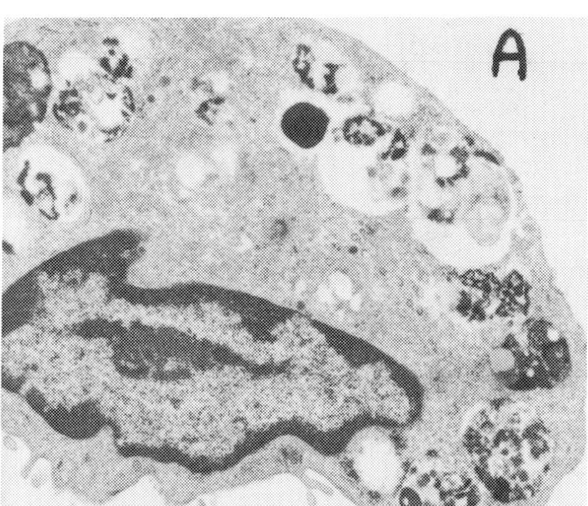

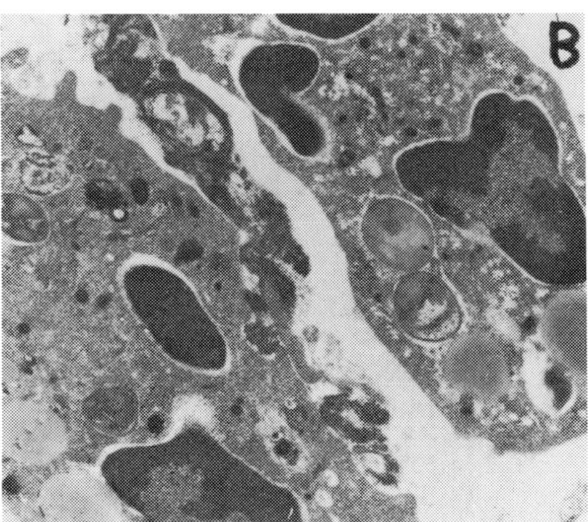

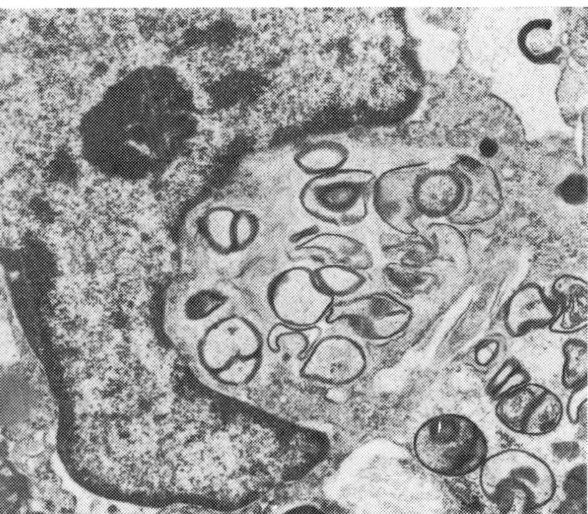

Fig. 3.
Degradation of staphylococci in joint lesion leukocytes.
A - 24 hrs following injection of intact bacteria; B - 24 hrs following injection of liquoid-grown bacteria; and C - 4 days following injection of intact bacteria. Note degradation of intact cells (A), stability of liquoid-bacteria (B), and accumulation of cell walls in macrophages (C).

staphylococci underwent very drastic morphological changes within phagosomes of rat granulocytes and macrophages in joint lesions 24-48 hrs following injection (Fig. 3A), the liquoid-grown cells appeared to be relatively intact (Fig. 3B), suggesting that liquoid stabilized the bacterial cells against attack by lysosomal enzymes. A very important observation made was that neither the control cells nor the liquoid-grown bacteria underwent any substantial loss of cell wall morphology 72 hrs to 14 days after injection (Fig. 3C), suggesting that autolytic wall enzymes may be shut down in *in vivo* milieu. Presumably this effect was caused by the accumulation in the inflammatory exudate of anionic polyelectrolytes which counteracted the activity of the autolytic wall systems (4, 11). The data represented suggest, therefore, that while cationic agents may be responsible for activating autolytic wall enzymes in bacteria, this process is tightly controlled by anionic polyelectrolytes which may also control cell wall lysis *in vivo*.

References

1. Rogers, H.J., Perkins, H.R., Ward, J.B.: Microbial Cell Walls and Membranes. Chapman and Hall, London, New York (1980).
2. Giesbrecht, P., Wecke, J.: Proc. 7th European Congress on Electron Microscopy, Vol. 2, The Hague, pp. 446-453 (1980).
3. Ginsburg, I., Sela, M.N.: Crit. Rev. Microbiol. 4, 249-332 (1976).
4. Ginsburg, I.: In: Lysosomes in Applied Biology and Therapeutics. Jacques P., Dingle, J. (eds.), North Holland Publ., Amsterdam, Vol. 6, 327-406 (1979).
5. Lahav, M., Ginsburg, I.: Inflammation 2, 165-177 (1977).
6. Ginsburg, I., Lahav, M., Giesbrecht, P.: Inflammation 6, 401-417 (1982).
7. Spitznagel, J.K.: J. Exp. Med. 114, 1079-1091 (1961).
8. Grossowicz, N., Ariel, M.: J. Bact. 85, 293-300 (1963).
9. Wesmacott, D., Perkins, H.R.: J. Gen. Microbiol. 115, 1-11 (1979).
10. Wecke, J., Lahav, M., Ginsburg, I., Giesbrecht, P.: Arch. Microbiol. 131, 116-123 (1982).
11. Ginsburg, I., Goultchin, J., Ne'eman, N., Lahav, M., Landstrom, L., Quie, P.G.: Agents and Actions, Suppl. 7, 260-270 (1980).

PART V

STRUCTURE AND FUNCTION OF PENICILLIN-BINDING PROTEINS

PENICILLIN BINDING PROTEINS AS TARGETS OF THE LETHAL EFFECTS OF β-LACTAM ANTIBIOTICS

Jack L. Strominger

Department of Biochemistry and Molecular Biology
Harvard University
7 Divinity Avenue
Cambridge, Massachusetts 02138, U.S.A

The fact that all bacterial species contain multiple penicillin-binding proteins, and that some of these at least are targets of penicillin lethality was discovered in 1972. How far have we come in the past decade in answering the following questions which were posed at that time?

1. What are the *in vivo* functions of the PBPs?

Do we know the functions of the PBPs in *any* microorganism? No, not even in *E. coli* are we sure of what the function of all 6 or 7 or 9 or 12 PBPs are. The numerology of how many PBPs there are depends on how the experiment is performed and what is counted as a PBP. For example, are PBP 1B-α, -β and -γ separate PBPs? Do they have some slightly different functions, or are they the result of an artifact that occured during the biosynthesis of these PBPs or during their isolation? Various additional PBPs called PBPs 3A, 4*, 5*, 7 and 8 have been observed, but we know almost nothing about them.

2. Are there multiple killing sites for penicillins in bacteria?

In fact several of the PBPs are killing sites, i.e. PBPs 1B, 2 and 3, as shown by Brian Spratt for example. Not all of the PBPs are essential; at least PBPs 4, 5 and 6 are not essential under the conditions in which we grow bacteria in the laboratory.

The Target of Penicillin

However, bacteria don't ordinarily live in the laboratory, so the question is: Are there ecological situations in which those PBPs which we believe are non-essential are really essential, for example, *E. coli* growing in the gastrointestinal tract? If not, why have those PBPs survived evolution? In particular, one is struck by the fact that the majority of the PBPs quantitatively in almost all microorganisms are these "non-essential" PBPs. For example, PBPs 5 and 6 of *E. coli* account for over 80% of the total PBPs of the organism.

3. Is binding of penicillins to PBPs sufficient to cause cell death?

As shown by Alex Tomasz and others, acylation of a PBP is not the only event which is necessary for cell death. Mechanisms involving bacteriolysins (autolysins) which operate after cell wall synthesis ceases lead to the death of the bacterial cell.

4. Are there any PBPs which escape detection by the methods used?

Some PBPs escape detection by labeling with penicillin G. One example is from the work of Gunnar Kleppe on *Bacillus subtilis*. Originally using 6-APA affinity columns only one PBP 2 was detected. In fact there are two components in the molecular weight range of PBP 2, now called PBP 2A and PBP 2B. One of them, PBP 2B, does not react with penicillin G or with 6-APA, but it reacts with cephalosporins. Chris Buchanan has shown that PBP 2B may have some importance in sporulation.

5. What enzymatic reactions are catalyzed by individual PBPs?

Several laboratories (Matsuhashi, Hirota and ourselves) have all begun to study the enzymatic reactions, particularly those catalyzed by the high molecular weight PBPs. However, the precise enzymatic reactions catalyzed by these PBPs are unknown, nor is it known if they catalyze the same reactions. If they catalyze the same reactions, then why is it necessary to have more than one, and how did more than one happen to evolve? It seems unlikely that they catalyze exactly the same biochemical reaction. With regard to the low molecular weight PBPs in *E. coli*, PBPs 4, 5 and 6, these all catalyze D-alanine carboxypeptidase reactions *in vitro*, and it is commonly believed that they are responsible for limitation of the extent of cross-linking of peptidoglycan strands, i.e. an acetylmuranyl-pentapeptide which is converted by one of these enzymes to acetylmuranyl-tetrapeptide could no longer be a donor for the usual type of transpeptidation reaction. However, there is no evidence to support this view, and a certain amount of evidence against it. At least one of these PBPs may act as a secondary transpeptidase *in vivo*.

6. What is the mechanism by which β-lactam antibiotics inhibit PBPs, and are β-lactam antibiotics really substrate analogues?

It has been clearly shown that PBPs are active site directed acylating agents. It seems most likely that they are substrate analogues. Crystallographic studies of PBPs which are in progress will give the answer to this question in detail.

7. What are the evolutionary relationships amongst the PBPs? What is the relationship of PBPs to β-lactamases?

Surely, the PBPs must be related to each other, but at the moment there is no firm information as to how much homology between different PBPs exists. Did ß-lactamases arise in evolution by modification of the PBPs so that the penicilloyl residue could be attacked by water? Although there is some homology between PBPs and ß-lactamases, it is very limited. If there is an evolutionary relationship, it is a distant evolutionary relationship. Yet the fact that there is some homology and that many PBPs catalyze a weak penicillinase reaction, i.e. they release the bound penicillin, certainly suggests that β-lactamases may have evolved from the PBPs.

So, a decade after the occurrence of multiple PBPs was detected, only incomplete answers to many of the questions which arose have been obtained. However, many of the critical features of the suggested mechanism of inhibition of transpeptidases by penicillins which was proposed in 1965 have now been shown to be correct. The bound form of penicillin is known to be the penicilloyl derivative of penicillin, and acyl enzymes in which the acyl group is derived from substrate have been shown to occur in the case of some PBPs, particularly in the case of the low molecular weight PBPs. However, the occurrence of acyl enzymes or intermediates in the transpeptidation reactions catalyzed by the high molecular weight PBPs has not yet been demonstrated. Definitive evidence that penicillins are substrate analogues, i.e. that the penicillin lies in the active site in the same way that substrate lies in the active site, will come from crystallographic studies. It seems difficult to imagine any other mode of action, but many things that have been difficult to imagine have turned out to be true. A penicilloyl enzyme would become a penicillinase if some modification occurs so that water can enter the active site and react with the penicilloyl residue. What kind of change in the catalytic site could produce that result? (See figure 1.)

We knew a long time ago that, although penicillin and substrate might be similar, they were not isosteric. In particular the distortion produced by the β-lactam ring around the C(=O)-N bond results in a considerably different bond angle than that in the normal substrate. Does the enzyme somehow bend the substrate into something like the conformation of the β-lactam ring before the peptide bond involved in transpeptidation is cleaved? That kind of question will certainly be answered in the next decade as we learn more about the mechanisms of these enzymes.

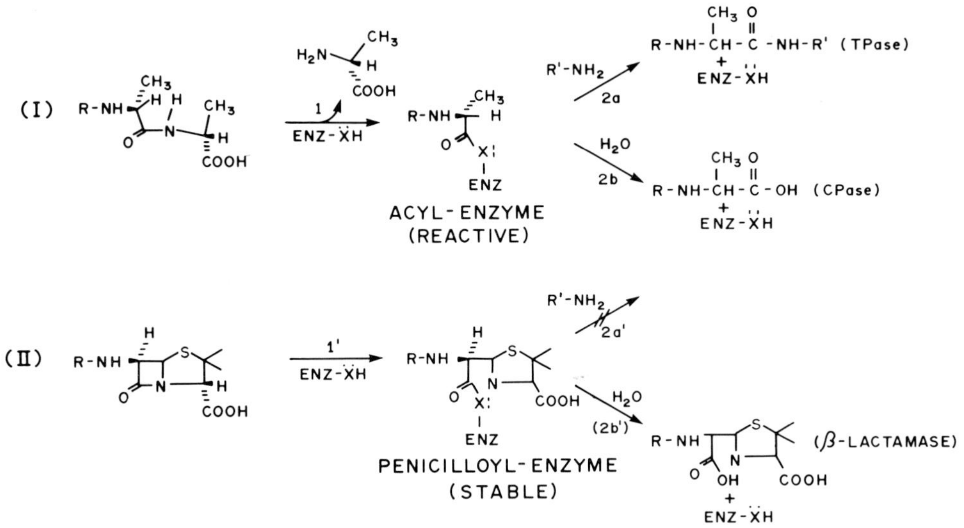

Fig. 1: Mechanism of enzymatic reactions catalyzed by PBPs and of their inhibition by ß-lactam antibiotics

Another extremely interesting question is the mechanism of the elimination reaction which some PBPs catalyze, i.e. namely with some PBPs the penicillin is slowly released as penicilloic acid (penicillinase activity), but with other PBPs, a cleavage of the C5-C6 bond in the penicilloyl-enzyme occurs (Fig. 2). As the result some compound related to dimethylthiazoline carboxylate and phenylacetylglycine are formed. The mechanism of that cleavage is not understood. In fact, we have recently found that some PBPs catalyze both of these reactions. For example, *E. coli* PBP 5 catalyzes both the release reaction to give penicilloic aicd and the cleavage reaction to yield phenylacetylglycine, and so does PBP 6. The proportion of the two is different with those two PBPs. What is the mechanism of that cleavage reaction?

The structure of the PBPs is also an interesting question, including the manner of their orientation in the cell membrane. We studied two in some detail, PBP 5 of *B. subtilis* and PBP 5

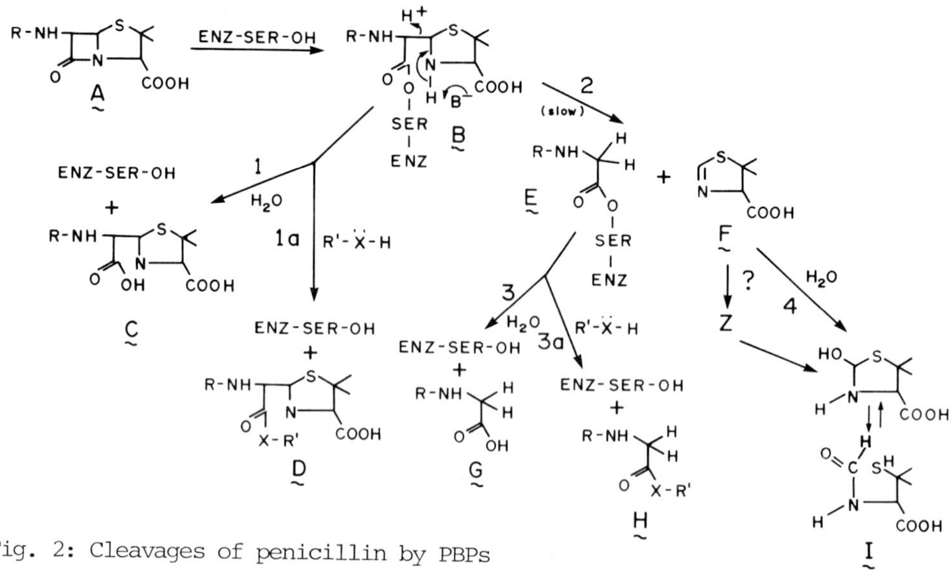

Fig. 2: Cleavages of penicillin by PBPs

of *B. stearothermophilus*. Both are D-alanine carboxypeptidases, at least *in vitro*. The catalytic sites of these molecules are near the amino terminus, and in fact in the case of these low molecular weight PBPs, 15,000 and 18,000 daltons proteolytic cleavage fragments retain the specific penicillin binding activity of these proteins. The catalytic activity in one case appears to become less stable to heat denaturation, so that it is possible that the mass of the protein evolved to stabilize the catalytic portion which is at the N-terminal end of the protein. At the carboxy terminus is a short "hydrophobic" region which is involved in anchoring these proteins in the membrane. PBP 5 of *E. coli* has a similar size to these *Bacillus* enzymes and catalyzes similar enzymatic reactions. However, as Brian Spratt showed, it does not have a particularly hydrophobic C-terminus. The sequences of the two C-termini from the bacilli are also not particularly hydrophobic. They contain several charged amino acids (lysine, aspartic acid) as well as several hydroxy amino acid residues (serine). Nevertheless, if the C-terminus of these proteins is removed, they no longer bind detergents, so the C-terminal region must be the region for insertion in the membrane. Molecular models of the C-termini in an α-helical configuration were constructed. These models showed that the C-termini have two faces, a hydrophobic face and a hydrophilic face. All of the residues on one face are hydrophobic, but on the other face all the hydrophilic residues occur. There would be an enormous energetic cost to insert such a structure into a lipid bilayer. How then does that "hydrophobic" region insert into a lipid bilayer? We have thought of two possibilities. There are probably more. They could pack as dimers or trimers into the membrane with the hydrophilic faces facing each other. Alternatively they may not go through the membrane; they could lie parallel to the membrane with the hydrophilic face facing the polar head groups of the lipid bilayer. Finally, nothing is known about where in membranes or in which direction in membranes these PBPs are located, or whether any of them can be located in the outer membrane as well as in the inner membrane of E. coli.

Isolation of acyl enzyme derivatives in which the acyl group is derived from substrate is another interesting question. A highly reactive ester substrate in which D-lactic acid replaces D-alanine (diacetyl-L-lysyl-D-alanine-D-lactate) was used to isolate acyl enzyme intermediates. That was all done with the low molecular weight PBPs of several bacteria. If the high molecular weight PBPs are used, this substrate does not work. A mixture of B. subtilis PBPs and isolated PBP 1 from B. subtilis were both used. A slow acylation occurs, but it stops, and only about 1% of these PBPs could be acylated. We do not presently understand why acyl enzyme intermediates from the high molecular weight PBPs cannot be trapped with this highly reactive substrate. Matsuhashi and his colleagues have shown that some of these proteins are double-headed enzymes, i.e. the high molecular weight PBPs catalyze transglycosylase reactions as well as transpeptidase reactions. Perhaps they are catalytically not very active until some appropriate substrate occupies the transglycosylase site. The synthetic ester substrate may simply be the wrong synthetic substrate.

A synthetic substrate with an appropriate carbohydrate moiety to occupy the transglycosylase site would be needed, much as acetyl-D-alanyl-D-alanine itself is not a substrate for the D-alanyl-D-alanine carboxypeptidases. However, both diacetyl-L-lysyl-D-alanyl-D-alanine and its ester analog are excellent substrates. Some kind of carbohydrate moiety may be required in these synthetic substrates for the high molecular weight PBPs. It is a constant struggle to isolate and keep good natural substrates for these enzymes; they are derivatives of C_{55}-isoprenyl alcohol. Real progress may not occur until synthetic substrates for these enzymes are available.

The enzymatic reactions catalyzed by the high molecular weight isolated PBPs are assayed under very unusual conditions. The activity of the isolated high molecular weight PBPs may be demonstrated in two ways. One way is to carry out the incubation on Whatman 3MM filter paper. In 1964 Pauline Meadow first described this effect of filter paper in describing the enzymatic synthesis of peptidoglycan in *S. aureus*. Matsuhashi and colleagues have found a way to make the reactions go in the test tube, viz. by addition of large amounts of solvent. With *B. stearothermophilus* or *B. subtilis* high molecular weight PBPs, the *in vitro* reaction goes optimally in the presence of 10% ethylene glycol, 10% glycerol and 5% methanol. All of the solvents used are polar hydrox solvents. In addition, high concentrations of N-acetylglucosamine in the absence of solvent, as well as several other sugars will also work. It is possible, therefore, that the filter paper effect and the solvent effect are a means of providing an accepto for the transglycosylation reaction, much as Neuhaus has describe that the melting of the cell wall fraction and the membrane fraction together activates these reactions. That is probably also an acceptor effect; it may introduce the acceptor into the proximity of the enzyme. In studying these cell-free reactions, the nature of the acceptor is as important as the nature of the donor substrate.

Finally another interesting problem is the dac A mutant of E. coli PBP 5. This mutant PBP 5 binds radioactive penicillin G and accumulates acyl enzyme intermediate, but it releases neither penicillin nor the acyl group. PBP 5 and PBP 6 of E. coli have quite different intrinsic D-alanine carboxypeptidase activities. In the wild type E. coli they both catalyze the carboxypeptidase reaction, but in the mutant strain PBP 5 has no carboxypeptidase activity, although it binds and is easily detected by binding of penicillin G. Although the wild type enzyme releases bound penicillin, the mutant enzyme does not release it at all in the presence of water. However, it can be released from the mutant enzyme in the presence of hydroxylamine; this strong nucleophile catalyzes an enzymatic release of the bound penicillin from the mutant PBP 5. This unusual mutation in PBP 5 dissects two different parts of the enzymatically catalyzed reaction; in the mutant PBP 5, k2, the acylation constant, is normal, but k3, the deacylation constant, is essentially zero. By studying this mutant enzyme, something of importance about the mechanism of the release reaction might be learned. In addition, although the mutant enzyme accumulates acyl enzyme, it does so very slowly.

Study of these and other residual problems is certain to provide much interesting information in the future.

References

This manuscript is an edited transcript of the presentation at the meeting. A fuller account and a complete list of references can be found in David J. Waxman and Jack L. Strominger, "Penicillin-Binding Proteins and the Mechanism of Action of β-Lactam Antibiotics", Annual Review of Biochemistry 52: 825-869 (1983).

This work was supported by research grants from NIH (AI-09152) and NSF (PCM-78-24129).

BINDING SPECIFICITIES OF PENICILLIN BINDING PROTEINS - A CONFORMATIONAL APPROACH

V.S.R. Rao

Molecular Biophysics Unit, Indian Institute of Science,
Bangalore 560 012, India.

Introduction

Penicillins and cephalosporins are widely used as antibacterial agents because of their broad-spectrum of activity and low toxicity. The widespread use of these antibiotics led to the emergence of resistant bacteria. These strains achieve their resistance mainly by producing β-lactamase enzymes which inactivate the antibiotic molecule by hydrolyzing the lactam peptide bond before it reaches the target enzyme. This suggests that at least one of the favoured conformations of the antibiotic may have either full or partial structural complimentarity to the binding site of the enzymes. Hence, a knowledge of various conformational manifolds of these antibiotics may throw light on the binding specificities of transpeptidases and β-lactamases.

Results and discussion

Crystal structure data (1-9) on several β-lactam compounds shows that the length of the C-N bond and the dihedral angle (ω) around C-N (Fig. 1) deviate significantly from their normal values. Depending on the nature of the rings fused, ω deviates by as much as 60°. We have recently carried out *ab initio* molecular orbital calculations on the β-lactam unit

The Target of Penicillin

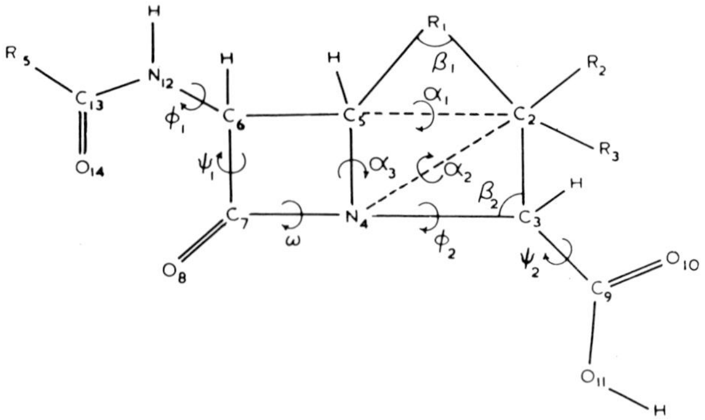

Fig. 1. Schematic representation of a β-lactam compound (penicillins). The dihedral angles are indicated.

Table 1. Calculated and observed C-N bond lengths in various β-lactams*

β-lactam antibiotics	Dihedral angle (ω^{o}) of the lactam peptide bond	STO-3G optimized C-N bond lengths (Å)
Δ^2-cepham	176	1.405 (1.350)[a]
Δ^3-cephams	162	1.415 (1.370)
Phenams	139	1.430 (1.390)
Thienamycins	120	1.445 (1.420)

* S. Vishveshwara and V.S.R. Rao (J. Mol. Struct., in press)
[a] The values in parentheses are taken from Refs. 1-9.

(Table 1). The optimised C-N bond length is 1.405 Å, 1.41 Å and 1.43 Å for ω around 176°, 162° and 139° respectively. This is in qualitative agreement with the experimental values.

The table shows that the planar nitrogen leads to a shorter peptide bond. The lactam peptide bond in cephalosporins and penicillins has about 1-2 kcal mol^{-1} and 5-6 kcal mol^{-1} respectively higher energy over the planar form. These studies suggest that non-planarity of nitrogen and the conjugation by inclusion of a double bond leads to weakening of the peptide bond. Thus, the peptide bond in penicillin is weakened mainly due to the pyramidal character of nitrogen and in cepham (active cephalosporin) due to resonance with the double bond. In highly active thienamycins the peptide bond is weakened by both the mechanisms.

There have been some attempts to correlate only the non-planarity of the lactam peptide bond to biological activity. However, our earlier conformational analysis (10, 11) on a number of β-lactam antibiotics showed that correlating only the non-planarity of lactam peptide bond to the antibacterial activity of the drug is an oversimplification. It was suggested that since the antibiotic should have a proper conformation to fit in the active site of the enzyme before the cleavage of the peptide bond, its overall shape is also important besides the non-planarity of the lactam peptide bond. The recent X-ray crystallographic studies (12) on a penicillin-sensitive D-alanyl carboxypeptidase-transpeptidase has also revealed that the antibiotic binding site overlaps the active site on the enzyme, thus supporting the substrate analog hypothesis of Tipper and Strominger (13). Virudachalam and Rao (14) have shown that the lactam ring is mainly responsible for the conformational similarity between the antibiotic and the D-ala-D-ala fragment of the substrate.

Δ^2-penems are the simplest among the β-lactam compounds which exhibit good antibacterial activity. The bicyclic ring system in these compounds favour a single conformation ($\alpha_1 \sim 5^o$, $\alpha_2 \sim 5^o$ and 180 , $\alpha_3 \sim 120^o$). The lactam peptide bond is significantly non-planar ($\omega = 120^\circ$) and the carboxyl group assumes an orientation with $\phi_2 = 75^\circ$. The high antibacterial activity of these compounds suggests that this conformation has high complimentarity to the active site of the receptor enzyme. If this is so, any deviation from these parameters will lead to improper fit in the active site and hence to weak activity.

In penicillins (10), the five-membered ring favours two different conformations, $C_2(\alpha_1, \alpha_2 \simeq 40^\circ, 10^\circ)$ and $C_3(\alpha_1, \alpha_2 \simeq -10^\circ, -25^\circ)$. The lactam peptide bond is non-planar. The carboxyl group favours $\phi_2 = 110^\circ$ and 150° in C_2 and C_3 conformations respectively. If penicillins (G, V or ampicillin) were to bind in the same mode of binding as Δ^2-phenems, these differences in the conformations of the bicyclic ring system may lead to some unfavourable interactions with the atoms in the active site of the enzyme. Hence these compounds may change their orientation of binding slightly from those of the thienamycins which may lead to weak activity. This explains the weak activity exhibited by the penicillin nucleus. However, the addition of suitable side groups as in penicillin G or V improves its antibacterial activity. This suggests that a side group with the proper conformation can act as a handle with which the enzyme can easily reorient the nuclear part of the molecule, not only to relieve any bad contacts but also for the proper positioning of the functional groups. Thus the 6-β side chain seems to be mainly responsible for the good antibacterial activity of penicillins.

Penicillin G, D-ampicillin and 3-pyridyl methyl penicillin which exhibit good activity against Gram positive bacteria favour similar (compact) conformations (Fig. 2) (15). This suggests that the compact conformation is either associated with biological activity or initiates the binding process and undergoes some conformational changes during binding.

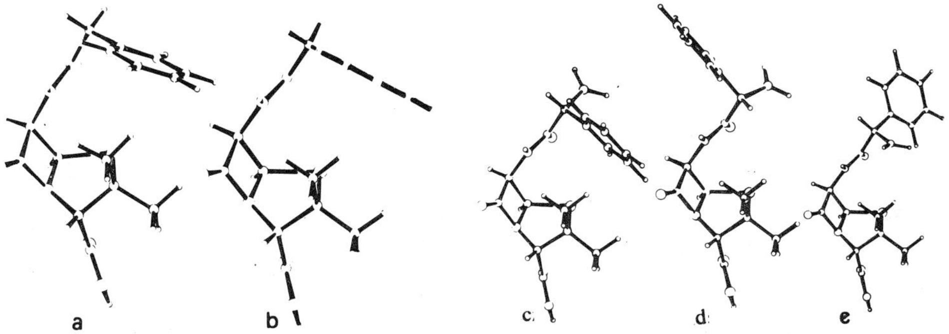

Fig. 2. Projections of minimum energy conformations of (a) penicillin G, (b) 3-pyridyl methyl penicillin, (c) D-ampicillin, (d) and (e) L-ampicillin, two conformations having energy (0.4 kcal mol^{-1}) higher than minimum.

The compact conformation of L-ampicillin has $\sim$2.8 kcal mol^{-1} higher energy. Even in this conformation the orientation of the amino group is different from that of D-ampicillin. On the other hand, in a conformation which has about 0.4 kcal mol^{-1} higher energy, the amino group has a similar orientation to that in D-ampicillin but in this conformer the phenyl ring assumes a different orientation (Fig. 2(d)). Since the side group is flexible, by rearrangement a compromise could be made in positioning the amino group and the aromatic group. Such an arrangement may lead to the expenditure of part of the binding energy and hence results in weak activity.

It is also interesting to look at the effect of the size of the 6-β side group on the antibacterial properties of Gram positive and negative bacteria. Some of the compounds which we have studied are displayed in Fig. 3. The first four compounds differ in the configuration at the C^{α} atoms of the side group; compounds 5, 6 and 7 differ in the nature of the group in the side chains. Compounds 1, 6 and 7 favour extended conformations (Fig. 4), though the nature of the

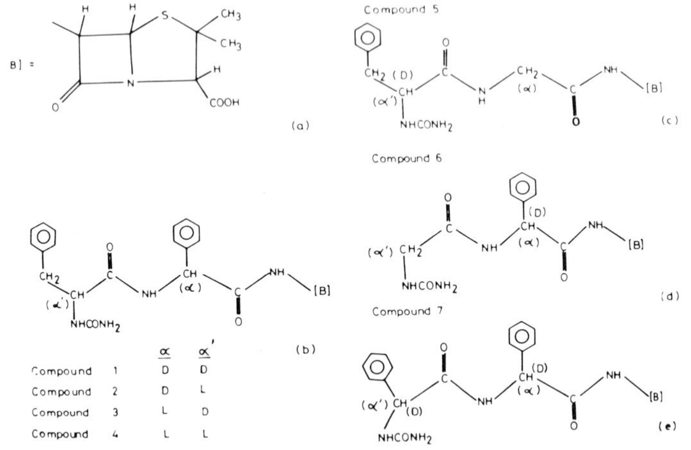

	α	α'
Compound 1	D	D
Compound 2	D	L
Compound 3	L	D
Compound 4	L	L

Fig. 3. Chemical structure of 6-[α-(α'ureido acylamino) acylamino] penicillanic acids.

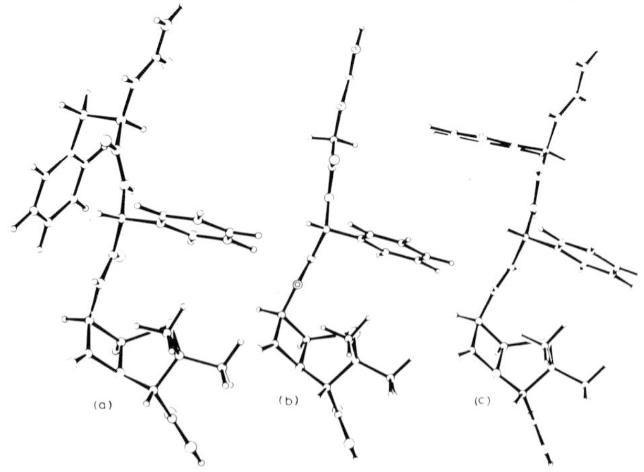

Fig. 4. Projections of minimum energy conformations of (a) compound 1 (b) compound 6 and (c) compound 7.

substituent at the second chiral centre C_{27} differs. The phenyl ring at the first chiral centre C_{17} folds over the bicyclic ring system, suggesting that this part of the molecule assumes a shape as favoured in penicillin G and D-ampicillin. Since these compounds (1, 6 and 7) exhibit good antibacterial properties, this suggests that the nature of the substituent at the second chiral centre, C_{27}, does not affect the antibacterial properties against Gram positive bacteria.

The favoured conformation of compound 2 (Fig. 5a) differs from 1 in the orientations of the group at C_{27}. The rest of the fragment in both has a similar conformation. The fact that both compounds 1 and 2 exhibit good antibacterial activity against Gram positive bacteria suggests that either cross-linking transpeptidases in this bacteria accommodate such changes or the part of the molecule beyond C_{17} might not interfere sterically in the active site of the cross-linking enzyme.

Compounds 3 and 4 favour conformations quite different from 1, 6 and 7. In the minimum energy conformer of compound 4

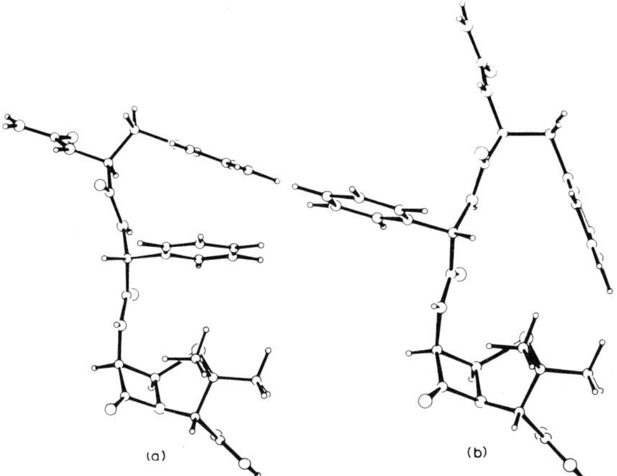

Fig. 5. Projections of minimum energy conformations of (a) compound 2 and (b) compound 4.

(Fig. 5b) and slightly higher energy (~0.24 kcal mol^{-1}) conformer of compound 3 (16), the fragment of the molecule upto the first chiral centre, C_{17}, resembles the favoured conformation of L-ampicillin. This resemblance perhaps explains their moderate activity against Gram positive bacteria. The minimum energy conformer of compound 5 (16) differs from 1. One of its higher energy (~2 kcal mol^{-1}) conformers has a close similarity to the favoured conformer of 1. Since the probability of occurrence of this conformer in solution will be low this may lead to its reduced activity.

Except 1, 6 and 7, therefore, the rest of the compounds show either poor or negligible antibacterial activity against Gram negative bacteria. This suggests that the changes in conformation that occur beyond the C_{17}-atom affect, to a great extent, the antibacterial activity against Gram negative bacteria. These results suggest that the cross-linking transpeptidases of Gram negative bacteria are much more specific than those of Gram positive bacteria.

References

1. Domiano, P., Nardelli, M., Balsamo, A., Macchia, B. and Macchia, F.: Acta Cryst. B35, 1363 (1979).
2. Hadgkin, I.D.C. and Maslen, E.N.: Biochem. J. 79, 395 (1961).
3. Slusarchyk, W.A., Applegate, H.E., Funke, P., Koster, W., Puar, M.S., Young, M. and Dolfini, J.E.: J. Org. Chem. 38, 943 (1973).
4. Van Meerssche, M., Germain, G., Declercq, J.P., Coeme, B. and Moreanch, C.: Cryst. Struct. Comm. 8, 281, 287 (1979).
5. Sweet, R.M. and Dahl, L.F.: J. Am. Chem. Soc. 92, 5489 (1970).
6. Paulus, Von. E.F.: Acta Cryst. B30, 1605, 1608, 2915, 2918 (1974).
7. Paulus, Von. E.F.: Acta Cryst. B33, 108 (1977).

8. Pfaendler, H.R., Gostcli, J., Woodward, R.B. and Rihs, G.: J. Am. Chem. Soc. 103, 4526 (1981).

9. Albers-Schonberg, G., Arison, B.H., Hensens, O.D., Hirshfield, J., Hoogsteen, K., Kaczka, E.A., Rhodes, R. E., Kahan, J.S., Kahan, F.M., Ratcliffe, R.W., Walton, E., Ruswinkle, L.J., Morin, R.B. and Christensen, B.G.: J. Am. Chem. Soc. 100, 6491 (1978).

10. Joshi, N.V., Virudachalam, R. and Rao, V.S.R.: Curr. Sci. 47, 933 (1978).

11. Vasudevan, T.K. and Rao, V.S.R.: Biopolymers, 20, 865 (1981).

12. Kelly, J.A., Moews, P.C., Knox, J.R., Frere, J.M. and Ghuysen, J.M.: Science, 218, 479 (1982).

13. Tipper, D.J. and Strominger, J.L.: Proc. Natl. Acad. Sci. U.S.A. 54, 1133 (1965).

14. Virudachalam, R. and Rao, V.S.R.: Biomolecular Structure, Function and Evolution, Vol. 2, Srinivasan, R., Pergamon Press, Oxford 1981.

15. Vasudevan, T.K. and Rao, V.S.R.: Int. J. Biol. Macromol. 4, 347 (1982).

16. Vasudevan, T.K.: Ph.D. Thesis, Indian Institute of Science, Bangalore, India (1981).

17. Ferres et al.: J. Antibiotics, 31, 1013 (1978).

THE ACTIVE SITES OF THE D-ALANYL-D-ALANINE-CLEAVING PEPTIDASES

Paulette Charlier, Otto Dideberg, Georges Dive, Jean Dusart, Jean-Marie Frère, Jean-Marie Ghuysen, Bernard Joris, Josette Lamotte-Brasseur, Mélina Leyh-Bouille and Martine Nguyen-Distèche

Université de Liège, Belgium. Service de Microbiologie appliquée aux sciences pharmaceutiques, Institut de Chimie, B6, B-4000 Sart Tilman; Service de Cristallographie, Institut de Physique, B5, B-4000 Sart Tilman; Service de Chimie analytique, Institut de Pharmacie, F1, Rue Fusch, 3-5, B-4000 Liège.

Introduction

Thousands of β-lactam compounds have been isolated or made, characterized and tested as antibacterial agents. Those few which have found a place in medicine came primarily from the exploitation of large screening programs and chance observations. "The reason for this has been a lack of knowledge of the three-dimensional active sites of the various enzymes with which β-lactam antibiotics react" (Abraham, 1981). This, of course, applies to β-lactamases and D-alanyl-D-alanine-cleaving peptidases (in short DD-peptidases).

Despite progress in different areas of structural research, crystallography is still the only way to determine the spatial structure of proteins (enzymes) to atomic resolution. The DD-peptidases are membrane-bound and membrane proteins are inherently difficult to obtain in a suitable crystalline form. In cases where the DD-peptidases are merely anchored to the membrane by a hydrophobic protein tail, one strategy may be to use proteases that split the enzymes into a hydrophilic and a hydrophobic part. The hydrophilic part (with, hopefully, the intact active site) can then be submitted to crystallization like a water-soluble globular protein. Another strategy makes use of the fact that some bacteria, especially strains of

The Target of Penicillin

Streptomyces and *Actinomadura*, excrete water-soluble DD-peptidases during growth. The G, R61 and R39 DD-peptidases have been isolated from culture filtrates. The G and R61 DD-peptidases have been crystallized. A fourth model DD-peptidase (the K15 enzyme) is membrane-bound. It has been purified almost to protein homogeneity with the help of the detergent N-cetyl-NNN-trimethylammonium bromide. Altogether, these four model enzymes well represent the many variations found among the bacterial DD-peptidases.

Enzymes are invariably able to distinguish between isomers. Like any peptidase, the DD-peptidases catalyse transfer of the electrophilic group $R-\overset{O}{\overset{\|}{C}}$ of aminoacyl amide ($R-\overset{O}{\overset{\|}{C}}-NH-R'$) or aminoacyl ester ($R-\overset{O}{\overset{\|}{C}}-O-R'$) substrates to a nucleophile HY. But the DD-peptidases are unique in that the scissile amide (ester) bond extends between two D centres and is in α position to a free carboxylate. These DD-peptidases also cleave the endocyclic amide bond of penicillins, cephalosporins, monobactams and other related compounds. However, the reaction flux stops at an abortive level, immobilizing the enzymes in an inactive form. β-Lactams are active site-directed inactivating reagents (or suicide substrates) of the DD-peptidases, but the inactivating efficacy very widely varies depending on the enzymes and the β-lactams. Low sensitivity of a DD-peptidase to β-lactam action is called intrinsic resistance.

Peptidase-catalysed cleavage of peptide (ester) $\overset{O}{\overset{\|}{C}}-N(O)$ bonds requires the concerted action of i) an electrophile (anion hole) which polarizes the bond C=O ; ii) a nucleophile which performs attack of the carbon atom; and iii) a proton donor which achieves proton donation to the nitrogen (oxygen) atom. These events can be achieved by different mechanisms. One metallo (Zn^{++}) and several serine DD-peptidases are known.

Initial Binding of Carbonyl Donor Substrates and β-Lactams to the DD-Peptidase Active Sites

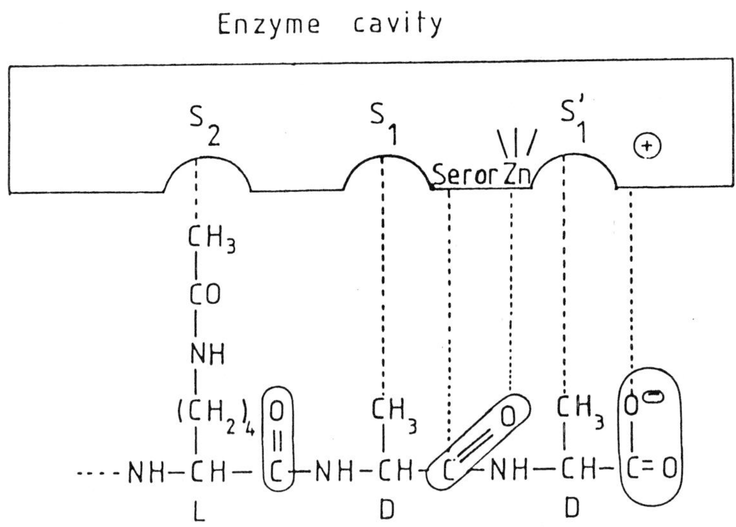

Initial recognition of the carbonyl donor substrate by a DD-peptidase, binding energy and proper alignment of the scissile bond with regard to the enzyme catalytically active functional groups rely on the complementation of at least three enzyme binding subsites. Subsites S'_1 and S_1 accommodate the two methyl groups in the D configuration that occur on both sides of the scissile bond. In addition, subsite S'_1 contains a cationic side chain involved in charge pairing with the C-terminal carboxylate of the substrate. Subsite S_2 accommodates a long side chain that protrudes from an L centre of the substrate backbone. Experimental evidence suggests that complementation of S_2 may not be essential for initial binding but is crucial for subsequent catalysis. Subsite S_2 is highly species-specific, a property which relates to the many structural variations found at the level of the L centre among the bacterial peptidoglycans.

D-Ala-D-Ala terminated peptides and β-lactams lack isosterism at least when the same sequences of atoms are compared. Yet, the D-Ala-D-Ala peptide bond in the carbonyl donor substrates and the endocyclic amide bond in the β-lactam inactivators are functionally equivalent in that they are predisposed to attack by the enzyme active site functional groups.

Looking at models, one sees the scissile $\underset{\overset{\|}{O}}{C}{-}N$ bond at a central position, flanked on one side by the C-terminal carboxylate and on the other, by another $\underset{\overset{\|}{O}}{C}{-}N$ amide bond. These three functional groups are well exposed on the α face of the molecules. They create, around the common backbone $\underset{\overset{\|}{O}}{C}{-}N{-}C{-}\underset{\overset{\|}{O}}{C}{-}N{-}C{-}\underset{\overset{\|}{O}}{C}{-}OH$, zones of positive and negative electrostatic potentials whose relative spatial disposition and strength must be important for the reactivity and orientation of the whole molecule within the enzyme active site. Examination of a set of peptide and β-lactam conformers shows common reactive properties around the carbonyl of the scissile bond. But, depending on the conformation of the peptide backbone, the presence of bulky side chains, the type of bicyclic framework in the β-lactams, the presence of ionized or electron-withdrawing substituents, etc., there are important variations in the electrostatic environments of the molecules. These variations suggest enzyme-ligand associations of widely varying complementarity and productiveness (whether the ligand is a carbonyl donor peptide or β-lactam).

Liganding Catalysis by the G DD-Carboxypeptidase

1. Geometry of the active site

The G DD-peptidase is a metallo (Zn^{++}) enzyme. Its chemical make-up has been elucidated. The molecule consists of one single polypeptide of 212 amino acid residues (or 213 if a Trp not detected in the sequence occurs between Arg^{52} and Phe^{53} as seen by X-ray diffraction analysis) with internal crosslinks

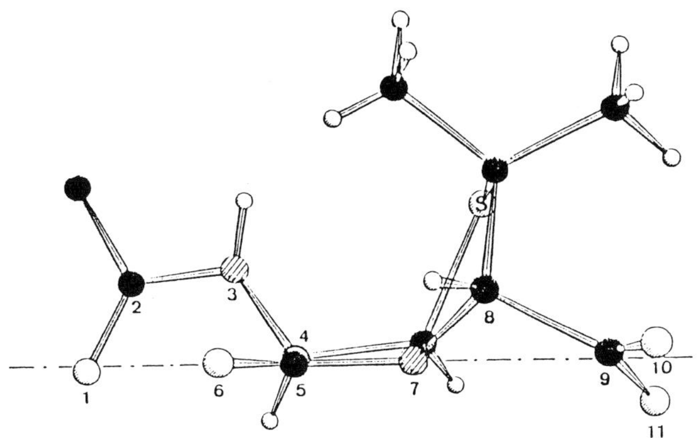

Model of penicillin, C_3 puckered conformer (BZPENK, Cambridge Data bank)
In this molecule, atoms O_1, O_6, C_5, N_7, C_9 and O_{10} of the backbone $\underset{O_1}{\overset{}{C}}{}_2-N_3-C_4-\underset{O_6}{C}{}_5-N_7-C_8-\underset{O_{10}^{\ominus}}{C}{}_9-O_{11}$ are virtually coplanar (α face).

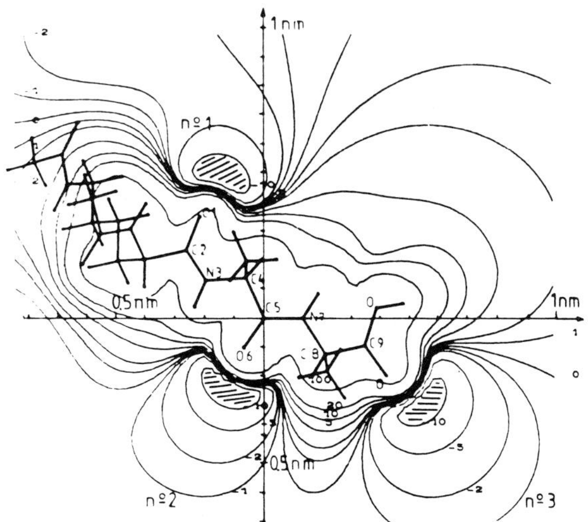

Potential electrostatic map of Ac_2-L-Lys-D-Ala-D-Ala.
The map shows contour of equipotential values and localized electrostatic potential wells (marks 1, 2 and 3). A well gives the energy charge, in Kcal/mol, that would occur if a proton was placed at this position. The pattern refers to a section made through the scissile bond $\underset{O}{C}$-N .

For more information, see Lamotte-Brasseur *et al.* (1983).

occurring in three places through S-S bridges. Its three-dimensional structure has been determined at 1.8 Å resolution, thus permitting comparison with the well-known zinc-containing thermolysin and carboxypeptidase A.

Thermolysin, carboxypeptidase A and the G DD-peptidase show complete lack of amino acid sequence homology and overall structural relatedness. Carboxypeptidase A is a one-domain protein of the α/β type structure; its dominant feature is one eight-stranded wall of β structure which constitutes the core of the molecule; its active site is a shallow depression on the surface of the domain. Thermolysin has two domains of the type $\alpha+\beta$ and all β, respectively; an α-helix runs through the centre of the molecule and the active site is located in the cleft at the junction between the two domains. The G DD-peptidase has two globular domains connected by a single link. The small N-terminal domain has three helices. The large C-terminal domain is of the α/β type structure and has three helices (the longest one being at the junction between the two domains) and five β strands. Remarkably, the enzyme active site (containing Zn^{++}) is located in the C-terminal domain; it appears as an open cleft, one side of which is a mixed sheet formed by the five β-strands; it has been identified as the binding site of the dipeptide Ac-D-Ala-D-Glu (a competitive inhibitor of the G DD-peptidase).

Models show that accommodation of the tripeptide substrate Ac_2-L-Lys-D-Ala-D-Ala in the enzyme active site leads to a close interaction between the C-terminal carboxylate and the guanidinium side chain of Arg^{137} involved in charge pairing and, simultaneously, to a positioning of the scissile $\overset{O}{\overset{\|}{C}}$-N bond such that the oxygen atom is oriented toward the zinc atom while the nitrogen atom is oriented toward the imidazole ring of His^{191} (apparently ready to donate a proton).

When the tripeptide is thus aligned, then the methyl group of the C-terminal D-Ala points to a large subsite S_1'. This

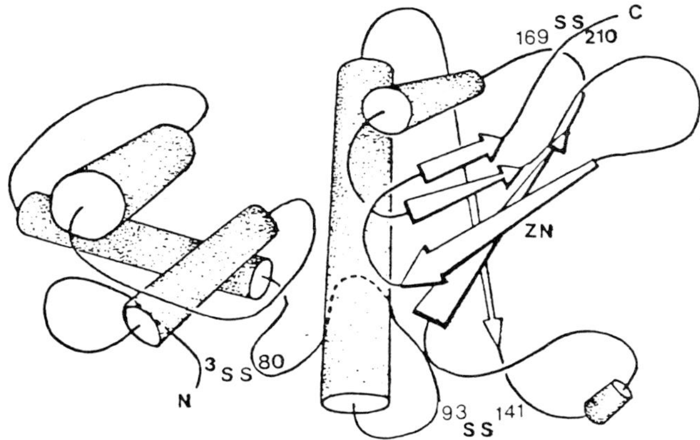

Schematic drawing of the model of the G DD-peptidase (from *Streptomyces albus* G) (Dideberg *et al.*, 1982). The numbering is that proposed by Joris *et al.* (1983).

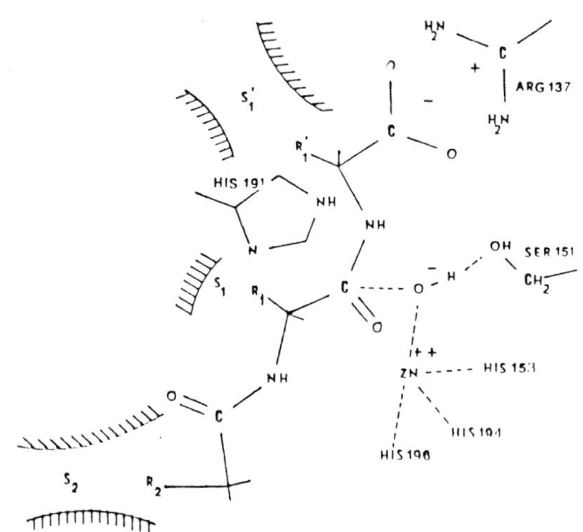

Schematic drawing of the active site of the G DD-peptidase showing the relative disposition of the bound peptide substrate and the enzyme functional groups.

subsite can accommodate side chains much larger than a methyl group, a property which is compatible with the peptidoglycan hydrolase activity of the G DD-peptidase. Changing the D configuration of the methyl group to an L configuration, however, suppresses all possible interaction with the enzyme cavity.

The methyl group of the penultimate D-alanine finds its place in a subsite S_1 of small size. Large side chains cannot be accommodated and a methyl group with an L configuration is excluded since it would collide with the zinc ligands and other residues of the active site. This property well explains the strict requirement shown by this DD-peptidase for the occurrence of a D-alanine at this position.

Finally, the side chain at the L centre of the carbonyl donor peptide complements subsite S_2. Subsite S_2 is a largely open cavity and Arg^{203}, located on the edge of the protein, offers a possible site of interaction. As already stated, substrate activity greatly depends on the interaction between a suitable L side chain of the bound substrate and subsite S_2.

2. Intrinsic resistance to β-lactams

Penicillins bind very loosely to the G DD-peptidase (dissociation constant $K \cong 150$ mM for phenoxymethylpenicillin). Cephalosporins have a somewhat more favorable K value (1-10 mM). Irrespective of the type of bicyclic framework and the nature of 6(7)substituent, the first order rate constant of enzyme inactivation is always very small ($\cong 1\text{-}5 \times 10^{-4} s^{-1}$ or less).

Kinetically, cephalothin and cephalosporin C act as noncompetitive inhibitors. Cephalosporin C destroys the crystal structure thus preventing any study by X-ray diffraction. Cephaloglycine and β-iodo-benzyl-7-amino-cephalosporanate act as competitive inhibitors and the crystal enzyme derivative formed with this latter β-lactam shows satisfactory isomorphism. Study at 4.5 Å resolution has provided direct evidence that the β-lactam binds to the enzyme active site.

With penicillins and cephalosporins, models show that it is not possible to align simultaneously both the β-lactam amide bond with the zinc ion and the C-terminal carboxylate with Arg^{137} because the fused ring system would collide with His^{191}. Intrinsic resistance to β-lactams can thus be seen before our eyes as being caused by the enzyme active site geometry which permits correct alignment and enzyme ligand association of high productiveness, only with carbonyl donor peptides.

6-β-Iodopenicillanate has a very short 6-β-substituent. When used at a high [6-β-iodopenicillanate]/[enzyme] molar ratio of about 8,000 , it also causes slow and irreversible inactivation of the G DD-peptidase (first order rate constant : $7 \times 10^{-4} s^{-1}$). The crystal enzyme derivative thus obtained is perfectly isomorphous. The difference Fourier synthesis at 2.8 Å shows that the inactivator molecule is located just in front of the zinc ion and superimposes His^{191} which has slightly moved toward the exterior of the cavity. This disposition suggests that, in the inactivation process, His^{191} acts as a nucleophile and undergoes alkylation by 6-β-iodopenicillanate with loss of the iodine.

3. Catalysis

The coordination around Zn^{++} involves two histidine residues and one glutamic acid residue in both thermolysin and carboxypeptidase A. Note, however, that His^{142}, His^{146} and Glu^{166} in thermolysin align with His^{196}, Glu^{72} and His^{69} in carboxypeptidase A. The fourth ligand is a water molecule.

Near Zn^{++}, there are two catalytic residues. Glu^{143} in thermolysin (or Glu^{270} in carboxypeptidase A) is thought to act as proton abstractor heightening the nucleophilicity of the zinc-bound water molecule. In turn, His^{231} in thermolysin (or Tyr^{248} in carboxypeptidase A) is thought to facilitate proton donation to the nitrogen atom at the level of the tetrahedral intermediate.

Following initial binding to thermolysin or carboxypeptidase A, the carbonyl donor substrate becomes the fifth ligand of Zn^{++} in a transient pentagonal complex in which Zn^{++} itself plays the role of an anion hole. Collapse of the intermediate by proton donation and re-entry of a water molecule permits release of the reaction products. At any time during the process, no covalent intermediate is formed and partitioning of the enzyme activity between alternate nucleophiles (H_2O and an amino compound) cannot occur. Thermolysin and carboxypeptidase A do not catalyse transpeptidation reactions; they are strict hydrolases.

The G DD-peptidase is also a strict hydrolase (DD-carboxypeptidase). Basically, its mode of action may be similar to that of thermolysin (or carboxypeptidase A), but there are marked differences between them. In the G DD-peptidase, the zinc protein ligands are three histidine residues (His^{194}, His^{196} and His^{153}). Moreover, to all appearances, His^{191} acts as proton donor while Ser^{151} would be somehow involved in water activation, a situation reminiscent to that found in the zinc-containing alcohol dehydrogenase (where a similar role is played by the dyad $His^{31}...Ser^{48}$).

A constant feature of the zinc-containing thermolysin, carboxypeptidases A and B, carbonic anhydrase and alcohol dehydrogenase, is that the hand and geometry of the zinc environment is invariant with regard to the protein ligands, water position, substrate binding site and proton abstractor (Argos *et al.*, 1978). The same disposition applies to the G DD-peptidase. In turn, the nucleophile in the environment of the carbon atom and the proton donor in the environment of the nitrogen atom of the scissile peptide bond determine the hand of the reactive intermediates (Argos *et al.*, 1978). In thermolysin and carboxypeptidase A, the nucleophile points to the carbonyl carbon from below the plane formed by the triad $\overset{\text{O}}{\overset{\|}{\text{C}}}\text{-N}$ while the proton donor points to the nitrogen atom from above that plane. Interestingly, a reverse disposition of these two catalytic groups occurs in the G DD-peptidase.

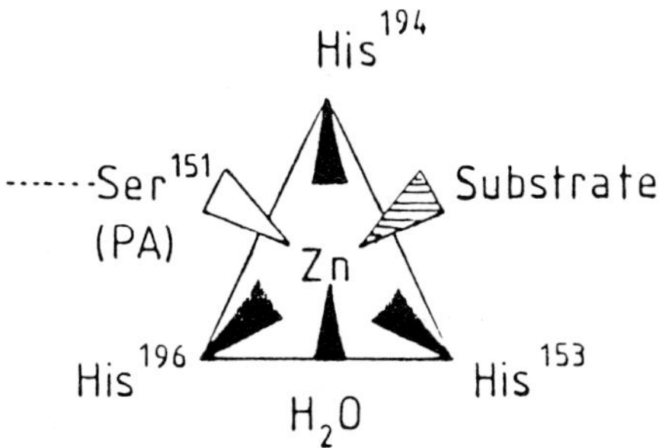

Hand and geometry of the zinc environment of the G DD-peptidase. PD = proton donor. The same disposition occurs in the zinc-containing thermolysin, carboxypeptidases A and B, carbonic anhydrase and alcohol dehydrogenase (Argos *et al.*, 1978).

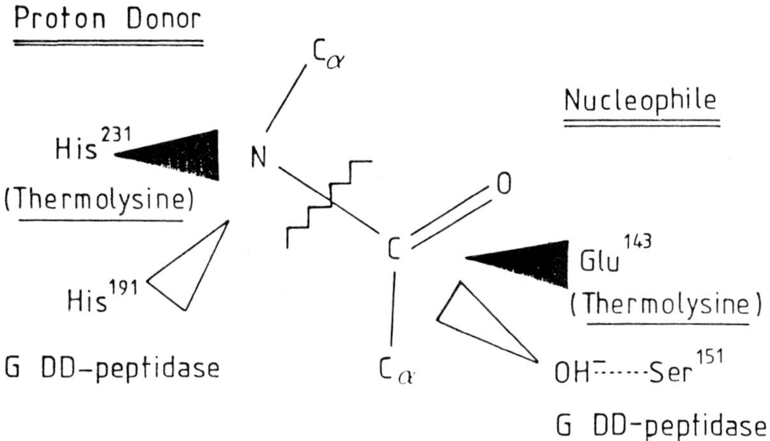

Disposition of the nucleophile and proton donor in thermolysin and the G DD-peptidase with respect to the carbon atom and nitrogen atom, respectively, of the scissile peptide bond.

The concept of bi-product analogues has led to the synthesis of specific and potent inhibitors of thermolysin, carboxypeptidases A and B, and the zinc-containing angiotensin converting enzyme (Ondetti *et al.*, 1981). Similarly, bifunctional compounds possessing both a C-terminal carboxylate function susceptible to interact with Arg^{137} and either a thiol, hydroxamate or carboxylate function susceptible to bind the zinc atom, are inhibitors of the G DD-peptidase. Inhibition is competitive. β-Mercaptopropionic acid ($Ki = 5 \times 10^{-9}$ M) and the L isomer of mercaptoisobutyric acid ($Ki = 1 \times 10^{-8}$ M) are very effective. Curiously, the D isomer is less potent ($Ki = 6 \times 10^{-8}$ M). Such small truncated and flexible molecules do not offer the possibility of complementing subsites S_1 and S_2 and, as the above observations suggest, subsite S_1' alone seems not to play any significant role in binding. Enzyme inhibition may be caused by bidentate interaction and formation of a penta-coordinate zinc atom. Obviously, specificity requires more elaborate structures that would be apt to interact with the S_1 and S_2 subsite targets.

Covalent Catalysis by the R61, R39 and K15 DD-Peptidases

The concept of acyl enzyme intermediate

α-Chymotrypsin (or any other serine peptidase)-catalysed cleavage of peptide (ester) bond proceeds through formation of a serine ester-linked acyl enzyme intermediate. Giving E-OH = the enzyme active serine residue; X = NH-R' or O-R'; K = dissociation constant, k_2 and k_3 = first order rate constants, the reaction can be written :

$$\text{E-OH} + \text{R-}\overset{\text{O}}{\overset{\|}{\text{C}}}\text{-X} \overset{K}{\rightleftharpoons} \text{E-OH}\cdot\text{R-}\overset{\text{O}}{\overset{\|}{\text{C}}}\text{-X} \xrightarrow[\text{HX product } P_1]{k_2} \underset{\text{acyl enzyme}}{\text{R-}\overset{\text{O}}{\overset{\|}{\text{C}}}\text{-O-E}} \xrightarrow[H_2O]{k_3} \text{E-OH} + \underset{\text{product } P_2}{\text{R-}\overset{\text{O}}{\overset{\|}{\text{C}}}\text{-OH}}$$

Release of products P_1 and P_2 proceeds through a Ping-Pong mechanism.

If acyl enzyme formation (k_2) is rate determining (as generally observed with amide substrates $R-\overset{O}{\overset{\|}{C}}-NH-R'$), the acyl enzyme does not accumulate detectably at the steady state and cannot be trapped. If acyl enzyme breakdown (k_3) is rate determining (as generally observed with ester substrates $R-\overset{O}{\overset{\|}{C}}-O-R'$), the acyl enzyme accumulates and can be trapped and isolated.

Assume the presence of an alternate amino nucleophile NH_2-R'' which does not bind to the free enzyme nor to the Michaelis complex but does react only after the leaving group P_1 has diffused away. Then, partitioning of the enzyme activity between the two nucleophiles, H_2O and NH_2-R'', occurs at the level of the acyl enzyme.

$$E-OH + R-\overset{O}{\overset{\|}{C}}-X \overset{K}{\rightleftharpoons} E-OH\cdot R-\overset{O}{\overset{\|}{C}}-X \xrightarrow[HX\ (P_1)]{k_2} R-\overset{O}{\overset{\|}{C}}-O-E \begin{cases} \xrightarrow{H_2O,\ k_3} R-\overset{O}{\overset{\|}{C}}-OH\ (P_2) + E-OH \\ \xrightarrow{NH_2-R'',\ k_3'} R-\overset{O}{\overset{\|}{C}}-NH-R''\ (P_3) + E-OH \end{cases}$$

If acyl enzyme formation (k_2) is rate determining, increasing concentrations of NH_2-R'' cannot increase the maximal rate of carbonyl donor consumption (Vmax) but aminolysis occurs at the expense of hydrolysis on a competitive basis. If acyl enzyme breakdown (k_3) is rate determining, increasing concentrations of NH_2-R'' may increase the rate of acyl enzyme breakdown and hence Vmax.

The R61 and R39 serine DD-peptidases

Although substantial progress in the determination of the three-dimensional structure of the R61 DD-peptidase is being made (Judith Kelly; this symposium), a precise picture of how this enzyme performs catalysis cannot yet be proposed. However, the enzyme catalyses concomitant hydrolysis and aminolysis of suitable carbonyl donor substrates (*i.e.* it functions both as carboxypeptidase and transpeptidase); with the ester substrate Ac_2-L-Lys-D-Ala-D-lactic acid (depsipeptide), an intermediate accumulates at the steady state; penicillin effectively immobilizes the enzyme in the form of a serine-ester linked penicilloyl derivative and experimental evidence strongly suggests that the same serine residue is involved in the formation of the intermediate that can be trapped during hydrolysis of the depsipeptide. To all appearances, the R61 DD-peptidase, as well as the R39 DD-peptidase, are serine enzymes analogous to α-chymotrypsin.

Using Ac_2-L-Lys-D-Ala-D-Ala as carbonyl donor and simple amino acids as acceptors (D-Ala; *meso*-A_2pm), hydrolysis and transpeptidation proceed as expected for an amide carbonyl donor substrate : the increase in the rate of transpeptidation and the decrease in the rate of hydrolysis caused by increasing concentrations of the amino acceptor, are commensurate. However, with more complex amino acceptors related to wall peptidoglycan, the picture is more complicated. The observed increase of the rate of transpeptidation is less than can be accounted for by the decrease of the rate of hydrolysis. Furthermore, at high concentrations of amino acceptor, both hydrolysis and transpeptidation are inhibited so that, eventually, the enzyme can be frozen in a catalytically inactive state. Kinetic studies also suggest that transpeptidation of Ac_2-L-Lys-D-Ala-D-Ala is an ordered pathway where the amino acceptor binds first to the enzyme. In addition, as shown with the R61 enzyme and complex amino acceptors, the enzyme can bind more than one molecule of amino acceptor leading to the formation of an unproductive quaternary complex [enzyme-donor-$(acceptor)_2$]. Finally, the R39 DD-peptidase possesses an additional peptide binding site distinct from the carbonyl donor and amino acceptor sites. Simple peptides (Gly-Gly-Gly, for example) that do not function as acceptors may produce extensive inhibition of the transpeptidation reaction while hydrolysis proportionally increases. Inhibition is noncompetitive *versus* the acceptor, implying that these "allosteric" inhibitors are not binding to the enzyme active site.

All these observations show that the carboxypeptidase and transpeptidase activities of the serine DD-peptidases are susceptible to exquisite modulation. Similar mechanisms might be involved in the control of the peptide crosslinking of peptidoglycan in living bacteria.

Much has been written on the inactivation of the serine DD-peptidases by the penicillins, cephalosporins and monobactams. Essentially, the underlying mechanism is that the corresponding acyl enzyme "intermediate" has, in most cases, a very long half-life. Since the scissile amide bond of the β-lactam ring is endocyclic, what should be regarded as the leaving group (P_1) during acyl enzyme formation cannot leave the enzyme active site which thus remains occupied. Enzyme deacylation, however, may slowly occur. In some cases, rupture of the C_5-C_6 linkage in the enzyme-bound penicilloyl moiety with formation of phenylacetylglycyl enzyme is the rate determining step of enzyme deacylation. Once formed, this new intermediate is immediately attacked by water or a suitable amino compound (on a competitive basis) with regeneration of an active enzyme

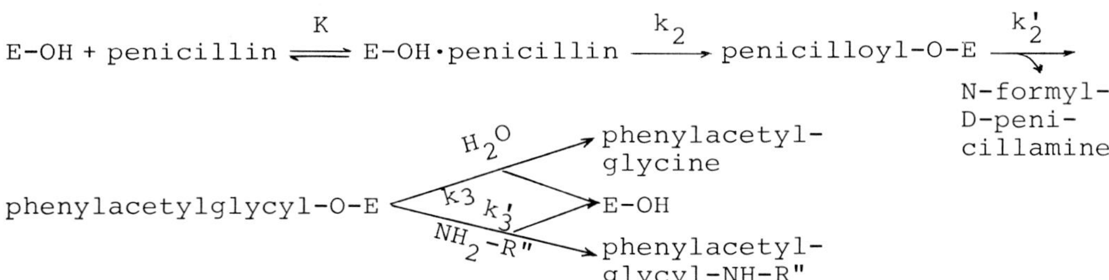

Since k'_2 is rate determinant, NH_2-R" does not accelerate the rate of enzyme reactivation. In other cases, enzyme deacylation may slowly occur without prior fragmentation of the bound acyl moiety. When this occurs, the DD-peptidase functions as a classical β-lactamase of very weak efficiency.

The higher the bimolecular rate constant of enzyme acylation (k_2/K), the more potent is the β-lactam as enzyme inactivator. Various side chains in homologous series of β-lactams can be accommodated by a given DD-peptidase but the k_2/K values thus generated may vary widely. The goodness of fit of the β-lactam molecule to the enzyme active site, rather than any other feature (intrinsic reactivity of the β-lactam ring, for example) is the primary parameter that governs the efficacy of enzyme inactivation.

The K15 DD-peptidase

The membrane-bound K15 DD-peptidase also performs covalent catalysis. Whether an active serine or another residue of the active site is involved in the process is not yet known. Whatever the case, the K15 enzyme strikingly differs from the R61 and R39 serine DD-peptidases. Indeed, the Vmax of consumption of the amide carbonyl donor Ac_2-L-Lys-D-Ala-D-Ala is very low in water (low carboxypeptidase activity).It is increased at least 30-fold in the presence of a suitable amino acceptor such as Gly-Gly (high transpeptidase activity). Hence, depending on whether the amino acceptor is present or not, the acyl enzyme, at the steady state, should be undetectable or should accumulate massively. Using $[^{14}C]Ac_2$-L-Lys-D-Ala-D-Ala at saturating conditions and SDS as trapping agent, attempts to estimate the acyl enzyme have been carried out by fluorography after gel electrophoresis in the presence of SDS. When Gly-Gly is present, no acyl enzyme can be trapped but, unexpectedly, when Gly-Gly is absent, no more than 10 % of the enzyme occurs in the form of acyl enzyme (indicating that the rate of acyl enzyme formation is much smaller than the rate of acyl enzyme breakdown). It thus seems that the effect of the amino acceptor on Vmax cannot be attributed to a simple partitioning at the level of the acyl enzyme but implies an acceleration of both acyl enzyme formation and breakdown.

This presumed effect of the amino acceptor on the rate of acyl enzyme formation is not observed when benzylpenicillin is used as carbonyl donor. Indeed, the k_2/K value (150 $M^{-1}s^{-1}$) is the same whether Gly-Gly is present or not. Similarly, Gly-Gly has no effect on the rate constant k_3 ($1 \times 10^{-4}s^{-1}$) of breakdown of the penicilloyl-enzyme since k_2' is rate determining (see above). It thus follows that, to all appearances, the K15 DD-peptidase does not require any "effector" to react with penicillin but reacts with a peptide carbonyl donor much more effectively in the presence of a suitable amino acceptor than in its absence. Such a behaviour may apply to the penicillin binding proteins (PBPs) 1A, 1B, 2 and 3 of *Escherichia coli*.

While the DD-peptidases R61 and R39 appear to be good models of the PBPs involved in secondary hydrolytic and transpeptidation reactions during wall peptidoglycan remodelling, the K15 DD-peptidase seems to be a good model of those PBPs involved in the primary transpeptidation reactions through which the nascent peptidoglycan undergoes attachment to the preexisting wall peptidoglycan.

Acknowledgement

Part of the work has been supported by *FRSM*, Brussels (contract n°3.4501.79), and an *Action concertée* with the Belgian Government (convention n°79/84-I1).
P. Ch. is *Aspirant du FNRS* and G.D. is *Chercheur qualifié du FNRS*.

References

Abraham, E.P. : Scientific American 244, 64-74 (1981).
Argos, P., Garavito, R.M., Eventoff, W., Rossmann, M.G. and Brändén, C.I. : J. Mol. Biol. 126, 141-158 (1978).
Ondetti, M.A., Cushman, D.W., Sabo, E.F., Natarajan, S., Pluscec, J. and Rubin, R. : *in* Molecular Basis of Drug Action. Singer and Ondarza, eds. Elsevier North Holland, 1981, pp. 235-246.
Ghuysen, J.M., Frère, J.M., Leyh-Bouille, M., Dideberg, O., Lamotte-Brasseur, J., Perkins, H.R. and De Coen, J.L. Penicillins and Δ^3-cephalosporins as inhibitors and mechanism-based inactivators of D-alanyl-D-Ala peptidases, *in* Topics in Molecular Pharmacology. Burgen and Roberts, eds. Elsevier North Holland Biomedical Press, 1981, pp. 63-97.
Frère, J.M., Kelly, J.A., Klein, D., Ghuysen, J.M., Claes, P. and Vanderhaeghe, H. : Biochem. J. 203, 223-234 (1982).
Dideberg, O., Charlier, P., Dive, G., Joris, B., Frère, J.M. and Ghuysen, J.M. : Nature 299, 469-470 (1982).

Kelly, J.A., Moews, P.C., Knox, J.R., Frère, J.M. and Ghuysen, J.M. : Science 218, 479-481 (1982).

Nguyen-Distèche, M., Leyh-Bouille, M. and Ghuysen, J.M. : Biochem. J. 207, 109-115 (1982).

Joris, B., Van Beeumen, J., Casagrande, F., Gerday, Ch., Frère, J.M. and Ghuysen, J.M. : Eur. J. Biochem. 130, 53-69 (1983).

Lamotte-Brasseur, J., Dive, G. and Ghuysen, J.M. : Eur. J. Biochem., submitted (1983).

This paper is a common contribution of several symposium participants

X-RAY STRUCTURE OF A PENICILLIN TARGET ENZYME

Judith A. Kelly

Biochemistry and Biophysics Section
Biological Sciences Group
University of Connecticut, Storrs, CT 06268 USA

Introduction

In 1965, Tipper and Strominger (1) and Wise and Park (2) independently demonstrated that β-lactams exerted their antibacterial effect in the final stage of cell wall synthesis during the crosslinking of the peptidoglycan strands. Transpeptidases which catalyze the formation of the peptide bridges are the targets of penicillins and cephalosporins. In the intervening years, substantial progress has been made toward describing at the molecular level the mechanism of action of penicillin (3). But to fully understand the data acquired from other studies (spectroscopic, kinetic, etc.), it would be valuable to have knowledge of the exact, atomic level structures of these penicillin targets and the complexes they form with antibiotics and substrate analogs.

I would like to report on progress to date on x-ray crystallographic studies underway on one particular penicillin target enzyme. Transpeptidases are, in general, membrane bound. However, some *Streptomyces* bacteria excrete water soluble D-alanyl-D-alanine peptidases during growth (4). J. M. Frère and J. M. Ghuysen of the Université de Liège have done extensive work characterizing these enzymes in order to use them as models of the membrane bound systems (5). Ghuysen and Frère have kindly provided our crystallographic laboratory at the University of Connecticut with pure samples of one of these exocellular enzymes, the D-D-carboxypeptidase/transpeptidase from *S.* R61. This afforded us the opportunity to crystallize a detergent-free, intact protein. Previous attempts had been made to solubilize membrane bound enzymes by

The Target of Penicillin

limited proteolysis or detergent extraction. Crystallization attempts on the penicillin-binding enzymes isolated by these methods failed (J. R. Knox, personal communication). However, with the exocellular R61 enzyme, we were successful (6).

Results

The solution of the enzyme's structure has proceeded to a resolution of 2.8 Å and the results of these studies were recently reported (7). The main structural characteristics of our model of the 37,400 dalton R61 enzyme are a clustering of the secondary structure elements into two regions (see Fig. 1). The region on the left contains five helices and on the right is a beta sheet of five strands with three helices protecting the faces of the beta sheet.

At that point, we reported observing via difference Fourier maps the binding site of cephalosporin C on the enzyme. In terms of the distribution of electrons, cephalosporin C is a rather symmetric V-shaped molecule at the resolution we are currently considering (2.8 Å).

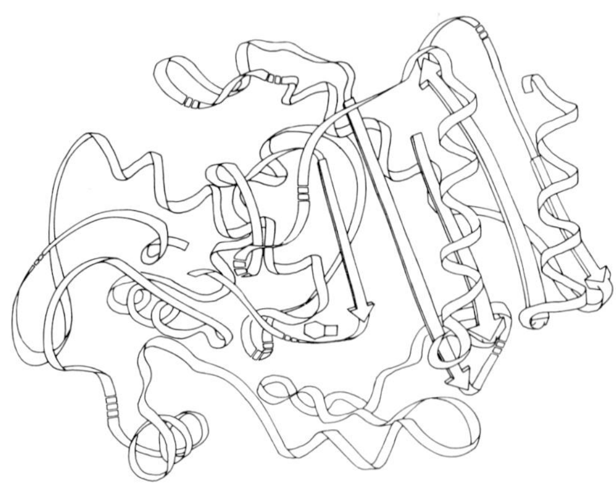

Fig. 1. Ribbon representation of R61 transpeptidase main chain. Cephalosporin C binding site is indicated.

The side chains of cephalosporin C also have a great deal of rotational freedom and hence density was not observed in the difference Fourier map for the ends of the C_7 and C_3 substituents. Because of these factors, we were unable to unambiguously orient the antibiotic in the observed electron density. With the aid of a second compound, we have been able to resolve this ambiguity and orient the cephalosporin C molecule in its electron density with respect to the protein. The second compound which was used is 6,6-dichloro-4-desaza-2,2-dides-methylpenicillanic acid, a desaza penicillin precursor. This compound was synthesized by G. J. Hite, A. Kumar and B. Tomczuk in the School of Pharmacy of the University of Connecticut. The final product of their synthesis, when C_2 has been methylated and C_6 has been functionalized, is intended to be a probe for the penicillin-destroying β-lactamases. At this stage, however, the dichloro precursor was a good candidate for identifying the position of a substituent extending out from the β-lactam ring. The desaza precursor had shown reversible kinetic behavior with the R61 enzyme with a K_I of 0.9 mM (J. M. Frère, personal communication). When the crystallographic experiments were done, the difference Fourier map clearly indicated the position of the chloride atoms. Using that information, we were able to orient the cephalosporin C and fit it into its electron density. Figure 2 shows the antibiotic positioned in its electron density and that portion of the enzyme which is adjacent. For comparison, Fig. 3 shows the strongest features of the desaza precursor difference density overlaid on the protein and cephalosporin C image. The three peaks shown correspond well with the C_6 dichloride substituent, the sulfur and the carboxyl groups of the thiazolidine ring. In order to appreciate the three-dimensional orientation of the cephalosporin C and the protein, a stereo view is shown in Fig. 4. While the density for the C_7 side chain of the antibiotic is not completely observed, that density which is present curves toward the β face further confirming our interpretation. In addition, on the β face near C_8 is a portion of electron density extending out toward the protein and not occupied by the cephalosporin C. The electron density for the protein is not perfectly clear in

Fig. 2. Cephalosporin C as fitted in its electron density with a portion of nearby residues of R61 enzyme including serine side chain.

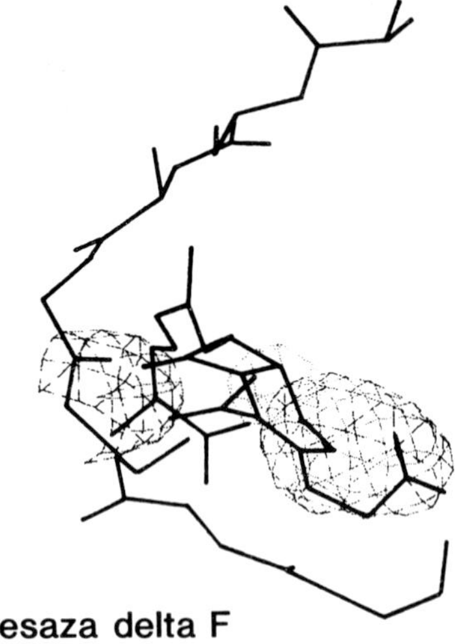

Fig. 3. Orientation of desaza penicillin precursor density with respect to cephalosporin C position and nearby residues of R61 enzyme.

Fig. 4. Stereographic view of cephalosporin C and R61 residues. Orientation as in Figs. 2 and 3.

this portion of the native map. However, it is possible to fit reasonably well a serine residue whose $O\gamma$ would be positioned for a β face attack of the β-lactam carbon atom. The serine side chain can be seen in Figs. 2, 3 and 4 with $O\gamma$ less than 3 Å away from carbonyl C_8 of the antibiotic. This would be consistent with the observation that certain cephalosporins undergo intramolecular nucleophilic attack which, because of structural constraints, must be from the β face (8). One additional area of unused density in the cephalosporin C map appears on the α face of the 6-membered dihydrothiazine ring. Not shown in Fig. 2 is a nearby negative peak which indicates a movement of the protein on binding the antibiotic. In the native protein map, there is density for a large side chain in this area, possibly an aromatic residue. This suggests that a protein side chain is moving down to interact with the dihydrothiazine ring of cephalosporin C.

In addition to binding studies with antibiotics and the desaza precursor, we have attempted to look at analogs of the peptidoglycan substrate of the R61 transpeptidase. In separate experiments, we have diffused into our crystals the following analogs: acetyl-D-alanyl-D-glutamine, provided by H. R. Perkins of The University of Liverpool; N-acetylmuramyl-L-alanyl-D-isoglutamine (MDP) and MDP-L-lysyl-L-alanine, provided by P. Lefrancier of Institut Choay; and α-t-Boc-ε-TFA-L-lysyl-D-glutamyl-D-alanine, provided by A. R. Zeiger of Jefferson Medical College. In no case was there adequate electron density in the difference Fouriers to warrant detailed interpretation. It is of note that in the case of the derivatized L-lysyl-D-glutamyl-D-alanine there is electron density that is coincident with the positions where S_1 and C_2 of cephalosporin C were placed. We continue to look for substrate analogs which will bind well to the R61 enzyme and result in a clearer image.

Detailed interpretation of interactions between the protein and inhibitors/ substrate analogs will be carried out as we refine our structure of this penicillin target enzyme and extend our map resolution to 1.8 Å. This work is currently in progress.

The author thanks G. J. Hite, J. R. Knox and P. C. Moews for helpful discussions and assistance in these studies. Also appreciated is the assistance provided by G. Petsko and D. Rose (MIT) in the use of their interactive computer graphics facility. This work was supported by the U.S. National Institutes of Health (AI 16702) and North Atlantic Treaty Organization (RG .082.81)

References

1. Tipper, D. J., Strominger, J. L.: Proc. Natl. Acad. Sci. U.S.A. 54, 1133-1141 (1965).

2. Wise, E. M., Jr., Park, J. T.: Proc. Natl. Acad. Sci. U.S.A. 54, 75-81 (1965).

3. Ghuysen, J.M.: Topics in Antibiotic Chemistry, P. G. Sammes, ed., Vol. 5, pp. 9-117, Ellis Horwood Ltd., Chichester (1980).

4. Leyh-Bouille, M., Coyette, J., Ghuysen, J. M., Idczak, J., Perkins, H. R. Nieto, M.: Biochemistry 10, 2163-2170 (1971).

5. Ghuysen, J. M.: J. Gen. Microbiol. 101, 13-33 (1977).

6. Knox, J. R., DeLucia, M. L., Murthy, N. S., Kelly, J. A., Moews, P. C., Frère, J. M., Ghuysen, J. M.: J. Molec. Biol. 127, 217-224 (1979).

7. Kelly, J. A. Moews, P. C., Knox, J. R., Frère, J. M., Ghuysen, J. M.: Science 218, 479-481 (1982).

8. Indelicato, J. M., Norvilas, T. T., Pfeiffer, R. R., Wheeler, W. J., Wilham, W. L.: J. Med. Chem. 17, 523-527 (1974).

THE FINE ARCHITECTURE AND FUNCTION OF THE GENE CODING FOR PBP-3 OF ESCHERICHIA COLI

Maruyama, I. N., Yamamoto, A., Maruyama, T. and Hirota, Y.

National Institute of Genetics, Mishima 411, JAPAN.

Introduction

The discovery of penicillin binding proteins (PBPs) in bacterial membrane (1) and the development of a sensitive method for detecting PBPs (2) have opened a new era of studies on the lethal target of penicillin action and murein biosynthesis. Seven major PBPs -1a, -1b, -2, -3, -4, -5 and -6 have been identified in E. coli (3,4). The PBP-3 has been proven to play an indispensable role in cell division, and to be a lethal target of penicillin action (4, 5, 6).

A synthetic ColE1 plasmid, pLC26-6, found in a gene bank (7) was shown to carry the gene (ftsI or pbpB) coding for PBP-3 (4, 5, 8). Using this DNA fragment, we determined the entire nucleotide sequence of the ftsI, consisting of 1,764 base-pairs which code a polypeptide of 588 amino acid residues and of a molecular weight 63,850 (9). In this symposium, we present new findings on the gene's fine structure and mechanism responsible for regulating the function of PBP-3.

Results and discussion

1) Promoter region of the ftsI gene.

Using a computer program, sequence regions likely to be a canonical Pribnow box were searched over a 480 nucleotide range in the upstream of the ftsI structural gene (Fig. 1). We found 9 potential candidates for Pribnow box, shown in Table 1. Of these, two candidates, GAAAATA and TATCGTA (nos. 2 and 4 in Table 1), are preceded by -35 regions showing a strong homology to TGTTGACA, which is known to exist frequently in -35

The Target of Penicillin

```
1         .         .         30        .         .         .        70
CAGCTGCGAGCACTAGGCAAGTTAATGCCGGGCGAAGAAGAGGTGGCTGAGAACCCTCGTGCCCGTAGTT
          .         .         100       .         .         .       140
CAGTTCTGCGTATTGCAGAGAGGACGAATGCATGATCAGCAGAGTGACAGAAGCTCTAAGCAAAGTTAAA
          .         .         170       .         .         .       210
GGATCGATGGGAAGCCACGAGCGCCATGCATTGCCTGGTGTTATCGGTGACGATCTTTTGCGATTTGGGA
                                      9
          .         .         240       .         .         .       280
AGCTGCCACTCTGCCTGTTCATTTGCATTATTTTGACGGCGGTGACTGTGGTAACCACGGCGCACCATAC
                         8        1
          .         .         310       .         .         .       350
CCGTTTACTGACCGCTCAGCGCGAACAACTGGTGCTGGAGCGAGATGCTTTAGACATTGAATGGCGCAAC
   3                                                      6
          .         .         380       .         .         .       420
CTGATCCTTGAAGAGAATGCGCTCGGCGACCATAGCCGGGTGGAAAGGATCGCCACGGAAAAGCTGCAAA
             7
          .         .         450       .         .         .       490
TGCAGACATGTTGATCCGTCACAAGAAAATATCGTAGTGCAAAAATAAGGATAAACGCGACGCATGAAAG
                        2    4             5                   MetLys
          .         .         520       .         .         .       560
CAGCGGCGAAAACGCAGAAACCAAAACGTCAGGAAGAACATGCCAACTTTATCAGTTGGCGTTTTGCGTT
AlaAlaAlaLysThrGlnLysProLysArgGlnGluGluHisAlaAsnPheIleSerTrpArgPheAlaLeu
```

Figure 1. Nucleotide sequence of the proposed promoter region of ftsI gene. Possible candidates for Pribnow box are enclosed in boxes with accompanying number indicating that given in Table 1.

regions of promoters (10). Therefore it is very likely that these two sequences are actually promoters of the ftsI. Spratt (3) has estimated that each E. coli cell contains about 50 molecules of PBP-3. This indicates that the promoter of the ftsI gene can be weak, and furthermore that other candidates listed in Table 1 can also be weak promoters of the ftsI gene. The initiation codon ATG is preceded by TAAGGA which begins 18 nucleotides upstream of the ATG. The location of this sequence for ribosomal binding is somewhat far (18bp) from the initiation codon, which is known to be usually 10 to 14 base pairs (12). There appears to be no alternative candidate, and therefore we assign this sequence for the ribosome binding site (Shine-Dalgarno sequence) of ftsI gene (13). This long distance between the ribosome binding site and initiation codon may be causing the low copy number of PBP-3 per cell.

2) Signal peptide of PBP-3 precursor.

We have reported that PBP-3 is first synthesized as a precursor having molecular weight of 63,850 daltons, and then processed to a mature form having molecular weight of 60,000 daltons (9). Here we put forward a

Table 1. Candidates for the promoter of ftsI gene.

	Position	Likelihood score	-35 Region ..TGTTGACA.TTT.	Pribnow box candidate
1	239	218	GGAAGCTGCCACTCTGCCTGTTCATTTGCAT	TATTTTG
2	445	210	CTGCAAATGCAGACATGTTGATCCGTCACAA	GAAAATA
3	284	208	TGACTGTGGTAACCACGGCGCACCATACCCG	TTTACTG
4	450	208	AATGCAGACATGTTGATCCGTCACAAGAAAA	TATCGTA
5	461	208	GTTGATCCGTCACAAGAAAATATCGTAGTGC	AAAAATA
6	337	207	CAACTGGTGCTGGAGCGAGATGCTTTAGACA	TTGAATG
7	363	203	AGACATTGAATGGCGCAACCTGATCCTTGAA	GAGAATG
8	236	197	TTGGGAAGCTGCCACTCTGCCTGTTCATTTG	CATTATT
9	179	190	TGGGAAGCCACGAGCGCCATGCATTGCCTGG	TGTTATC

a. Numerals indicate the position of first nucleotide of corresponding Pribnow box candidate.

b. Likelihood score is evaluated using the following rule. Basing on the sequence data of 57 known Pribnow boxes (10,11), usage frequencies of the nucleotides at each of the 7 sites were tabulated.

	1st	2nd	3rd	4th	5th	6th	7th
A	1	49	13	36	35	0	17
C	2	0	11	2	13	0	6
G	3	3	5	9	4	0	18
T	51	5	28	10	5	57	16

We assume that likelihood is proportional to the usage frequency at each of the corresponding site. The likelihood score represents the sum of the 7 usage frequencies given above. For example, in the case of TATTTTG, 51(T)+49(A)+28(T)+10(T)+5(T)+57(T)+18(G)=218.

tentative hypothesis that the precursor has a signal peptide consisting of about 40 residues and located at its amino-terminus, and this signal peptide is cleaved off to make mature PBP-3. We have several lines of evidence supporting the assertion. a) Using a method of Segrest and Feldmann (14), hydrophobicity index (HI) has been calculated along the entire sequence of the precursor to find regions having structure required for a signal peptide. The results presented in Fig. 2 reveal that the precursor possesses several regions showing strong hydrophobicity which is one of the requirements for a signal peptide. One of such regions is located near the amino terminus (24th to 40th residues) and is preceded by a hydrophilic region which is another requirement. This hydrophobic area has a HI value (2.42) that is close to the mean HI (2.51) of 20 previously

```
 +   +  +.++ --+   .  +       .        .+        .-   +- - +  +    .
MKAAAKTQKPKRQEEHANFISWRFALLCGCILLALAFLLGRVAWLQVISPDMLVKEGDMRSLRVQQVSTS  70
HHHHHHHCCCHHHHHHTHHHHHTTTTTTHHHHHHHHHHHHHHEEEEECCTHHHHHHHHHHHEEEEECCC
+    -+ +        +.   - +- +-        -+.+        . -     +  .  + +    .
RGMITDRSGRPLAVSVPVKAIWADPKEVHDAGGISVGDRWKALANALNIPLDQLSARINANPKGRFIYLA 140
TTEEETTTCCCCEEEEEEHHHHHHHTHHHHTTTCEECTTHHHHHCCCCEEEHHEEEEETCCCTTEEEEEE
+    -   -. ++ +    + +-- ++  .  -   +   .    -    -  -+  -+ .    -+ +
RQVNPDMADYIKKLKLPGIHLREESRRYYPSGEVTAHLIGFTNVDSQGIEGVEKSFDKWLTGQPGERIVR 210
TTCCCTHHHHHHHCCCCHECHHTTTTTCCTCEEEEEEEEECCCCCCCCEHHHHTHTTTCCCCCCCEEEE
+-+  +  --    -    +       --+.    +-  .      + - .      -  .-        .
KDRYGRVIEDISSTDSQAAHNLALSIDERLQALVYRELNNAVAFNKAESGSAVLVDVNTGEVLAMANSPS 280
TTTTCEEEEEECCCCHHHHHHHHHHHHHHHHHHHHEEHHHHHHHHHHHHHHHHHEEEEEECHHHHHEEECCTT
         .+-  + +  .-   -       +         +   +-     .     +   +- +-  +  -.
YNPNNLSGTPKEAMRNRTITDVFEPGSTVKPMVVMTALQRGVVRENSVINTIPYRINGHEIKDVARYSEL 350
CCTTTCCCCCHHHHHHTTCEEEECTTTCCEEEEEEEHHHHEEEEEETTTEEEEETTCCCCCEHHHHHHHHH
        + .    +   .         - . +    +  .      -+  .   + +  -.-+     .
TLTGVLQKSSNVGVSKLALAMPSSALVDTYSRFGLGKATNLGLVGERSGLYPQKQRWSDIERATFSFGYG 420
EEEEEEEECCTECETEEHHHHHHHHEEEEEEETTCCCCCEEEEEEETTTCCCTTTTCCCCETTTTTTTC
         +         . +     + -     -+  .-    +    +. -       .  +   +   +
LMVTPLQLARVYATIGSYGIYRPLSITKVDPPVPGERVFPESIVRTVVHMMESVALPGGGGVKAAIKGYR 490
CEEEEHHHEEEEEEETTEEEEEEEEEEEEECCCCCTEEEECTHHHHHHHHHEHHHECCTTCCEEEEEEHHEE
   +   ++  - +  +           .     +    .   -     +.          .        +.
IAIKTGTAKKVGPDGRYINKYIAYTAGVAPASQPRFALVVVINDPQAGKYYGGAVSAPVFGAIMGGVLRT 560
EEEEECCCTEECTTTTEEEEEEEEEEEECCCTCCCTEEEEEEEEECCTTTEETTEEEEEEEEEEEECECEEE
   - -   . -+ -    . -    +
MNIEPDALTTGDKNEFVINQGEGTGGRS                                           588
ECCCCHHHHCCCCCEEEECCCCCCCCCC
```

Figure 2. Predicted secondary structure and hydrophobic region of PBP-3. The areas enclosed in boxes show the sequences for which hydrophobicity index (HI) is higher than 2.0. Plus and minus charged amino acids are denoted by + and -. H, E, T, and C stand for α-helix, extended or β-sheet, reverse turn, and aperiodic or coil structures, respectively. Secondary structure was predicted according to the directional method of Garnier et al. (17) using a decision constant equal to zero.

studied cases (15). It is also worth mentioned that the sizes of these 20 previously studied cases vary from 9 to 24 residues, and the present case of 17 residues falls well in the permissible range. It is known that most signal peptides begin by a polar segment at the amino terminus. The precursor of PBP-3 also contains a positive charge (+5) which is followed by the hydrophobic core. Evidence has been presented for that the positively charged amino-terminal region of a signal peptide plays an important role in efficient protein secretion across the membrane (16). Combining all these evidence, we are quite certain that this amino terminus region is in fact a signal peptide of the precursor. b) The secondary structure of the PBP-3 has been determined by the method of Garnier et al. (17) and the results are shown in Fig. 2. The amino-terminal sequence of the precursor begins with an α-helix region and the hypothetical signal peptide of the precursor has high content of α-helix. The presence of this secondary structure at the amino-terminus of the PBP-3 precursor coincides with the secondary structure of known signal peptides (18). These results confirm that PBP-3 is synthesized as a precursor having a signal peptide and then processed to its mature form.

PBP-3 might be an ectoprotein; a part of the molecule is integrated into membrane and the rest is exposed to periplasmic space. The signal peptide of PBP-3 might play an essential role for PBP-3 to function as an ectoprotein in _E. coli_ cell membrane. The amino acid residue at the cleavage site of signal peptide is known to have a side chain containing at most one carbon (19). Based on this rule and the difference of molecular weights of the precursor and mature PBP-3, the cleavage of the signal peptide seems to occur between any of the following pairs of residues; 40th glycine and 41st arginine, 43rd alanine and 44th tryptophan, or 49th serine and 50th proline. Therefore, the PBP-3 precursor must have a signal peptide consisting of 40 to 49 amino acid residues. This is considerably longer than other known signal peptides which consist of 15 to 34 residues. Structure of the signal peptide of the precursor appears to be unique in that a long amino-terminal hydrophilic portion is followed by a hydrophobic core. An intriguing question is how this signal peptide serves for PBP-3 to be established as a functional ectoprotein. This remains to be solved.

3) Penicillin binding site of PBP-3.

In order to find a penicillin binding site of PBP-3, we have compared amino acid sequences of four class A β-lactamases (20), two D-alanine carboxypeptidases (CPases) (21) and PBP-3. Following three categories of homology have been considered. a) Homology with 25 amino acid residues of penicillin binding sites of CPases and β-lactamases (21): Regions with more than 25% homology were searched allowing a single residue insertion or deletion. Nine regions of PBP-3 were found to be homologous to these binding sites under the criterion. b) Homology with the 52 invariable

Table 2. Amino acid sequence homology between PBP-3 and other penicillin binding enzymes.

Amino acid residues compared with PBP-3	Identical /Total	Homology(%)
Entire residues of		
β-lactamase, E. coli	33/254	13.0
S. aureus	33/253	13.0
B. licheniformis	28/245	11.4
B. cereus	35/253	15.4
Invariable residues among 4 β-lactamases	19/52	36.5
Invariable residues among 2 CPases and 4 β-lactamases	4/5	80.0

a. Amino acid sequence of PBP-3 around the predicted penicillin binding site was compared with three categories of amino acid residues of other penicillin binding enzymes, 4 class A β-lactamases and 2 carboxypeptidases (CPases).

amino acid residues previously established among 4 β-lactamases (20): We found 39 regions of PBP-3 having more than 10% homology (6 or more identical residues among the 52 invariable residues). c) Homology with the 5 invariable amino acid residues which are known to be common among 4 β-lactamases and 2 CPases (21): We found 7 regions having strong homology for the 5 invariable residues. These regions contain a site of serine responsible for the penicillin binding, and they have at least two other common sites among the 5 invariable residues.

However, only one of these regions, containing the serine at 307th residue, satisfies all the three criteria stated above. The hypothetical penicillin binding sites of PBP-3, 4 β-lactamases and 2 CPases are presented in Fig. 3. The data of homology between PBP-3 and β-lactamases, and between PBP-3 and CPases are summarized in Table 2. Homology between the 254 residues of β-lactamases and the PBP-3 was 13%. The homology between the 52 invariable residues of the 4 β-lactamases and PBP-3 showed 36.5%. Furthermore, 4 out of the 5 invariable residues of β-lactamases and CPases were found in the hypothetical penicillin binding

```
                               2 2     2     2     3     3     3     3   3     3
                               6 7     8     9     0     1     2     3   4     5
                               6 0     0     0     0     0     0     0   0     0
1. PBP-3, E.coli               DVNTGEVLAMANSPSYNPNNLSGTP//RNRTITDVFEPGSTVKPMVVMTALQRGVVRENSVINTIPY//IKDVARYSEL

2. β-lactamase, E.coli         kvkdaedqlgarvgyieldlnsgkilesfrpeerfpmmstfkvllcgavlsrvdagqeqlgrrihysqndlveyspv
3.              S.aureus       elndlekkymahigvyaldtksgke-vkfnsdkrfayastskainsailleqvpynklnkkvhi--nkddivayspi
4.              B.lichen.      dfakleeqfdaklgifakdtgtnrt-vayrpderfafastikaltvgvllqqksiedlnqrity--trddlvnynpi
5.              B.cereus       efsqlekkfdarlgvyaidtgtnet-isyrpdqrfafastykalaagvllqqnsidslnevlgl--tkedlvdyspv

6. CPase, B.subtilis             asdpidinas-aaimieassgkilysknadkrlpiasmtkmmteyllleaidqgkvkwdqty--tpdxyvxqxs
7.        B.stearoth.            esapldirad-aailvdaqtgkilyekniidtvlgiasmtkm


          3     3      3     3    4      4     4     4     4     4
          6     7      8     9    0      1     2     3     4     5
          0     0      0     0    0      0     0     0     0     0
1.   TLTGVLQKSSNVGVSKLALAMPSS--ALVDTYSRFGLGKATNLGLV//SGLYPQKQRWSDIERATFSFGYGLMVTPLQLARVYATIGSYGIYRPLSITKVD

2.   tekhlt-dgmtvrelcsaaitmsdntaanllttiggpkeltaflhnmgdhvtrldrqepelneaipn-derdttmpaama-ttlrklltgelltlasrq
3.   lekyvg-kditlkalieasmtysdntannkiikeiggikkvkqrlkelgdkvtnpvryeielnyyspk-skkdtstpaafg-ktlnkliangklskenkk
4.   tekhvd-tgmtlkeladaslrysdnaaqnlilkqiggpeslkkelrkigdevtnperfepelnevnpg-etqdtstaralv-tslrafaledklpsekre
5.   tekhvd-tgmklgeiaeaavrssdntagnlifnkiggpkgyekalrhmgdritmsmrfetelneaipg-dirdtstakaia-tnlkaftvgnalpaekrk
```

Figure 3. Amino acid sequence alignment of homologous regions of PBP-3, class A β-lactamases and carboxypeptidases (CPases). Numbering of amino acid sequence is identical to that of PBP-3 amino acid sequence shown in Figure 2. The insertion and deletion of amino acid residue to obtain optimal matching are presented by / and -, respectively. Amino acids identical with those of the *E. coli* PBP-3 are in boxes. x shows unknown amino acid.

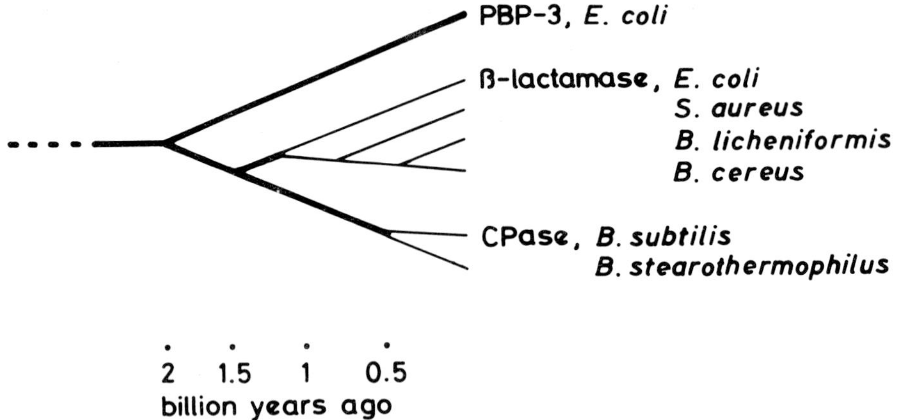

Figure 4. Phylogenic tree representing evolution of penicillin binding proteins. The rate of amino acid substitution (22) is estimated to be 4.9×10^{-10}/year from the equation, $p=(19/20)\exp[-2(20/19)\lambda t]+1/20$ where P is homology between two amino acid sequences compared and t is the time of divergence measured in unit of years (23). Here we assumed that the time of divergence between gram negative and positive bacteria is 1200 million years ago and that between B. subtilis and B. stearothermophilus is 400 million years ago (24). Thick lines represent the divergence among proteins and thin lines the divergence among bacterial species.

region of PBP-3 mentioned above (alanine 276, serine 307, lysine 310 and leucine 318 in Fig. 3), that is 80% homology for these 5 invariable sites.

Based on these results, a phylogenic tree (dendrogram) was constructed (Fig. 4) (22). We assume that PBP-3 is an ancestral protein from which β-lactamases and CPases have diverged, since only the PBP-3 among three kinds of enzymes shown in the figure is essential for survival of E. coli (4). The divergence of bacterial species, E. coli, Bacillus spp. and Staphylococcus aureus have occurred after the divergence of the enzymes. Assuming that the time of divergence between E. coli and gram positive bacteria (Bacillus and Staphylococcus) is approximately 1200 million years ago and that among Bacillus species is 400 million years ago (24), the rate of amino acid substitution of the penicillin binding region has been estimated to be 4.9×10^{-10} per year. This estimate indicates that this region surrounding the penicillin binding site is more conservative than the known average rate (about 10^{-9}) among various proteins, but is slightly less than cytochrome c (22). This finding is consistent with the well known fact that proteins playing essential role in life tend to have lower rate of evolution. The time of divergence between PBP-3 and β-lactamase-CPase, and that between β-lactamases and CPases are approximately 2000 and 1400 million years ago, respectively.

This work was supported by a grant from the Ministry of Education, Science and Culture of Japan to Y. H. (No. 56122005).

References

1. Blumberg, P.M., Strominger, J.L.: Bacteriol. Rev. 38, 291-335 (1974).
2. Spratt, B.G., Pardee, A.B.: Nature 254, 516-517 (1975).
3. Spratt, B.G.: Eur. J. Biochem. 72, 341-352 (1977).
4. Suzuki, H., Nishimura, Y., Hirota, Y.: Proc. Natl. Acad. Sci. USA 75, 664-668 (1978).
5. Nishimura, Y., Nishimura, A., Suzuki, H., Inouye, M., Hirota, Y.: Plasmid, 1, 67-77 (1977).
6. Spratt, B.G.: Proc. Natl. Acad. Sci. USA 72, 2999-3003 (1975).
7. Clarke, L., Carbon, J.: Cell 9, 91-99 (1976).
8. Tamura, T., Suzuki, H., Mizoguchi, J., Hirota, Y.: Proc. Natl. Acad. Sci. USA 77, 4499-4503 (1980).
9. Nakamura, M., Maruyama, I. N., Soma, M., Kato, J., Suzuki, H., Hirota, Y.: Mol. Gen. Genet., in press.
10. Rosenberg, M., Court, D.: Ann. Rev. Genetics 13, 319-353 (1979).
11. Siebenlist, U., Simpson, R.B., Gilbert, W.: Cell 20, 269-281 (1980).
12. Steitz, J. A., Jakes, K.: Proc. Natl. Acad. Sci. USA 72, 4734-4738 (1980).
13. Shine, J., Dalgarno, L.: Proc. Natl. Acad. Sci. USA 71, 1342-1346 (1974).
14. Segrest, J.P., Feldmann, R.J.: J. Mol. Biol. 87, 853-858 (1974).
15. Austen, B.M.: FEBS Lett. 103, 308-313 (1979).
16. Inouye, S., Soberon, X., Franceschini, T., Nakamura, K., Itakura, K., Inouye, M.: Proc. Natl. Acad. Sci. USA 79, 3438-3441 (1982).
17. Garnier, J., Osguthorpe, D.J., Robson, B.: J. Mol. Biol. 120, 97-120 (1978).

18. Garnier, J., Gaye, p., Mercier, J.-C., Robson, B.: Biochimie 62, 231-239 (1980).

19. Inouye, M., Halegoua, S.: CRC Crit. Rev. Biochem. 7, 339-371 (1980).

20. Ambler, R.P.: Phil. Trans. R. Soc. Lond. B289, 321-331 (1980).

21. Waxman, D.J., Strominger, J.L.: J. Biol. Chem. 255, 3964-3976 (1980).

22. Dayhoff, M.O. (ed.) Atlas of Protein Sequence and Structure. National Biomedical Research Foundation, Washington, D.C. 1972

23. Jukes, T.H., Cantor, C.R.: Mammalian Protein Metabolism, 3rd ed., ed. by Munro, H.N., Academic Press, N. Y. 1969

24. Hori, H., Osawa, S.: Proc. Natl. Acad. Sci. USA 76, 381-385 (1979).

SEQUENCE OF PENICILLIN-BINDING PROTEIN 5 OF *ESCHERICHIA COLI*

Jenny Broome-Smith, Alex Edelman and Brian G. Spratt

Microbial Genetics Group, School of Biological Sciences
University of Sussex, Brighton BN1 9QG, UK

Introduction

The determination of the structure of penicillin-binding proteins (PBPs; refs 1,2) is essential to an understanding of the molecular details of the interaction of β-lactam antibiotics with their killing targets. This type of information will come from X-ray crystallographic studies and it is therefore of interest to know how similar PBPs are to each other and therefore whether the solution of the structure of one PBP will allow a rapid solution to the structure of others. A second interest in the structure of PBPs relates to the suggestion that β-lactamases evolved from penicillin-sensitive enzymes involved in cell wall synthesis.
The relationships amongst PBPs, and between PBPs and β-lactamases, should emerge from a comparison of their amino acid sequences. At present little is known of the amino acid sequences of PBPs. A small amount of sequence from the NH_2-terminus of PBP 5 of *Bacillus subtilis* and *B.stearothermophilus* and from the NH_2-terminus of PBPs 5 and 6 of *E.coli* has been reported (3,4) and recently the complete sequence of PBP 3 of *E.coli* has been obtained from the nucleotide sequence of the *fts*I gene (5). We have recently obtained the sequence of PBP 5, and of its signal peptide, from the nucleotide sequence of the *dac*A gene (6) and will describe here its sequence relationship to other PBPs and to β-lactamases.

The Amino Acid Sequence of PBP 5 of *E.coli*.

PBP 5 of *E.coli* is a major D-alanine carboxypeptidase but the precise role of this enzyme in peptidoglycan synthesis is unknown. The *dac*A gene has been located on a 1.6-kb *Eco*RI-*Bam*HI fragment (7). We have sequenced both

The Target of Penicillin

strands of this fragment using the M13 phage/dideoxynucleotide method (6). Waxman et al. (4) have identified the NH_2-terminal 28 amino acids of PBP 5 by protein sequencing, and this information has allowed us to align the start of the mature protein with the amino acid sequence derived from the DNA sequence. Figure 1 shows the sequence of the 374 amino acids of mature PBP 5. The molecular weight of PBP 5 calculated from the sequence is 41,340 which agrees well with the reported value of 42,000 (1). The NH_2-terminal sequence of Waxman et al. (4) differs from that derived from the nucleotide sequence at only one position where residue 27 is tentatively assigned as proline but is shown to be serine from the DNA sequence.

Fig. 1. The Amino Acid Sequence of PBP 5 of Escherichia coli

```
               -29                                                -1
                MNTIFSARIMKRLALTTALCTAFISAAHA

1                  20                  40
DDLNIKTMIPGVPQIDAESYILIDYNSGKVLAEQNADVRR

41                 60                  80
DPASLTKMMTSYVIGQAMKAGKFKETDLVTIGNDAWATGN

81                100                 120
PVFKGSSLMFLKPGMQVPVSQLIRGINLQSGNDACVAMAD

121               140                 160
FAAGSQDAFVGLMNSYVNALGLKNTHFQTVHGLDADGQYS

161               180                 200
SARDMALIGQALIRDVPNEYSIYKEKEFTFNGIRQLNRNG

201               220                 240
LLWDNSLNVDGIKTGHTDKAGYNLVASATEGQMRLISAVM

241               260                 280
GGRTFKGREAESKKLLTWGFRFFETVNPLKVGKEFASEPV

281               300                 320
WFGDSDRASLGVDKDVYLTIPRGRMKDLKASYVLNSSELH

321               340                 360
APLQKNQVVGTINFQLDGKTIEQRPLVVLQEIPEGNFFGK

361          374
IIDYIKLMFHHWFG
```

Comparison of the Amino Acid Sequences of D-Alanine Carboxypeptidases, β-Lactamases and Peptidoglycan Transglycosylase/transpeptidases.

Figure 2a shows our alignment of the NH_2-terminal region of PBP 5 with the available sequences of D-alanine carboxypeptidases. The alignment of the E.coli and Bacillus sequences differs markedly from that suggested previously (4). All four sequences can be aligned without the introduction of any gaps and, although the information is still fragmentary, the degree of similarity between the NH_2-terminal regions of these D-alanine carboxypeptidases is such that they may form a homologous group (Figure 2a). For example, PBP 5 of E.coli and PBP 5 of B.subtilis have identical amino acids at 25 of the 65 residues for which data is available for the two proteins (39% identity) and structurally similar amino acids are present at several of the other positions.

Serine-36 has been identified as the residue acylated by both penicillin and substrate in PBP 5 of B.subtilis and B.stearothermophilus (3) and the sequence similarities in this region allow serine-44 to be identified as the corresponding active site residue in PBP 5 of E.coli (Figure 2a). As expected, there was no significant homology between PBP 5 of E.coli and the penicillin-insensitive zinc-requiring D-alanine carboxypeptidase from Streptomyces albus G (8).

It has often been suggested that β-lactamases may have arisen from penicillin-sensitive enzymes involved in bacterial cell wall synthesis and certainly there are several similarities between these two groups of enzymes (2,3,4). Waxman et al. (3,4) have pointed out sequence similarities between the class A β-lactamases (e.g. TEM β-lactamase) and the available NH_2-terminal amino acid sequences of D-alanine carboxypeptidases. Figure 2b shows the alignment of the NH_2-terminal region of TEM β-lactamase with the NH_2-terminal region of the D-alanine carboxypeptidases. Within the NH_2-terminal 65 amino acids there is 23% identity between TEM β-lactamase and PBP 5 of both E.coli and B.subtilis. At several positions identical, or structurally similar, amino acids are present in all of the D-alanine carboxypeptidases and class A β-lactamases that have been examined (e.g. asp/glu-24, ser/thr-27, asp/glu-37, ser-44, lys-47). Maximum similarity between the D-alanine carboxypeptidases and class A β-lactamases occurs immediately to the NH_2-terminal side of the active site serine residue and the similarity decreases rapidly on the COOH-terminal side of this residue. Comparison of the entire sequence of PBP 5 of E.coli with those of class A β-lactamases shows that no further significant similarities exist outside of the NH_2-terminal region shown in Figure 2b.

Whilst sequence similarities do exist between the D-alanine carboxypeptidases and class A β-lactamases the two classes of proteins are clearly not closely related. If they have evolved from a common ancestral protein, then the divergence in their biological function and cellular location has been paralleled by extensive diverence in their primary sequences, and the region of similarity close to the NH_2-terminus may reflect the conservation of a sequence that is part of the binding site for β-lactam antibiotics. It should be stressed that other classes of β-lactamase exist.

The chromosomal β-lactamase of E.coli (a class C enzyme) has recently been sequenced but shows little similarity to either class A β-lactamases (9) or D-alanine carboxypeptidases.

Comparison of the amino acid sequence of E.coli PBP 5 with that of E.coli PBP 3 (5), which is a transglycosylase/transpeptidase (10) and a killing target for β-lactam antibiotics (1), shows that these two proteins are completely unrelated except for one small region of significant similarity (Figure 3). Interestingly this region in PBP 5 is the region that is most highly conserved within the D-alanine carboxypeptidases and between the carboxypeptidases and class A β-lactamases. Whereas this sequence occurs close to the NH_2-terminus of the D-alanine carboxypeptidases it occurs towards the middle of the larger transglycosylase/transpeptidases. This raises the possibility that the sequence is part of the penicillin-binding region of PBP 3 and that the transglycosylase domain of PBP 3 may be NH_2-terminal and the penicillin-sensitive transpeptidase domain may be COOH-terminal.

```
A)  E. coli PBP 6                          NH2-AEQTVEAPSVDASAWFLMDYAXGKV---
                                                   *  **     * **  ***
    E. coli PBP 5                   NH2-DDLNIKTMIPGVPQIDAESYILIDYNSGKVLAEQNADVRRDPASLTKMMTSYVIGQAMKAGKFKETDLVTIG---
                                                   * *   * *   *** *   *** *   ** ***** *     *   ** *     *
    B. subtilis PBP 5                       NH2-ASDPIDINASAAIMIEASSGKILYSKNADKRLPIASMTKMMTEYLLLEAIDQGKVKWDQTYTPD---
                                                * * ** * ***    *  ***** ** *   * *******
    B. stearothermophilus PBP 5             NH2-ESAPLDIRADAAILVDAQTGKILYEKNIDTVLGIASMTKM---


B)  E. coli PBP 5                   NH2-DDLNIKTMIPGVPQIDAESYILIDYNSGKVLAEQNADVRRDPASLTKMMTSYVIGQAMKAGKFKETDLVTIG---
                                             *          ** * **** *        *    *  *              **
    TEM β-lactamase                NH2-HPETLVKVKDAEDQLGARVGYIELDLNSGKILESFRPEERFPMMSTFKVLLCGAVLSRVDAGQEQLGRRIHYS---
                                                        *     ***** *      * *  *  *        *   * *
    B. subtilis PBP 5                       NH2-ASDPIDINASAAIMIEASSGKILYSKNADKRLPIASMTKMMTEYLLLEAIDQGKVKWDQTYTPD---
```

Fig. 2. A) Alignment of NH_2-terminal sequences of D-alanine carboxypeptidases. B) Alignment of D-alanine carboxypeptidases and TEM-β-lactamase.

```
A A I L V D A Q T G K I L Y E K N
  + : + + +   : + +   : +       +
S A V L V D V N T G E V L A M A N
+   : + : +   + : +   + + +     +
S Y I L I D Y N S G K V L A E Q N
```

Figure 3. A region of similarity between D-alanine carboxypeptidases and a bifunctional peptidoglycan transglycosylase/transpeptidase. The top line shows residues 11 - 27 of PBP 5 of B. stearothermophilus, the middle line residues 218 - 234 of PBP 3 of E. coli, and the bottom line residues 19 - 35 of PBP 5 of E. coli. Identical residues are marked + and structurally related residues :. The single-letter notation of amino acids is used.

Organisation of PBP 5 of E.coli in the Cytoplasmic Membrane.

PBPs catalyse reactions on the exterior surface of the cytoplasmic membrane and therefore have to transport portions of their polypeptide chain across the membrane. In B.subtilis and B.stearothermophilus PBP 5 appears to be an ectoprotein which can be cleaved from the membrane with trypsin to produce a water-soluble active enzyme that has lost a small COOH-terminal hydrophobic membrane anchoring peptide (11). In eukaryotes, proteins with this organisation within the cytoplasmic membrane are made as pre-proteins with NH_2-terminal signal peptides. We have previously shown that both PBP 5 and PBP 6 of E.coli are also made as pre-proteins (12). Analysis of the DNA sequence upstream of the start of the mature protein identifies a single significant translation initiation sequence. Assuming that this translation start is used in vivo, PBP 5 contains a 29 amino acid signal peptide (Figure 1). This is the only reported signal peptide of a processed cytoplasmic membrane protein, encoded by a bacterial gene, and the sequence shows all of the features characteristic of the signal peptides of periplasmic and outer membrane proteins, e.g. basic residues at the NH_2-terminus followed by an uninterrupted stretch of hydrophobic amino acids and alanine at the splice point (6).
There is no indication from the amino acid sequence that PBP5 is inserted in the cytoplasmic membrane by a hydrophobic COOH-terminal membrane anchoring peptide. The protein is not particularly hydrophobic (polarity index of 45%) and contains no extensive hydrophobic stretches characteristic of membrane anchoring peptides. The organisation of PBP 5 of E.coli in the cytoplasmic membrane may therefore be different from that of PBP 5 of Bacillus.

References

1. Spratt, B.G.: Phil. Trans. Roy. Soc. B289, 273-283 (1980).
2. Spratt, B.G.: J. Gen. Microbiol. 129, in press (1983).
3. Waxman, D.J., Yocum, R.R., Strominger, J.L.: Phil. Trans. Roy. Soc. B289, 257-271 (1980).
4. Waxman, D.J., Amanuma, H., Strominger, J.L.: FEBS Letters 139, 159-163 (1982).

5. Maruyama, I.N., Nakamura, M., Soma, M., Nishimura, Y., Hirota, Y.: Abstract, EMBO meeting on β-lactam antibiotics, El Escorial (1982).

6. Broome-Smith, J.K., Edelman, A., Spratt, B.G.: In preparation (1983).

7. Markiewicz, Z., Broome-Smith, J.K., Schwarz, U., Spratt, B.G.: Nature 297, 702-704 (1982).

8. Joris, B., Van Beeumen, J., Casagrande, F., Gerday, C., Frere, J-M., Ghuysen, J-M.: Eur. J. Biochem. 130, 53-69 (1983).

9. Jaurin, B., Grundstrom, T.: Proc. Natl. Acad. Sci. USA. 78, 4897-4901 (1981).

10. Matsuhashi, M., Nakagawa, J., Tomioka, S., Ishino, F., Tamaki, S.: In: Drug Resistance in Bacteria - Genetics, Biochemistry and Molecular Biology. pp. 297-310. Ed. S. Mitsuhashi, Japan Scientific Societies Press, Tokyo 1982.

11. Waxman, D.J., Strominger, J.L.: J. Biol. Chem. 254, 4863-4875 (1979).

12. Pratt, J.M., Holland, I.B., Spratt, B.G.: Nature 293, 307-309 (1981).

This work was supported by the Medical Research Council.

PURIFICATION AND IDENTIFICATION OF THE *Pseudomonas aeruginosa* DD-ENDOPEPTIDASE AS A PENICILLIN-BINDING PROTEIN

Juan-Carlos Montilla, Alfredo Rodríguez-Tébar and David Vázquez
Instituto de Bioquímica de Macromoléculas, Centro de Biología Molecular, CSIC and UAM, Cantoblanco, Madrid-34, Spain

Introduction

Elongation and septation of *P. aeruginosa* is a dynamic process that requires the participation of a number of enzymes acting on the synthesis or hydrolysis of peptidoglycan. One of these hydrolytic enzymes is the DD-endopeptidase that splits the D-alanyl-(D)*meso*-2,6-diaminopimelyl linkages that crosslink the peptidoglycan peptides.

Results

1) Solubilization and partial purification of DD-endopeptidase activity.
The molecule

GlcNAc-MurNAc-L-Ala-γ-D-Glu-*meso*-(^{3}H)A_2pm-D-Ala
GlcNAc-MurNAc-L-Ala-γ-D-Glu-meso-(^{3}H)A_2pm-D-Ala

(C3) (1) is used as a substrate in routine assays of DD-endopeptidase activity and the product of the reaction is GlcNAc-MurNAc-L-Ala-γ-D-Glu-*meso*-(^{3}H)A_2pm-D-Ala (C6) (1) when the purified enzyme is used. However, in the purification procedure of DD-endopeptidase there is a contamination with NAcMuramyl-L-Alanine amidase along the first three steps and therefore the end product of the reaction is not C6 but the tetrapeptide L-Ala-D-Glu-*meso*-A_2pm-D-Ala.
When cell suspensions of *P. aeruginosa* NCTC 10 662

The Target of Penicillin

TABLE I

Purification of DD-endopeptidase from *P. aeruginosa*

Preparation	Total protein (mg)	Specific activity	Purification (X fold)	% Yield
Crude extract	9,000	8.8	—	100
Fraction I	4,505	15.0	1.7	85
Fraction II	1,664	36.0	4.1	75
Fraction III	712	84.0	9.5	75 (a)
Fraction IV	18	1201.0	136.5	68

a) This yield was obtained by KBr treatment. However, only 50-60% yield was obtained after subsequent removal of the salt by dialysis (see Results) since there was some precipitation of proteins. KBr increased 80% the specific activity.

(200 mg/ml) were broken by 5x1 min pulses of ultrasonic treatment and centrifuged at 100 000 xg for 90 min, most DD-endopeptidase activity is present in the supernatant fraction (Fraction I). However, when this Fraction I was centrifuged at 100 000 xg for 16 h the DD-endopeptidase activity was pelleted. Nonionic detergents such as Triton X-100, Genapol or Nonidet are not able to solubilize the DD-endopeptidase activity from this precipitate. However, the enzyme is solubilized when the pellet was resuspended and treated with 1.0 M KBr and centrifuged at 100 000 xg for 16 h. When this supernatant was dialyzed, the DD-endopeptidase is partially precipitated but 50-60% of the enzyme remains in the supernatant (Fraction III). By chromatographic treatment of Fraction III in a DEAE-Sepharose column, the DD-endopeptidase passed through the column (Fraction IV), whereas the amidase activity was retained in the column (Table I). The DD-endopeptidase in this Fraction IV is very unstable and further purification was not possible.

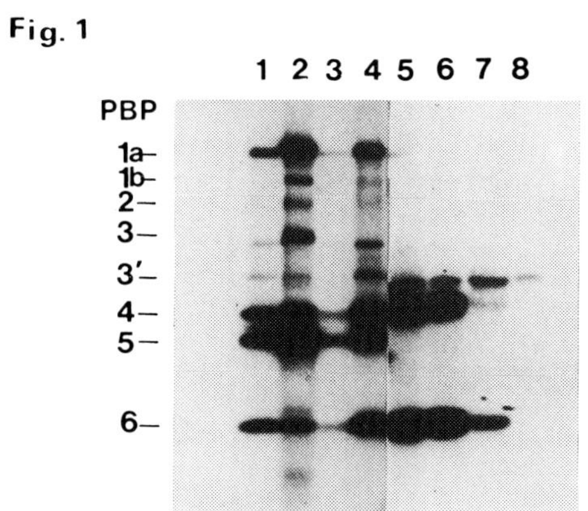

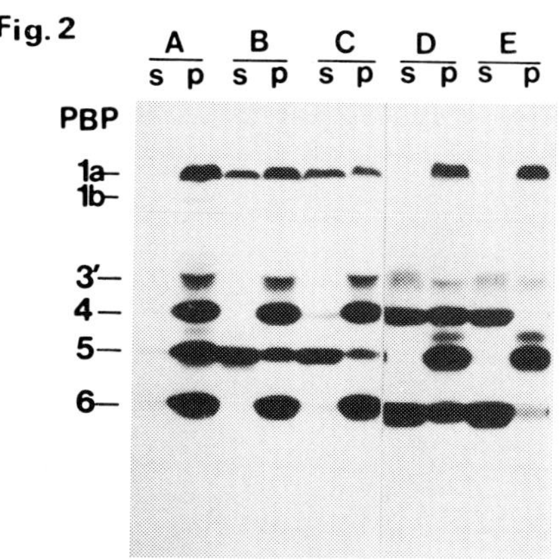

Figure 1. Patterns of the PBPs from P. aeruginosa in different steps of purification of the DD-endopeptidase (for nomenclature see the text).
1) Fraction I; 2) precipitate of 90 min centrifugation; 3) supernatant of 16 h centrifugation; 4) Fraction II; 5) fraction extracted with KBr; 6) precipitate after KBr removal; 7) Fraction III; 8) Fraction IV.

Figure 2. Solubilization of PBPs from Fraction II. Fraction II (final protein concentration 10 mg/ml) was treated for 30 min at 30°C with: A) nothing; B) 0.25% Triton X-100; C) 1.0% Triton X-100; D) 0.25 M KBr; E) 1.0 M KBr. Samples were then centrifuged at 100 000 xg for 16 h. s) supernatants, p) precipitates.

---

2) Properties of DD-endopeptidase.

The DD-endopeptidase of Fraction IV has an apparent Km = 25 µM for the subtrate C3 and is a Mg^{++}-dependent enzyme, the optimum of Mg^{++} concentration beeing 0.5-1.0 mM. Between pH 6.0 and pH 8.5 there is a linear increase of activity, but doubts remain whether this increase is an intrinsic characteristic of the enzyme or merely an effect of the protonation of the substrate C3. The enzyme is inhibited by heparin, retained in a Poly(U)-Sepharose column (see below) and fully blocked by moderately high concentrations of benzylpenicillin (ID_{50}= 200 µg/ml).

3) DD-endopeptidase and PBPs.

Since DD-endopeptidase is a benzylpenicillin-sensitive enzyme, it might be conceivable that the enzyme was one of

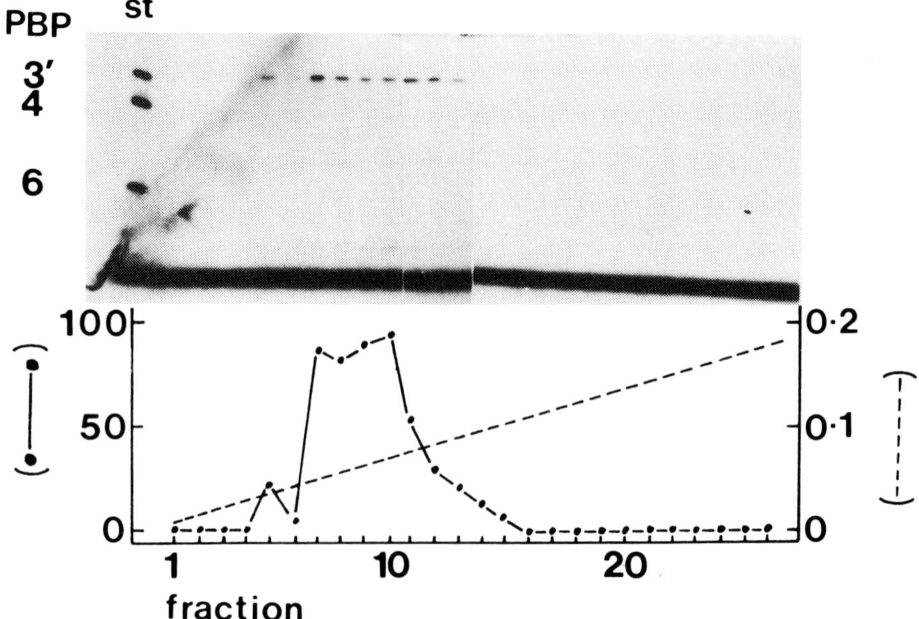

Figure 3. Chromatography of Fraction III on Poly(U)-Sepharose. Fraction III (in 10 mM sodium phosphate, pH 7.0) was passed through a column of Poly (U)-Sepharose. Proteins retained were eluted with a linear gradient of phosphate (between 10 and 200 mM). (o——o) DD-endopeptidase activity (arbitrary units). (- - -) Phosphate concentration (M). st: PBP pattern of Fraction III.

---

the PBPs of P. aeruginosa. Indeed, there are a number of data supporting this hypothesis. Figure 1 shows the distribution of PBPs throughout the purification procedure of DD-endopeptidase clearly showing that only PBP 3' is present in the fractions containing the DD-endopeptidase activity.
PBP 3' from P. aeruginosa has been detected by some workers (2) but not by many others since it does not sediment with the membrane fraction of sonicated cells by the standard procedure of centrifugation at 100 000 xg for 60-90 min. Nevertheless, PBP 3' is present in the pellet obtained by centrifugation at 100 000 xg for 16 h (Fraction II). When PBPs are solubilized with nonionic detergents or salts, DD-endo-

peptidase activity is always obtained in the same fractions containing PBP 3' (Figure 2).
Attempts to purify PBP 3' and DD-endopeptidase by affinity chromatography on ampicillin-Sepharose were not successfull since no enzymic activity was detected and the enzyme lost 90% of its capacity to bind β-lactam antibiotics. However, when Fraction I was chromatographed on Sepharose 6B (not shown) and Fraction III was chromatographed on either Poly-(U)-Sepharose (Figure 3) of Sephacryl S-200 (not shown), DD-endopeptidase was always associated with fractions containing PBP 3', further suggesting that PBP 3' is the DD-endopeptidase.

Conclusions

Data presented here show that *P. aeruginosa* membranes are provided of a DD-endopeptidase. The enzyme is contaminated in the initial steps of purification with a NAcMuramyl-L-Alanine amidase as previously reported in Streptomyces (3).
Our purified DD-endopeptidase is sensitive to benzylpenicillin as previously reported for some bacterial DD-endopeptidases (4,5) but is also blocked by heparin as previously described by other type of DD-endopeptidases insensitive to β-lactams (6,7).
Our data show an association of PBP 3' and DD-endopeptidase in all preparations and fractions and strongly suggest that PBP 3' itself is the DD-endopeptidase enzyme.

References

1. Primosigh, J., Pelzer, H., Maass, D., Weidel, W.: Biochim. Biophys. Acta 46, 68-80 (1961).

2. Noguchi, H., Matsuhashi, M. and Mitsuhashi, M.: Eur. J. Biochem. 100, 41-49 (1979).

3. Katayama, T., Matsuda, T., Kato, K. and Kotani, S.: Biken J. 19, 75-91 (1976).

4. Tamura, T., Imae, Y. and Strominger J. L.: J. Biol. Chem. 261, 414-423 (1976).

5. Keck, W.: Phil. Dissertation. Eberhard-Karls Universität Tübingen, G.F.R. (1980).

6. Keck, W. and Schwarz, U.: J. Bacteriol. 139, 770-774 (1979).

7. Tomioka, S. and Matsuhashi, M.: Biochem. Biophys. Res. Comm. 84, 778-784 (1978).

PURIFICATION OF PENICILLIN BINDING PROTEINS FROM STREPTOCOCCUS PNEUMONIAE

Regine Hakenbeck
Max-Planck-Institut für molekulare Genetik, Abt. Trautner,
D-1000 Berlin 33

Introduction

Penicillin binding proteins (PBPs) in pneumococci have gained interest since it has been discovered that penicillin resistance in these bacteria (intrinsic resistance) is coupled with a change in affinity towards penicillin in the high molecular weight (hmw) PBPs (PBP 1a, 1b, 2a and 2b), but not in the PBP with the lowest molecular weight (lmw), PBP3. In addition, at a high level of resistance, novel PBPs appear (1, 2, 3). Since all the resistant strains (clinical isolates) grow normally at 37°C, one has to assume that the change in the altered PBPs affects only the binding to ß-lactams and not (or not substantially) their enzymatic function. It is obvious that isolation of the PBPs is one prerequisite to understand their function as well as the mechanism of penicillin resistance. In this paper, sone aspects of the isolation and characterization of pneumococcal PBPs from the penicillin-sensitive wild type will be presented. Details of the chromatographic procedures will be published elsewhere, or are documented in (4).

Results and Discussion

PBPs were purified from solubilized membrane proteins. Triton X100 (0.1 %) was used, and this amount of detergent was present during all further steps. PBP2b binds to the membrane very strongly, and could not completely be solubilized (Fig. 1) suggesting that it contains a very hydrophobic region, whereas all of PBPs 1a and 1b and most of PBP3 were liberated by

The Target of Penicillin

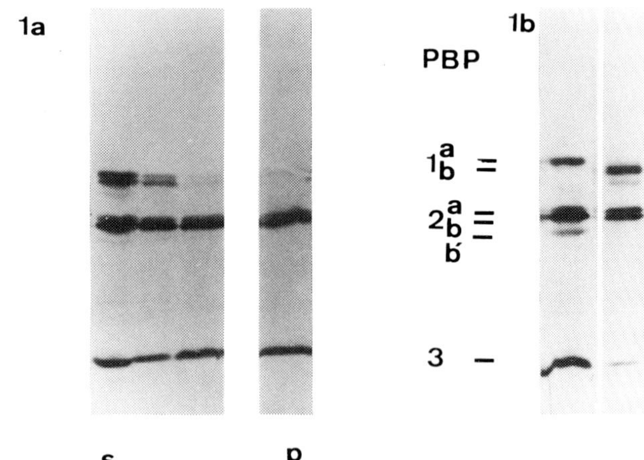

Fig. 1: PBPs after solubilization and phosphocellulose chromatography. 1a: Fluorogram of PBPs after three successive extractions of membranes with 0.1 % Triton X100 in the solubilized fractions (s) and the residual membrane pellet (p). 1b: PBPs in the PC non-adsorbed (left) and PC-adsorbed (right) fraction. Chromatography was performed in 40 mM Na-phosphate buffer pH 6.6. All buffers used for chromatographies contained 0.1 % Triton X100, 10 % glycerol, 5 mM ß-mercaptoethanol and 0.5 mM EGTA.

this mild extraction procedure. For separation of the individual PBPs, three types of chromatographic methods were used. I: ion exchange columns, phosphocellulose followed by DEAE-sephacel; II: separation according to the isoelectric point by chromatofocusing; III: affinity chromatography on poly-U sepharose and Blue sepharose. Ampicillin sepharose was avoided for several reasons: not all PBPs can be easily eluted, and separation between different PBPs may not be achieved by this method. Furthermore, it is not clear whether the conditions needed for elution affect the enzymatic activity of some of the PBPs. For monitoring penicillin-binding activity, a filter binding assay has been developed recently (5). Although the individual PBPs still have to be analyzed on SDS-PAGE and fluorography, the binding assay economizes the testing by rapidly discriminating between PBP-containing and non-containing fractions. Also, quantification of the binding activity is easily possible. For radioactive labelling, ^{3}H-propionyl-ampicillin (^{3}H-PA) was used routinely (6).

Fig. 1b shows that the phosphocellulose (PC) chromatography results in separation of PBP 1a and 3 (PC non-adsorbed) from PBP 1b (PC adsorbed). PBP 2b and 2a are present in both fractions. All PBPs bound to DEAE-sephacel and all of them eluted below 0.3 M NaCl. At this stage, a binding component that sometimes appeared already in crude membranes was concentrated and could be analyzed on a two-dimensional gel system (Fig. 2). It clearly originated from PBP 2b and was consequently named PBP 2b'.

Major separation between the individual PBPs was achieved by chromatofocusing. All PBPs eluted at a pH below 6 (Fig. 4) in agreement with their behaviour on 2D-gels with isoelectric focusing in the first dimension (Fig. 3). PBP 2b eluted (and focused) at two different pH: a small portion at pH 5.4, and the bulk of the protein at pH 4.8.

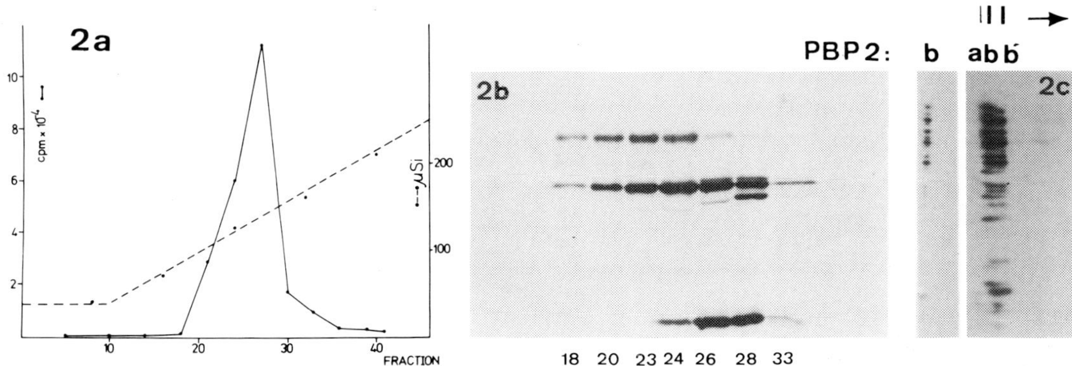

Fig. 2: DEAE-sephacel chromatography of the PC non-adsorbed fraction. 2a: penicillin binding activity during elution with a linear NaCl gradient (20 - 700 mM) in 20 mM Tris-HCl pH 8.3. 2b: Fluorogram of PBPs in the active fractions. 2c: two-dimensional analysis of fraction 28. After labelling with ^{3}H-PA, PBPs were separated on SDS-PAGE as shown in 2b. In the second dimension, PBPs were digested during SDS-electrophoresis using SV8-protease in the stacking gel (Ellerbrok and Hakenbeck, this issue). For comparison, a fluorogram of proteolytic products derived from PBP 2b is shown on the left.

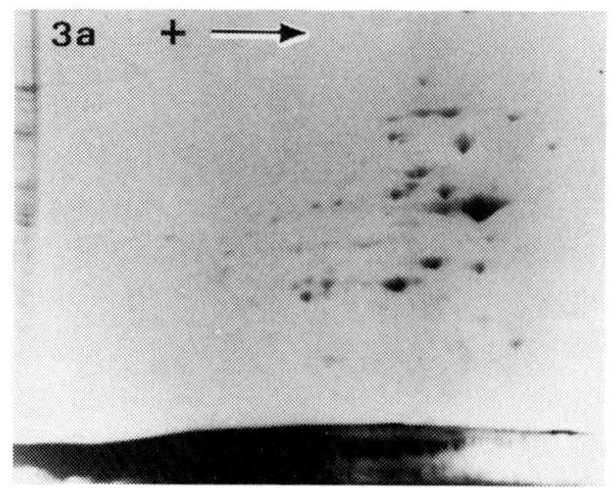

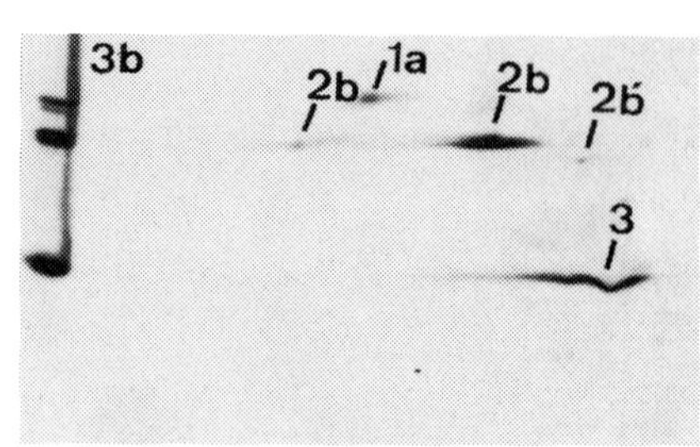

Fig. 3: Two-dimensional electrophoresis of membrane bound PBPs. 150 µg of membrane proteins were incubated with ^{3}H-PA and separated by isoelectric focusing in the first dimension and on SDS-PAGE in the second dimension (9). 3a: Coomassie blue stain; 3b: Fluorogram. BPB 1b and 2a cannot be identified under these conditions.

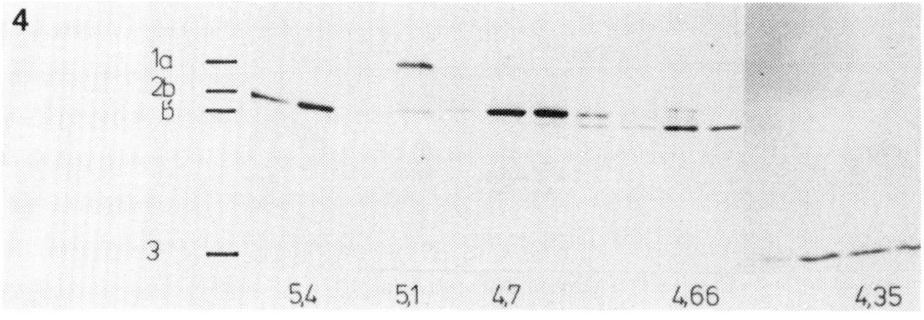

Fig. 4: Chromatofocusing of PBPs. Fraction 26 - 33 of the DEAE-chromatography shown in Fig. 2 were separated on a chromatofocusing column according to the procedure described by Pharmacia for a pH-range from 6 - 4. The active fractions were analyzed after fluorography on SDS-PAGE; the pH of the fractions is indicated.

Partial proteolysis of these two fractions revealed no difference. In addition, PBP 2a and 2b occurred as somewhat smaller molecules, called PBP 2a' and 2b' (see also Fig. 2). Whether some specific modification or a processing-like cutting of the molecules is responsible for that variation remains to be clarified. Final purification of PBP3 has been published (4) and will be briefly summarized here: the protein shows weak interactions with Blue sepharose, so that all contaminating proteins did not adsorb to the column, whereas PBP3 eluted very late after extensive washing with the same buffer. PBP 3 contained penicillin-sensitive D,D-carboxypeptidase activity, thereby resembling lmw PBPs of other species. Its isoelectric point is the most acidic among the pneumococcal PBPs (4.35). PBP 1a and 2b could be purified after affinity chromatography on poly-U sepharose and Blue sepharose, for the latter 2M NaCl and more were required for elution. SDS-PAGE of the purified PBPs is shown in Fig. 4. PBP 2b still contains other non-PBP proteins, whereas PBP 1a and 3 appear over 95 % pure. For PBP3, a recovery of 11 % and for PBP 1a of 6 % was calculated. The proteins were kept at -20° C in 10 % glycerol. No loss of penicillin binding activity occurred over a period of at least 10 months. Binding kinetics resembled those as shown for native, membrane bound PBPs (5): saturation of the purified PBP3 occurred at the lowest, and of PBP 1a at the highest ^{3}H-PA concentration. Therefore it is likely that the penicillin sensitive enzymatic functions are still actively contained in the proteins. To our surprise, PBP 1a and even more PBP 2b, but not PBP 3, showed high affinity to ds DNA. Binding was apparent using 50 - 100 ng protein and 50 ng DNA (calf thymus, 300 base pairs) and optimum binding occurred at 0.1 M NaCl. At this DNA concentration no interference with penicillin binding activity could be detected. The specificity of this reaction is currently under investigation.

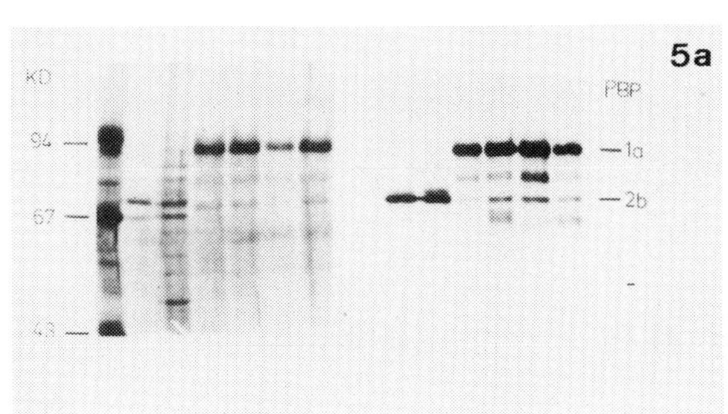

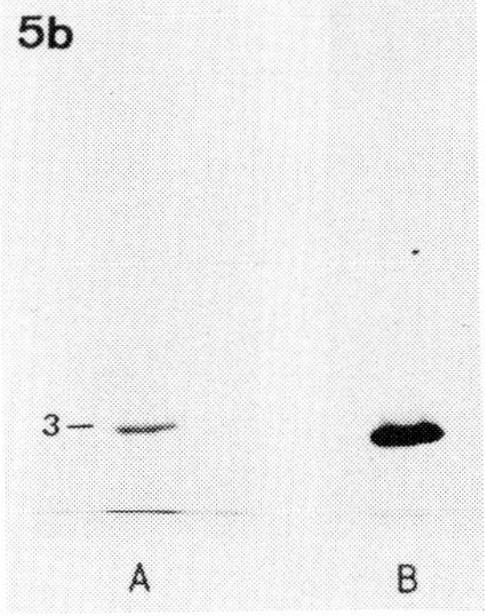

Fig. 5: SDS-PAGE of the purified PBPs. 5a: silver stain (10) of PBP 2b and 1a (left) and fluorogram (right) after Blue-Sepharose and final concentration on DEAE-sephacel. 5b: coomassie blue stain (A) and fluorogram (B) of PBP3.

Interestingly, affinity to polynucleotides has also been documented for another enzyme acting on murein, the trans-glycosylase in *E. coli* (7). One might speculate that the DNA binding capacity reflects a regulatory device needed for correlation of DNA segregation and septal wall synthesis. Another, equally attractive but also equally speculative hypothesis can be formulated in analogy to a eucaryotic membrane protein complex functioning in processing of secretory proteins (8): a specific complex between PBPs in association with a polynucleotide might be required for coordinate function.

References

1. Percheson, P. B., Bryan, L. E.: Antimicrob. Agents Chemother 12, 390-396 (1980).
2. Hakenbeck, R., Tarpay, M., Tomasz, A.: Antimicrob. Agents Chemother 17, 364 - 371 (1980).
3. Zighelboim, S., Tomasz, A.: Antimicrob. Agents Chemother. 17, 434 - 442 (1980).
4. Hakenbeck, R., Kohiyama, M.: Eur. J. Biochem. 127, 231 - 236 (1982).
5. Hakenbeck, R., Kohiyama, M: FEMS Microb. Lett. 14, 241 - 245 (1982).
6. Schwarz, U., Seeger, K., Wengenmayr, F., Strecker, H.: FEMS Microb. Lett. 10, 107 - 109 (1981).
7. Kusser, W., Schwarz, U.: Eur. J. Biochem. 103, 277 - 281 (1980).
8. Walter, P., Blobel, G.: Nature 299, 691 - 698 (1982).
9. Ferro-Luzi Ames, G., Nikaido, K.: Biochemistry 15, 616 - 623 (1976).
10. Wray, W., Boulikas, T., Wray, V. P., Hancock, R.: Anal. Biochem. 118, 197 - 203 (1981).

CHARACTERIZATION OF PENICILLIN BINDING PROTEINS FROM _STREPTOCOCCUS PNEUMONIAE_ BY PROTEOLYSIS

Heinz Ellerbrok and Regine Hakenbeck

Max-Planck-Institut für molekulare Genetik
D-1000 Berlin 33

Introduction

The binding process of ß-lactam antibiotics to PBPs has been studied mostly with D,D-carboxypeptidases, using these low molecular weight (lmw) PBPs as model enzymes. The recent isolation of two high molecular weight (hmw) PBPs (PBP1a and 2b) (Hakenbeck, this issue) and the lmw PBP3, a D,D-carboxypeptidase (1) of pneumococci has enabled us to study the ß-lactam binding site of these proteins in proteolytic digests of unfractionated membranes. Analysis of the binding site is of interest also in respect to the understanding of the intrinsic resistance phenomenon: e.g. in pneumococci, resistant strains show a correlating alteration in the affinity for penicillin in four out of five PBPs (2, 3, 4). In the present communication we have determined the smallest native peptide(s) which contain a functional ß-lactam binding site of pneumococcal PBPs. In addition we tried to get some information about the orientation of the binding site in these membrane components.

Results and Discussion

PBPs were routinely labelled in their ß-lactam binding site with ^{3}H-propionyl-ampicillin (^{3}H-PA; 5). Fig. 1 shows the radioactive peptides derived from denatured PBPs after _Staphylococcus aureus_ V8 (SV8) protease digestion, using a

Fig. 1: Comparison of PBPs by 2D SDS-PAGE. a) Fluorogram of ^{3}H-PA labelled membranes after SDS-PAGE (1st dimension). b) The gel slice was layered on top of a second PAGE containing SV8-protease (0.2 µg/ml) in the stacking gel (2nd dimension).

Fig. 2: Fluorography of solubilized ^{3}H-PA labelled trypsin digested PBPs. a: undigested control; c, e, g, i: labelled after, and b, d, f, h: labelled prior to trypsin treatment. Trypsin action (37° C) was stopped by adding an excess of trypsin inhibitor. Trypsin concentrations (µg/ml): b, c 50; d, e 200; (15 min); f, g 200; h, i 500; (2.5 hrs). Indicated are the relatively stable proteolytic end products. 1a, 2b and 3 refer to the origin of these fragments. X, Y and Z could not be identified.

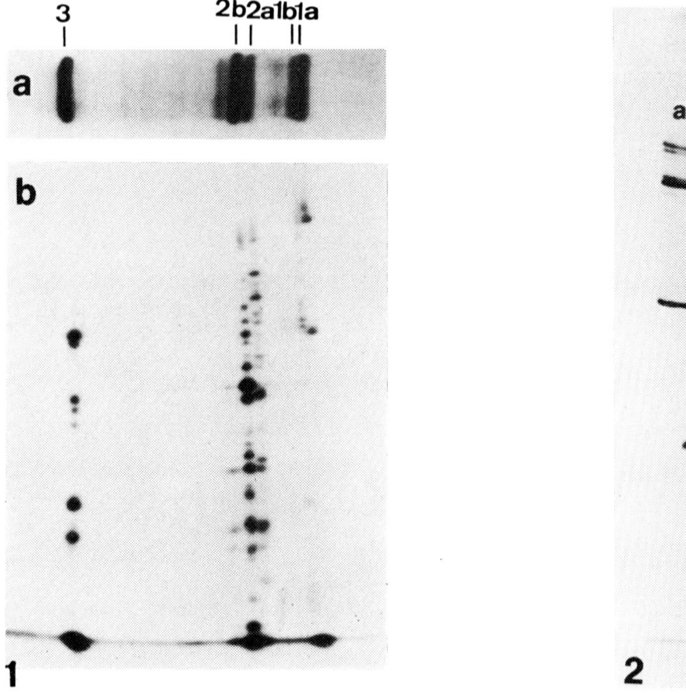

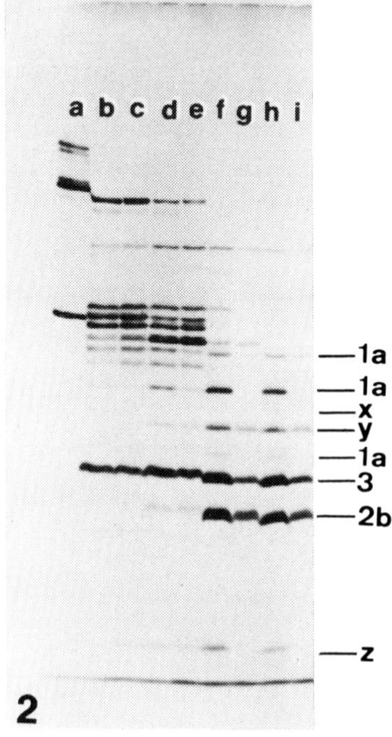

two-dimensional gel system for analysis. It can be seen that each PBP gave rise to a characteristic peptide pattern; no similarity between the different PBPs could be detected by this method. This kind of analysis represents a convenient approach if PBPs are to be compared in different unfractionated samples, e.g. in order to determine the origin of the novel PBPs that have often been found in intrinsic resistant strains. A similar system has been used recently for investigating PBPs during the growth cycle of B. subtilis (6).

When we compared native membrane bound or solubilized membrane proteins that were labelled with ^{3}H-PA and treated with trypsin, both fractions showed a similar pattern of radioactively labelled peptides. The purified PBPs were considerably more sensitive to trypsin, suggesting that in the crude fractions other proteins might protect the PBPs against tryptic action. In all cases relatively stable end products of proteolysis were obtained, some of them could be identified using the purified trypsin treated PBPs as references as indicated in Fig. 2. Tryptic digestion of denatured PBPs showed further degradation suggesting that the stability of the fragments mentioned above is not due to a lack of trypsin cleavage sites but originates from the three dimensional structure of the ß-lactam binding site (data nat shown).

In order to determine whether the peptides still contain the ability to bind ^{3}H-PA, PBPs from different fractions were labelled with the antibiotic after various times of trypsin digestion. As can be seen in Fig. 2, almost all PBP-fragments contained still an active binding site, although in some cases the binding activity decreased considerably. Preliminary results showed that some of the fragments bind and release ^{3}H-PA with the same kinetics as the corresponding native, undigested PBPs.

Waxman and Strominger reported, that carboxypeptidases in B. subtilis (50 KD) and in B. stearothermophilus (46 KD) are attached to the membrane via a small hydrophobic region at the C-terminus, having a size of 2.5 KD and 3 KD, respectively (8, 9). We have investigated whether also pneumococcal PBPs are similarly anchored into the membrane and determined the largest peptide containing the binding site that could be

Fig. 3: Tryptic digests of PBPs in spheroplasts, membranes and solubilized membranes. PBPs in the different fractions were labelled with ^{3}H-PA and trypsin treated (5 μg/ml, 15 min, 37° C). Purified PBP 1a and 2b (0.25 μg/ml trypsin, each) and PBP 3 (10 μg/ml) were used as references. Labelled fragments from membranes and spheroplasts were separated into solubilized (S) and membrane bound (P) peptides by high speed centrifugation. Solubilized fragments derived from PBP 2b (o) and 3 (o) are indicated. Solubilized fragments derived from PBP 1a have only been tentatively identified from this and other PAGEs ( ); fragment 8 could only be detected in the crude fractions when higher trypsin concentrations were used. Arrow ( ): peptide not present in spheroplast sample.

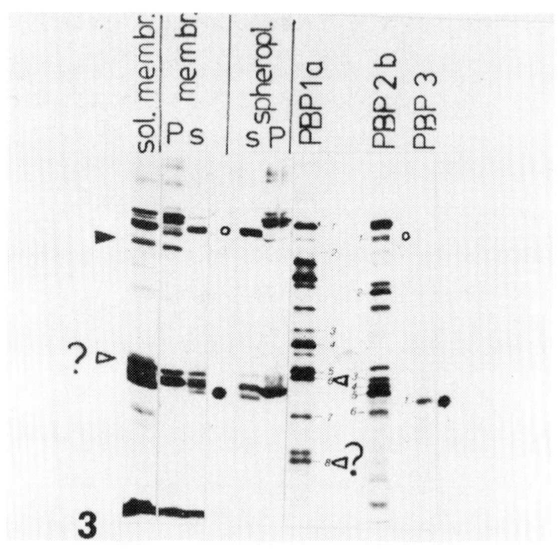

solubilized upon trypsin treatment. Furthermore, by comparing tryptic, ^{3}H-PA labelled peptides that were obtained from spheroplasts (where only that part of the protein can be attacked that faces the outside) to those derived from membranes (where the protease can act on both sides of the membrane) we tried to get some information about the orientation of the PBPs in the cytoplasmic membrane. Again, peptides that were solubilized by trypsin were separated from the membrane bound protein part by centrifugation prior to analysis on SDS-PAGE. The result of these experiments is summarized in Fig. 3 (details will be published elsewhere). From spheroplasts as well as from membranes, soluble fragments from PBP 2b and 3 were obtained having a maximum size of 74 KD, respectively, therefore leaving a membrane bound fragment of 4 KD and 2 KD (35 and 16 amino acids). Surprisingly, one fragment was present only in the membrane digest (67 KD) and not in the corresponding spheroplast fraction. The most plausible explanation is that one PBP loops through the membrane so that one (or more) cleavage sites can be reached only from the cytoplasmic site of the membrane. We have tentatively identified this fragment as being generated from PBP 1a. Since the ß-lactam binding sites of all PBPs could be attacked from the outside of the cell, these binding sites are facing the cell wall in the native stage. This is to be expected from their supposed function namely to be engaged in the final steps of murein biosynthesis. Since the penicillin binding activity resides in relatively small fragments, a second enzymatic function could be located in the residual protein part of PBP 1a and 2b. This situation would resemble the finding of Matsuhashi et al. for *E. coli* hmw PBPs (10).

References

1. Hakenbeck, R., Kohiyama, M.: Eur. J. Biochem. 127, 231-236, (1982).
2. Zighelboim, S., Tomasz, A.: Antimicrob. Agents Chemother. 17, 434-442 (1980).
3. Hakenbeck, R., Tarpay, M., Tomasz, A.: Antimicrob. Agents Chemother. 17, 364-371 (1980).
4. Percheson, P. B., Bryan, L. E.: Antimicrob. Agents Chemother. 12, 390-396 (1980).
5. Schwarz, U., Seeger, K., Wengenmayr, F., Strecker, H.: FEMS Microbiol. Lett. 10, 107-109 (1981).
6. Cleveland, D. W., Fischer, S. G., Kirschner, M. W., Laemmli, U. K.: J. Biol. Chem. 252, 1102 - 1106 (1977).
7. Todel, J. A., Ellar, D. J.: Nature 300, 640-643 (1982).

8. Waxman, D. J., Strominger, J. L.: J. Biol. Chem. 254, 4863-4875 (1979).

9. Waxman, D. J., Strominger, J. L.: J. Biol. Chem. 256, 2067-2077 (1981).

10. Matsuhashi, M., Fumitoshi, I., Shigeo, T., Nakajima-Iljima, S., Tomioka, S., Nakagawa, J., Hirata, A.: "Trends in Antibiotic Research", Japan Antibiotics Research Association, Tokyo 1982.

DISTRIBUTION OF PENICILLIN-BINDING PROTEINS WITHIN THE CELL ENVELOPE OF *Escherichia coli*

Alfredo Rodríguez-Tébar, Julio A. Barbas and David Vázquez
Instituto de Bioquímica de Macromoléculas, Centro de Biología Molecular, C.S.I.C. and U.A.M., Cantoblanco, Madrid-34, Spain

Introduction

A number of penicillin-binding proteins (PBPs) are enzymes that catalyze the last stages in peptidoglycan biosynthesis (1, review). Some crucial events in the cell cycle, such as elongation and septation in which peptidoglycan structure is drastically modified, take place at certain places in *E.coli* and other bacterial rods. Therefore, it is very important to study the localization of PBPs and other enzymes involved in peptidoglycan biosynthesis, since their function has to be modulated in the place where they are needed during the cell cycle. Thus, studies on localization of PBPs have been performed in the past (2,3) and are the main interest of this contribution.

Results

The separation of inner membrane (IM) and outer membrane (OM) of *E.coli* W7, growing exponentially, was achieved on sucrose discontinuous gradients (4). Separation of both structures was very neat and clean as judged by visual observation, clearly distinguishing both bands in the gradient. Furthermore, purity of both fractions was assessed by measurement of membrane markers (KDO for OM and succinic dehydrogenase for IM) and examination of protein patterns in electrophoregrams stained with Coomassie blue. Gradients were analyzed into 30-40 fractions

The Target of Penicillin

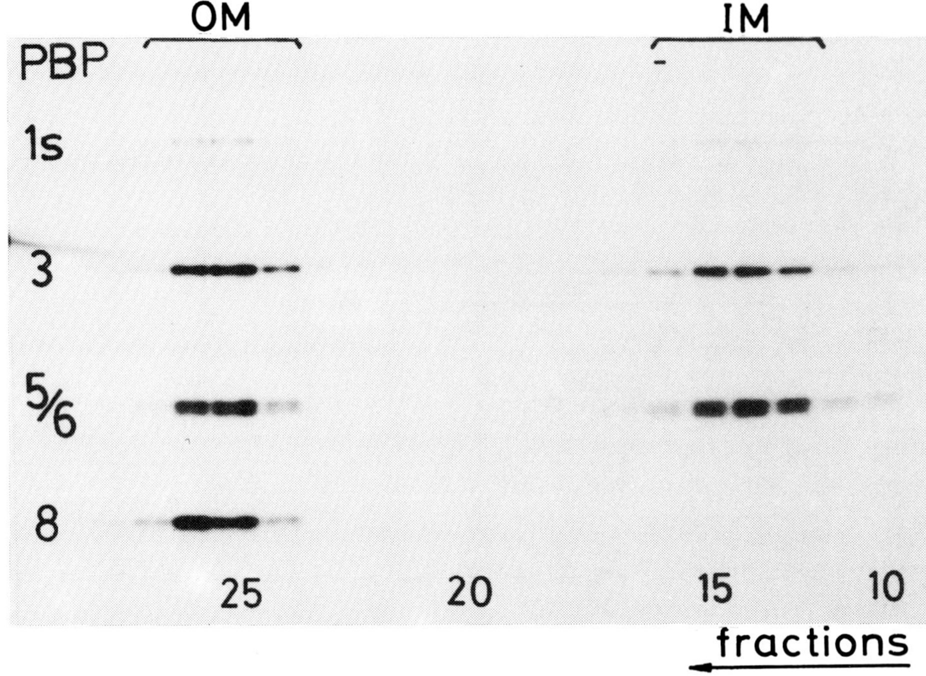

Figure 1. Outer and inner membranes were separated in sucrose gradients according to the procedure of Schnaitman (4). Gradients were fractioned and PBPs from each fraction were detected as described in the text.

and PBPs were detected following methods previously described (5,6)(Figure 1). Standard PBPs are distributed in both inner and outer membrane fractions with the only exception of PBP 8 (MW 34 700) that is only present in the OM. This reproducible distribution of PBPs was also studied in E.coli W7 in the steady state and in a temperature sensitive strain of E.coli PAT 84 growing at 30°C, or in its filamentous form growing at 42°C (Table I). It can be observed that PBP 8 is in all cases in the OM and that both PBPs 3 and 8 totally disappeared in the steady state. Although it is certain that PBPs in the IM are soluble in 1% Sarkosyl as previously described (5), we have observed that 20-30% of the total protein of the OM is also soluble in 1% Sarkosyl, obviously including the PBPs present in this structure. In fact a number of proteins present in the OM are also soluble in detergents even milder than Sarkosyl.

Table I: Distribution of the PBPs in IM and OM from E.coli.

Strain and conditions	PBPs in IM (%) 1s	2	3	5/6	8	PBPs in OM (%) 1	2	3	5/6	8
W7,exponential	27	26	44	54	0	73	74	56	46	100
W7,steady st.(a)	30	0	-	31	-	70	100	-	69	-
PAT 84, 30°C	37	33	42	52	0	63	67	58	48	100
PAT 84, 42°C	35	62	43	48	0	65	38	57	52	100

Results were obtained from densitometric measurements of the autoradiographical plates. They only show the percentage of radioactivity for each PBP found in IM and OM and do not imply that the absolute amount of one PBP is maintained under the different conditions used. Actually, the percentage of PBPs 3 and 8 with respect to the total amount of PBPs changed according to the conditions:

Strain and conditions	(total quantitative data,b) PBP 3	PBP 8
W7,exponential	21	20
PAT 84, 30°C	26	20
PAT 84, 42°C	14	8

a) Similar results were obtained after 12 and 24 h of steady state; under these conditions PBPs 3 and 8 disappeared from the envelopes.

b) Percentage of total radioactivity associated with PBPs.

In further experiments, we have separated the IM and OM fractions of E.coli W7 by following the method of Osborn et al (7) in which lysozyme is used prior to separation of the fractions in sucrose gradients. Separation of IM and OM fractions is also accomplished as shown by IM and OM markers and other criteria indicated above, but there is always and intermediate band (IntM) containing some 5% of the total protein of the bacterial envelope. Studies of PBP patterns in the different fractions show again that PBP 8 is always associated with the OM (Figure 2). It is most striking that, by following this fractionation, PBP 3 and, to a lesser extent PBPs 4/5, appear highconcentrated in the intermediate fraction.

In order to isolate some specific fragments of the membrane, we also prepared the known M-band containing membrane portions associated to DNA by using the Mg^{++}-Sarkosyl technique (8). By using this technique a number of proteins, including some of

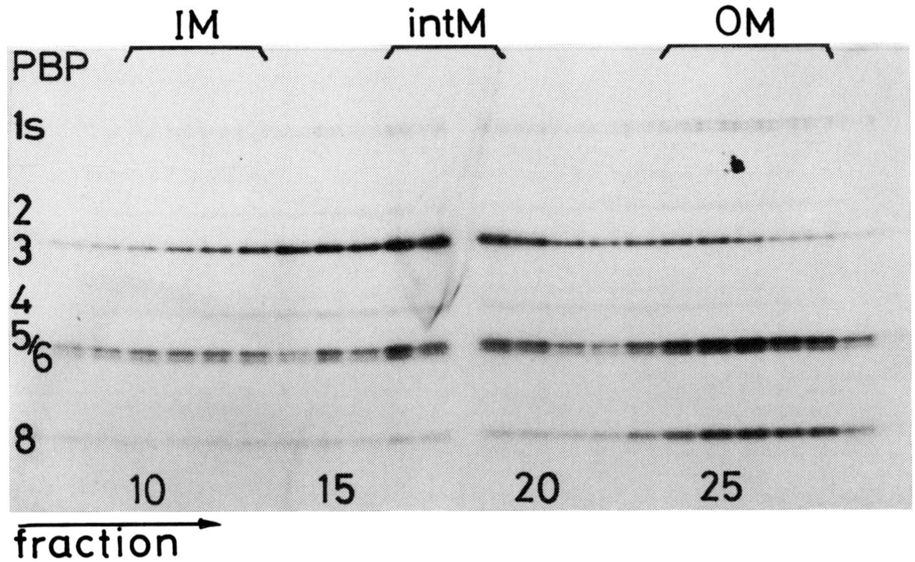

Figure 2: Cell envelope from E.coli W7 was fractionated according to Osborn et al (7). PBPs were detected as in figure 1.

the OM have been reported to be associated to DNA (9, and references therein). Indeed, we have identified electroforetically some of the M-band proteins associated to DNA, showing that a number of them are OM proteins (not shown). Furthermore, our M-band preparations contain some PBPs (Figure 3). Densitometric quantification of these data (not shown) revealed that M-band from exponentially growing E.coli W7 or PAT 84 at 30°C, contained a 4-fold enrichment in PBPs 1s, 3 and 8 compared to the total content of these PBPs in the whole envelope. On the other hand, the proportion of the other standard PBPs was much smaller (except PBPs 5/6 in PAT 84) when compared with their amounts in the entire envelope. M-band PBPs were also studied in E.coli in which the capacity of the bacteria to divide was impaired in a number of ways, such as: a) W7 in the steady state, or b) E.coli W7 in filamentous forms produced by treatment with cephalexin, or c) E.coli PAT 84 in filamentous forms produced by growing at 32°C. In the steady state PBPs 3 (10) and 8 are not observed in the whole envelope, whereas in fila-

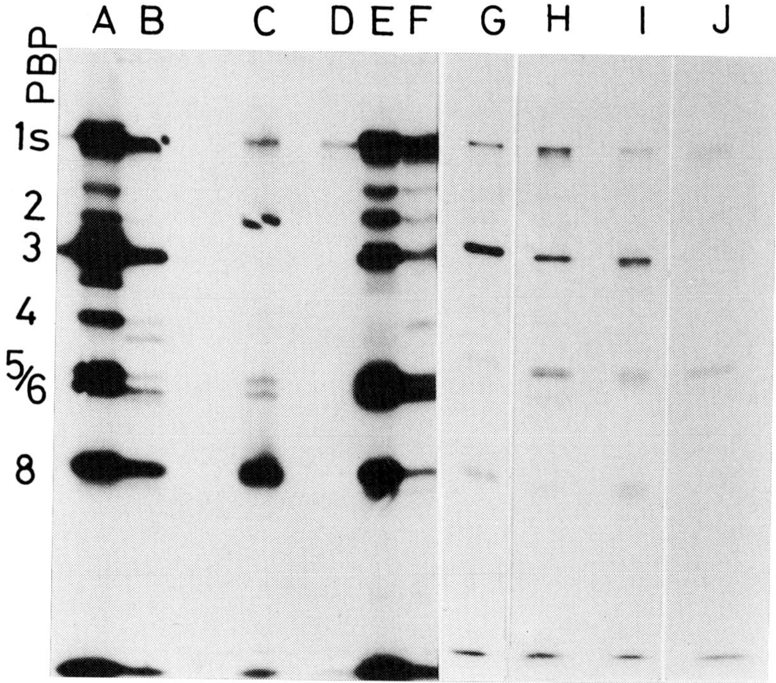

Figure 3: Analysis of PBPs in M-band fractions of E.coli grown under different conditions: E.coli W7 grown exponentially, A) whole envelope, B) M-band; E.coli W7 24 h in the steady state, C) whole envelope, D) M-band; E.coli W7 growing exponentially and then treated with 25 ug/ml cephalexin, E) whole envelope, F) M-band; E.coli PAT 84 growing exponentially at 30°C, G) whole envelope, H) M-band; E.coli PAT 84 grown at 42°C, I) whole envelope, J) M-band.

mentous forms of PAT 84, PBPs 3 and 8 appeared in the whole envelope but disappeared from the M-band. Essentially similar results are obtained in the filamentous forms of E.coli W7 produced by cephalexin, but in this case PBP 3 is partially masked by the antibiotic; however the antibiotic does not interfeared with the analysis of PBP 8 since it does not bind to this protein at the concentration used.

Conclusions

Data presented here suggest that, unlike reports of previous findings, standard PBPs are essentially found not only in the IM but also in the OM with the exception of PBP 8 that is only present in the OM.

PBP 3 appeared to be selectively located in the intermembrane fraction probably associated with intermembrane fusion sites (11). PBPs 3 and 8 are specifically associated to DNA and they both appeared to be involved somehow in the process of cell division and septation.

References

1. Mirelman, D. in Bacterial Outer Membranes, M. Inouye (ed), pp. 115-166. John Wiley & Sons. New York 1979
2. Buchanan, C. E.: J. Bacteriol. 145, 1293-1298 (1981).
3. Goodell, E. W. and Schwarz, U.: Eur. J. Biochem. 81, 205-210 (1979).
4. Schnaitman, C. A.: J. Bacteriol. 104, 882-889, ib. 890-901 (1970).
5. Spratt, B. G.: Eur. J. Biochem. 72, 341-352 (1977).
6. Schwarz, U., Seeger, K., Wengenmayer, F. and Strecker, H.: FEMS Microbiol. Lett. 10, 107-109 (1981).
7. Osborn, M. J., Gander, J. E., Parisi, E. and Carson, J.: J. Biol. Chem. 247, 3962-3972 (1972).
8. Earhart, C. F., Tremblay, G. Y., Daniels, M. J. and Schaechter, M.: Cold Spring Harbor Symp. Quant. Biol. 707-710 (1968).
9. Hendrickson, W. G., Kusano, T., Yamaki, H., Balakrishnan, R. King, M., Murchie, J. and Schaechter, M.: Cell 30, 915-923 (1982).
10. delaRosa, E. J., dePedro, M. A. and Vázquez, D.: FEMS Microbiol. Lett. 14, 91-94 (1982).
11. Bayer, M. H., Costello, G. P. and Bayer, M. E.: J. Bacteriol. 149, 758-767 (1982).

LABELLING AND CROSS-LINKING OF *E. coli* PBPs WITH BIS-ß-LACTAM AND PHOTOREACTIVE DERIVATIVES OF ß-LACTAM ANTIBIOTICS

Vicente Arán, Alfredo Rodríguez-Tébar and David Vázquez

Centro de Biología Molecular CSIC-UAM, Canto Blanco, Madrid-34, Spain

Introduction

PBPs have been detected by using the few radiolabelled ß-lactams that are readily available. These are mostly benzyl-(^{14}C)penicillin (1,2) and more recently (^{3}H)benzyl-penicillin (3) and either ^{125}I- or ^{3}H-labelled derivatives of ampicillin (4-8). PBP patterns depend on the labelled antibiotic used, and hence doubts still remain as to whether other PBPs might exist to which the radioactive ß-lactams used so far do not appear to bind, by the standard methods used, due to their low affinity. Furthermore, not only the number but also the localizations of the PBPs are very relevant to study crucial processes of the cell cycle that only occur in discrete places of the cell envelope such as elongation of the sacculus, formation of the septum and cell-lysis caused by ß-lactams (9-11). New compounds are required in order to have ß-lactams which bind to PBPs of *E. coli* with a higher affinity to resolve some possibly undetected PBPs and to study some aspects of their topography in the bacterial membrane. Thus we have synthesized a number of radiolabelled (a) photoreactive ß-lactams that react and form permanent covalent bonds with PBPs and (b) bis-ß-lactam antibiotics which are nearly symmetrical dimers of well known ß-lactam antibiotics and act as bifunctional specific crosslinking reagents for the PBPs.

The Target of Penicillin

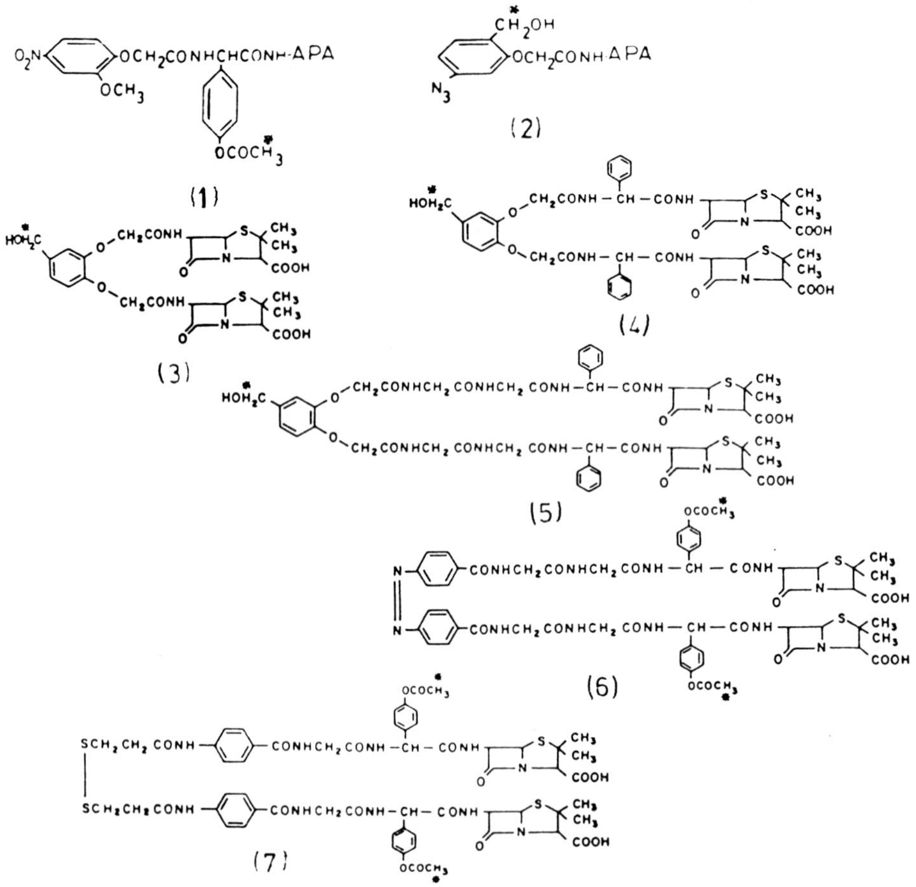

Fig. 1. Chemical structures of our radioactively labelled (*) bis-ß-lactams and photoreactive ß-lactams

Results

1) Radiolabelled photoreactive ß-lactams. The most interesting radiolabelled photoreactive ß-lactams of our series are compounds 1 and 2 (Fig 1) (12). When binding experiments were performed in the dark, both photoreactive ß-lactams displayed affinity for all PBPs that are labelled with

benzyl(^{14}C)penicillin. Quantitative data showed higher affinities of both photoreactive ß-lactams 1 and 2 for the PBPs than their parental ß-lactams. This is more remarkable in the case of lower (PBP 8 of 34.7 KD) and higher molecular weight (protein 170 KD) (Table 1)proteins that are labelled with rather high concentrations of benzyl(^{14}C)penicillin and lower concentrations of photoreactive ß-lactams 1 and 2. In all these cases the corresponding ß-lactam was removed from the labelled protein with hydroxylamine, further suggesting the specificity of the ß-lactam-protein interaction through a nucleophilic attack, just as other PBPs do.

When membrane samples were irradiated at 270 nm in the presence of the radiolabelled photoreactive ß-lactam 2 and electrophoresed, a number of new protein-labelled bands were observed including one of 190 KD and another one of 105 KD. Proteins of 190 KD and 105 KD appear to be dimers of PBP 1a and PBP 3 respectively cross-linked by our photoreactive ß-lactam 2; indeed PBP 1a and PBP 3 disappear when proteins 190 KD and 105 KD respectively appear. Similarly, by irradiating membrane samples at 340 nm in the presence of the labelled photoreactive ß-lactam 1, a dimer of PBP 1b (protein 145 KD) and a dimer of PBP 3(protein 110 KD) are formed (Table 2). In the case of membranes irradiated in the presence of either compound 1 or 2 of the radiolabelled photoreactive ß-lactams there were new labelled protein bands located between PBP 4 and PBP 5 and the label was not removed by hydroxylamine treatment suggesting that ß-lactams bound to membrane proteins, other than PBPs, through the photoreactivable group. Therefore the use of radiolabelled photoreactive ß-lactams as protein affinity labels and cross-linkers can be very useful to elucidate the topology of a number of proteins in the membrane.

Table 1

Heavy monomeric proteins labelled with radioactive bis-ß-lactam and photoreactive ß-lactam antibiotics

ß-lactam	ß-lactam carbonyl distance	Heavy monomeric protein bands radiolabelled (KD)			
		190	170-175	145	125
Compound 1	-	-	+	-	-
Compound 2	-	-	+	-	-
Compound 3	15 Å	-	+	-	-
Compound 4	22 Å	-	+	-	-
Compound 5	34 Å	+	+	+	+
Compound 6	40 Å	+	+	+	+
Compound 7	44 Å	+	+	+	+

Table 2

PBPs cross-linked with radioactive bis- ß-lactam and photoreactive ß-lactam antibiotics

Protein bands	Components
190 KD	ß-lactam 2 cross-linked PBP la-dimer
145 KD	ß-lactam 1 cross-linked PBP lb-dimer
110 KD	ß-lactam 1 cross-linked PBP 3-dimer
105 KD	ß-lactam 2 cross-linked PBP 3-dimer
115 KD	ß-lactam 4 or 3 cross-linked PBP 3-dimer
W(175-178 KD)	Bis-ß-lactam 7 cross-linked PBP lc-dimer
X(165-168 KD)	Bis-ß-lactam 7 cross-linked PBP lb-dimer
Y(102-105 KD)	Bis-ß-lactam 6 or 7 cross-linked PBP 3-dimer
Z(57 KD)	Bis-ß-lactam 6 or 7 cross-linked PBPs 7 and(or) 8

2) Radiolabelled bis-ß-lactam molecules. They are derived from very common ß-lactam antibiotics: compound 3 from 6-amino-penicillanic acid, compounds 4 and 5 from

ampicillin and compound 6 and 7 from amoxycillin (Fig 1) (13). Bis-ß-lactams, in general, bound to the PBPs more strongly than their parent ß-lactams, and the most remarkable increase in affinity corresponds to PBP 1a. Since compounds 3 and 4 were rather poor cross-linkers we developed compounds 5,6 and 7 in which the theoretical distances, according to the models, between the carbonyl groups of the ß-lactams are longer and therefore chances for cross-linking are higher. Although it is known that aliphatic chains are usually shorter than calculated from the models, our bis-ß-lactams 5,6 and 7 have a number of polar groups that will partially avoid these phenomena. Furthermore, our bis-ß-lactam 7 can be broken by disulfide-cleaving reagents and therefore monomeric and dimeric protein bands labelled with this compound can be easily differenciated by 2-mercaptoethanol treatment. As many as four monomeric protein bands heavier than PBP 1a (190 KD, 170-175 KD, 145 KD and 125 KD) are consistently labelled with our bis-ß-lactam antibiotics (Table 1), as well as the previously described PBPs 1a,1b,1c,2,3,4,5,6,7 and 8 (data not shown). We have also observed that some of our bis-ß-lactams cross-link two molecules of either PBP 1b, or PBP 1c or PBP 3 (Table 2). Moreover we have observed dimers of PBP 4 and also PBP 7 and(or) PBP 8 in some experiments but the results were not so consistent and reproducible. We have also detected some protein bands moving very slowly, which appear to be dimers or cross-linking species of the new high molecular weight PBPs described above. The only cases of cross-linking of heterologous PBPs that we have possibly detected are (a) between PBP 2 and one of the PBPs 1 not yet elucidated and observed only with our compound 7, and (b) perhaps the Z band (Table 2), not yet identified.

References

1. Blumberg, P.M., Strominger, J.L.: Proc. Nat. Acad. Sci. USA 69, 3751-3755 (1974).

2. Spratt, B.G. Eur. J. Biochem. 72, 341-352 (1977).

3. Dougherty, T.J., Koller, A.E., Tomasz, A. Antimicrob. Agents Chemother. 20, 109-114. (1981).

4. Schwarz, U., Seeger, K., Wengenmayer, F., Strecker, H. FEMS Microbiol. Lett. 10, 107-109. (1981).

5. Rosa, E. de la, Pedro, M.A. de, Vázquez, D. FEMS Microbiol. Lett. 14, 91-94. (1982).

6. Rodríguez-Tébar, A., Rojo, F., Vázquez, D. Eur. J. Biochem. 126, 161-166. (1982).

7. Rodríguez-Tébar, A., Rojo, F., Dámaso, D., Vázquez, D. Antimicrob. Agents Chemother. 22, 255-262. (1982).

8. Berenguer, J., Pedro, M.A. de, Vázquez, D. Eur. J. Biochem. 126, 155-159. (1982).

9. Ryter, A., Hirota, Y., Schwarz, U. J. Mol. Biol. 78, 185-195. (1973).

10. Schwarz, U., Asmus, A., Frank, H. J. Mol. Biol. 41, 419-429. (1969).

11. Goodell, E.W., Schwarz, U. J. Gen. Microbiol. 86, 201-209. (1975).

12. Arán, V., Rodríguez-Tébar, A., Vázquez, D. FEBS Letters (In press).

13. Rodríguez-Tébar, A., Arán, V., Vázquez, D. (Submitted for publication).

TOPOLOGICAL INTERRELATION OF PENICILLIN-BINDING PROTEINS IN E. COLI - A STUDY USING CLEAVABLE CROSS-LINKERS

Ibrahim Mohamed Said and Joachim-Volker Höltje
Max-Planck-Institut für Virusforschung, Abteilung Biochemie, D-7400 Tübingen

Introduction

Penicillin-binding proteins (PBPs) are characterized by their ability to bind ß-lactam antibiotics covalently. It has been shown that in E. coli there are at least 8 PBPs present in the cytoplasmic membrane (1,2). It is well established that these proteins are involved in murein-biosynthesis (1-5), which has to be carried out in a precisely co-ordinated fashion to maintain the specific shape of the bacterium and to avoid lysis of the cell. This leads to the assumption that some of the proteins involved in this intricate process may be organized in functional complexes. In an attempt to demonstrate such complexes, we tried to cross-link the PBPs in the membrane with cleavable cross-linking agents.

Results

1) Analysis of cross-linked complexes by symmetrical two-dimensional SDS-polyacrylamide electrophoresis.

Isolated membranes were labelled with ^{125}J-ampicillin (7) followed by treatment with dithiobis(succinimidylpropionate) (6) and solubilisation with Sarcosyl. As a control, membranes were treated exactly the same way except that the cross-linking agent was omitted. The samples were analysed by symmetrical two-dimensional SDS-polyacrylamide gel

The Target of Penicillin

electrophoresis (6). The dithiobis-specific complexes were cleaved by the reducing agent mercaptoethanol in an agarose layer on top of the second slab gel. A diagonal line of penicillin-binding proteins was obtained in non-treated control samples (Fig. 1A'). In contrast, cross-linked samples (Fig. 1A) which were cleaved prior to the development in the second dimension show, among others, 3 prominent off-diagonal PBP spots arising from a 140 000 dalton complex. Comparison with the diagonal line allows identification of these proteins as PBPs 1, 3, and 5.

In addition to this stochiometric complex, PBPs 3 and 5 form intramolecular cross-linking products which migrate faster than the unaltered molecules, and appear as off-diagonal line bands to the left of the corresponding PBP. A cross-linking product arising from the PBP 1b with an apparent higher molecular weight appeared on the right of PBP 1b.

Mutants defective in PBPs 3 and 6 (strain JE5703 kindly provided by Dr. Hirota) (Fig. 1B) and in PBPs 1b and 4 (strain 3-19) (Fig. 1C) were cross-linked under identical conditions, but showed no indication of complexes of PBPs, although the intramolecular cross-linkage products and the modified product of PBP 1b were formed, proving that cross-linkage did occur in the samples.

2) Analysis of cross-linking products by isoelectric focusing in the first dimension and SDS-polyacrylamide gel electrophoresis in the second dimension.

The pattern of PBPs was also analysed, after treatment of membranes with dithiobis(succinimidylpropionate), by two-dimensional electrophoresis, wherein the first dimension proteins were separated according to their isoelectric point and in the second dimension according to their molecular weight. As can be seen from the autoradiograms (Fig.2) this system has some disadvantages, namely tailing and the

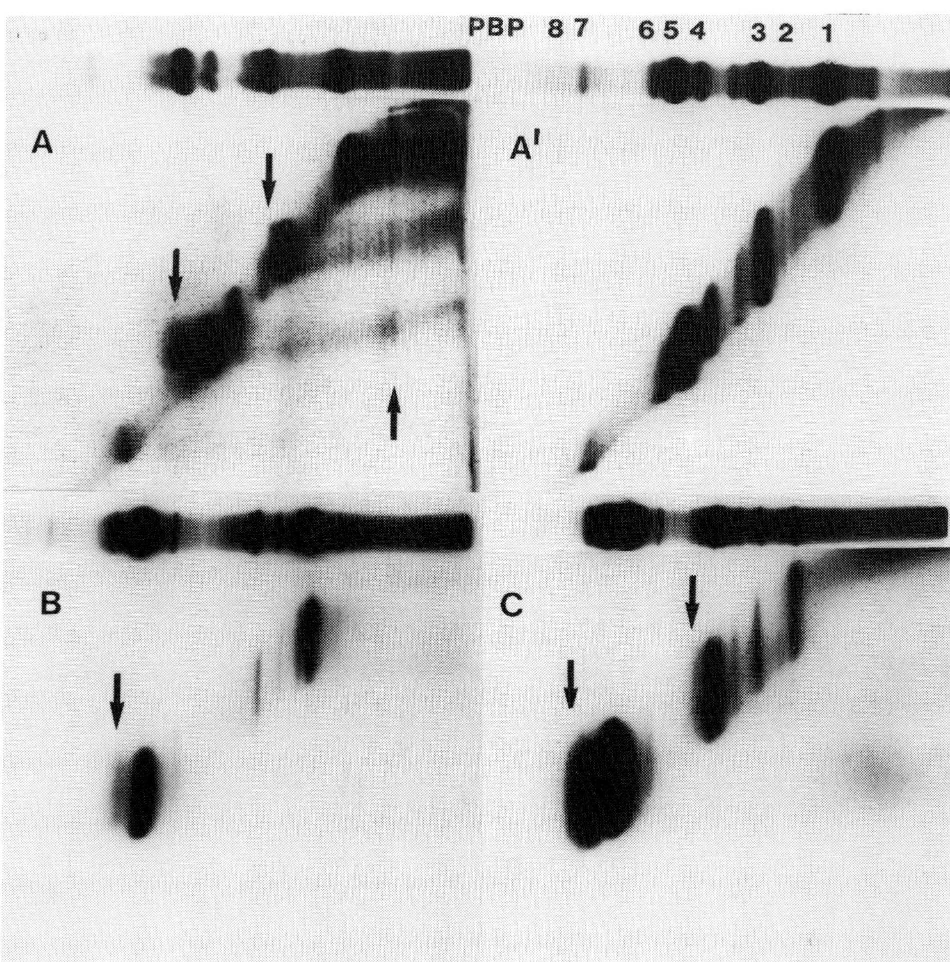

Fig. 1: Autoradiograms of symmetrical two-dimensional SDS-polyacrylamide gels. Envelopes from (A) E. coli PA3092 (wild type), (B) JE5703 (ftsI) and (C) 3-19 (ponB, dacB) were prepared as described (2). Penicillin-binding proteins (0.3 mg protein/ 30 μl) were labelled with 3 μCi ^{125}J-Ampicillin (7) and cross-linked with 6 μg dithiobis(succinimidylpropionate) for 1 min at RT (6). Membrane proteins were solubilized with 1% Sarcosyl NL-97 and resolved on a SDS-10% polyacrylamide slab gel in two dimensions under identical conditions (3,5 h at 25 mA). The cross-linked products were cleaved with 2% 2-mercaptoethanol before performing the second dimension. (↑) complex of PBP 1, 3, and 5; (↓) intramolecular cross-linking products.

fact that PBPs 1 cannot be detected. Presumably, the conditions (e.g. the pH at the site of sample application) of this technique is too harsh to preserve the PBP 1-penicillin complexes. However, a native complex of PBPs 3 and 5 is clearly seen in untreated samples. Cross-linking which would stabilize this complex modifies its behaviour in the isoelectric focusing step. Thus, the complex migrates to a different isoelectric point.

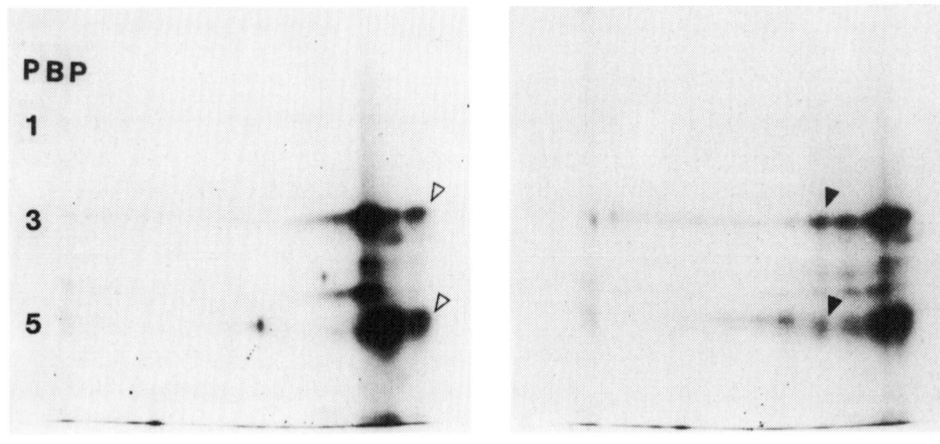

Fig. 2: Autoradiograms of two-dimensional gel electrophoresis with isoelectric focusing in the first dimension and SDS-polyacrylamide electrophoresis in the second dimension. Samples were prepared as described in Fig. 1 except that membrane proteins were solubilized with Triton X-100 and urea according to O'Farrell (8). Proteins were first fractionated on a horizontal isoelectric focusing gel (9) using a 3.5 - 10 pH gradient for 16 h at constant power (0.6 Watt). Cleavage of cross-linked proteins and SDS-polyacrylamide gel electrophoresis in the second dimension was as described in Fig. 1. (∇) native complex; (▼) complex formed by cross-linking.

Discussion

Our data indicate that it is possible to cross-link PBPs 1, 3, and 5 to form a specific stochiometric complex as well as a number of random complexes which may also include other proteins. This result could be obtained by two-dimensional SDS-gel electrophoresis. Using isoelectric focusing in the first dimension, participation of PBPs 3 and 5 in a native complex without chemical cross-linkage could be demonstrated. However, with this technique participation of PBP 1 in this complex could not be demonstrated. From the chemical nature of the cross-linking agent it can be deduced that the PBPs 1, 3, and 5 have a maximal distance of 12 Å from one other in the native complex. The PBPs 3 and 5 form intramolecular cross-linking products which migrate faster than the unaltered molecules. Similarly, the stochiometric complex seems to contain intramolecular cross-bridges which result in a faster movement of the complex. From the position in the gel, an apparent molecular weight of about 140 000 daltons has been calculated instead of 190 000 daltons which would correspond to the sum of the molecular weights of PBPs 1, 3, and 5. In mutants missing either of the three PBPs, no complex could be detected under identical conditions. We conclude that PBPs 1, 3, and 5 assemble in the membrane and, depending on the presence of all 3 proteins, form a complex which very likely represents a functional one. The physiological relevance of this finding should be further investigated.

References

1. Spratt, B.G.: Proc.Natl.Acad.Sci. USA 72, 2999-3003 (1975).

2. Spratt, B.G.: Eur.J.Biochem. 72, 341-352 (1977).

3. Suzuki, H., Nishimura, Y., Hirota, Y.: Proc.Natl.Acad.Sci. USA 75, 664-668 (1978).

4. Matsuhashi, M., Nakagawa, J., Ishino, F., Nakajima-Iijima, S., Tomioka, S., Doi, M., Tamaki, S.: In: Beta-lactam antibiotics (Mitsuhashi, S., ed.) pp. 203-223, Japan Scientific Press, Tokyo, and Springer-Verlag, Berlin, Heidelberg, New York 1981).

5. Nakagawa, J., Tamaki, S., Matsuhashi, M.: Agric.Biol. Chem. 43, 1379-1380 (1979).

6. Reithmeier, R., Bragg, P.D.: Biochim.Biophys.Acta 466, 245-256 (1977).

7. Schwarz, U., Seeger, K., Wengenmayer, F., Strecker, H.: FEMS Microbiol.Lett. 10, 107-109 (1981).

8. O'Farrell, P.H.: J.Biol.Chem. 250, 4007-4021 (1975).

9. Wallenfels, B.: Proc.Natl.Acad.Sci. 76, 3223-3227 (1979).

Acknowledgement

We thank Uli Schwarz for his advice and helpful discussions.

ELECTROFOCUSING OF INTEGRAL BACTERIAL MEMBRANE PROTEINS WITH SPECIAL REGARD TO PBPs.

Giuseppe A. Botta and Alessandro Costa.
Istituto di Microbiologia dell'Università di Genova
16132 Genova, Italy.

Introduction

Isoelectrofocusing on polyacrylamide gels (PAGIF) allows separation of a protein mixture on the basis of the different isoelectric points (pI) thus providing an alternative approach to SDS-PAGE membrane protein characterization.Very little experience has been gained in the analysis of envelope proteins by this procedure.We present here the results of our efforts in trying to define the optimal conditions for PAGIF analysis of the proteins in the bacterial envelopes.Aim of our approach is to use PAGIF to investigate PBPs in a condition in which the native state of the proteins is preserved by treatment with mild detergents.We assume that this method would be valuable in exploring the possibility that:a) some PBPs work as a functional complex *in vivo*;b) mutants do exist in which a PBP maintains its MW but migrates at a different pI;and finally c) to study division mutants in which the PBPs profile in SDS-PAGE is not consistent with the observed behaviour.Preliminary results of PBPs separation in PAGIF utilizing different PBPs deficient mutants are presented.

The Target of Penicillin

Materials and Methods

Bacterial strains and growth conditions - The strains used were: B.subtilis W23, E.coli KN126 and its pbpA Ts mutant SP45 (1), E. coli BUG6 (2), E.coli PA 3092 and its mutants JE10528 (ponA), JE10704 (ponB) and JE10730(ftsI)from the Hirota's col= lection (3).Strains were grown in Antibiotic medium N° 3(Difco) at 30° or at 42°C.Bacteroides distasonis and B. ovatus were grown in Schadler broth enriched with haemin (0.5 µg/ml) and vit.K (0.1 µg/ml).

Protein solubilization - After cell sonication, proteins from isolated envelopes were solubilized with different procedures: a) according to O'Farrell (4) with minor modifications, b) according to Ames and Nikaido (5) and c) envelopes were resuspended in Tris-HCl containing $MgCl_2$ (1mM), Triton X-100 (1.8% final concentration), zwittergent 3-10 (0.2%final concentration) and urea (8 M)(100 µl/80 mg of envelopes), incubated at RT for 30 min and centrifuged.

Binding with radiolabelled penicillin - Washed envelopes were incubated for 15 min at 30° or 42°C with saturating concentrations of ^{35}S-benzylpenicillin (spec. act. 4.7Ci/mmol;N.E.N.). Reaction was stopped by adding 80 µl of solution c), containing 1 mg/ml of cold penicillin.After extraction for 30 min at RT and centrifugation, samples of the supernatant were loaded on the PAGIF gel.

For PBPs analysis only the extraction procedure c) was used; with or without urea.When urea was not included in the solubilizing mixture was also omitted from the gels.

Gel preparation - Slab gels (1 mm thick) were prepared in the following way: acrylamide 28.8%) 10 ml, bis-acrylamide(1.2%)

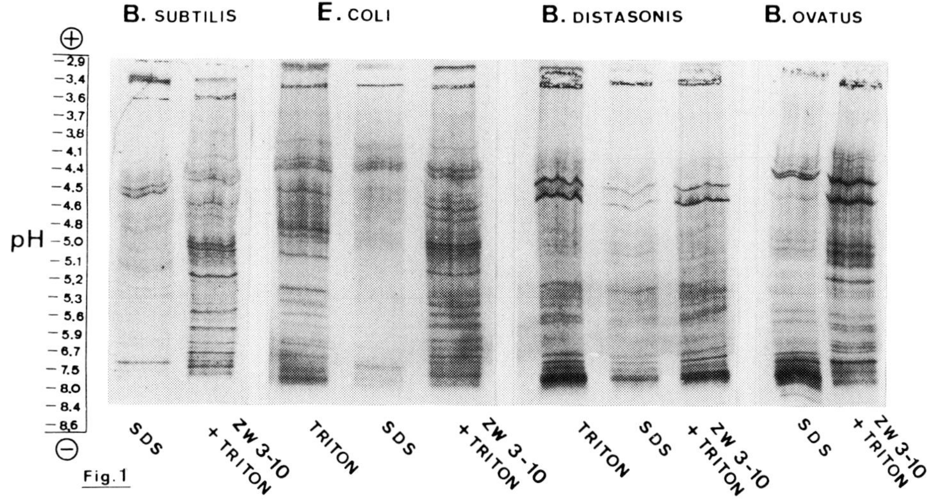

Fig. 1

10 ml, urea (8M,final concentration),Triton X-100 (1.8% final conc.),zwittergent 3-10 (0.2% final conc.)ampholytes 3-10 (0.6 ml),5-7 (0.9 ml),3-5 (1.5 ml),distilled water to 60 ml.To the mixture,degassed under vacuum,0.24 ml of 10% ammonium persulfate and 20 µl TEMED were added and slab gels poured.After 20min to allow polymerization,gels were transferred at 4°C until needed.Separation was achieved running the gels at constant power (17W) for two hrs;final voltage 600V.Electrodes were NaOH 1M (catode) and H_3PO_4 1M (anode);pG gradient ranged from 2.9 to 8.6.Gels were fixed in TCA 20% and stained with Serva Blau R-250.PBPs were detected by fluorography and autoradiography as described for SDS-PAGE.

Results and Discussion

In Fig.1 are shown the PAGIF profiles of membrane proteins of different bacterial species solubilized with the procedures above described.Best results were obtained when method c) was utilized.Indeed addition of zwitterionic detergent 3-10 to the solubilization mixture and to the gels,proved to be a significant improvement.Zwitterionic detergents,possessing both cationic and anionic groups,preserve the net charge of solubilized proteins and they are generally non denaturating (6,7). Among several zwittergents tested,differing in chain lenght, compound 3-10 was selected for further experiments because of beter resolution of bands in the pH range utilized.Separation of PBPs from different E. coli strains was achieved in two different gel systems.The results obtained using Triton X-100, zwittergent 3-10 and urea are shown in Fig.3.From this autoradiogram,5 bands are clearly evident in the parental strains, located in the region ranging from pH 4.8 to 5.9.The possible existence of bands in the basic region requires further analysis with a different pH range.The three major bands possess a pI value of 4.8 (IF-1),5.6 (IF-4) and 5.9 (IF-5).The analysis of the profile of the pon A mutant shows a remarkable decrease in the intensity of IF-5.In this strain the splitting of IF-1 in two bands of equal intensity was consistently observed.Interestingly enough,in the pon B mutant a decrease in the intensity of IF-4 is clearly evident.The profile of the PBPs obtained with the thermosensitive mutant SP 45 when grown at 42°C,shows both a decrease in the intensity of the PBP with pI 5.6 (IF-4) and the appearance of two bands with a more basic pI(between 7.5 and 8.0).No clear conclusions can be drawn

pH
8.8
8.4
8.2
8.1
8.2
7.9
7.7
7.5
7.4
7.1
6.9
6.7
6.6
6.3
6.1
5.9
5.8
5.6
5.4
5.2
5.1
4.9
4.8
4.6
4.4
4.3
4.1

Fig. 2

KN 126
SP 45
42°c
30°c
KN 126
PA 3092
ponA
fts I
42°
30°
ponB

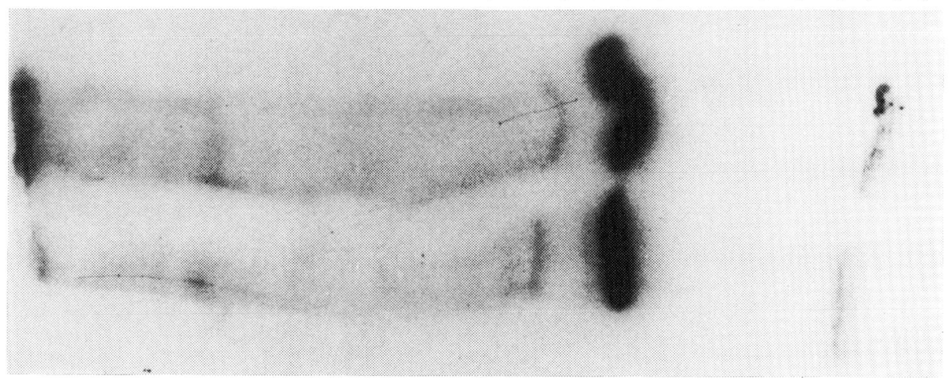

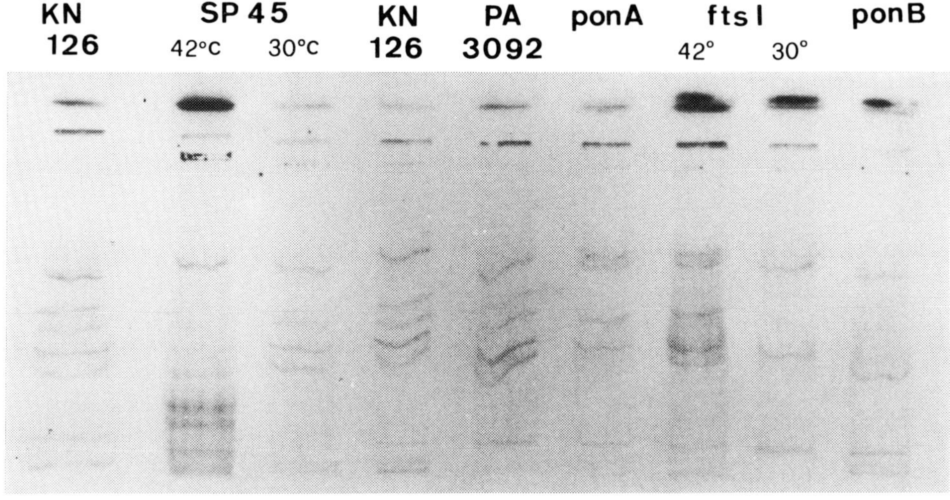

Fig. 3

from the analysis of PBPs in the ftsI mutant when grown at 42° C or at 30° C.When PBPs were extracted and separated in the absence of urea,a complete different profile was obtained (Fig. 4).In this gel only two bands can be detected: a major one with pI 5.6 and a second one ,of lower intensity,with pI 6.0. No other bands are evident in these experimental conditions.

Two major findings characterize this report.The first one is the description of a method for solubilizing membrane proteins which is more suitable for PAGIF analysis thanpreviously described procedures.The second one is the detection by PAGIF of a limited number of PBPs compared to that observed in SDS-PAGE. Although these results are highly suggestive for PBPs complexes in the native state we think that further work is required for conclusive evidence (analysis of greater number of mutants, competitive binding with antibiotics possessing specific affinity,binding of PBPs after separation).PAGIF looks very promising in that it can provide completely new information about PBPs in their native state and could represent a valid tool in their purification.

References

1. Spratt,B.G. :Eur. J. Biochem. 72,341-352 (1977).
2. Botta,G. and Park,J.T. : J. Bacteriol.145,333-340 (1981)
3. Suzuki,H.et al.:Proc. Natl. Acad. Sci. USA 75, 664-668(1978)
4. O'Farrell P.H.: J. Biol. Chem. 250, 4007-4021 (1975)
5. Ames,G.F. and Nikaido,K. : Biochem. 15, 616-623(1976)
6. Gonenne,A. and Ernst,R. : Anal. Biochem. 87, 28-33 (1978)
7. Hjelmeland,L. et al. : Anal. Biochem. 95, 201-207 (1979)

CHARACTERIZATION OF THE CHANGES IN PENICILLIN-BINDING PROTEINS THAT OCCUR DURING SPORULATION OF *BACILLUS SUBTILIS*

Christine E. Buchanan, Margaret O. Sowell

Department of Biology, Southern Methodist University
Dallas, Texas 75275

Introduction

While inhibition of bacterial septum formation and normal wall growth are widely-recognized consequences of penicillin treatment, it is sometimes overlooked that penicillin can also inhibit sporulation in many gram-positive bacilli. The changes that occur in the penicillin-binding proteins (PBPs) during this developmental sequence could provide some important clues about the physiological roles of these proteins in both growing and sporulating cells. We have recently compared the PBPs from *Bacillus subtilis* 168T in different stages of sporulation with the PBPs from nonsporulating cells (1). During sporulation the vegetative PBPs 1, 2A, 4 and 5 decrease in amount, PBPs 2B and 3 increase at specific times, and two new PBPs 4* and 5* (60,000 daltons and 42,000 daltons, respectively) appear. The results are essentially the same whether the nutrient exhaustion or the nutrient downshift technique is used to induce sporulation. These changes, except for the appearance of PBP 4*, also occur in other sporulating strains of *B. subtilis*, but never in the exponentially-growing cells. From a comparison of the PBP activities in sporulating cells and two asporogenous (stage 0) mutants, it was concluded that the increases in PBPs 2B and 3, and the appearance of PBPs 4* and 5* are likely to be sporulation-related events rather than the consequences of stationary-phase aging.

The Target of Penicillin

Further characterization of the PBP changes described above is necessary before their significance can be properly evaluated. For example, it would be useful to know if the sharp decline in the amounts of PBPs 1, 2A, 4 and 5 after initiation of sporulation (t_0) is due to an early sporulation event or to non-specific inactivation or degradation. The stability of the individual vegetative PBPs has not been previously examined either in vivo or in vitro. It would also help to know if the apparent increase in the amounts of the vegetative PBPs 2B and 3 reflects de novo synthesis or unmasking of preexisting protein. And finally, we need to know whether the new PBPs 4* and 5* are truly new proteins, or if they are derived instead from other PBPs or prePBPs. Our approach to answering these questions has been to treat sporulating cells with a protein synthesis inhibitor for various time intervals to determine which of the PBP changes require protein synthesis during specific stages of sporulation. We have also determined the in vivo stability of the PBPs and their susceptibility to sporulation-specific proteases.

Results and Discussion

<u>In vivo stability of the PBPs</u>. The in vivo stability of the membrane-bound PBPs of <u>B</u>. <u>subtilis</u> 168T was measured by a technique described previously (2). Exponential-phase cells in Difco sporulation medium (3) were treated for 4 hours with either 100 µg of chloramphenicol (CAM) per ml or 10 µg of tetracycline (TET) per ml. During these 4 hours no protein synthesis was possible so loss of any unstable proteins was obvious when the treated cultures were compared with a parallel untreated one (now in its stationary phase and beginning to sporulate), or compared with an exponential-phase sample harvested at the time of addition of the inhibitors (Fig. 1). The results are summarized in Table 1. PBP 2A was clearly the most unstable of the six vegetative PBPs while PBPs 3 and 5 were

TABLE 1. Relative Amount of Each Vegetative PBP

PBP	Stat. phase/ exp. phase	CAM-treated/ exp. phase	TET-treated/ exp. phase
1	43 ± 22[a]	58 ± 29	59 ± 18
2A	36 ± 2	16 ± 6	17 ± 6
2B	66 ± 5[b]	58 ± 20	60 ± 23
3	141 ± 20	96 ± 29	102 ± 37
4	64 ± 14	49 ± 21	45 ± 19
5	66 ± 1	90 ± 15	99 ± 27

[a]The average of data obtained from 3 pairs of cultures ± one standard deviation.

[b]The increase in PBP 2B that normally occurs in stationary phase is not apparent here because the sampling time did not correspond to the time of the protein's peak amount.

remarkably stable. The stabilities of the other PBPs fell between these two extremes. Because CAM and TET had similar effects on the PBPs, it is unlikely that synthesis of any PBP was uniquely resistant to one of the inhibitors.

Effect of CAM on PBP changes during sporulation. After induction of sporulation by a nutrient downshift, CAM was added to separate flasks of sporulating cells for the following time intervals: t_0 to t_4, t_2 to t_4, and t_4 to t_6. It was known from previous studies that stage II formation of the asymmetric forespore septum occurred at about t_3 with this strain under these conditions, while stage IV cortical peptidoglycan synthesis began near t_6 (1). Thus, the t_4 to t_6 interval roughly corresponded to the period of membrane proliferation during stage III engulfment of the forespore. The results from CAM-treatment of four downshifted cultures are summarized below (manuscript in preparation).

CAM completely inhibited the sporulation-related increase in the amounts of vegetative PBPs 2B and 3 as well as the appear-

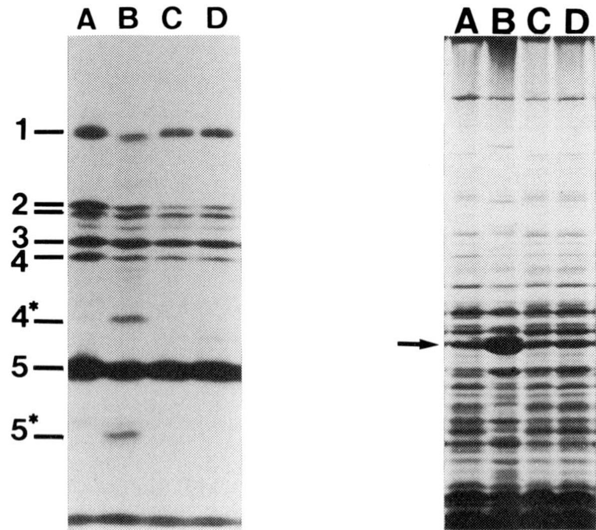

FIG. 1. Left: PBPs of B. subtilis analyzed by SDS-polyacrylamide gel electrophoresis and fluorography. A, PBPs from exponential-phase cells in Difco sporulation medium; B, PBPs from stationary-phase cells; C, PBPs from cells after 4 hours with CAM; D, PBPs from cells after 4 hours with TET. Right: Coomassie brilliant blue-stained gel corresponding to the fluorograph on the left. Approximately the same amount of membrane protein was added to each slot. An arrow indicates the location of PBP 4*.

ance of the two new PBPs 4* and 5*. A similar effect was also observed after sporulation was induced by the nutrient exhaustion technique (Fig. 1, left). Furthermore, if CAM was added at t_4, when the levels of the four proteins were usually still increasing, not only did their synthesis cease immediately but the various proteins began to rapidly decline in amount. This indicates that at least some synthesis of PBPs 2B, 3, 4* and 5* occurred simultaneously with their turnover and that the actual amounts of these proteins synthesized during sporulation were probably much greater than originally described (1).

The decline in PBPs 1, 2A, and 4 that normally begins at t_0 is probably not related to an early sporulation event because these proteins were relatively unstable even in growing cells (Table 1), and CAM added at t_0 had little effect on their subsequent loss. PBP 2B, which usually increases after t_0, also steadily declined in the presence of CAM. In contrast to PBPs 1, 2A, 2B and 4, the early loss of PBP 5 may be related to sporulation since it was a particularly stable PBP under vegetative conditions (Table 1). This tentative conclusion is supported by the observation that no loss of PBP 5 occurred if CAM was added at t_0.

Throughout these studies we observed a protein band on Coomassie blue-stained gels that appeared to correspond to PBP 4* (Fig. 1, arrow). That is, it moved with the same electrophoretic mobility as PBP 4* and visibly increased in amount at the time when PBP 4* appeared in fluorographs. This protein seemed to be present in all strains of _B. subtilis_, but it did not significantly increase during sporulation except in those strains that produced PBP 4*. The increase in this relatively major membrane protein was susceptible to inhibition by CAM and TET (Fig. 1, right). Two explanations for the presence of this protein in vegetative cell membranes are possible: either there is inactive PBP 4* in the membranes during growth which is not detectable by fluorography, or there is another membrane protein that happens to move to the same relative position as PBP 4* on gels.

Treatment of B. subtilis membranes with extracts of sporulating cells. Although the data do not support the notion that the new PBPs 4* and 5* originated directly by processing of the PBPs present at t_0 (1), these PBPs could have appeared as a result of additional synthesis of the vegetative proteins during sporulation which was followed by their rapid specific conversion to the new forms. Alternatively, they could have been formed by processing of undetectable prePBPs located in the membrane. We tested these possibilities by incubating a presporulation (t_0) membrane sample with extracts of sporulating

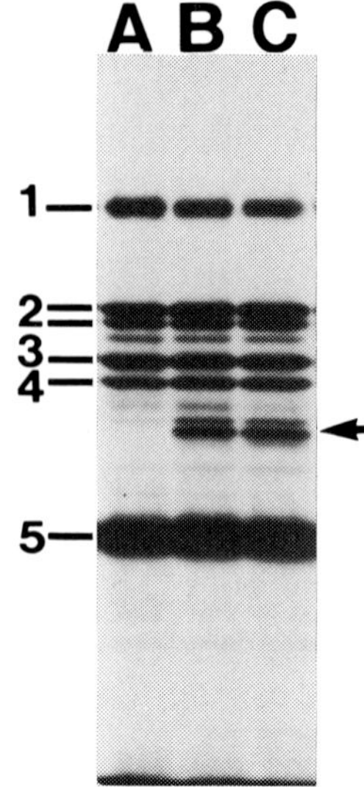

FIG. 2. PBPs from B. subtilis cells harvested at t_0. A, untreated; B, after exposure to an extract of t_3 cells for 30 min.; C, after exposure to an extract of t_6 cells for 30 min. The arrow indicates the two PBPs generated by this treatment.

cells harvested at t_3 and t_6. The extracts were prepared by a method designed to avoid inactivation of the sporulation-specific intracellular protease and contamination with the extracellular enzymes (4).

The appearance of the extract-treated membrane samples on a stained gel provided clear evidence of proteolytic activity. There was significantly less protein in these samples after incubation for 30 min. at 37°C with either extract (not shown). In contrast to the bulk of the membrane proteins, most of the PBPs, including some that were fairly unstable in vivo, were affected very little by this treatment. Neither PBP 4* nor PBP 5* was produced, even after the incubation time was extended to 60 min. Proteolysis did yield two other PBPs (Fig. 2) whose relation to the vegetative PBPs is not known. These same two proteins were also generated by treatment of membranes from an asporogenous mutant and by treatment of t_6 membranes from sporulating cells. They did not appear when membranes were incubated with various extracts that had no measurable proteolytic activity (unpublished results).

Summary

The immediate decline in the amounts of PBPs 1, 2A, and 4 after induction of sporulation is probably unrelated to sporulation because these proteins were also unstable under vegetative conditions. However, the decline in PBP 5, which is normally quite stable, may be the result of an early CAM-sensitive event in sporulation. The increase in PBPs 2B and 3 and the appearance of PBPs 4* and 5* are apparently due to de novo synthesis of each protein, rather than to processing of preexisting protein by sporulation-specific proteases.

This research was supported by NSF grant PCM-7921673.

References

1. Sowell, M.O., Buchanan, C.E.: J. Bacteriol. 153, in press.
2. Buchanan, C.E.: FEMS Microbiol. Letters 7, 253-256 (1980).
3. Schaeffer, P., Millet, J., Aubert, J.-P.: Proc. Natl. Acad. Sci. U.S.A. 54, 704-711 (1965).
4. Reysset, G., Millet, J.: Biochem. Biophys. Res. Comm. 49, 328-334 (1972).

PATTERN OF PENICILLIN-BINDING PROTEINS DURING THE LIFE CYCLE OF *ESCHERICHIA COLI*

Frans B. Wientjes and Uli Schwarz, Department of Biochemistry, Max-Planck-Institut für Virusforschung, Tübingen, West Germany

Tom J.M. Olijhoek and Nanne Nanninga, Department of Electronmicroscopy and Molecular Cytology, University of Amsterdam, The Netherlands

Introduction

The *Escherichia coli* cell contains about 10 penicillin-binding proteins (PBPs), which play roles in cell-wall growth and shape maintenance of cells. For example, PBP2 is necessary for maintenance and/or production of the rod-form of the cells whereas PBP3 has a function in cell division (1) and might be directly involved in murein synthesis during septum formation (2). Furthermore, it has been found that cell-wall growth in *E. coli* occurs predominantly in growth zones (3) and that murein biosynthesis oscillates during the cell cycle (4). Therefore, the cells must control the activities of the enzymes involved in murein metabolism in space and/or time. It is not known whether this is done by controlling the presence of the enzymes in time (i.e. by switching the synthesis of the enzymes on and off), or in space (i.e. a restricted location of the proteins) or by controlling their activities only (i.e. by activation and deactivation). To test the first possibility we investigated the PBP pattern in synchronously growing *E. coli* cells. The cells were synchronized by the elutriation technique (5) and the pattern of PBPs was determined with an iodinated derivative of ampicillin. We did not find evidence that any of the PBPs is present exclusively or mainly in one part of the cell cycle. This makes it unlikely that the cells exert their control on murein metabolism on the level of enzyme synthesis.

The Target of Penicillin

Materials and Methods

The *E. coli* K12 strain PA3092 was used. The cells were grown in minimal citrate medium at 30°C. The doubling time was 70 min.
Synchronously growing cells were obtained by elutriation (5).
Membranes were prepared by sonication of the cells and were isolated by centrifugation through 30% (w/v) sucrose in 50 mM phosphate buffer, pH 7.0 (100 000 xg_{av}, 1h, 4°C). The membranes were resuspended in 20 µl of buffer and used immediately.
The penicillin-binding assay was done as described by Spratt (6). The total volume of the assay was 20 µl with 10-20 µg membrane protein (or the corresponding amount of intact cells). Each test tube contained 10 µCi $[^{125}I]$ampicillin (prepared according to Schwarz et al. (7)) and incubation was for 10 min at 30°C. The reaction was stopped by addition of an excess of cold phosphate buffer. Intact cells were washed with this buffer and then boiled in sodium dodecylsulphate-containing sample buffer for gel electrophoresis. Labeled membranes were washed free from unbound $[^{125}I]$ampicillin by sucrose cushion centrifugation and were then resuspended in sample buffer and boiled. Gel electrophoresis, staining and exposure was with standard techniques. Quantitation was done by scanning the films with a densitometer (Joyce-Loebl) and measuring the peak surfaces with a planimeter.

Results

Exponential increase of PBPs. Figs. 1 and 2 show that in synchronously growing *E. coli* the PBPs increase in an approximately exponential way. This experiment was done with membranes. Fig. 1 shows the X-ray film and Fig. 2 shows the increase of PBPs 1, 3 and 5. The other PBPs gave similar results except PBP 7. However, the latter result was not reproducible (cf. Fig. 3) and was probably an artefact.

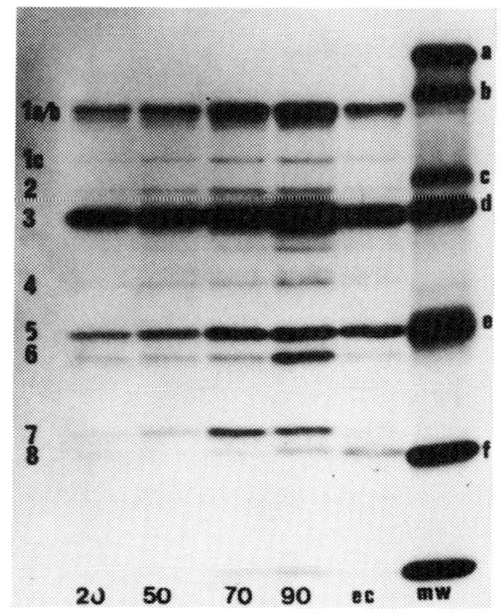

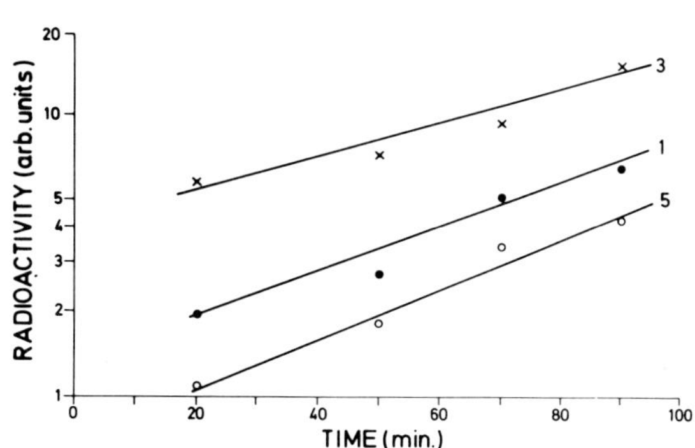

Fig. 1. PBP pattern after labeling of membranes from synchronously growing E. coli with [^{125}I]ampicillin. 20, 50, 70,90: minutes of synchronous growth. Division is at about 70 min; ec, exponential culture; mw, molecular weight marker proteins: a, 116 000; b, 96 000; c, 68 000; d, 60 000; e, 40 000; f, 30 000.

Fig. 2. Exponential increase in PBPs 1, 3 and 5 in synchronously growing E. coli. The X-ray film of Fig. 1 was scanned and peak areas were measured.

Constant ratio of PBPs. In a second series of experiments, samples were taken during two division cycles and care was taken to apply equal protein amounts to the test mixtures to minimize differences in the test conditions. To this purpose the bacteria were grown in the presence of ^{3}H-amino acids and an equal amount of ^{3}H-radioactivity was applied to each assay. Since losses of material could not completely be avoided, leading to differences in the amount of material applied to the gels, the relative intensities of the PBPs were calculated. Fig. 3 shows the result of a series of experiments with intact cells. The values are percentages [^{125}I]ampicillin bound to the particular PBP in relation to the total amount of ^{125}I-label in the sample and are averages of 4-6 determinations with the standard deviation. The data show that none of the PBPs is predominantly present at any stage of the

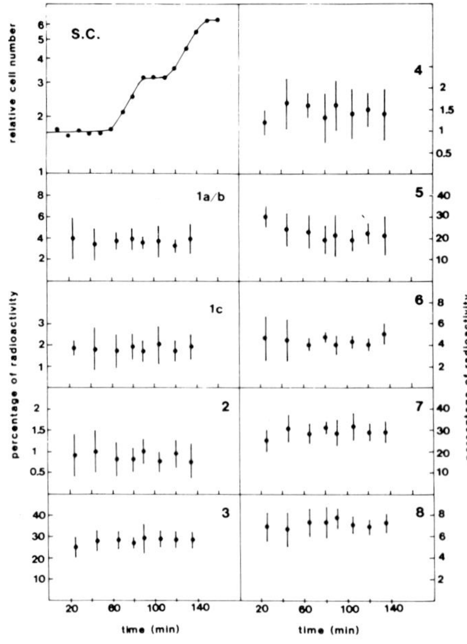

Fig. 3. Relative intensities of PBPs after labeling of intact cells from synchronous cultures with [^{125}I]ampicillin. Average values of percentages of radioactivity in each PBP band ± S.D. are shown. S.C., synchronization curve; the numbers 1-8 refer to the different PBPs.

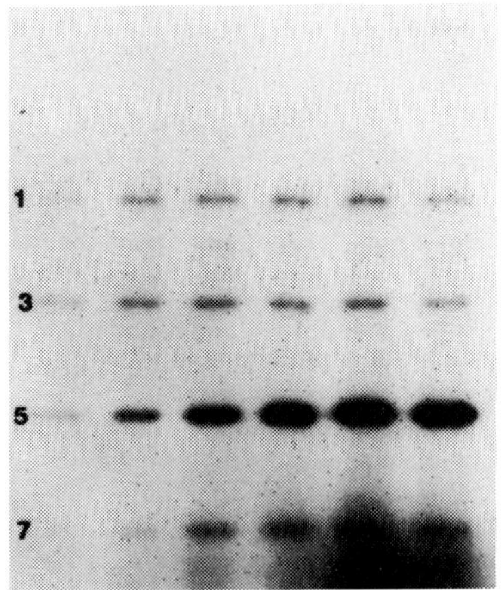

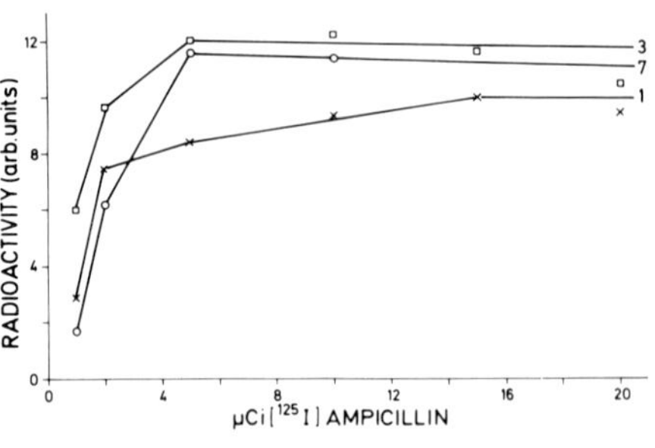

Fig. 4. Saturation of PBPs in E. coli with [^{125}I]ampicillin. Membranes (2 µg protein) were labeled with increasing amounts of [^{125}I]ampicillin (1-20 µCi). X-ray film exposed during 3 days.

Fig. 5. Same experiment as Fig. 4; saturation curves for PBPs 1, 3 and 7. The curves for the other PBPs were similar except that the film had to be exposed for a shorter or longer time.

division cycle. Also the total amount of PBPs did not fluctuate: the total bound radioactivity was proportional to the protein amount in the samples. Similar results were obtained with isolated membranes from synchronously growing cells.

Saturation of PBPs with [^{125}I]ampicillin. The chemical concentration of ampicillin in the assay is too low to saturate the PBPs. To exclude the possibility that this obscured eventual fluctuations in the ratio of the PBPs, we investigated whether the PBPs could be saturated with [^{125}I]ampicillin. To this purpose smaller amounts of membrane protein were assayed (2 μg). Saturation of the PBPs was observed at 5-10 μCi [^{125}I]ampicillin (Figs. 4 and 5). This result was obtained with intact cells and with membranes. Also under the saturating conditions described here, no changes were observed in the relative intensities of the PBPs during the cell cycle (not shown).

Discussion

The standard deviations of the data for the relative PBP intensities are rather high but the method is sufficiently accurate to detect major changes in the PBP pattern. Our results show that when there is any fluctuation of the labeled PBPs during the cell cycle, these fluctuations are small. The absence of oscillations is especially striking for PBPs 2 and 3 (and remarkably clear in the latter case, see Fig. 3). For these proteins an oscillation could be expected on the basis of their roles in elongation and septation, respectively (1).
We conclude that the concentration of PBPs in *E. coli* is constant during the cell cycle and that murein synthesis is probably not regulated by controlling the amount of PBPs. Cell-wall growth and cell-shape maintenance are apparently regulated by another mechanism. One possibiliy is that the activity of the enzymes involved in cell-wall metabolism is controlled. For example, such a control might be exerted on the level of substrate availability or via macroscopic parameters such as osmotic pressure. Alternatively,

the enzymes might function under topological control. This would mean that PBP3 is present mainly or exclusively in the region of the cell where the polar cap is formed, whereas PBPs la, lb and 2 would be located at the longitudinal growth zones. Studies are being initiated in our laboratories to determine the totographical distribution of several of the relevant enzymes over the cell envelope of E. coli.

References

1. Spratt, B.G.: Proc. Natl. Acad. Sci. USA 72, 2999-3003 (1975).
2. Ishino, F. and Matsuhashi, M.: Biochem. Biophys. Res. Commun. 191, 905-911 (1981).
3. Schwarz, U., Ryter, A., Rambach, A., Hellio, R. and Hirota, Y.: J. Mol. Biol. 98, 749-759 (1975).
4. Olijhoek, A.J.M., Klencke, S., Pas, E., Nanninga, N. and Schwarz, U.: J. Bacteriol. 152, 1248-1254 (1982).
5. Figdor, C.G., Olijhoek, A.J.M., Klencke, S., Nanninga, N. and Bont, W.S.: FEMS Lett. 10, 349-352 (1981).
6. Spratt, B.G.: Eur. J. Biochem. 72, 341-352 (1977).
7. Schwarz, U., Seeger, K., Wengenmayer, F. and Strecker, H.: FEMS Lett. 10, 107-109 (1981).

GROWTH DEPENDENT MODIFICATIONS OF THE PENICILLIN BINDING PROTEINS OF *Escherichia coli* : INFORMATION SUPPORTING THE GENERALITY OF THE PROCESS AND THE MULTIPLICITY OF MECHANISMS.

Enrique J. de la Rosa, Miguel A. de Pedro, David Vázquez.
Instituto de Bioquímica de Macromoléculas, C.S.I.S.-U.A.M.
Madrid, Spain.

Introduction.

In *Escherichia coli* the pattern of binding of radioactive β-lactams to the penicillin binding proteins (PBPs) is modified considerably when the cells stop active growth (1,2), as could be expected from its crucial participation in murein biosynthesis, a strictly growth-dependent process. Although the capacity of binding of most PBPs is modified during the transition to stationary phase, hardly anything is known about the mechanism by which these alterations take place, clearly an important point for the understanding of the regulation of the PBPs.

In this communication, we report information on the generality of most of the alterations in the PBPs previously reported (2), as well as observations suggesting the involvement of different mechanisms in the modifications of the PBPs.

Results.

Table 1 shows the results of a comparative analysis of the modifications in the binding of N[3-(4-hydroxy 5-[^{125}I]iodophenyl)propionyl] ampicillin ([^{125}I]-ampicillin) (6) to the PBPs of several unrelated strains of *E. coli*, in stationary phase cultures, grown in two different media, L-broth (7) and mini-

The Target of Penicillin

mal-citrate (8). With the notable exception of PBPs 5 and 6, the alterations in the PBPs were similar in all the strains. The modifications of PBPs 5 and 6, however, showed strong strain-dependence.

The nature of the growth medium clearly influenced the changes in the pattern of binding of $[^{125}I]$-ampicillin to most PBPs. Apparently when L-broth was used the modifications in the pattern of binding were accelerated. A peculiar medium-dependent modification was detected in the strain E. coli W7 . In stationary phase cells of W7 , grown in L-broth, a new $[^{125}I]$-ampicillin binding element, with an electrophoretical mobility slightly faster than that of PBP 1B, became detectable. This protein, named PBP 1B* (2), was apparently related to PBP 1B, as indicated by the fact that in a pon B derivative of W7 (PBP $1B^-$), obtained in our laboratory, no PBP 1B* could be detected in stationary phase cultures grown in L-broth. When active growth was stopped by other methods, like deprivation of some essential nutrient, the changes in the pattern of binding of $[^{125}I]$-ampicillin to the PBPs of the strains tested (MC6 and W7) were similar, in general, to those found in stationary phase cultures, grown in minimal-citrate medium. It is interesting to point out the very high reproducibility of the modification of PBP 3. Whenever cell growth was stopped, the level of binding to this protein decreased rather quickly. However, the modifications to the other PBPs showed a certain variability as to speed.

The inhibition of protein synthesis by chloramphenicol at the end of exponential growth, had no effect on the changes in the binding capacity of the PBPs, except in the cases of PBP 3 and, in the strain W7, PBPs 5 and 6. In fact, the characteristic reduction in the level of PBP 3 was prevented by chloramphenicol. In the two strains so far tested, W7 and MC6, we found that the level of PBP 3 in the control cultures was below 40% while in the chloramphenicol treated cultures it was still higher than 85% of the value found in exponentially growing cells. The behaviour of PBPs 5 and 6 in the strain W7, was also affected by

TABLE 1. Modification of the pattern of binding of ^{125}I ampicillin to the PBPs in stationary phase cells from three strains of E.coli grown in L-broth or minimal-citrate media.

Penicillin Binding Protein	Strain and growth media					
	L-broth			Minimal-citrate		
	W7	MC6	AB1157	W7	MC6	AB1157
1A	38	45	48	114	83	103
1B	81(1B*)	67	78	117	92	123
1C	6	51	77	38	92	95
2	57	45	50	124	99	92
3	25	26	41	47	25	82
5	162	46	71	149	93	107
6	334	72	79	170	83	80

Samples of 4.10^{10} cells were obtained from exponentially growing or 12 hour old stationary phase cultures of the strains W7 (3),MC6 (4) and AB1157(5) grown in L-broth or minimal-citrate media. The purification of cell envelopes and the $[^{125}I]$ ampicillin binding assays were as described(2). The autorradiographies were quantified by densitometry and the results are expressed as percent relative to the values found in the exponentially growing cells.

chloramphenicol whose presence, in stationary phase cultures, blocked the increase in the level of binding of $[^{125}I]$-ampicillin to these proteins. The effect of chloramphenicol on the modification of PBP 3 suggests the involvement of an active process induced at the end of exponential growth. Since the rel A gene is known to participate in the regulation of protein degradation, specially at the end of active growth (9), we have studied the effect of a rel A mutation on the modification of the PBPs. The results obtained with MC6 and its rel A derivative indicated that, whereas PBP 3 might belong to the group of proteins actively degraded at the end of exponential growth by a rel A-regulated mechanism, the behaviour of the other PBPs was unaffected by the rel A mutation (manuscript sent for publication).

Discussion

The modification of the pattern of binding of radioactive antibiotics to the PBPs of E.coli, beginning at the end of active growth, is, in its main features, reproducible when unrelated strains are compared, with the remarkable exception of PBPs 5 and 6. The apparent level of these proteins increased considerably in stationary phase cells of W7, although, in two different K-12 strains (MC6 and AB1157) it was reduced with respect to the level in exponentially-growing cells. The behaviour of PBPs 5 and 6 in the latter cannot be considered a general characteristic of K-12 strains. According to Buchanan and Sowell (1),in E.coli X975 and CP78, two K-12 strains, the level of PBP 6 increased in stationary phase cells, whereas that of PBP 5 remained at a constant level. These results suggest, therefore, that the modifications of PBPs 5 and 6 are strain-specific.

The nature of the growth medium, seems to affect considerably the behaviour of the PBPs in all the strains tested. In minimal-citrate medium the modifications ocurred at a slower rate than in L-broth,although the general tendency of the changes was maintained.

The effect of chloramphenicol and of a rel A mutation on the modifications of the PBPs suggest that at least two different mechanisms are involved: one sensitive to chloramphenicol, and probably controlled by rel A, and a second one insensitive to the inhibition of protein synthesis and independent of rel A.

References.

1. Buchanan, C.E., Sowell, M.O.: J. Bacteriol. 151, 491-494 (1982).
2. de la Rosa, E.J., de Pedro, M.A., Vázquez, D.: FEMS Microbiol. Lett. 14, 91-94 (1982).
3. Hartmann, R., Höltje, J., Schwarz, U.: Nature 235, 426-429 (1972).

4. de Pedro, M.A., Llamas, J.E., Cánovas, J.L.: J. Gen. Microbiol. 91, 307-314 (1975).

5. Takeda, Y., Nishimura, A., Nishimura, Y., Yamada, M., Yasuda, S., Suzuki, H., Hirota, Y.: Plasmid 5,86-98 (1981).

6. Schwarz, U., Seeger, K., Wengenmayer, F., Strecher,H.: FEMS Microbiol. Lett. 10, 107-109 (1981).

7. Lennox, E.S.: Virology 1, 190-206 (1955).

8. Vogel, H.J., Bonner, D.M.: J. Biol. Chem. 218, 97-106 (1956).

9. Goldberg, A.L., St. John, A.C.: Ann. Rev. Biochem. 45, 748-801 (1976).

RECOMBINANT PLASMIDS CARRYING PENICILLIN-BINDING PROTEIN/CELL SHAPE GENES FROM THE *lip-leuS* REGION OF THE *ESCHERICHIA COLI* CHROMOSOME

Brian G. Spratt, Jenny Broome-Smith, Alex Edelman

Microbial Genetics Group, School of Biological Sciences
University of Sussex, Brighton BN1 9QG, UK

Neil G. Stoker

Department of Genetics, University of Leicester
Leicester LE1 7RH, UK

Introduction

The *lip-leuS* region of the *Escherichia coli* chromosome contains the genes *pbpA*, *rodA* and *dacA* (1). All of these genes are believed to be involved in peptidoglycan synthesis. The *pbpA* gene is the structural gene for penicillin-binding protein 2 (PBP 2). Inactivation of PBP 2, either by β-lactam antibiotics or by mutation, results in the growth of *E.coli* as spherical cells and subsequent cell death (2). The *rodA* gene encodes a minor cytoplasmic membrane protein (M_r=31,000) and inactivation of this gene by mutation also results in growth of *E.coli* as spherical cells (1,3). The *dacA* gene encodes PBP 5, a major D-alanine carboxypeptidase. PBP 5 is not essential for cell growth as a strain of *E.coli* with a deletion of the entire *dacA* gene grows without obvious impairment (4). However, slight overproduction of PBP 5 results in spherical morphology and high levels of overproduction are lethal (5). The in vivo function of PBP 5 is unknown. Genetic mapping of the region using a series of lambda transducing phages has shown the gene order to be *lip-dacA-rodA-pbpA-leuS* (1). Analysis of the role of each of these genes in the biosynthesis of the peptidoglycan is complicated by their close linkage. We describe here the sub-cloning of this region to produce plasmids that carry the *pbpA*, *rodA* and *dacA* genes separate from each other.

The Target of Penicillin

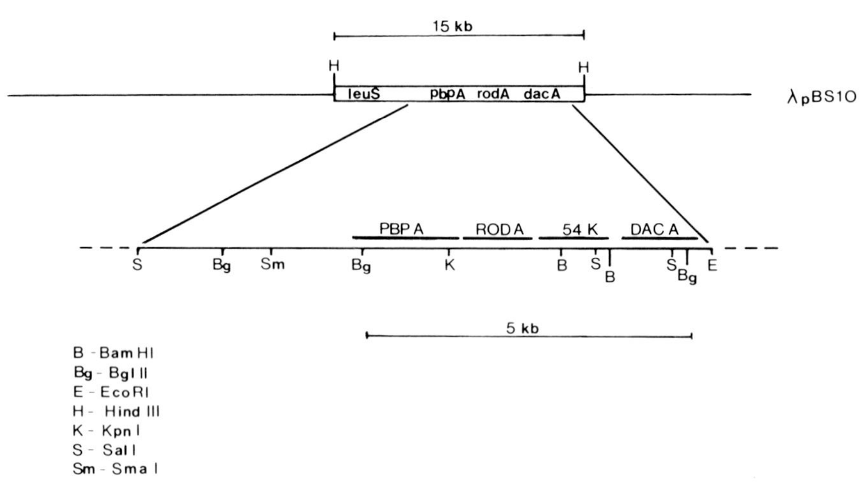

Fig. 1. The structure of λpBS10.

Plasmids Carrying the pbpA, rodA and dacA Genes

Figure 1 shows the structure of λpBS10 which contains a 15-kb HindIII fragment carrying the leuS, pbpA and rodA genes. Deletion mapping of λpBS10 and subcloning of fragments into plasmid vectors has delineated the position of these genes. Both the pbpA and rodA genes are present on a 7-kb SalI fragment (1, Figure 1). In order to clone the pbpA gene separate from the rodA gene we used λpBS10Δ13, a derivative of λpBS10 that has a 1.2-kb deletion within the 7-kb SalI fragment. This deletion removes the BamHI site within the 7-kb SalI fragment and inactivates rodA. The deleted SalI fragment (5.8-kb) was cloned between the SalI sites of the low copy number pSC101-based vector pLG318 (6) to give pBS47 (Figure 2). The pbpA gene was further localised to a 3.7-kb SmaI-SalI fragment which was cloned between the SmaI and XhoI sites of the pSC101-based vector, pLG338 (6), to produce pBS60 (Figure 2).
The rodA gene was found to be on a 1.6-kb KpnI-BamHI fragment and this was cloned between the KpnI and BamHI sites of pLG318 to produce pLG346 (Figure 3). The dacA gene has been shown to be located on a 1.6-kb EcoRI-

BamHI fragment (5) and this was cloned between the EcoRI and BamHI sites of pLG338 to produce pBS59.

Neither the pbpA gene nor the dacA gene could be inserted stably into high copy number plasmids. The rodA gene has been inserted into high copy number plasmids (e.g. pACYC184 and pBR328), and cells carrying the resulting plasmids grew normally.

The pbpA, rodA and dacA genes, as well as a gene that encodes a cytoplasmic membrane protein of molecular weight 54,000, map within a 5.6-kb region and are probably contiguous.

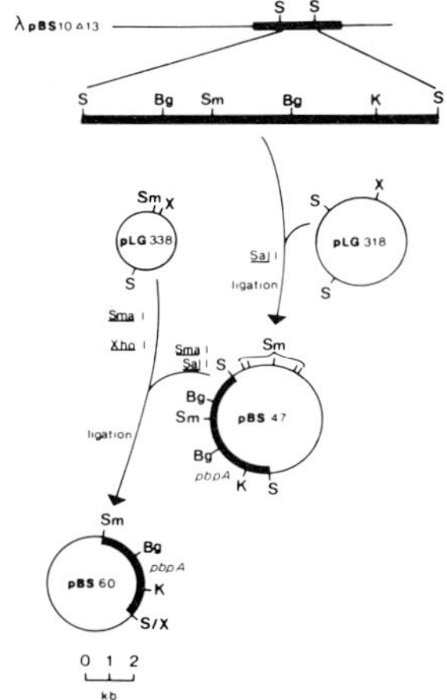

Fig. 2. Subcloning of the pbpA gene. X, XhoI; Sm, SmaI; S, SalI; K, KpnI; Bg, BglII; S/X, SmaI/XhoI hybrid.

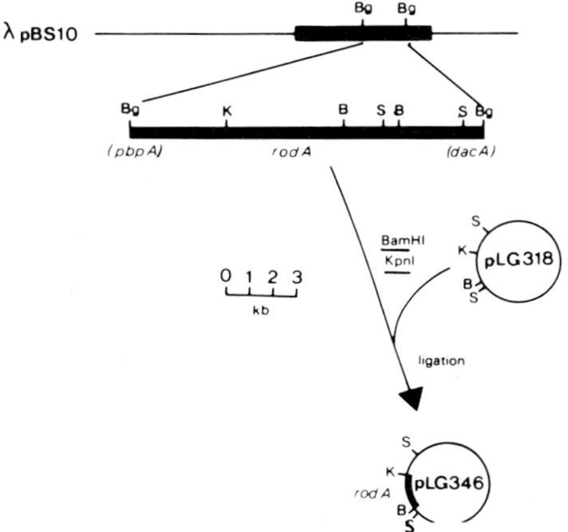

Fig. 3. Subcloning of the rodA gene. Bg, BglII; K, KpnI; B, BamHI; S, SalI.

References

1. Spratt, B.G., Boyd, A., Stoker, N.G.: J. Bacteriol. 143, 569-581 (1980).

2. Spratt, B.G.: Phil. Trans. Roy. Soc. London B289, 273-283 (1980).

3. Stoker, N.G., Pratt, J.M., Spratt, B.G.: Submitted for publication (1983).

4. Spratt, B.G.: J. Bacteriol. 144, 1190-1192 (1980).

5. Markiewicz, Z., Broome-Smith, J.K., Schwarz, U., Spratt, B.G.: Nature 297, 702-704 (1982).

6. Stoker, N.G., Fairweather, N.F., Spratt, B.G.: Gene 18, 329-335 (1982).

This work was supported by the Medical Research Council. N.G. Stoker was supported by Hoechst (UK).

SOME ASPECTS OF THE MECHANISM OF EXPRESSION OF PBP 3 IN E.coli

II. Differential Behaviour of Transducing Phages Complementing the pbpB Thermosensitive Mutation sep_{2158}.

Juan A. Ayala, Miguel A. de Pedro and David Vázquez

Instituto de Bioquímica de Macromoléculas
Centro de Biología Molecular. C.S.I.C. - U.A.M.
Universidad Autónoma de Madrid
Canto Blanco, Madrid-34, Spain.

Introduction

Penicillin binding protein 3 (PBP 3) of Escherichia coli is a membrane protein involved in septum formation during the process of cellular division (1,2,3). In spite of its importance, little is known about the mechanisms controlling the expression and activity of this protein, coded for by the gene pbpB mapping at 1.9 min. on the E.coli genetic map (4). To study the mechanisms controlling the expression of PBP 3, we have analyzed two groups of λ phages carrying bacterial DNA inserts able to complement pbpB ts mutations. The first group of phages studied were obtained in our laboratory, using λ540 as vector and DNA from W3110 (5). The hybrid molecules were constructed by integration of Hind III fragments of bacterial DNA into the single Hind III site present on λ540. The recombinant phages were selected by their ability to complement the ts mutation $ftsI_{730}$ (6). Among the recombinant complementing phages λ31 was of particular interest. When JE 10730 (6) was lysogenized by λ31 it became thermoresistant, and produced an apparently normal PBP 3, althoughit does not seem to contain a copy of the structural gene coding for PBP 3. A more detailed description of λ31 has been previously communicated (5).

The Target of Penicillin

In this article, we report some interesting characteristics observed in a different group of vectors, Walker's phages λsep^+46 and λsep^+82 (7). Both phages contain the structural gene coding for PBP 3 and complement the thermosensitive mutation sep_{2158} in the strain Ax655/ λ^+ (7). However, they differ in the size of their respective bacterial DNA inserts. Whereas λsep^+46 contains a 26.8 Kbp insert carrying the genes from leuA to envA, the insert in λsep^+82 is shorter (10.8 Kbp) and contains the genes leuA and pbpB. The bacterial strains and phages used in this work were kindly provided by Dr. M. Vicente.

Results

Analysis of the capacity of binding of PBP 3 to radioactive β-lactams in membranes obtained from the strain Ax655/ λ^+ and the dilysogens Ax655/ λ^+, λsep^+46 and Ax655/ λ^+, λsep^+82, showed the existence of striking differences between both dilysogens. The radioactive antibiotic used for the detection of the PBPs was N[3-(4-hydroxy-5- ^{125}I iodophenyl)propionyl] ampicillin (^{125}I-ampicillin)(2000 Ci/mMol)(8), an antibiotic with an extremely high affinity for PBP 3. In saturation experiments performed under standard conditions (250 µg of protein, 10 min. of incubation at 37°C) the half saturation value obtained was 5-10 nMolar. Our results showed that in the membranes from Ax655/ λ^+, no PBP 3 could be detected with ^{125}I-ampicillin, even in membranes from cells grown at 30°C, when cell division was normal (fig.1). The analysis of membranes obtained from the strains Ax655/ λ^+, λsep^+46 and Ax655/ λ^+, λsep^+82, both diploids for the gene pbpB, showed a clear difference in the capacity of binding of their respective PBP 3s to ^{125}I-ampicillin. In the former, no binding of ^{125}I-ampicillin to PBP 3 could be detected, irrespective of the temperature of growth (30°C or 42°C). However, when Ax655/ λ^+, λsep^+82 was grown under the same conditions, a normal binding of ^{125}I-ampicillin to PBP 3 was observed at both temperatures (fig.1). In contrast, when binding experiments were performed with membranes obtained after induction of the prophages (U.V. irradiation and thermal shock) a very heavily labelled band, in the position of PBP 3, became the predominant feature in the autoradiographies for both strains, Ax655/ λ^+, λsep^+46 and Ax655/ λ^+, λsep^+82 (fig.1). The amount of labelled PBP 3 detected under the above conditions indicated an overproduction by a factor of 10 with respect to the amount detected in membranes of uninduced Ax655/ λ^+, λsep^+82. Although the behaviour of both dilysogens was apparently similar after induction of the prophages, the analysis by bidimensional slab-gel electrophoresis (isoelectric focusing as first dimension and sodium dodecyl sulphate electrophoresis as second dimension) showed that the ^{125}I-ampicillin labelled protein appearing at the position of PBP 3 in membranes from induced Ax655/ λ^+, λsep^+46 had a far more acidic isoelectric point (pI=6.2) than that of the PBP 3 from Ax655/ λ^+, λsep^+82 or the genetically unrelated W-7 (9)(pI=7.6).

The study of the morphological effects of β-lactams with high affinity for PBP 3 like azthreonam (ID_{50} for PBP 3 in Ax655/ λ^+ = 0.5 μg/ml) on the dilysogens Ax655/ λ^+, λsep^+82 and Ax655/ λ^+, λsep^+46 showed that both strains were equally sensitive to filament-inducing β-lactams (fig.2), suggesting that the lack of binding of ^{125}I-ampicillin to the PBP 3 from uninduced cells of Ax655/ λ^+, λsep^+46 does not reflect a general lack of interaction of this protein with β-lactam antibiotics. Similar results were obtained with other β-lactams with high affinity for PBP 3 like furazlocillin(10) or MCH3K(Lafarquim laboratories)(manuscript in preparation)

Discussion

In both strains Ax655/ λ^+, λsep^+46 and Ax655/ λ^+, λsep^+82, the structural gene coding for the thermoresistant PBP 3 is, ap-

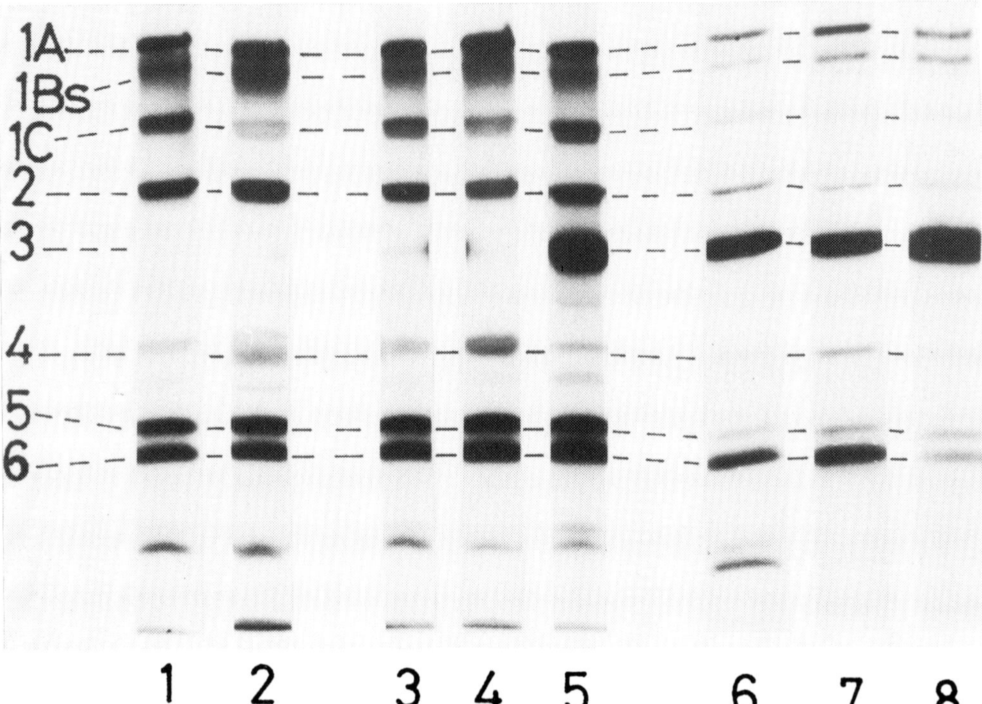

Fig.1.- Pattern of ^{125}I-ampicillin binding to membranes of the different strains. Membranes of the strains Ax655/ λ^+(lanes 1,2), Ax655/ λ^+, λsep$^+$46 (lanes 3,4,5) and Ax655/ λ^+, λsep$^+$82(lanes 6,7,8) were obtained from exponentially growing cells at 30°C (lanes 1,3,6), 42°C (lanes 2,4,7) or after prophage induction (60 sg. at 400 erg/mm^2 and 30 min. at 42°C)(lanes 5,8). The binding assay was as described(12). ^{125}I-ampicillin(2000 Ci/mMol specific activity) was used at a final activity of 30 μCi/ml, equivalent to 15 nMolar.

parently, the same; so, the variations observed in the behaviour of PBP 3 must be due to some other differential element between the two dilysogens. Since the only known difference between these two strains is the diploidity of Ax655/ λ^+, λsep$^+$46 for the gene cluster murE-envA, we suggest that the duplication of some genetic element in this region might be responsible for the differences observed in the behaviour of PBP 3. This hypothesis suggests the existence of elements

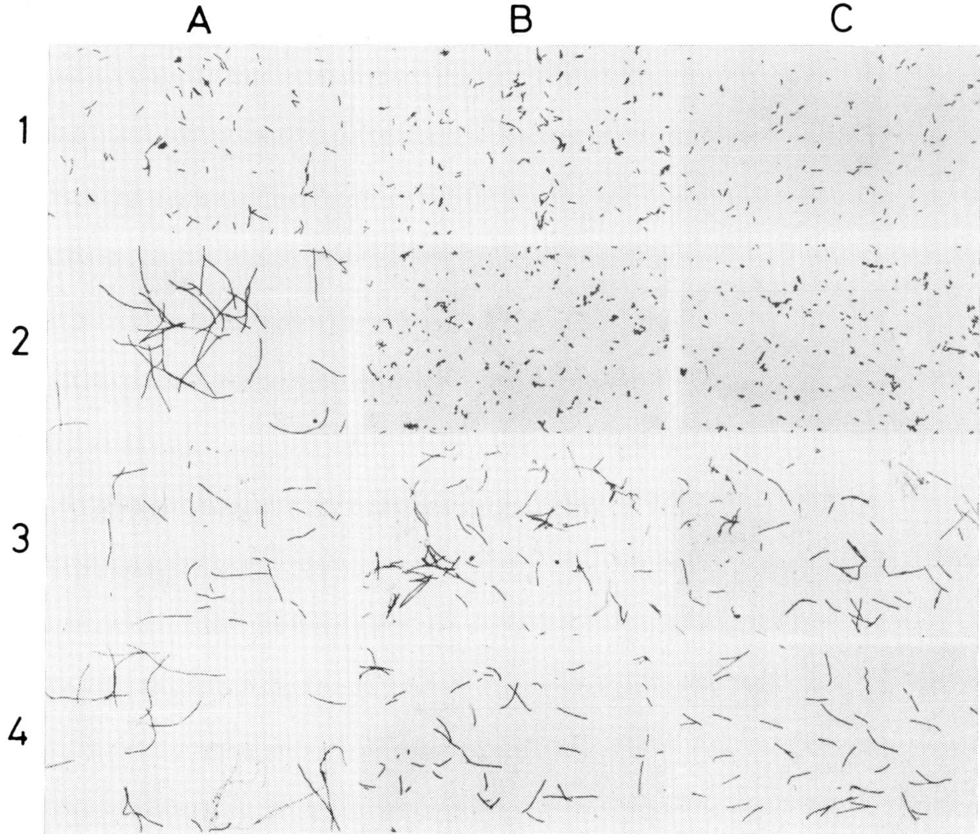

Fig.2.- Effect of azthreonam on exponentially growing cells of the different strains. Cells of strains Ax655/ λ^{+}(A), Ax655/ λ^{+}, λsep^{+}46(B) and Ax655/ λ^{+}, λsep^{+}82(C) were grown in medium LB at 30°C(1 and 3) or at 42°C (2 and 4) in the presence(3 and 4) or in the absence(1 and 2) of azthreonam at 1 μg/ml.

able to modify the physico-chemical characteristics of PBP 3; either by direct action on the protein itself, as suggested by the drastic change observed in the isoelectric point of the PBP 3 from induced cells of Ax655/ λ^{+}, λsep^{+}46, or by an indirect effect, via modification of the cell membrane, as could be expected from the duplication of genes like ftsA or envA involved in the control of cell growth (11). In order to check this hypothesis we are at present working to obtain a series of λsep^+ vectors derived from λsep^+46 by progressive shortening of its bacterial DNA insert.

References

1. Spratt B.G.: Proc. Nat. Acad. Sci. USA 72,2999-3003 (1975).
2. Ishino F. and Matsuhashi M.:Agric. Biol. Chem. 43,2641-2642 (1979).
3. Ishino F. and Matsuhashi M.: B.B.R.C. 101,905-911 (1981).
4. Bachmann B.J. and Brooks Low K.: Microbiological Reviews 44, 1-56 (1980).
5. Ayala J.A., de Pedro M.A.,Martinez-Salas E.,Vicente M. and Vazquez D.:Book of Abstracts. EMBO workshop on Mode of Action of β-lactam Antibiotics. El Escorial, Madrid (1982).
6. Suzuki H.,Nishimura Y. and Hirota Y.:Proc. Natl. Acad. Sci. USA 75,664-668 (1978).
7. Fletcher G.,Irwin C.A.,Henson J.M.,Fillingim C.,Malone M.M. and Walker J.R.:Journal of Bacteriology 133,91-100 (1978).
8. Schwarz U.,Seger K.,Wengenmayer F. and Strecker H.: FEMS Microbiol. Lett. 10,107-109 (1981).
9. Hartmann R.,Höltje J.V. and Schwarz U.: Nature (London) 235,426-429 (1972).
10. Schmidt L.S.,Botta G. and Park J.T.: Journal of Bacteriology 145,632-637 (1981).
11. Donachie W.D.,Begg K.J.,Lutkenhaus J.F.,Salmond G.P.C., Martinez-Salas E. and Vicente M.:Journal of Bacteriology 140,388-394 (1979).
12. Spratt B.G.: Eur. J. Biochem. 72,341-352 (1977).

THE INTERDEPENDENCE OF BACTERIAL SHAPE AND SIMULTANEOUS BLOCKING OF PBP 2 AND 3 IN ESCHERICHIA COLI SP 45 AND ESCHERICHIA COLI SP 63

K. Seeger and E. Schrinner, Hoechst AG, Frankfurt

Introduction:

Thermosensitive mutants, such as *Escherichia coli* SP 45 lacking PBP 2, when grown at 42 °C, form spherical bacteria with a more or less intact murein layer. *E. coli* with normal PBPs, when grown in the presence of Mecillinam show the same shape with PBP 2 blocked (Spratt 1975, 1977). *E. coli* lacking PBP 3, as for example *E. coli* SP 63, when grown at 42 °C, forms filaments. The same effect is achieved with *E. coli* K12, grown in the presence of a ß-lactam-antibiotic binding preferentially to PBP 3 (Spratt 1975). What is the morphology of *E. coli* SP 45, when PBPs 2 and 3 are simultaneously blocked?

Material and Methods:

Escherichia coli SP 45 kindly supplied by B. Spratt
Escherichia coli SP 63
Mueller-Hinton-Broth (Difco)
Cefotaxime (Hoechst AG,Frankfurt) used in final concentrations of 5 µg/ml;
Mecillinam (Leo, Ballerup) 25 µg/ml; *E. coli* SP 45 and *E. coli* SP 63, overnight culture was used in a dilution 1 : 100, 0.05 ml Suspension/ml medium. The antibiotic was added to this dilution containing approximately 10^5 CFU after incubation for 2 h at 42 °C.

The Target of Penicillin

For Scanning electron microscopy (SEM), the samples were prepared as follows:
Samples were diluted (1 : 1) with glutardialdehyde solution in phosphate buffer (1 %; pH 6.88) and kept for 16 h at 4 °C. Centrifugation followed by fixation with glutardialdehyde solutions (1 - 5 %), for either 2 hr or 16 hr at 4 °C was done. The samples were dehydrated in ethanol of increasing concentrations (30 % to absolute ethanol) and finally transferred to a cover glass and air dried. Coating with gold in a sputter-coater (Fa. Balzer) was performed. Pictures were taken in a Scanning Electron microscope (Stereoscan 150, Cambridge).

Results and Discussion:

As can be seen from Fig. 1, E. coli SP 45 grown at 42 °C was not completely homogenous, spheres and rod-like bacteria were present. Possibly residues of PBP 2 were still functioning. If a suspension as shown in Fig. 1 was grown for 2 hr in 25 µg Mecillinam/ml broth, a more homogenous population was obtained with spheroids almost exclusively seen (Fig. 2). It remains to be confirmed, whether this is due to a selection of round shaped, Mecillinam-resistant, PBP 2-lacking mutants or whether residual amounts of PBP 2 are still present in E. coli SP 45 when grown at 42 °C. If sheroid E. coli SP 45 was further cultivated at 42 °C in solutions of ß-lactam-antibiotics, binding preferentailly to PBP 3 (e.g. Cefotaxime) the spheroids grew out to forms rods, even short filaments could be found. Growth of cylindrical murein in some individual bacteria can be seen starting from the spheroid, in other cases the starting point for further growth cannot be identified. When E. coli SP 63, preincubated at 42 °C, was grown in 25 µg Mecillinam/ml broth, the filaments showed dilation in diameter near to one pole.

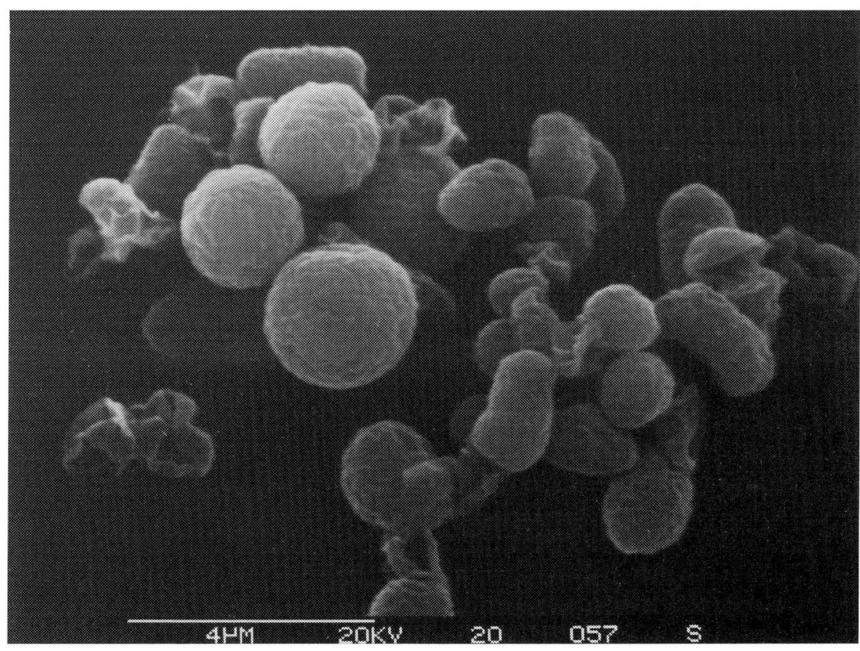

Fig. 1
Scanning-electron-micrograph of E. coli SP 45, grown in Mueller-Hinton Broth at 42 °C (Magnification: 10,000).

If we consider the bacterial rod as being composed of a cylindrical part and two half-spheroids as poles. Lack or inhibition of PBP 2 and stimulation of PBP 5 inhibit growth of the cylindrical part of the murein or stimulate growth of that part of murein responsible for septum- and poleformation (Markiewicz et al., 1982). When PBP 3 is inhibited or lacking, the cylindrical part grows, but few or no septa are formed and no separation of daughter cells occurs. Where PBPs 2 and 3 are together inhibited, the sequence of inhibition can influence the outcome. If PBP 2 is first inhibited followed by inhibition of PBP 3 rods grow from the spheroids.

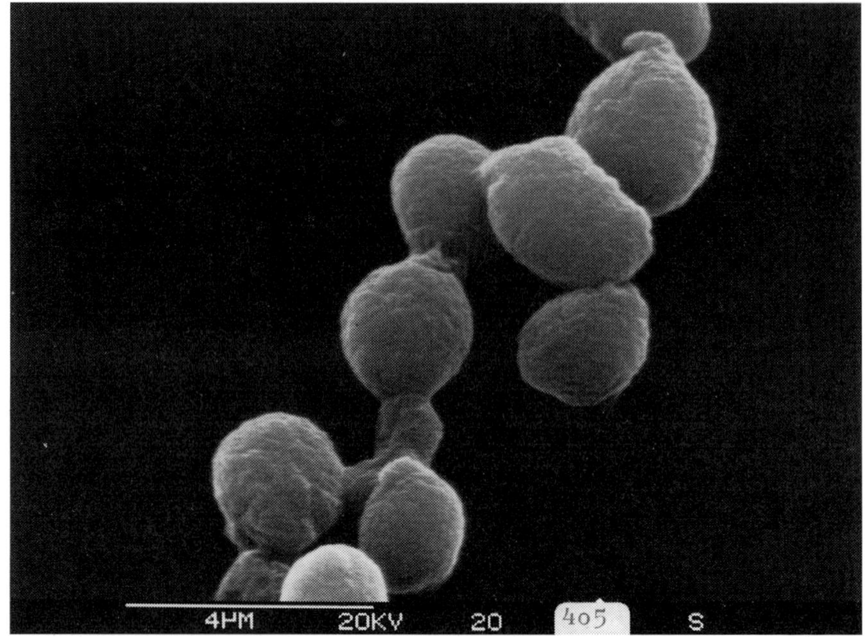

Fig. 2
Scanning-electron-micrograph of E. coli SP 45, grown in Mueller-Hinton Broth at 42 °C, containing 25 µg Mecillinam/ml (Magnification 10,000).

If PBP 3 is first inhibited, followed by inhibition of PBP 2 club-shaped filaments arise, similar to the findings of Otsuki (1981). The diameter of the part grown under influence of Mecillinam is enlarged. If PBP 5 prevails or if PBP 2 is inhibited causing a higher degree of cross-linkage of murein, then this higher cross-linked murein could be responsible for curved spheroid growth of normal E. coli and for the spheroid extension of the sacculi in growth zones.
Blockade of PBP 2 first and PBP 3 later allows a more balanced growth, daughter cells from spheres grow again as rods. Obviously inhibition of PBP 3 reverts spheroid growth into cylindrical growth.

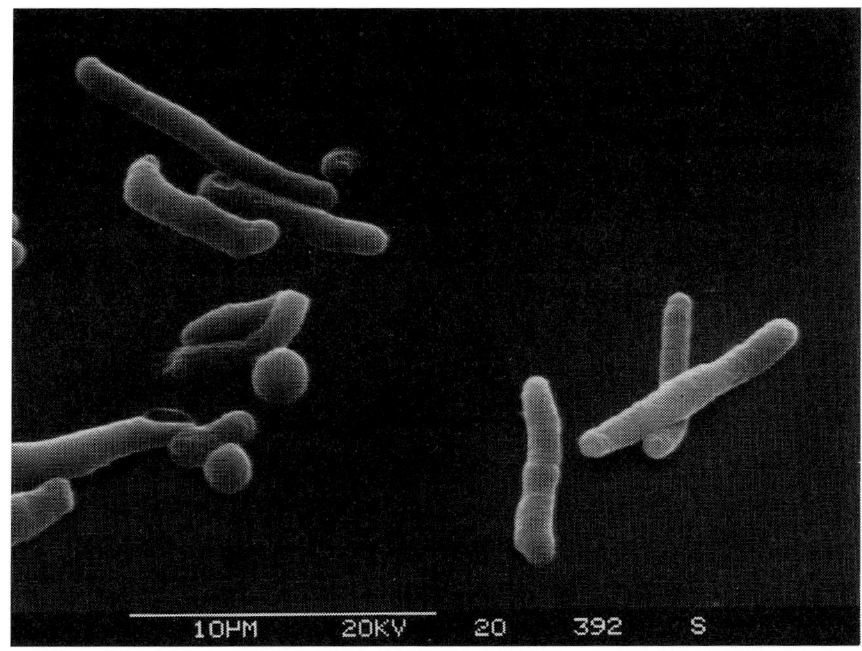

Fig. 3
E. coli SP 45, grown at 42 °C, containing 5 μg Cefotaxime/ml (Magnification 5,000).

References:

1. Chase, H.A., Fuller, C., Reynolds, P.E.: Eur. J. Biochem. 117, 301 - 310 (1981)
2. Otsuki, M. J.: Journ. Antibiot. 34, 739 - 752 (1981)
3. Markiewicz, Z., Broome-Smith, J.K., Schwarz, U., Spratt, B.G.: Nature, 297, 702 - 704 (1982)
4. Matsuhashi, M., Ishino, F., Tamaki, S., Nakajima-Iijima, S., Tomioka, S., Nakagawa, J., Hirata, A., Spratt, B.G., Tsuruoka, T., Inonye, S., Yamada, Y. in Trends in Antibiotic Research, Umezawa, H. et al. (Ed) Tokyo 1982

5. Spratt, B.G.: Proc. Nat. Acad. Sci. USA 72, 2999 - 3003 (1975)
6. Spratt, B.G.: Eur. J. Biochem. 72, 341 - 352 (1977)
7. Spratt, B.G.: Phil. Trans. R. Soc. Lond. B 289, 273 - 283 (1980)

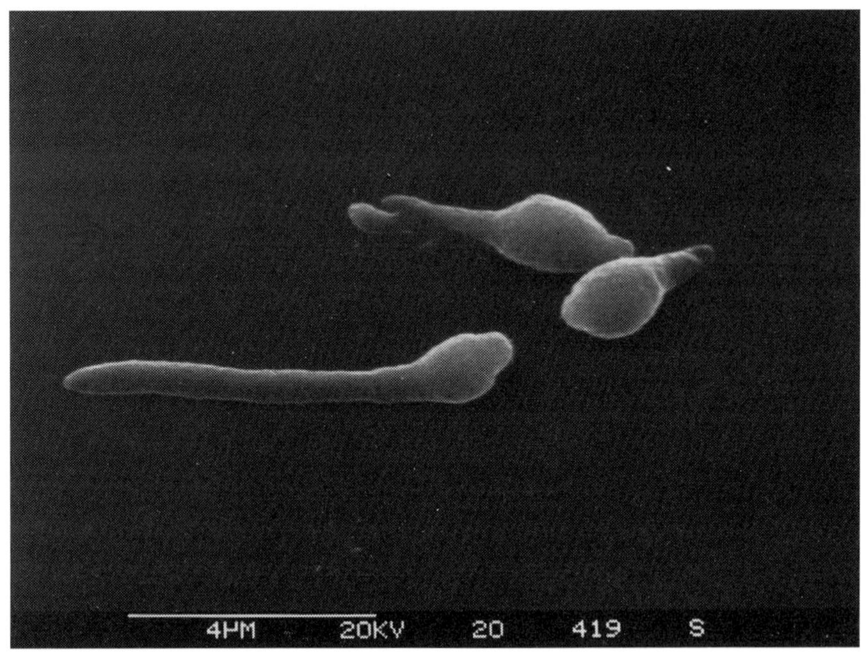

Fig. 4
E. coli SP 63, grown at 42 °C, containing 25 µg Mecillinam/ml (Magnification 10,000).

BENZYLPENICILLIN-RESISTANT MUTANTS OF *CLOSTRIDIUM PERFRINGENS*

Russell Williamson

Laboratoire de Microbiologie,
Institut Biomédical des Cordeliers, Paris, France.

Introduction

In contrast to most Gram-positive bacilli, *Clostridium perfringens* grows as filaments in the presence of low concentrations of benzylpenicillin (1,2). The organism contains six penicillin-binding proteins (3), and the four lowest molecular weight PBPs become virtually saturated at concentrations of benzylpenicillin that cause inhibition of septation (4). Since this morphological response can be obtained without any inhibition of the increase of cell mass, and that there was no measurable binding of the antibiotic to the high molecular weight PBPs, it was suggested that PBPs 1 and/or 2 could have transpeptidase activity essential for the incorporation of peptidoglycan and continued growth of the bacteria (4). There have been no reports of β-lactamase activity in *C. perfringens* and it could therefore be expected that mutants resistant to benzylpenicillin might have high molecular weight PBPs with reduced affinity for the antibiotic as the sole basis for resistance. The isolation and some characteristics of such mutants are reported in this presentation.

Results

Wild-type bacteria were not able to grow on agar with concentrations of benzylpenicillin above 0.015μg/ml. However, with a large inoculum, some colonies were able to grow at

The Target of Penicillin

0.03μg/ml. One of these spontaneous slightly resistant mutants was then exposed to increasing concentrations of the antibiotic in liquid medium, and more resistant mutants were obtained which were able to grow at a two-fold higher concentration. In this stepwise selection procedure, a range of benzylpenicillin-resistant strains was obtained as spontaneous mutants, each derived from the preceding less-resistant strain. In each case the strains were isolated, purified, and the resistance level determined before selection of the next mutant in the series. In addition, only strains that grew with normal morphology (rods) in the absence of benzylpenicillin were used in the selection procedure. As shown in Table 1, the mutants became increasingly more resistant to all the β-lactam antibiotics tested. In contrast, each strain was as sensitive to chloramphenicol, rifampicin, and vanco-

Table 1 Antibiotic susceptibilities of wild-type and benzylpenicillin-resistant mutants of *C.perfringens*

	Antibiotic MIC (μg/ml) [a]				
Strain	Pen G	Carb	Thin	Mox	Gen.time [b]
667	0.03	0.25	0.25	0.06	27
P0.03 [c]	0.06	0.5	0.5	0.125	29
P0.06	0.125	1.0	1.0	1.0	30.5
P0.125	0.25	1.0	2.0	2.0	31
P0.25	0.5	1.0	4.0	4.0	32
P0.5	1.0	2.0	8.0	8.0	34
P0.75	1.5	4.0	16.0	16.0	37
P1.0	2.0	8.0	16.0	16.0	41
P1.25	2.5	4.0	16.0	16.0	44
P1.5	2.5	4.0	32.0	16.0	47
P2.0	3.0	4.0	32.0	16.0	51

(a) MICs were determined on BHI agar.
(b) Generation times (min) were measured in liquid medium without antibiotics.
(c) Strain designation indicates the highest concentration of benzylpenicillin in which the mutant could grow.

mycin as the wild-type, parental strain. No β-lactamase activity was detected in any strain, whether grown with a β-lactam antibiotic or not. It was found that as the resistance of the strains increased to benzylpenicillin, the growth rate of cultures in the absence of any antibiotic decreased in comparison with that of the wild-type (Table 1). The most resistant mutant (P2.0) grew at half the rate of the parental strain (667).

All of the strains formed filaments when grown in medium containing benzylpenicillin at half the respective MIC of this antibiotic for each mutant. Moreover, each mutant also grew as filaments when grown in this antibiotic at half the MIC for the wild-type, parental strain, and this effect was reversed by subsequent addition of penicillinase. Similar morphological results were obtained when the strains were grown in the appropriate concentrations of carbenicillin, cephalothin, or moxalactam.

When the PBPs in whole organisms were examined after exposure of the mutants to a sufficient quantity of (^{3}H)-benzylpenicillin to saturate the PBPs of the wild-type bacteria, the apparent amount of PBP1 decreased as the resistance of the strain increased (Fig. 1). None of the other five PBPs appeared changed in any of the mutants in comparison with the parental strain. In fact the relative amounts of these PBPs and their affinities for benzylpenicillin were virtually identical in each strain, indicating that no changes had occurred in these proteins in the resistant mutants. In a few cases, the apparent amount of PBP2 differed by about two-fold within mutants at a particular resistance level, but this did not correlate to increased resistance. In contrast, when the amount of benzylpenicillin was increased to detect the PBPs, it was evident that the total amount of PBP1 present in the resistant strains was essentially the same as that in the wild-type parent (Fig. 1).

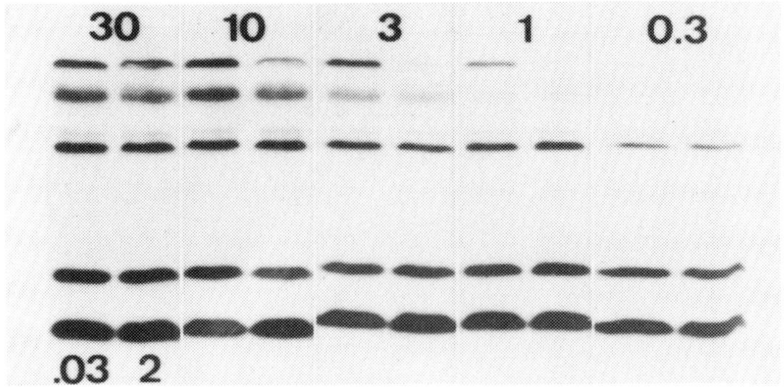

Fig. 1 PBPs of wild-type (0.03) and benzylpenicillin-resistant (2.0) strains of C. perfringens.
1ml. volumes of bacteria were incubated with various concentrations (μg/ml) of benzylpenicillin for 10min at 37°C.

The calculated 50% saturation values (μg/ml) for PBP1 from each of the strains are as follows: 667, 0.9; P0.03, 0.85; P0.06, 0.95; P0.125, 1.1; P0.25, 1.35; P0.5, 3.2; P0.75, 5.8; P1.0, 6.2; P1.25, 6.8; P1.5, 7.4; and P2.0, 7.6.

During the isolation of the resistant mutants, about a hundred isolates at each successive level were screened for abnormal growth in medium without benzylpenicillin. The majority (95%) all grew as normal rods, but a few grew as chains of about 10 bacilli or as short filaments about 5-10 times normal length, and in some cases both of these were also curly. There were no detectable PBP differences in each of these strains when compared with normal-morphology organisms at the respective resistance level.

Discussion

Resistance of Gram-positive bacteria to β-lactam antibiotics by an apparent reduction in affinity of one or more PBPs has been reported for *Staphylococcus aureus* (5), *Streptococcus pneumoniae* (6), *Bacillus subtilis* (7), and *B. megaterium* (8). In the case of mutants of *B. subtilis* and *B. megaterium* high-molecular weight PBPs were involved in the resistance, and the PBP1 of *B. licheniformis* is also an important target of β-lactam antibiotics (9). PBP1 of *C. perfringens* appears to be a similar target, since this protein had significantly reduced affinity for benzylpenicillin in the resistant mutants. However, it is important to note that the affinity of a single target decreased as the resistance increased, and this is in contrast with the multiple PBP changes in resistant strains of *S. pneumoniae* (6), and some other changes in *B. subtilis* (7) or *B. megaterium* (8). Hence, the PBP1 of *C. perfringens* may be the primary, β-lactam antibiotic-sensitive transpeptidase. Since the total amount of this protein appears to be the same in each strain, the decreasing growth rate as the resistance increases may indicate that the activity of this PBP may become rate limiting for growth of the strain.

The lack of additional changes in the lower molecular weight PBPs of the resistant strains of *C. perfringens* presumably indicates that these proteins are not of primary importance to the organism. The fact that each strain grew as filaments in the same concentration of benzylpenicillin as the parental organism, as well as at half the respective MIC for each mutant, suggests that the function(s) of these proteins were identical in all strains and that they could be inhibited to the same extent to cause inhibition of septation.

References

1. Gardner, A.D.: Nature (London) 146, 837-838 (1940).
2. Crofts, T.E., Evans, D.G.: Brit. J. Exp. Pathol. 31, 550-561 (1950).
3. Murphy, T.F., Barza, M., Park, J.T.: Antimicrob. Agents Chemother. 20, 809-813 (1981).

4. Williamson, R., Ward, J.B.: J. Gen. Microbiol. 128, 3025-3035 (1982).

5. Hartman, B.J., Tomasz, A.: Antimicrob. Agents Chemother. 19, 726-735 (1981).

6. Zighelboim, S., Tomasz, A.: Antimicrob. Agents Chemother. 17, 434-442 (1980).

7. Buchanan, C.E., Strominger, J.L.: Proc. Natl. Acad. Sci. U.S.A. 73, 1816-1820 (1979).

8. Giles, A.F., Reynolds, P.E.: Nature (London) 280, 167-168 (1979).

9. Chase, H.A., Reynolds, P.E., Ward, J.B.: Eur. J. Biochem. 88, 275-285 (1978).

STUDIES OF β-LACTAM RESISTANCE IN STREPTOCOCCUS FAECIUM

L. Daneo-Moore, Michael Pucci, Edward Zito, Maria Ferrero
Department of Microbiology and Immunology
Temple University School of Medicine
Philadelphia, Pennsylvania, USA

Introduction

Enterococci differ from other streptococci by a relatively high level of β-lactam sensitivity, having minimal inhibitory concentrations (MICs) of 3-10 μg/ml for penicillin G. This relatively high MIC does not appear to be due to a difficulty in drug access, or to a low level of permeability (1,2). In addition, penicillin-sensitive (MICs of 0.04-0.08 μg/ml) enterococci can be isolated (3; Fig. 1). These sensitive derivatives do not appear to differ from the parent organism in penicillin-binding activity (3; our results not shown here). Recently, a number of high-level penicillin-resistant strains of Streptococcus faecium have been examined, obtained in the laboratory (4,5) or from clinical material (5).

Results

Preliminary studies reported in ref. 5 and here indicate the following:

1) Stepwise penicillin-resistant derivatives (MICs of 8-80 μg/ml) exhibit no qualitative changes in their penicillin-binding proteins (5). There appears to be an increase in the binding activity of PBP 5, which is roughly related to the level of penicillin resistance.

The Target of Penicillin

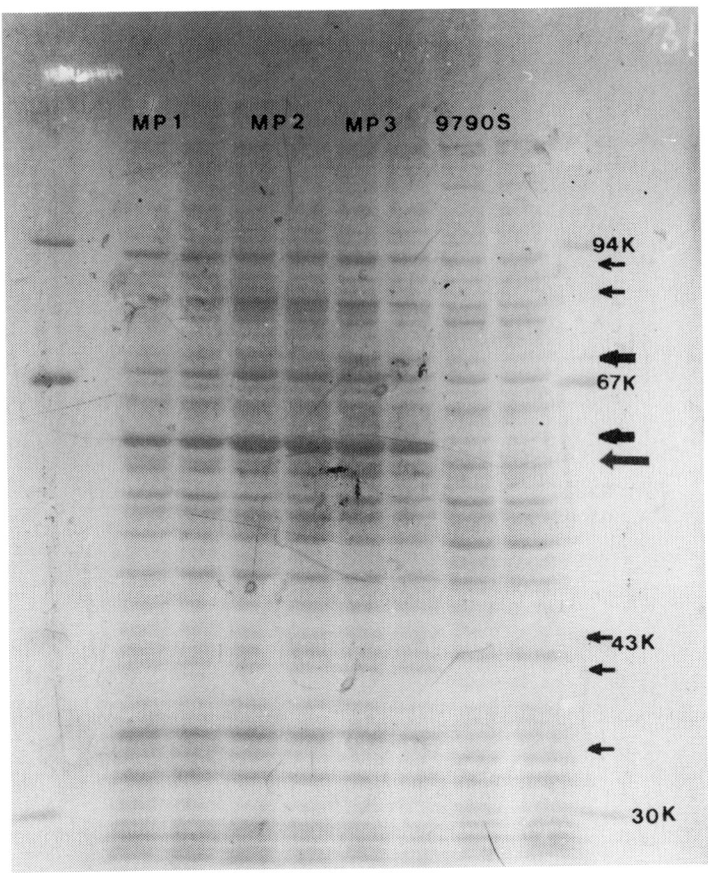

Fig. 1. Membrane protein patterns of _S. faecium_ ATCC 9790 WT, MP1, MP2, MP3. Membrane proteins were subjected to SDS-polyacrylamide gel electrophoresis. 9790S is equivalent to _S. faecium_ ATCC 9790 WT. Strains were grown in Todd-Hewitt glucose medium and run in duplicate. 200 µg protein was loaded per well.

2) At least seven membrane proteins detected on staining with Coomassie Blue are increased in staining intensity in the laboratory-derived, penicillin-resistant isolates (Fig. 1). Two of these proteins increase in staining intensity with increased level of penicillin resistance (Fig. 1, heavy arrows). The estimated molecular weights of these proteins are 86, 84, 76, 58, 43, 39, and 35 Kd. At least one 50-Kd membrane protein decreases in intensity in the laboratory-derived, penicillin-resistant organisms (Fig. 1, long arrow).

3) The time course-dependent increase in penicillin-binding activity of PBP 5 can be partially prevented by the presence of PMSF (1 mM) during membrane preparation and incubation (Fig. 2). The kinetics of binding of the other PBPs are not affected by the presence of PMSF.

4) High-level penicillin-resistant clinical isolates differ in PBP-binding activity from our laboratory strain both qualitatively and quantitatively (5). Interestingly, the Coomassie Blue-stained membrane proteins differ from the laboratory strain in the same bands as do the laboratory isolates, as well as in other bands. These common differences in penicillin-resistant strains hold for both the proteins that are increased, and the protein that is decreased in staining intensity.

5) In preliminary experiments we have obtained derivatives that resemble the clinically isolated resistant organisms qualitatively in the absence of one (or more) PBP bands. These isolates are relatively resistant to thienamycin and cefoxitine, respectively, but are only slightly more resistant than the parent to penicillin G.

6) The tentative preliminary conclusion from these studies is that enterococci possesses two classes of penicillin

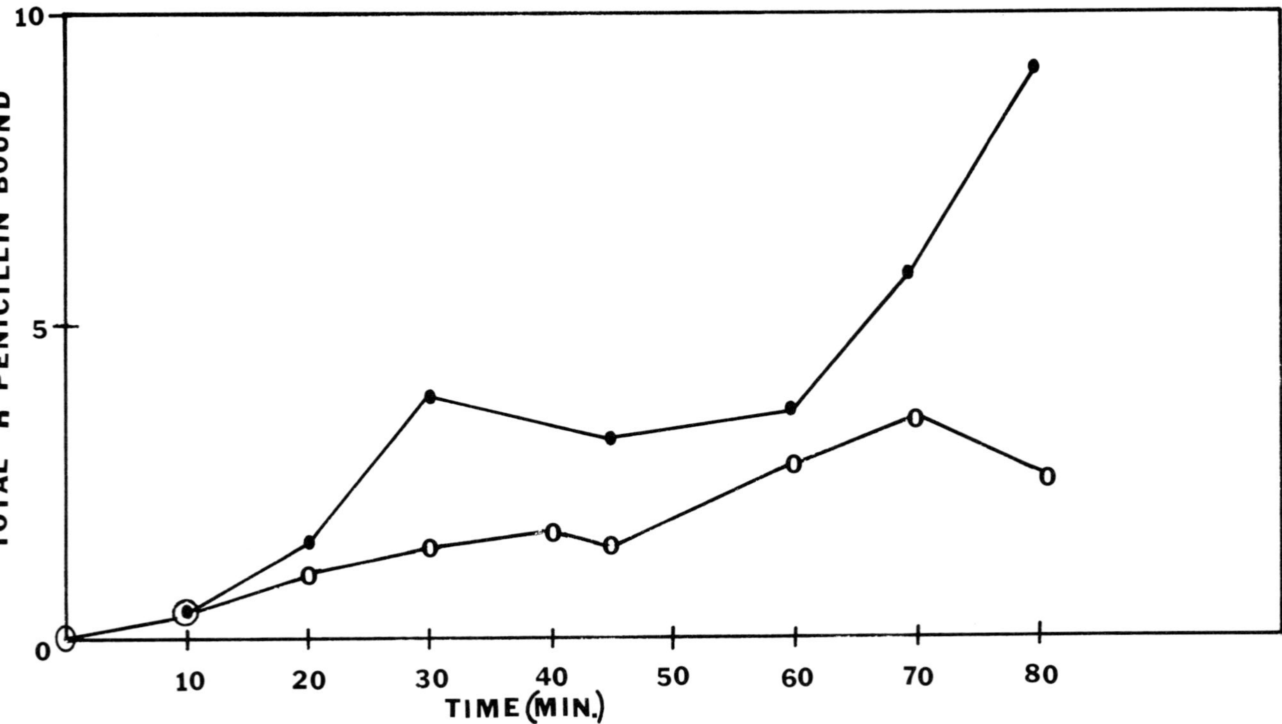

Fig. 2. Percent of total radiolabeled penicillin bound associated with PBP 5 of *S. faecium* ATCC 9790 over time. Growth of cells and binding were also carried out in the presence of PMSF (1 mM, final concentration). Symbols: ●——●, -PMSF (control); O——O, +PMSF.

resistance mechanisms. One, which is relatively conventional is accompanied by alterations in PBP affinity and mobility. The second, which appears to be novel, consists of alterations in the staining intensity of several specific membrane proteins and by an increase in the binding of PBP 5 (4,5). Fontana and her associates (this volume) have proposed that in enterococci a slow-binding kinetics PBP, such as PBP 5, can confer resistance by taking over the function(s) of other PBPs. The relationship between the change in membrane proteins described here and the accompanying penicillin resistance is unclear at this time.

References

1. Coyette, J.I., Ghuysen, J.M., Fontana, R.: Eur. J. Biochem. 110, 445 (1980).
2. Williamson, R., Calderwood, S.B., Moellering, R.C., Jr., Tomasz, A.: J. Gen. Microbiol. (in press).
3. Eliopoulos, G.M., Wennersten, C., Moellering, R.C., Jr.: Antimicrob. Agents Chemother. 22, 295 (1982).
4. Fontana, R., Cerini, R.: in Current Chemotherapy and Infectious Disease, Am. Soc. Microbiol., Washington, D.C. (1982).
5. Pucci, M., Daneo-Moore, L.: in Microbiology - 1982, Am. Soc. Microbiol., Washington, D.C., p. 199 (1982).

BIOSYNTHESIS OF MUREIN (PEPTIDOGLYCAN)

Michio Matsuhashi, Jun-ichi Nakagawa, Shigeo Tamaki, Fumitoshi Ishino, Shigeo Tomioka, Wan Park

Institute of Applied Microbiology, The University of Tokyo, Bunkyo-ku, Tokyo, 113 Japan

Introduction

Peptidoglycan in bacterial cell walls and septa is produced by three main processes: [1] synthesis of UDP-N-acetylmuramyl (MurNAc)-pentapeptide and UDP-GlcNAc by cytoplasmic enzymes; [2] synthesis of disaccharide-pentapeptide (repeating unit) linked to undecaprenol-pyrophosphate by hydrophobic enzymes in the cytoplasmic membrane; and [3] extension of glycan chain by addition of repeating units linked to undecaprenol-pyrophosphate (transglycosylase reaction) and formation of crosslinkage with concomitant removal of one terminal D-alanine molecule (transpeptidase reaction)(1-3); the third process is also performed by cytoplasmic membrane-bound proteins. In Gram-positive Staphylococci and Micrococci, the two reactions in this process seem to be carried out by separate enzymes(2,4), transglycosylase and β-lactam-sensitive transpeptidase, which is one of the so-called penicillin-binding proteins(PBPs). In Gram negative *Escherichia coli* (and probably in all Enterobacteria and *Pseudomonas* strains), these two reactions are carried out mainly by bifunctional PBPs that possess two domains for activities of transglycosylase and β-lactam sensitive transpeptidase(2,3)(Fig. 1). The last process is biologically most interesting because it has been established that multiple enzymes are involved in this process, and the differentiation is correlated to the different steps of the cell cycle, i.e., cell elongation, determination of the rod-shape of the cell and cell division.

The Target of Penicillin

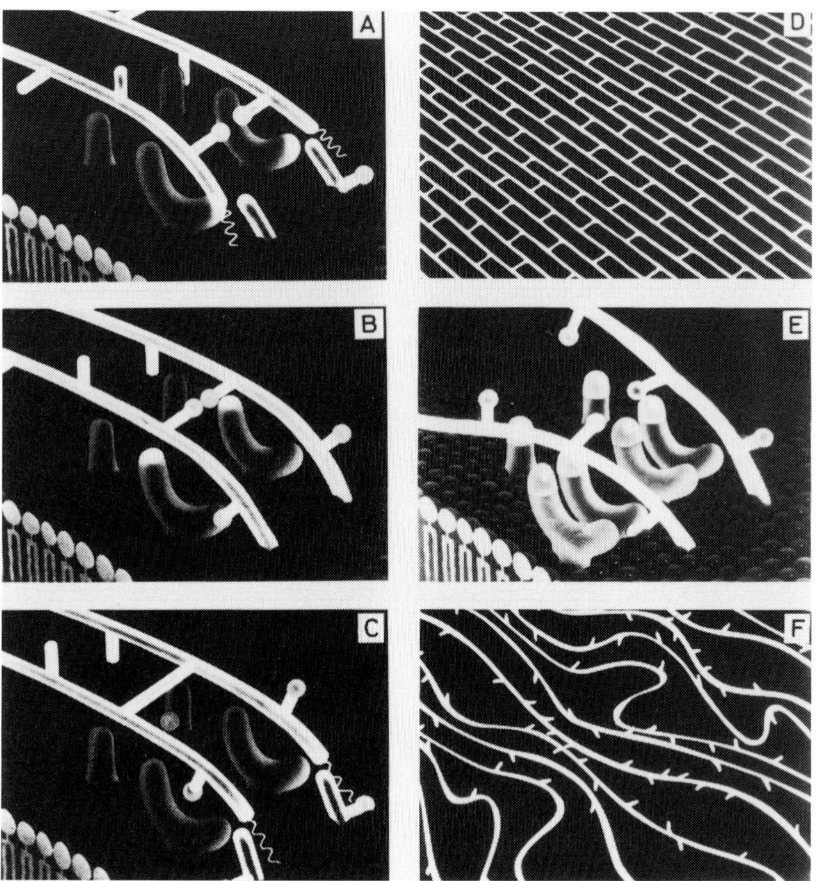

Fig. 1. Proposed model of peptidoglycan synthesis by penicillin-binding proteins in E. coli. A: Glycan strands are extended at the transglycosylase site of a PBP (for instance, 1B) by addition of repeating units, GlcNAc-MurNAc-pentapeptide, which are linked to undecaprenol pyrophosphate in the cytoplasmic membrane. In this figure peptidoglycan strands are also linked to undecaprenol pyrophosphate (zigzag at the end of the strands). B: Crossbridge between two pentapeptide side chains (short bars each with a ball at the end) is formed at the transpeptidase site of a PBP with release of one terminal D-alanine (ball) per crossbridge. C: DD-carboxypeptidase, presumably PBP-5 or 6, removes residual terminal D-alanine molecules that have not been involved in crosslinking reaction. D: Network of peptidoglycan formed. E: If a β-lactam antibiotic binds to the transpeptidase site of both PBPs 1A and 1B (and PBP-5 and 6, too), the crosslinking reaction is inhibited. F: Uncrosslinked strands of peptidoglycan are formed and the bag-shaped network of peptidoglycan is broken. (Reproduced from reference 3).

These steps must be initiated and terminated under a strict regulatory mechanism and the biosynthetic enzyme reactions causing cell-cycle-specific formation of peptidoglycan are the most probable candidates for the above regulation.
In the following chapters we will devote most of our description to the last steps of peptidoglycan biosynthetic reactions in *E. coli*.

Peptidoglycan Synthesis by PBPs in *E. coli*

Seven major PBPs are known in *E. coli* and among them four higher molecular weight PBPs are enzymes that synthesize the network structure of peptidoglycans(3). Enzymatic activities of PBPs were demonstrated in purified protein preparations. Surprisingly, the highly purified preparations of PBP-1B which consist of at least three proteins with similar molecular weights, showed two enzymatic activities of transglycosylase and penicillin-sensitive transpeptidase, as deduced from their ability to form crosslinked peptidoglycan from lipid-linked precursor(5). Subsequently the two enzyme activities were also demonstrated in PBP-1A and then in PBP-3, which, unlike the PBP-1B triplet, were isolated as a single band in sodium dodecylsulfate(SDS) - polyacrylamide gel electrophoresis(PAGE) (6,7). The triplet proteins of PBP-1B were separated from each other by SDS-PAGE, guanidine-HCl treated and reactivated. Each protein showed both enzymatic activities of transglycosylase and penicillin-sensitive transpeptidase. This provides strong evidence for the location of the two enzymatic activities in each single protein molecule(8).
After treatment with trypsin, PBP-1B is degraded into fragments with molecular weights of about 50,000 (50K fragments, probably doublet proteins), which retained [^{14}C]benzylpenicillin bound to the protein prior to the trypsin treatment, and a smaller amount of non-penicillin-binding 32K fragment which was further

degraded to form much smaller fragments(9). Extended trypsin treatment, which caused complete conversion of PBP-1Bs to 50K and to other smaller fragments did not appreciably alter the binding of [^{14}C]benzylpenicillin to the 50K fragments(9). Moreover, by careful experimentation, even the two biosynthetic enzyme activities, i.e., transglycosylation and transpeptidase activity, were found to be unimpaired after extensive trypsin treatment(10). These results indicate that activities of the transglycosylase and the transpeptidase both reside in the 50K fragment. Preliminary results of DNA-recombination experiments suggested that the domain for the transpeptidase activity is located closer to the C-terminal of the 50K fragment. DNA- and amino acid-sequence determination and X-ray crystallography are required for exact location of the active domains.

Based on the results of genetic experiments, PBP-1A and PBP-1B both function in the process of cell elongation(11-13). Differences in the properties of these two PBPs with respect to biological functions will be discussed elsewhere(14).

PBP-3 septum-peptidoglycan synthetase also has the activities of transglycosylase and transpeptidase(7). The transpeptidase activity is very sensitive to the β-lactam antibiotics that inhibit septum formation of the cell, such as benzylpenicillin, apalcillin, acylureidopenicillins and cephalexin, but is insensitive to nocardicin A, which does not induce filamentous cells (3). Mutants with thermosensitive PBP-3 which form filamentous cells at higher temperatures have been previously isolated by Spratt(15) and others.

Enzymatic activities of purified PBP-2 have not yet been demonstrated. However, membranes from the cells that produce an excessively large amount of PBP-2 were very active in synthesizing crosslinked peptidoglycan from UDP-linked precursors(16). The crosslinking reaction was insensitive to 6-α-methoxy-cephalosporins(cephamycins) but was very sensitive to an amidinopenicillin mecillinam(3,16). Mecillinam binds to PBP-2 and induces round cells(15).but cephamycins do not bind to this protein. Probably PBP-2 functions in peptidoglycan synthesis in the process of determining the rod-shape of the cell. The mecillinam-sensitive transpeptidase activity is certainly due to PBP-2 but it is still unknown if a transglycosylase activity is also possessed by this protein. Structure of the cross-linked product is under the investigation (with collaboration of Dr. U. Schwarz and his colleagues).

DD-Carboxypeptidase activities of smaller PBPs have been reviewed previously(3).

Peptidoglycan Transglycosylase in Staphylococcus aureus and Micrococcus luteus.

Peptidoglycan transglycosylase activity was eluted from membranes of S. aureus and M. luteus and partially purified on a column of DEAE cellulose. The transglycosylase activities were eluted from the column differently from the major PBPs of the two bacteria, and also was not adsorbed on the ampicillin-Sepharose affinity column. The transglycosylase activity of S. aureus was very sensitive to the antibiotics moenomycin and macarbomycin but that of M. luteus was resistant to these antibiotics. Transpeptidase activity supposedly due to some of the PBPs(17) has not yet been demonstrated in any fraction. Presence of the two enzyme activities in a single PBP has also been suggested recently in a Gram positive Bacillus (J. L. Strominger, personal communications).

Acknowledgments: This work was partially supported by a Grant-in-Aid for Special Project Research (57111006) from the Ministry of Education, Science and Culture of Japan.

References

1. Strominger, J.L., Izaki, K., Matsuhashi, M., Tipper, D.J.: Federation Proc. 26, 9-22 (1967).

2. Matsuhashi, M., Nakagawa, J., Tomioka, S., Ishino, F., Tamaki, S.: Drug Resistance in Bacteria — Genetics, Biochemistry and Molecular Biology (ed.: Mitsuhashi, S.), Japan Scientific Societies Press, Tokyo, and Thieme-Stratton Inc., New York 1982, pp. 297-310.

3. Matsuhashi, M., Ishino, F., Tamaki, S., Nakajima-Iijima, S., Tomioka, S., Nakagawa, J., Hirata, A., Spratt, B.G., Tsuruoka, T., Inouye, S., Yamada, Y.: Trends in Antibiotic Research — Genetics, Biosyntheses, Actions & New Substances (eds.: Umezawa, H., Demain, A., Hata, T., Hutchinson, C.R.), Japan Antibiotics Research Association, Tokyo 1982, pp. 99-114.

4. Park, W., Matsuhashi, M.: Manuscript in preparation.

5. Nakagawa, J., Tamaki, S., Matsuhashi, M.: Agric. Biol. Chem. (Tokyo) 43, 1379-1380 (1979).

6. Ishino, F., Mitsui, K., Tamaki, S., Matsuhashi, M.: Biochem. Biophys. Res. Commun. 97, 287-293 (1980).

7. Ishino, F., Matsuhashi, M.: Biochem. Biophys. Res. Commun. 101, 905-911 (1981).

8. Nakagawa, J., Matsuhashi, M.: Biochem. Biophys. Res. Commun. 105, 1546-1553 (1982).

9. Nakagawa, J., Matsuhashi, M.: Agric. Biol. Chem. (Tokyo) 44, 3041-3044 (1980).

10. Nakagawa, J., Matsuhashi, M.: Manuscript in Preparation.

11. Tamaki, S., Nakajima, S., Matsuhashi, M.: Proc. Natl. Acad. Sci. USA 74, 5472-5476 (1977).

12. Suzuki, H., Nishimura, Y., Hirota, Y.: Proc. Natl. Acad. Sci. USA 75, 664-668 (1978).

13. Spratt, B. G., Jobanputra, V., Schwarz, U.: FEBS Lett. 79, 374-378 (1977).

14. Tomioka, S., Ishino, F., Tamaki, S., Sudo, A., Nakagawa, J., Matsuhashi, M.: These Proceedings.

15. Spratt, B.G.: Proc. Natl. Acad. Sci. USA 72, 2999-3003. (1975).

16. Ishino, F., Tamaki, S., Spratt, B.G., Matsuhashi, M.: Biochem. Biophys. Res. Commun. 109, 689-696 (1982).

17. Hayes, M.V., Curtis, N.A.C., Wyke, A.W., Ward, J.B.: FEMS Microbiol. Lett. 10, 119-122 (1981).

ENZYMATIC SYNTHESIS OF PEPTIDOGLYCAN-CROSSBRIDGES IN *E. COLI*

Shigeo Tomioka, Fumitoshi Ishino, Shigeo Tamaki, Atsushi Sudo, Jun-ichi Nakagawa, Michio Matsuhashi

Institute of Applied Microbiology, The University of Tokyo, Bunkyo-ku, Tokyo, 113 Japan

Introduction

The rod-shape of bacilli-formed bacterial cells is determined mainly by the chemical structure of peptidoglycan (murein according to Weidel and Pelzer(1)) network. In Gram-negative *Escherichia coli* and related bacilli, the peptidoglycans seem to cover the whole cell surface with a single layer of network structure. The differences in chemical structure at specific positions of the bag-shaped peptidoglycan sacculus must then be determined by the length of glycan chains, and extent and type of crossbridges. The main determinants of these two important factors are penicillin-binding proteins(PBPs) (2). PBP-3 undoubtedly determines the network structure that distinguishes hemispherical parts of *E. coli* sacculi, originally having been septa in cell division, from other cell walls. PBP-2, probably in cooperation with a few other proteins, e.g., *rodA*(*mrdB*) gene product, may synthesize parts of the sacculi that enable the extension of new peptidoglycan networks on the long axis of the cell. PBP-1A and 1B apparently function in the last step, the step of sacculi extension. As suggested by genetic observations, PBP-1A and -1B seem to compensate, at least partly, for the lack of each other, although the two proteins are fairly different in properties as well as in enzyme activities(3). In this paper we will describe the properties of the crosslinking reaction by purified PBP-1A and the apparent contribution of this protein to the formation of sacculi.

The Target of Penicillin

To identify the structure of the crosslinked repeating units, we examined the complete lysozyme digests of peptidoglycans by two dimensional thin-layer chromatography. Fig. 1 indicates the lysozyme-digested fragments obtained from normal E. coli peptidoglycan. Besides the major C3, C5 and C6 spots(1), a number of minor spots appeared. The spot with the smallest Rf1 is probably the crosslinked trimer of repeating units(4). Among spots with higher Rf1 and lower Rf2, there may be dehydrated compounds which are formed from peptidoglycans by action of anhydromuramidase (transglycosylase according to Höltje et al.(5)). The chromatographic pattern was reproducible by a skillful hand and may be applicable to the study of the fine structures of peptidoglycan.

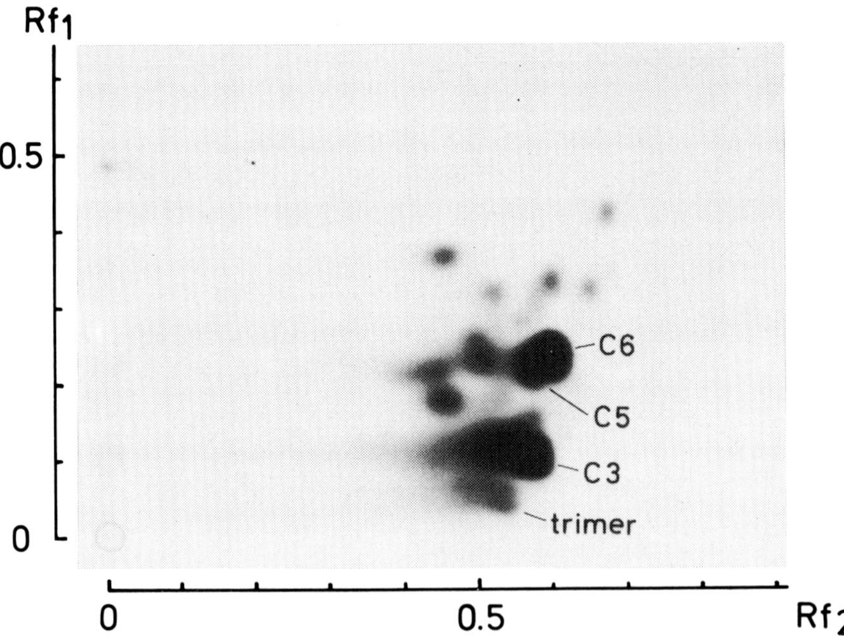

Fig. 1. Two dimensional thin-layer chromatographic pattern of lysozyme-treated E. coli peptidoglycan (labeled in meso-[^{14}C]-diaminopimelic acid). An autoradiogram is shown. Solvent 1: isobutyric acid - 1M ammonia (1:0.6); solvent 2; 95% ethanol - 1M ammonium acetate, pH 7 (1.5:0.6).

Formation of Hypercrosslinked Peptidoglycan with Multiple Crosslinkages by PBP-1A

Purified PBP-1A catalyzes formation of crosslinked peptidoglycan from the lipid-linked precursor(6,7). Unlike PBP-1B, which formed crosslinked peptidoglycan to up to about 25%, the extent of the crosslinkage by PBP-1A increased with time of incubation to about 40%(3). After lysozyme digestion, PBP-1A yielded disaccharide-peptide compound(s) and crosslinked bis-(disaccharide-peptide) compound(s), designated as spots C6* and C3*, respectively (Fig. 2), and a third spot, C2* (Fig. 2), which was assumed to be tris-(disaccharide-peptide) compound(s). These starred compounds apparently have one more D-alanine molecule than the correlating compounds from the normal peptidoglycan, since PBP-1A has no DD-carboxypeptidase activity. C3* and C2* did not appear from the product of PBP-1A reaction formed in the presence of β-lactams. On two-dimensional chromatograms, not shown due to space limitation, a new spot between C3* and C6* (Rf1 0.2 and Rf2 0.6, correcting to Fig. 1) was found, which also disappeared when benzylpenicillin was

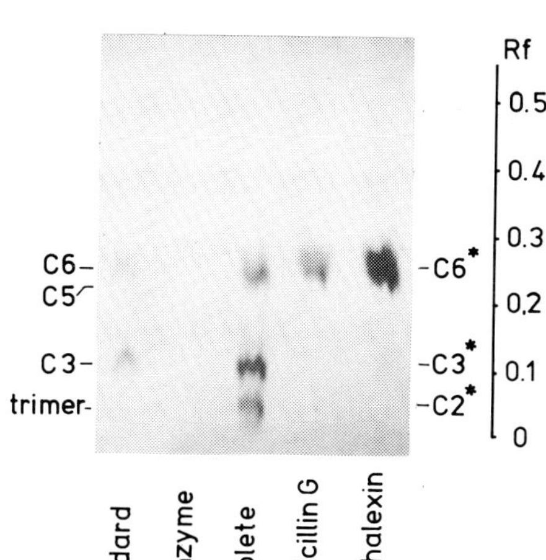

Fig. 2. Lysozyme products of peptidoglycans formed by PBP-1A in the absence and presence of β-lactams. An autoradiogram of the one dimensional paper chromatogram is shown. Solvent: isobutyric acid - 1 M ammonia (1:0.6). For experimental conditions see reference 8. The figure was reproduced from reference 8.

present in the reaction mixture. A faint doublet spot (Rf1 0.35∿ and Rf2 0.65) and a very faint single spot (Rf1 0.55 and Rf2 0.55) were present in samples obtained with or without benzylpenicillin.
Compounds C2* and C3* were eluted from the paper and subjected to digestion with DD-endopeptidase (penicillin-insensitive), which splits the D-alanyl-meso-diaminopimelyl linkage(9). C3* was converted into C6* and C2* was reduced to C6* and C3* which was further converted to C6*. By N-terminal determination, 1/2 and 1/3 equivalent of ε-amino terminal in C3* and C2*, respectively, was found to be free. Thus C3* was deduced to be the crosslinked dimer and C2* to be the doubly-crosslinked trimer of the repeating unit(8).

Possible roles

PBP-1B is probably a generally functioning peptidoglycan-synthetic enzyme in E. coli, as can be assumed from its enormous heat stability (60°C), resistance to anionic detergents, such as Sarkosyl, and from the fact that mutants lacking this protein are often thermosensitive (42°C)(10). Moreover, the relatively low sensitivity of the transpeptidase activity of PBP-1B to many β-lactam antibiotics may be favorable for the survival of the cells. PBP-1B is very hydrophobic and is bound tightly to the cytoplasmic membrane, its elution being possible only in 1% Triton in the presence of 1 M NaCl. The tight binding of this protein to the cytoplasmic membrane is compatible with the apparent function of this protein in synthesizing long-chain peptidoglycan with a constant extent of crosslinkage (25% in optimal conditions), which may ensure the four-fold helical structure of peptidoglycan (reference 11, see also Fig. 1 of reference 7).
On the other hand PBP-1A is a much more unstable protein and is also extremely sensitive to most β-lactam antibiotics. This protein is, moreover, loosely bound to the cytoplasmic membrane.

The peptidoglycan-synthetic reactions of this protein also have characteristics different from those of PBP-1B. First, the transglycosylase reaction of PBP-1A has sigmoid kinetics with respect to reaction time and enzyme concentration (see also reference 3). It requires incubation on a filter paper in the presence of a certain range of concentrations of non-ionic detergent(6). Under certain conditions it forms uncrosslinked, lipid-linked oligomers of repeating units. The transglycosylase reaction and the transpeptidase reaction are not tightly linked in PBP-1A: the peptidoglycan formed at early times of incubation is poorly crosslinked but that formed after a long incubation is hyper-crosslinked(3). The real function of PBP-1A in E. coli is probably different from that of PBP-1B, in that the former may cause diversion of the peptidoglycan, and the latter may synthesize a rather constant structure of peptidoglycan. A definite answer of this question may be obtained from reconstitution experiments of synthetic membranes.

In Fig. 3 we show the two-dimensional thin-layer chromatogram of the lysozyme-products of peptidoglycans in different E. coli cells which lack each one of the PBPs, 1A or 1B. Unfortunately no significant difference in the chromatographic pattern could

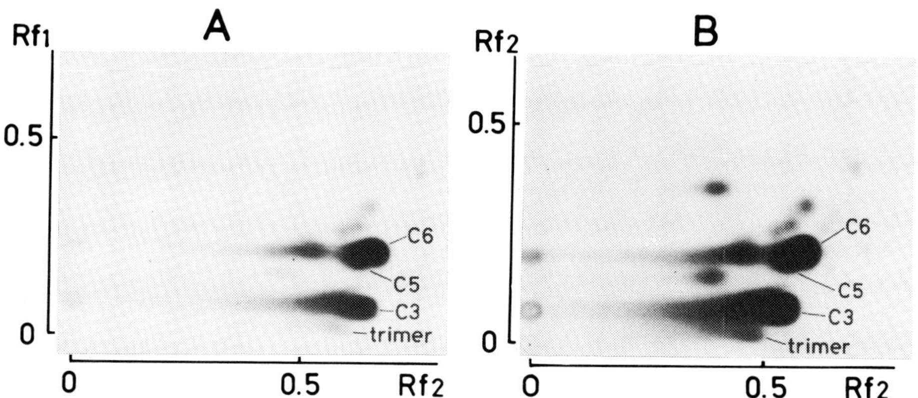

Fig. 3. Lysozyme products of peptidoglycans from PBP-1A⁻ (A) and PBP-1B⁻ (B) cells.

so far be identified. In the living cells, a detour mechanism or salvation mechainsm may be so tightly operating that the defect can not be seen in such simple experiments.

Acknowledgments: This work was partially supported by a Grant-in-Aid for Special Project Research (57111006) from the Ministry of Education, Science and Culture of Japan.

References

1. Weidel, W., Pelzer, H.: Adv. Enzym. 26, 193-232 (1964).

2. Matsuhashi, M., Ishino, F., Tamaki, S., Nakajima-Iijima, S., Tomioka, S., Nakagawa, J., Hirata, A., Spratt, B.G., Tsuruoka, T., Inouye, S., Yamada, Y.: Trends in Antibiotic Research — Genetics, Biosyntheses, Actions & New Substances (eds.: Umezawa, H., Demain, A., Hata, T., Hutchinson, C.R.), Japan Antibiotics Research Association, Tokyo 1982, pp. 99-114.

3. Matsuhashi, M., Nakagawa, J., Tomioka, S., Ishino, F., Tamaki, S.: Drug Resistance in Bacteria — Genetics, Biochemistry and Molecular Biology (ed.: Mitsuhashi, S.), Japan Scientific Societies Press, Tokyo, and Thieme-Stratton Inc., New York 1982, pp. 297-310.

4. Gmeiner, J.: J. Bacteriol. 143, 510-512 (1980).

5. Höltje, J.V., Mirelman, D., Sharon, N., Schwarz, U.: J. Bacteriol. 124, 1067-1076 (1975).

6. Ishino, F., Mitsui, K., Tamaki, S., Matsuhashi, M.: Biochem. Biophys. Res. Commun. 97, 287-293 (1980).

7. Matsuhashi, M., Nakagawa, J., Tamaki, S., Ishino, F., Tomioka, S., Park, W.: These Proceedings.

8. Tomioka, S., Ishino, F., Tamaki, S., Matsuhashi, M.: Biochem. Biophys. Res. Commun. 106, 1175-1182 (1982).

9. Tomioka, S., Matsuhashi, M.: Biochem. Biophys. Res. Commun. 84, 978-984 (1978).

10. Tamaki, S., Nakajima, S., Matsuhashi, M.: Proc. Natl. Acad. Sci. USA 74, 5472-5476 (1977).

11. Burge, R.E., Fowler, A.G., Reaveley, D.A.: J. Mol. Biol. 117, 927-953 (1977).

NEW INSIGHTS IN THE PROCESS OF IN VITRO MUREIN SYNTHESIS IN E. COLI REVEALED BY MUREIN-ANALYSIS WITH HIGH-PRESSURE-LIQUID-CHROMATOGRAPHY

Werner Kraus, Bernd Glauner, and Joachim-Volker Höltje
Max-Planck-Institut für Virusforschung, Abteilung Biochemie, D-7400 Tübingen

Introduction

The biosynthesis of the murein sacculus is an extremely complex process that is not well understood. Several membrane-associated enzymes, mainly the penicillin-binding proteins (PBPs), catalyse the final stages of murein synthesis in *E. coli* (1,2,3). Little is known about the specific functions of these proteins in the assembly of the murein sacculus. To unravel the mechanisms involved, it has been proven helpful to study separate aspects of murein synthesis in better controllable *in vitro* systems. The development of a highly sensitive method of murein analysis using high-pressure-liquid-chromatography enabled us to study the products of *in vitro* synthesis with increased resolution.

Results and Discussion

1) Differences between native and *in vitro* synthesized murein.

A comparison of the murein synthesized by crude cell envelope preparation with the *in vivo* murein reveals differences in their compositions (Fig. 1). Several components are shown to be characteristic of the murein synthesized *in vitro*. Because they are hardly or not at all found in the *in vivo* murein, these muropeptides may represent pre-

The Target of Penicillin

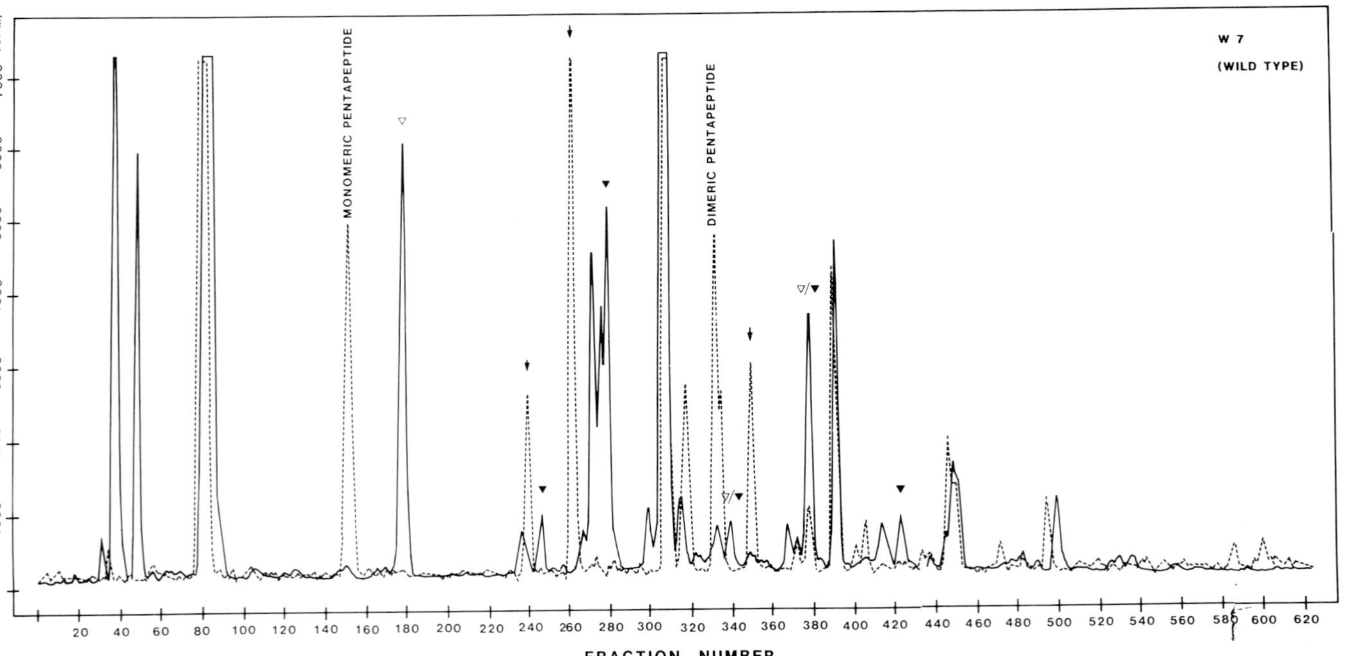

Fig. 1: High-pressure-liquid-chromatography of muramidase digests of native murein and in vitro synthesized murein. Native murein was prepared from cells grown in the presence of $^{3}H-A_2pm$ (2 mCi/ml; 25 Ci/mMol) as has been described previously (4). In vitro murein synthesis was performed according to Izaki et al. (5). Envelopes (0.2 mg protein) were incubated at 30°C for 1 h in the presence of 80 nmol UDP-N-acetylmuramylpentapeptide and 0.8 nmol UDP-N-acetyl-^{14}C-glucosamine (247 mCi/mmol) in 0.05 Tris-HCl buffer, 0.01 M $MgCl_2$, 0.001 M 2-mercaptoethanol, pH 8. Murein was digested with Chalaropsis muramidase, reduced with sodium borohydride and fractionated by HPLC as described in the contribution by Glauner and Schwarz. (—) native murein; (••••) in vitro murein; (↓) muropeptide present predominantly in murein synthesized in vitro; (▽) Arg-Lys containing muropeptide; (▼) dimeric muropeptide with direct A_2pm-A_2pm bond.

cursors of the native murein. Furthermore, in the *in vitro* system lipoprotein is not attached to murein and none of the recently discovered direct A_2pm-A_2pm cross-linkages of the native murein are synthesized. These findings are in agreement with other results indicating a correlation between these two reaction steps (see contribution Glauner and Schwarz). Although direct A_2pm-A_2pm bonds found in the *in vivo* murein are not formed *in vitro* normal A_2pm-Ala bonds are. It may be concluded that the two types of cross-linkages are formed by different mechanisms which can be uncoupled.

2) Muropeptide pattern in various carboxypeptidase IA mutants.

The analysis of murein synthesized *in vitro* by mutants with defects in PBP 5 shows a high increase in pentapeptide components (Table 1).

MUTANTS IN PBPs 4 AND 5

STRAIN	RELEVANT GENOTYPE	HIGH MOLECULAR WEIGHT MUREIN (CPM)	CROSSLINKAGE (%)	PENTAPEPTIDES IN % OF TOTAL	
				MONOMERIC	DIMERIC
PA 3092	WILD TYPE	84 463	18.4	6.1	2.1
DL 64	DAC B 12	85 569	25.3	8.0	3.2
SP 5003	DAC A	109 937	28.7	13.4	12.7
JE 5631	FTS I 730 DAC B 12 DAC A 1191	128 750	28.0	23.0	18.7

Table 1: In vitro murein synthesis products of carboxypeptidase IA mutants. Murein synthesis and analysis by HPLC was performed as described in Fig. 1. Mean values of two independent assays are presented. Cross-linkage is calculated using the formula 100x(1/2 Dimer + 2/3 Trimer)/Monomer + Dimer + Trimer.

This is in accordance with the fact that PBP 5 exhibits carboxypeptidase IA activity (6,7). PBP 4, which also has carboxypeptidase IA activity (8), seems to be a minor factor in the observed increase of pentapeptides as indicated by only a small increase of these muropeptides in a dacB mutant. In a double mutant defective in PBP 4 and PBP 5 the majority of the muropeptides carries pentapeptide side chains. The carboxypeptidase activity of PBP 6 seems to be of secondary importance, at least in the *in vitro* system. This is in agreement with other results showing that PBP 6 has only about 10% of the carboxypeptidase IA activity of PBP 5 (9).

3) *In vitro* murein synthesis in mutants defective in cell division.

Temperature-sensitive mutants having a defect in cell division, including PBP 3 and ftsA mutants, turned out to be severely hampered in synthesizing murein *in vitro* when grown at non-permissive temperature until filamentation was complete (Table 2).

A): MUTANTS IN PBP 3

STRAIN	RELEVANT GENOTYPE	GROWTH TEMPERATURE (°C)	HIGH MOLECULAR WEIGHT MUREIN (CPM)	MUREIN SYNTHESIS IN % OF CONTROL	CROSS-LINKAGE (%)
SP 63	TS IN PBP 3	30	89 376	100	22.6
		42	16 774	18.8	14.6
KN 126	WILD TYPE	30	84 947	100	22.8
		42	102 850	121	18.2
JE 5703	FTS I 730	30	88 393	100	18.9
		42	50 851	57.4	14.1
PA 3092	WILD TYPE	30	84 463	100	18.4
		42	71 659	84.8	14.3
B): TEMPERATURE SENSITIVE DIVISION MUTANT PAT 84					
PAT 84	FTS A	30	70 898	100	20.1
		42	30 238	42.8	14.5

Table 2: In vitro murein synthesis products of mutants defective in cell division. (Details as given in Fig. 1 and Table 1).

In contrast, murein is synthesized *in vivo* at an almost normal rate, although no septum is formed (10). Changes in the muropeptide composition are not observed either *in vivo* or *in vitro*.

4) Participation of PBP 1a and 1b in *in vitro* murein synthesis

The cross-linkage of the *in vitro* murein synthesis product in PBP 1b mutants decreases by 50% when compared with the wild type (Table 3). Various PBP 1a mutants show a normal degree of cross-linkages. Thus, PBP 1b exhibits high transpeptidase activity, at least under the *in vitro* conditions. The degree of overall murein synthesis is drastically reduced in PBP 1b mutants (by 95%) but only moderately decreased (by 40%) in PBP 1a mutants. While *in vivo* these two enzymes can compensate for each other (11) *in vitro* PBP 1a cannot replace the lacking PBP 1b. This could mean that in the *in vitro* system, the PBP 1b is mainly functional.

MUTANTS IN PBP'S 1A AND 1B

STRAIN	RELEVANT GENOTYPE	GROWTH TEMPERATURE (°C)	HIGH MOLECULAR WEIGHT MUREIN (CPM)	MUREIN SYNTHESIS IN % OF CONTROL	CROSS-LINKAGE (%)
JE 5637	PON B 1085	30	4 304	4.9	9.0
SP 61	PON A	30	51 792	58.4	20.9
JE 5686	PON B 353, DAC B 12, DAC A 1191	30	58 900	67.4	14.0
JE 5683	PON A 980, DAC B 12, DAC A 1191	30	109 377	125.2	30.1

Table 3: In vitro murein synthesis products of mutants in penicillin-binding proteins 1a and 1b (details as given in Fig. 1 and Table 1; overall murein synthesis is calculated as compared with the data of the corresponding wild types given in Table 2).

Acknowledgement

We thank U. Schwarz for his advice and helpful discussions and Y. Hirota and B. Spratt for kindly providing strains.

References

1. Tamaki, S., Nakajima, S., Matsuhashi, M.: Proc.Natl.Acad. Sci. USA 74, 5472-5476 (1977).
2. Suzuki, H., Van Heijenoort, Y., Tamura, T., Mizoguchi, J., Hirota, Y., Van Heijenoort, J.: FEBS Lett. 110, 245-249 (1980).
3. Tamura, T., Suzuki, H., Nishimura, Y., Mizoguchi, J., Hirota, Y.: Proc.Natl.Acad.Sci. USA 77, 4499-4503 (1980).
4. Höltje, J.-V., Mirelman, D., Sharon, N., Schwarz, U.: J. Bacteriol. 124, 1067-1076 (1975).
5. Izaki, K., Matsuhashi, M., Strominger, J.L.: J.Biol.Chem. 243, 3180-3192 (1968).
6. Matsuhashi, M., Tamaki, S., Curtis, S.J., Strominger, J.L.: J. Bacteriol. 137, 644-647 (1979).
7. Nishimura, Y., Suzuki, H., Hirota, Y., Park, J.T.: J. Bacteriol. 143, 531-534 (1980).
8. Matsuhashi, M., Takagaki, Y., Maruyama, I.N., Tamaki, S., Nishimura, Y., Suzuki, H., Orino, U., Hirota, Y.: Proc. Natl.Acad.Sci. USA 74, 2976-2979 (1977).
9. Amanuma, H., Strominger, J.L.: J.Biol.Chem. 255, 11173-11180 (1980).
10. Mirelman, D., Yashouv-Gan, Y., Schwarz, U.: Biochem. 17, 1781-1790 (1976).
11. Suzuki, H., Nishimura, Y., Hirota, Y.: Proc.Natl.Acad.Sci. USA 75, 664-668 (1978).

PENICILLIN-BINDING PROTEINS AND PEPTIDOGLYCAN BIOSYNTHESIS IN BACILLUS MEGATERIUM

Peter E. Reynolds

Department of Biochemistry, University of Cambridge
Tennis Court Road, Cambridge CB 2 1 QW, U.K.

Introduction

The protoplast membrane of vegetative cells of an asporogenous strain of *Bacillus megaterium* contains five major penicillin-binding proteins (PBPs) with trace amounts of a sixth component of molecular weight intermediate between PBP 3 and PBP 4 (1). Purification of four of the five PBPs has been carried out but only one of the purified proteins had D,D-carboxypeptidase activity when tested against the cell wall precursor, UDP-MurNAc-pentapeptide (2). Circumstantial evidence, based on a comparison of the rate of loss of benzylpenicillin from all PBPs with the rate of recovery of transpeptidation activity in a wall/membrane preparation presaturated with benzylpenicillin indicated that PBP 1 was probably a transpeptidase (3). Additional evidence for the importance of PBP 1 was obtained by binding experiments : treatment of growing cells of *B. megaterium* with benzylpenicillin at the MIC value indicated that PBP1 was the sole PBP to contain covalently-bound antibiotic and was thus essential for survival (3). These experiments did not indicate that PBP 1 was the only lethal target of β-lactam antibiotics in this organism and more recent experiments have shown that certain cephalosporins have low affinity for PBP 1 but nevertheless inhibit cell growth at low concentrations (4). This paper reports the results of investigations into the inhibitory effects of cephalexin and cefaclor on transpeptidation in toluenised cells of *B. megaterium* and relates these effects to the affinities of the antibiotics for PBPs in membrane preparations and in whole cells. It is concluded that one or more PBPs other than PBP 1 also has transpeptidase activity.

The Target of Penicillin

Results and Discussion

Peptidoglycan synthesis in toluenised cells. Nascent peptidoglycan synthesised from the two cell wall precursors UDP-MurNAc-peptapeptide and UDP-GlcNAc is attached to pre-existing peptidoglycan by a penicillin-sensitive transpeptidase in toluenised cells of B. megaterium (5,6). The attachment is assayed by following the incorporation of radioactive precursors into trichloroacetic acid precipitable, hot sodium dodecyl sulphate insoluble, material. The process was inhibited by concentrations of cefaclor, cephalexin and cephradine which did not result in binding to PBP 1 in membrane preparations. The effect of these antibiotics on the formation of dimers and higher oligomers was studied by digesting the peptidoglycan synthesised in a cell free incubation with a muramidase and separating the products chromatographically. In the control incubation the percentage incorporation into monomers, dimers and higher oligomers was 47, 38 and 15 respectively. These values approximate to the ratios of these components in digested cell walls of this organism. At the lowest concentration of β-lactam antibiotics having an observable effect on peptidoglycan synthesis cell wall dimers were synthesised at a similar rate as in control incubations without antibiotic but the formation of higher oligomers was inhibited. At higher concentrations both dimer and higher oligomer formation were affected (Table 1).

Table 1. Effect of Cefaclor on the Formation of Dimers and Higher Oligomers by Toluenised preparations of B. megaterium and Reversal of its Inhibitory effects.

	Control	Cefaclor (μg/ml)		Cefaclor 1 μg/ml at -10 min, followed by β-lactamase at t=0			
		0.1	1	7.5min	15min	45min	120min
Monomers	47	54	70	62	55	49	47
Dimers	38	37	26	31	36	39	39
Higher oligomers	15	9	4	7	9	12	14

The values in the table represent the percentage incorporation into the cell wall fractions.

In further experiments the ability of toluenised preparations to recover from the inhibitory effects of pretreatment with cefaclor, cephalexin or cephradine was tested by destroying excess antibiotic with β-lactamase following initial binding for 10 min and then adding the radioactive cell wall precursors. The ability to attach nascent to existing peptidoglycan was recovered very rapidly with all three antibiotics with a half-life less than 5 min at 23°. This is not necessarily a true indication of the half-life of any PBP which catalyses the attachment of nascent peptidoglycan to pre-existing wall since it only requires a single enzymic event to link a long chain of nascent peptidoglycan to the wall and such a reaction could occur if the enzyme was only partially active.

A kinetic investigation of the production of dimers and higher oligomers under the same conditions revealed that the recovery of the ability to form cell wall dimers occurred rapidly. The percentage of radioactivity incorporated into dimers had reached the value in the control incubation within 15 min. However, it required a considerably longer time for higher oligomer formation to approach the value in the control (Table 1). The results with cephradine and cephalexin were similar to those obtained with cefaclor.

These results suggest that two different transpeptidases may be involved in the synthesis of dimers and higher oligomers - one, involved in higher oligomer formation, is fairly sensitive to β-lactam antibiotics and takes longer to recover from the inhibitory effects of β-lactams than the other, involved in dimer formation, which is less sensitive and recovers rapidly from inhibition when excess antibiotic is removed. It is possible that the second enzyme is the attachment transpeptidase.

It is also possible to explain the results on the basis of a single transpeptidase. The formation of cell wall oligomers is almost certainly dependent on the prior formation of dimers and the consequence of interfering with the relative topology of the newly incorporated glycan chains and a transpeptidase enzyme might mitigate against higher oligomer synthesis until dimer formation had completely recovered from inhibition. If this were so one would expect higher oligomer formation to be more sensitive to β-lactam antibiotics than dimer formation and to recover more slowly from inhibition. It will require pulse-chase experiments carried out _in vivo_ to distinguish between these two possibilities.

Affinity of penicillin-binding proteins for cefaclor and cephalexin. A previous investigation had indicated that the relative affinities of PBPs for β-lactams in protoplast membranes and in growing cells were greatly dependent on the experimental conditions, particularly the concentration of membrane protein and the presence of β-lactamase in intact cells (7). The present studies confirmed these observations (Table 2). Labelling with cefaclor or cephalexin indicated that binding to PBP 1 was unlikely to be responsible for the cessation of cell growth since the MIC value was considerably lower than the affinity of PBP 1 in intact cells. Labelling of protoplast membranes in thick suspension (comparable to the concentration of membranes in toluenised cell preparations) indicated that PBP 1 had a much lower affinity than in intact cells both to cefaclor and cephalexin (Table 2). At the concentrations

Table 2. Binding of Cefaclor and Cephalexin to Penicillin-binding Proteins of _Bacillus megaterium_ and comparison with MIC values.

	Cefaclor		Cephalexin	
	Intact cells	Membranes	Intact cells	Membranes
50 % Binding (µg/ml)				
to: PBP 1	0.1	> 1	1	300
PBP 2	0.1	0.003	0.3	3
PBP 3	0.01	0.03	0.03	0.3
PBP 4	<0.003	0.003	0.01	0.1
PBP 5	>10	> 1	>100	>300
MIC (µg/ml)	0.01 - 0.03		0.03 - 0.1	

50% binding values were determined by incubating growing cells or membranes with varying concentrations of unlabelled cephalosporin for 10 min followed by saturation of the remaining PBP molecules with [^{3}H]-benzylpenicillin.
MIC values were determined on exponentially-growing cells. The lower value given permitted growth at 50% of the control rate. The higher value completely inhibited growth.

of these antibiotics which partially inhibited the formation of higher oligomers and of dimers in toluenised preparations no binding to PBP 1 would have occurred. It therefore appears likely that at least one of the other PBPs, 2, 3 and 4 has transpeptidase activity which functions in the synthesis of higher oligomers. PBP 5 is excluded from consideration on the basis of its low affinity for cephalosporins (Table 2). The high affinity of cefaclor for PBP 4 in intact cells and in membrane preparations, with 50% binding values lower than the concentrations which inhibit transpeptidation, make it unlikely that this protein is a transpeptidase that is being studied in these experiments.

Location of Penicillin-binding Proteins. In order for a membrane-bound PBP to function as a transpeptidase in peptidoglycan biosynthesis it is likely that the active site will be located in a protein domain outside the membrane. Consequently, labelling studies with impermeable labelling reagents should label such PBPs if they contain reactive groups in their external domains that are avilable to the reagent. Labelling of protoplasts with the iodine/lactoperoxidase method followed by fractionation of the membrane proteins on sodium dodecylsulphate polyacrylamide gels indicated that PBPs 1 and 4 had areas exposed to the exterior of the cell. This observation was confirmed by labelling intact cells with isethionyl acetimidate - the same two PBPs were labelled together with a number of other membrane proteins (Soar and Reynolds, unpublished observations). Labelling of PBPs 2, 3 and 5 could not be detected in intact cells or protoplasts though small amounts of label reacted with PBPs 3 and 5 when protoplast membranes were treated with the reagents. The labelling of these two proteins was only detected by subjecting a detergent-solubilised extract of isethionyl acetimidate labelled membranes to covalent affinity chromatography on penicillin-sepharose. PBPs 1 and 4 were heavily labelled and were not absorbed to the affinity column but PBPs 3 and 5 bound covalently to the column and were subsequently eluted with neutral hydroxylamine. It cannot be assumed that PBPs 3 and 5 do not project beyond the external surface of the cell membrane: indeed there is evidence to the contrary for PBP 5 in that the majority of the total D,D-carboxypeptidase activity can be detected in stabilised protoplasts and thus the active site must be outside the cell membrane. PBPs 3 and 5 may not possess many available reactive sites or it is possible that they are located in the region of the cell septum in which case they might be protected from labelling in intact cells. It remains a possibility that PBP 3 is a transpeptidase as proposed by Rodriguez-Tebar et al. (4).

References

1. Chase, H.A., Shepherd, S.T., Reynolds, P.E.: FEBS Lett. 76, 199-203 (1977).
2. Chase, H.A.: J. Gen. Microbiol. 117, 211-224 (1980).
3. Reynolds, P.E., Shepherd, S.T., Chase, H.A.: Nature 271, 568-570 (1978).
4. Rodriguez-Tebar, A., Rojo, F., Vazquez, D.: Eur. J. Biochem. 126, 161-166 (1982).
5. Schrader, W.P., Fan, D.P.: J. Biol. Chem. 249, 4815-4818 (1974).
6, Giles, A.F., Reynolds, P.E.: FEBS Lett. 101, 244-248 (1979).
7. Chase, H.A., Reynolds, P.E.: FEMS Microbiol. Lett. 10, 285-289 (1981).

FUNCTION OF PENICILLIN-BINDING PROTEIN 3 IN *STREPTOCOCCUS FAECIUM*

Jacques Coyette, Anne Somzé, Jean-Jacques Briquet, Jean-Marie Ghuysen

Université de Liège, Faculté de Médecine, Service de Microbiologie, Institut de Chimie, B6, B-4000 Sart Tilman, Liège, Belgium

Roberta Fontana

Università di Padova, Istituto di Microbiologia, I-35121 Padova, Italy

Introduction

The membranes of *Streptococcus faecium* (*S. faecalis* ATCC 9790) contain a maximum of eight penicillin-binding proteins (PBPs) (1). Among them, only the protein with the smallest molecular weight (43,000 Mr) has been shown to possess an enzymatic activity. It is a DD-carboxypeptidase (2). Up to now, no physiological function could be attributed to any of these membrane proteins.

The results presented in this paper tend to show that PBP3 is involved in cell septation and PBP2 and/or PBP6 are implicated in pole formation.

Results

At 37°C in a rich medium, cefotaxime, a recently introduced cephalosporin (3), has the same inhibitory potency on *S. faecium* as benzylpenicillin (MIC ≅ 4 μM) but these two β-lactams have different MBC values : 420 μM for cefotaxime and 8 μM for benzylpenicillin. Hence, under these growth conditions, *S. faecium*

The Target of Penicillin

shows tolerance for cefotaxime. When the growth temperature is reduced (32°C), the MIC value of cefotaxime is increased to 1000 µM and the tolerance phenomenon disappears. Under identical conditions, benzylpenicillin shows almost unchanged MIC and MBC values (8 and 16 µM , respectively).

At 37°C in a rich medium, cefotaxime, at the MIC, inhibits completely *S. faecium* growth only 2 hrs after the beginning of the treatment. The inhibition is temporary and growth resumes. At a concentration corresponding to 20 × MIC, cefotaxime induces a slow lysis.

Morphology is deeply influenced by cefotaxime treatment. At concentrations around the MIC (5 µM) or below (0.1 to 1 µM), the normal coccal cells (Fig. 1,A) are transformed in bacilliform cells whose length increases as the duration of the treatment increases (Fig. 1,B). The diameter of the treated cells is also increased. Thin sections of such cells, examined by transmission electron microscopy, show that septation is inhibited (Fig. 2). Septa are initiated but never reach completion. In addition, constrictions similar to those seen in normal cells (Fig. 3) are not observed.

Affinity of the PBPs (of isolated membranes and intact cells) for cefotaxime has been determined by direct binding experiments using [^{14}C]cefotaxime or by competition experiments between non radioactive cefotaxime and radioactive [^{14}C]benzylpenicillin. Affinity is defined as the concentration permitting 50 % saturation. Both techniques yield essentially the same results (Table 1). Cefotaxime binds preferentially to the three highest molecular weight proteins. PBPs 2 and 3 are about 100-fold more sensitive than PBP1 and therefore must be involved in cell septation. This is confirmed by time course experiments using cefotaxime at subinhibitory concentration (1 µM). From cell samples, collected 30 and 60 min after addition of cefotaxime, membranes are isolated, labelled with saturating [^{3}H]benzylpenicillin and examined by fluorography (Fig. 4,B). Under these conditions, only PBPs 2 and 3 are saturated by

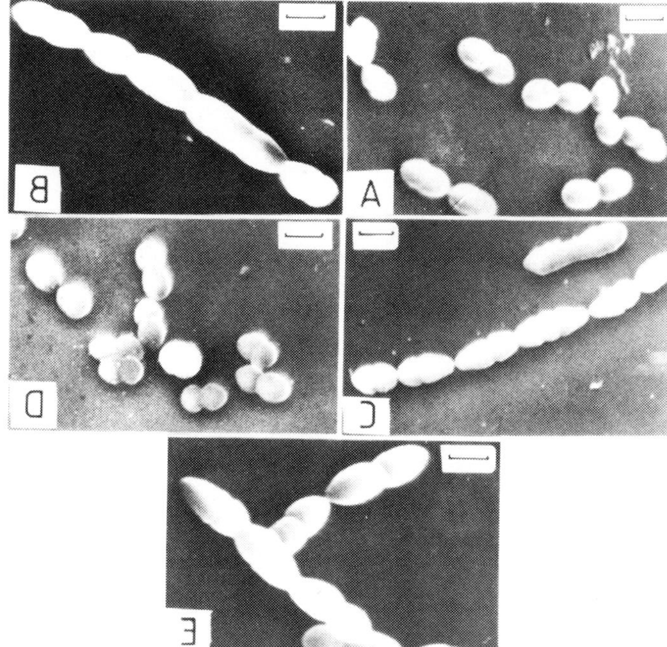

Figure 1. Scanning electron micrographs of untreated wild type cells (A); cefotaxime-treated (120 min) wild type cells (B); cefoxitin - treated (120 min) wild type cells (C); untreated $NT_1/20$ mutant cells (D); cefotaxime-treated (120 min) $NT_1/20$ mutant cells (E). The bar represents 1 μM .

Table 1. Affinity of the PBPs

PBPs	Cefotaxime	Cefoxitin	Benzylpenicillin
1	11.0	4.8	0.4
2	0.05	0.05	0.24
3	0.13	3	0.08
4	1300	215	1.2
5	—	85	> 10
4*	1300	—	1.0
6	2500	0.50	1.3

Results expressed as the concentrations (μM) required to achieve 50 % binding to the proteins. They represent the mean value of several experiments.

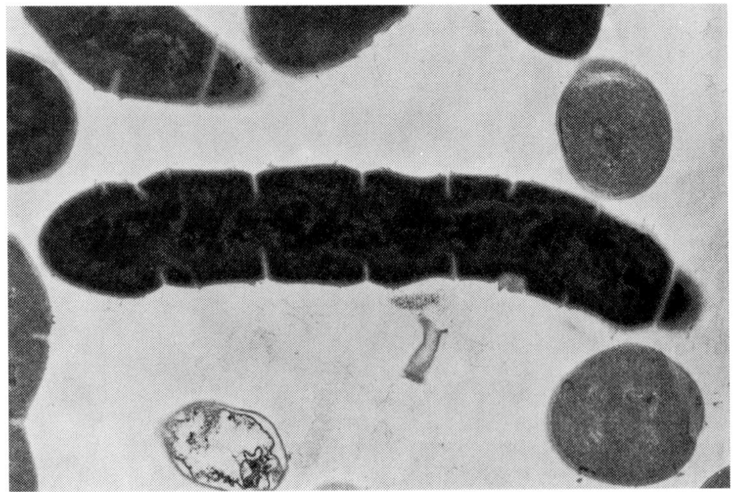

Figure 2. Thin sections of cefotaxime treated (90 min) wild type cells of *S. faecium*.
The bar represents 1 µM .

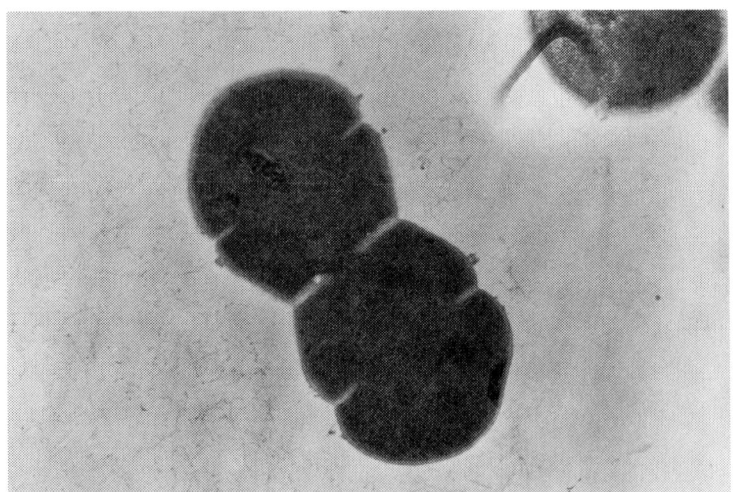

Fig. 3. Thin sections of untreated wild type cells of *S. faecium*.
The bar represents 1 µM .

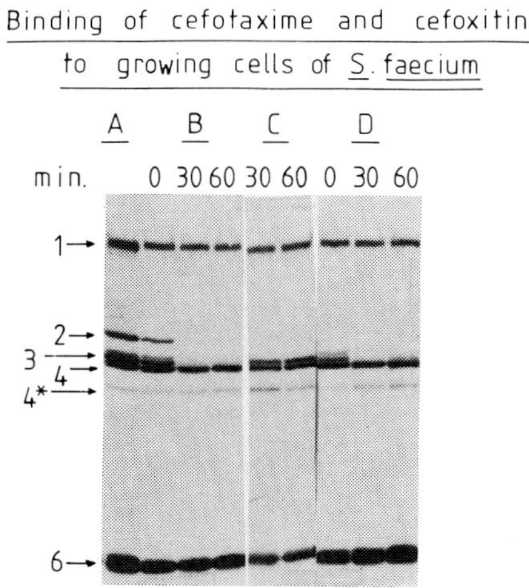

A : isolated membrane
B : wild type cells + cefotaxime
C : wild type cells + cefoxitin
D : mutant cells + cefotaxime

Figure 4. A : isolated membranes; B : wild type cells + cefotaxime; C : wild type cells + cefoxitin ; D : $NT_1/20$ mutant cells + cefotaxime.

cefotaxime. At this stage, no distinction can be made between the two proteins; either both or only one of them may be involved in cell division.

As shown by binding experiments, carried out on both isolated membranes (1) and intact cells (Table 1), PBP2, PBP6 and then PBPS 3 and 1 have, in the indicated order, decreasing affinities for cefoxitin. PBPs 5 and 4 are the most resistant ones. The MIC and MBC values are 50 and 500 μM , respectively (1). At the MIC, cefoxitin inhibits completely cell growth and induces a very slow lysis. Below the MIC, growth inhibition decreases gradually as the antibiotic concentration decreases. Above the MIC, cell lysis is more rapidly induced as the cefoxitin concentration increases.

The morphological abnormalities induced by cefoxitin, at a concentration (0.5 μM) **below** the MIC, differ from those induced by cefotaxime (1 μM). As shown by scanning and transmission electron microscopy, the cefoxitin-treated cells have an increased diameter and are slightly elongated (Fig. 1,C). But the most stricking alteration is the frequent presence of conical poles contrasting with the round poles observed in control cells (Fig. 1,A).

Analysis of membranes isolated from cells grown for 30 and 60 min in the presence of a subinhibitory concentration of cefoxitin (0.5 μM) shows that PBPs 2 and 3a are completely saturated while PBP6 is partially saturated. The other PBPs are not affected. Thus, the morphological alterations observed in cefoxitin-treated cells can be attributed only to the inhibition of the functioning of one or several of these 3 PBPs.

Elongated cells similar to those obtained with cefotaxime are never found in cultures treated with cefoxitin at concentrations sufficient to saturate PBP2. As already suggested above, PBP6 does not play any direct role in the inhibition of cell septation. The main difference between cefotaxime- and cefoxitin-treated cells is that PBP3 is saturated by cefotaxime but completely untouched by cefoxitin. Thus, septation inhibition must be due to the interaction of cefotaxime with PBP3.

A constitutive (NT_1/20) mutant of *S. faecium* devoid of PBP2 grows at 32°C more slowly than the wild type strain (generation times : 42 and 36 min , respectively). At 37°C, the generation time is 60 min and the mutant does not grow at all at 42°C. It may be a double mutant because revertants capable of growing at 42°C as the wild type does have been isolated. They still lack PBP2. At 32°C, the NT_1/20 mutant has a normal morphology but it is slightly smaller than the wild type cells (Fig. 1,D). Thus the absence of PBP2 seems not to cause any important morphological alteration. Consequently, PBP2 appears not to have any vital function in *S. faecium*.

When treated with cefotaxime (1 μM), the $NT_1/20$ mutant grows as elongated cells similar to those observed with wild type treated cells (Fig. 1,E). Analysis of the PBPs in cefotaxime-treated cultures of the $NT_1/20$ mutant shows that only PBP3 is saturated by the antibiotic and is uncapable of binding [^{3}H]benzylpenicillin (Fig. 4,D).

Discussion

PBP3 is critical for cell division probably at the level of cell septation. Its function can be related to that of PBP3 in *Escherichia coli* (4) which is a DD-transpeptidase specifically required for cell septation (5).

PBP2 has no obvious role although it may be somehow involved in cell size control. Its function could be taken over by one or several of the other PBPs (6). When the functioning of PBP2 and/or PBP6 is inhibited by a β-lactam such as cefoxitin, cells of *S. faecium* are induced to form conical poles.

Acknowledgements

This work has been supported in part by the US National Institutes of Health (grant AI 13364-05); the FRSM, Brussels (grant 3.4501-79), the Belgian Government (*Action concertée* 79/84-I1) and the FNRS, Brussels (*crédit aux chercheurs*).
The $NT_1/20$ mutant is a gift of Dr. P. Canepari from the Istituto di Microbiologia, Università di Genova, Genova, Italy.

References

1. Coyette, J., Ghuysen, J.M., Fontana, R. : Eur. J. Biochem. 110, 445-456 (1980).
2. Coyette, J., Ghuysen, J.M., Fontana, R. : Eur. J. Biochem. 88, 297-305 (1978).
3. Schrinner, E., Limbert, M., Heymes, R., Dürckheimer, W. : *in* β-Lactam Antibiotics (S. Mitsuhashi, ed.), Japan Scientific Societies Press, Tokyo, pp. 120-127 (1981).

4. Spratt, B.G. : Eur. J. Biochem. 72, 341-352 (1977).
5. Ishino, F., Matsuhashi, M. : Biochem. Biophys. Res. Commun. 101, 905-911 (1981).
6. Hayes, M.V., Curtis, N.A.C., Wyke, A.W., Ward, J.B. : FEMS Microbiol. Lett. 10, 119-122 (1981).

THE ROLE OF A PROTEIN THAT BINDS PENICILLIN WITH SLOW KINETICS IN PHYSIOLOGY AND RESPONSE TO PENICILLIN OF *STREPTOCOCCUS FAECIUM* ATCC 9790.

Roberta Fontana
Istituto di Microbiologia, Università di Padova, Padova, Italy.

Piero Canepari
Istituto di Microbiologia, Università di Genova, Genova, Italy.

Giuseppe Satta
Istituto di Microbiologia, Università di Cagliari, Cagliari, Italy.

Introduction

The beta-lactams are known as the least toxic and most valuable antibiotics available. Their use is limited by both the production of beta-lactamases and the emergence of resistant strains non producers of these enzymes (intrinsically resistant).
The development of beta-lactamase resistant beta-lactam derivatives is currently an area of intensive and successful research. The number of clinical isolates which are intrinsically resistant to beta-lactams is continously increasing particularly in *staphylococci*, *neisseriae*, *pseudomonas*, *pneumococci*, and *group D streptococci* (1 - 6). For this reason the development of beta-lactams capable of overcoming both acquired and natural intrinsic resistance is desirable. The first step towards this goal is the clear understanding of the biochemical basis of intrinsic resistance. In this paper we show that in *Streptococcus faecium* ATCC 9790 intrinsic resistance is due to an overproduction of a penicillin binding protein (PBP) that has the novel property of binding the antibiotic with slow kinetics (SKPBP) and can take over the function of other PBPs with a higher affinity for the antibiotic.

The Target of Penicillin

Results

General properties of penicillin resistant mutants. Spontaneous mutants resistant to different penicillin concentrations were isolated from *S.faecium* ATCC 9790 by serial passages in medium containing increasing antibiotic concentrations. Such mutants were designed as R5, R20, and R40. Table 1 reports some properties of these strains which had been described elsewhere (R. Fontana, submitted for publication). These mutants apparently showed no differences in respect to the parent in PBP electrophoretic patterns, permeability and PBP affinity for penicillin.

Identification of PBP 5 as a slow kinetics penicillin binding protein. In order to obtain more complete information on PBPs, we have modified the technique previous described (7) which did not allow, in our experiments, PBP 5 detection (see also Table 1). By increasing incubation time of membranes with the antibiotic we have found that PBP 5 bound penicillin with kinetics much slower than the other *S.faecium* PBPs, being saturated

Table 1. Some properties of penicillin resistant mutants.

Strain[1]	MIC[2] μg/ml	MBC[3] μg/ml	Total amount of ^{3}H-penicillin bound by PBPs[4] in		Concn. (μM) of ^{3}H-penicillin required to bind 50% of PBP				
			membranes	whole cells	1	2	3	4	6
ATCC	0.25	5	914	925	0.9	0.5	0.06	2.1	1.6
R5	20	20	980	960	1.0	0.5	0.06	2.1	1.5
R20	40	40	920	915	1.1	0.5	0.07	2.1	1.8
R40	80	80	975	960	1.1	0.6	0.07	2.1	1.6

1. All strains were grown in SB medium (7) at 37°C.
2. Minimum inhibitory concentration (MIC) was determined by a two fold dilution technique.
3. Minimum bactericidal concentration (MBC) was the minimum concentration which killed 99.9% of the cells after 18 h incubation at 37°C.
4. PBP detection was performed by using the technique described by Coyette, Ghuysen and Fontana (1980). ^{3}H-benzyl-penicillin, ethylpiperidinium salt (31 Ci/m mol) was used instead of ^{14}C-benzyl-penicillin. Under these conditions PBP 5 was not found. The amount of radioactivity bound by PBPs was calculated by measuring the band intensity with a densitometer. Values were expressed in arbitrary OD units.

only after 60 min of incubation. For this reason it was called SKPBP. When a binding time of 60 min was used, all the resistant strains showed an increased amount of PBP 5 which paralled the increase in penicillin resistance (Fig. 1). No differences were found in PBP penicillin affinity of resistant strains in respect to the parent by using the modified technique. PBP 5 resulted to be the less sensitive of S.faecium PBPs (data not shown).

Interaction of penicillin with PBPs in growing cells. To confirm a possible role of PBP 5 overproduction in penicillin resistance of S.faecium, we have carried out an experiment in which exponentially growing bacteria were first exposed to various concentrations of penicillin and subsequently to a saturating dose of ^{3}H-penicillin. Fig. 2A shows that in parental strain the penicillin MIC saturated PBPs 1, 2 and 3 but not PBPs 4, 5 and 6. In contrast, in R40 a penicillin concentration much lower than the MIC (5 µg/ml) saturated all PBPs, except PBP 5. In resistant strain PBP 5 was saturated only by a penicillin concentration equal to the MIC (Fig. 2B). Neverthless both in ATCC and in R40 the concentration that saturated PBP 5 was equal to MBC (see also Table 1).

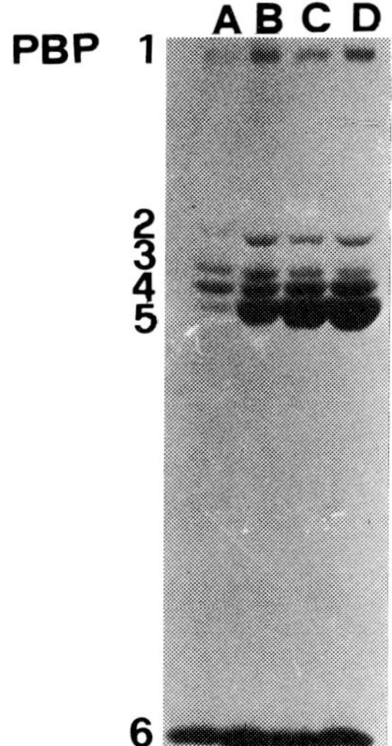

Fig. 1. PBPs of penicillin resistant mutants.
PBPs were detected in isolated membranes essentially as described previously (7) with the only exception that membranes were incubated with ^{3}H-penicillin for 60 min. A: ATCC 9790; B: R5; C: R20; D: R40.

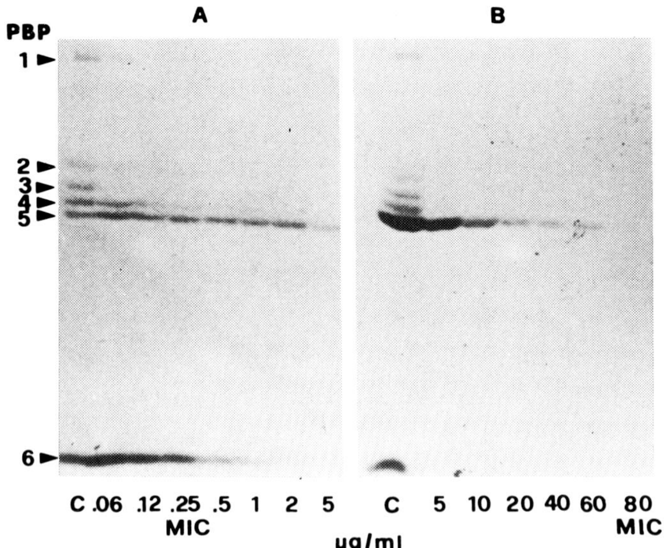

Fig. 2. Interaction of penicillin with PBPs in growing cells. Exponentially growing cells (10 ml, about 1×10^8 bacteria/ml) were incubated with various penicillin concentrations at 37°C for 60 min, centrifuged and resuspended in 100 μl of 0.01 M phosphate buffer pH 7. Cell suspension were then incubated with an amount of ^{3}H-benzyl-penicillin sufficient to saturate the PBPs of untreated organisms (100 μM) for another 60 min. The control samples were preincubated with no antibiotic before exposure to the radioactive penicillin. Membrane isolation, electrophoresis and PBP detection were then carried out as described (7). A: ATCC 9790. B: R40.

Discussion

This study has lead to the discovery that PBP 5, a PBP which shows very peculiar properties, is responsible for penicillin intrinsic resistance in S.faecium. This conclusion is supported by the fact that resistance to penicillin is consistently associated with the overproduction of PBP 5 and that this protein is the target for growth inhibition in resistant mutants, but not in the parent. In fact in the parent the penicillin MIC saturated PBPs 1, 2 and 3 but not PBPs 4, 5 and 6. On the contrary the resistant strain continued to grow normally in the presence of penicillin concentrations which saturated all PBPs, other than PBP 5, and stopped growing only in the presence of concentrations saturating PBP 5. In addition, the finding that in both strains MBC corresponded to the minimum concentration required to saturate PBP 5 supports the hypothesis that this protein is the killing target for the drug in *S.faecium*.

The fact that cells continue to grow when all PBPs other PBP 5 are saturated by penicillin suggests that either the binding of penicillin does not necessarily inhibit PBP functions or that under certain conditions the activity of one single PBP is sufficient for the cell to grow. These results confirm previous observations showing that the target of penicillin inhibitory action in *S.faecium* can vary depending on conditions of cell growth (8, R. Fontana: J. Bacteriol., 1983, in press) and strongly support the previous proposal that each PBP does not consistently perform a specific given function, but that they interact in a complex way and the function they perform can be influenced by the physiological status of the cells.

Based on the data shown here we propose that the overproduced PBP 5 takes over the essential cell growth functions that in sensitive strains are performed by other PBP(s). Inhibition of growth by penicillin requires the saturation of PBP 5. Due to both the very slow kinetics of binding and the very low affinity of this PBP for penicillin, high concentrations of antibiotic are required to stop the growth of cells overproducing it. This is the biochemical basis of penicillin intrinsic resistance in *S.faecium*.

Resistant strains which overproduce one of the PBPs have already been described in *Bacillus megaterium*, *Staphylococcus aureus*, and Pseudomonas (4, 9, 10). In the former case it was suggested that the overproduced PBP takes over the function of a PBP previously identified as the penicillin lethal target. Some similarities exist between such previously proposed mechanism of intrinsic resistance and that here suggested for *S.faecium*.

Bacteria that are killed by a penicillin concentration many times higher than the MIC are considered to be tolerant to the drug. If, as in *S.faecium*, also in other bacteria, the targets for growth inhibition and that for killing by penicillin reside in different PBPs, then these bacteria could show tolerance for the drug when penicillin affinity or binding kinetics of the two targets are very different.

Acknowledgements

We wish to thank Dr. J.D. Cassidy of Merck and Co., Inc., Rahway, N.J. for the very generous gift of ^{3}H-benzyl-penicillin.

References

1. Sabath, L.D.: Microbiology - 1979, D. Schlessinger, American Society for Microbiology, Washington D.C., 1979.

2. Hartman, B., Tomasz, A.: Antimicrob. Agents Chemother. 19, 726-735, 1981.

3. Dougherty, T.J., Koller, A.E., Tomasz, A.: Antimicrob. Agents Chemother. 18, 730-737, 1980.

4. Godfrey, A.J., Bryan, L.E., Rabin, H.R.: Antimicrob. Agents Chemother. 19, 705-711, 1981.

5. Hakenbeck, R., Tarpay, M., Tomasz, A.: Antimicrob. Agents Chemother. 17, 364-371, 1980.

6. Moellering, R.C., Jr., Krogstad, D.J.: Microbiology - 1979, D. Schlessinger, American Society for Microbiology, Washington D.C., 1979.

7. Coyette, J., Ghuysen, J.M., Fontana, R.: Eur. J. Biochem. 110, 445-456 1980.

8. Fontana, R., Canepari, P., Satta, G., Coyette, J.: Nature, 287, 70-72, 1980.

9. Giles, A.F., Reynolds, P.E.: Nature 280, 167-168, 1979.

10. Brown, F.J., Reynolds, P.E.: FEBS Lett. 122, 275-278, 1980.

TRANSPEPTIDATION IN STAPHYLOCOCCUS AUREUS WITH INTRINSIC RESISTANCE TO β-LACTAM ANTIBIOTICS ("METHICILLIN RESISTANCE")

Derek F.J. Brown

Clinical Microbiology and Public Health Laboratories,
New Addenbrooke's Hospital, Cambridge, England

Peter E. Reynolds

Department of Biochemistry, University of Cambridge,
Cambridge, England

Introduction

Strains of S. aureus with heteroresistance to β-lactam antibiotics produce large amounts of a penicillin-binding protein (PBP) with low affinity for β-lactam antibiotics (1). This PBP, which has an M_r on SDS-PAGE gels similar to PBP3 in sensitive strains and may be an altered PBP3, is produced only when the organism is grown under conditions in which resistance is expressed and is not produced under any conditions in sensitive strains. A similar low-affinity PBP has been reported in a strain of S. aureus which was methicillin-resistant under all growth conditions (2). PBPs are thought to be enzymes involved in metabolism of peptidoglycan although PBP4 is the only purified PBP in S. aureus shown to have transpeptidase activity (3). There is also evidence that PBP4 has transpeptidase activity in-vivo (4). Transpeptidation in S. aureus with intrinsic resistance to β-lactam antibiotics might be expected to have similar sensitivity to β-lactams as the "resistant" PBP3 produced in these strains. In addition, differences in sensitivity of transpeptidation to β-lactam inhibition might be expected between organisms grown in conditions favourable or unfavourable for the expression of resistance.

The Target of Penicillin

Methods

Penicillinase-negative, methicillin-resistant ($13136p^{-}m^{+}$) and methicillin-sensitive ($13136p^{-}m^{-}$) strains of S. aureus were examined. Growth conditions favourable for the expression of resistance were incubation at 30°C in medium containing 5% (w/v) NaCl, and conditions unfavourable for the expression of resistance were incubation at 40°C with no NaCl in the medium. In 'transpeptidase' assays (5) chloramphenicol (50 µg/ml) was added to late log-phase cultures. After 10 min [^{14}C]-glycine was added to the cultures with or without β-lactam antibiotics. At intervals samples were treated with an equal volume 10% (w/v) trichloroacetic acid at 90°C for 10 min. Precipitated material was collected by centrifugation, resuspended in 4% (w/v) sodium dodecyl sulphate and boiled for 20 min. The residue, cross-linked peptidoglycan, was collected by filtration on Whatman GF/C paper, and the radioactivity counted. This assay measures newly-synthesized peptidoglycan that has been covalently attached to 'old' wall material, presumably by transpeptidation since incorporation in sensitive strains is completely inhibited by low concentrations of β-lactam antibiotics.

Results and discussion

Transpeptidation in β-lactam-sensitive S. aureus $13136p^{-}m^{-}$ was largely inhibited by 0.1 µg benzyl penicillin/ml when the organism was grown in conditions favourable (Figure 1c) or unfavourable (Figure 1d) for the expression of resistance. The MIC of benzyl penicillin for the strain and the affinities of PBPs 1, 2 and 3 for benzyl penicillin (1) were similar to the concentration inhibiting transpeptidation. Similarly with methicillin and cephaloridine (data not shown), the concentrations of antibiotics inhibiting transpeptidation were similar to the concentrations inhibiting growth of the

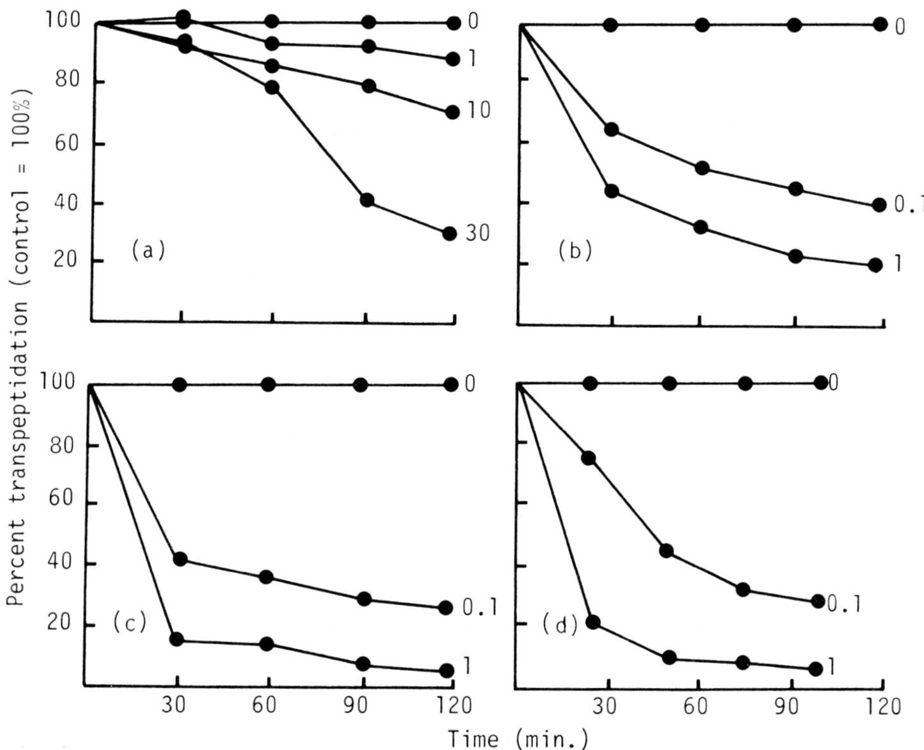

Figure 1. Inhibition of transpeptidation by benzyl penicillin. S. aureus 13136p$^-$m$^+$ grown under conditions favourable (a) and unfavourable (b) for the expression of resistance. S. aureus 13136p$^-$m$^-$ grown under conditions favourable (c) and unfavourable (d) for the expression of resistance. Concentrations of benzyl penicillin (μg/ml) are indicated against curves.

organisms, and the affinities of PBPs 2 and 3 for the antibiotics.

When S. aureus 13136p$^-$m$^+$ was grown under conditions favourable for the expression of resistance, transpeptidase activity was resistant to inhibition by benzyl penicillin (Figure 1a) to a degree similar to the MIC of benzyl penicillin for the organism (16 μg/ml). When grown under these conditions S. aureus 13136p$^-$m$^+$ produces a PBP3 with an affinity for benzyl penicillin >10 μg/ml. When grown under conditions where resistance is not expressed transpeptidase activity in S. aureus 13136p$^-$m$^+$ (Figure 1b) was almost as sensitive to inhibition by benzyl penicillin as in the sensitive strain. Under these conditions the MIC and the affinities of PBPs for benzyl penicillin were similar to those for the sensitive strain. Similar results were obtained with S. aureus 13136p$^-$m$^+$ with methicillin and cephaloridine (results not shown), although the concentrations required to inhibit transpeptidation when the organism was grown under conditions favourable for the expression of resistance were about 4-fold higher than the MICs of the antibiotics. This discrepancy may be due to the concentration of organisms in the transpeptidase assays being higher than those used in the MIC determinations.

The results suggest that the PBP3 detected in S. aureus 13136 p$^-$m$^+$ only when the organism is grown under conditions favourable for the expression of resistance is probably the enzyme responsible for the β-lactam-resistant transpeptidase activity described here. The limited resistant transpeptidase activity detected when S. aureus 13136p$^-$m$^+$ was grown under conditions unfavourable for the expression of resistance may reflect the production of a small amount of the enzyme insufficient to be detected on fluorograms or to allow growth of the organism in the presence of higher concentrations of β-lactam antibiotics.

Transpeptidation in the presence of chloramphenicol would be predominantly due to cell wall thickening. If the β-lactam-resistant PBP3 is the enzyme responsible for the β-lactam-resistant transpeptidation described here it might also be expected to be a septal transpeptidase as it is the only PBP detected which is not saturated by β-lactam antibiotics at concentrations well above those inhibiting the growth of sensitive strains. Using methods not requiring chloramphenicol-inhibition of cultures, Smith and Wilkinson (6) reported results which suggested that septal and wall-thickening peptidoglycan synthesis are distinct in the methicillin-resistant strain of

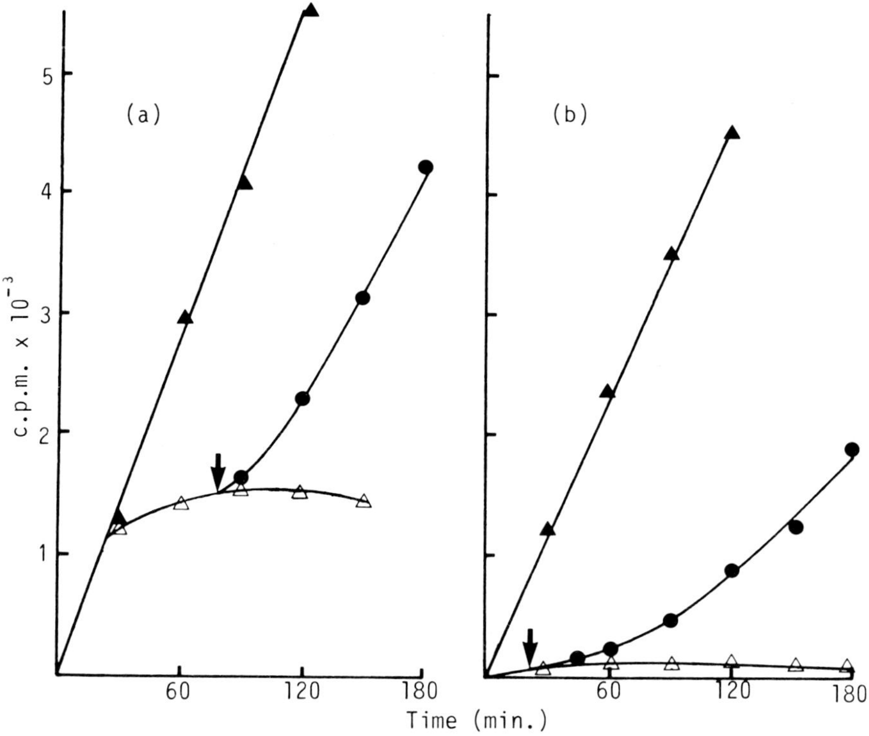

Figure 2. Recovery of transpeptidation in S. aureus strains 13136p⁻m⁺ (a) and 13136p⁻m⁻ (b) grown under conditions favourable for the expression of resistance. ▲ control, no benzyl penicillin; △ benzyl penicillin 30 μg/ml (a) and 10 μg/ml (b); ● recovery from inhibition by benzyl penicillin following the addition of β-lactamase at the point indicated by the arrow.

S. aureus that they studied. They indicated that the septal transpeptidase might be resistant to inhibition by methicillin while the wall-thickening transpeptidase remained sensitive. "Intermediate" results were obtained with chloramphenicol-inhibited cultures.

If transpeptidation in chloramphenicol-inhibited organisms is inhibited by a β-lactam antibiotic and the antibiotic subsequently removed by centrifugation and resuspension of cells in β-lactam-free medium and/or the addition of β-lactamase, transpeptidation recovers as the PBP-β-lactam complex breaks down. Recovery was not always complete in that the cells did not regain the same rate of $[^{14}C]$-glycine incorporation as the β-lactam-free control. When grown under conditions where resistance was expressed the resistant strain 13136p^-m^+ recovered from inhibition of transpeptidase activity by benzyl penicillin considerably more rapidly than the sensitive strain 13136p^-m^- (Figure 2). Similar results were obtained with methicillin and cephaloridine (data not shown). Rapid dissociation of the PBP-β-lactam complex might ensure that a proportion of the transpeptidase present remains free to perform its enzymic function even when higher concentrations of β-lactam antibiotics are present.

Thus it appears that when grown under conditions favourable for the expression of resistance the intrinsically resistant strain of *S. aureus* investigated produces a transpeptidase which, relative to that produced by its sensitive variant, is resistant to inhibition by β-lactam antibiotics. It also dissociates rapidly from β-lactam antibiotics.

References

1. Brown, D.F.J., Reynolds, P.E.: FEBS Lett. 122, 275-278 (1980).
2. Hayes, M.V., Curtis, N.A.C., Wyke, A.W., Ward, J.B.: FEBS Microbiol. Lett. 10, 119-122 (1981).
3. Kozarich, J.W., Strominger, J.L.: J. Biol. Chem. 253, 1272-1278 (1978).
4. Wyke, A.W., Ward, J.B., Hayes, M.V., Curtis, N.A.C.: Eur. J. Biochem. 119, 389-393 (1981).
5. Reynolds, P.E., Chase, H.A.: β-lactam Antibiotics, Ed. M. Salton and G.D. Shokman, Academic Press, New York, 153-158 (1981).
6. Smith, P.F., Wilkinson, B.J.: J. Bact. 148, 610-617 (1981).

PENICILLIN-SENSITIVE ENZYMES IN *STAPHYLOCOCCUS AUREUS*

Anne W. Wyke
National Institute for Medical Research, Mill Hill, London NW7 1AA, U.K.

J.Barrie Ward and Michael V. Hayes
Glaxo Group Research, Greenford, Middlesex UB6 OHE, U.K.

Membranes of *Staphylococcus aureus* contain 4 (or possibly 5) penicillin-binding proteins (PBPs). Mutants lacking either PBP1 or 4 have been isolated (1,2) suggesting that neither of these is a lethal target for β-lactam antibiotics and that PBPs 2 and/ or 3 must be the penicillin-sensitive enzymes responsible for the primary transpeptidation reaction in cell wall biosynthesis. No morphological changes have been described in the mutants lacking PBP1 or 4 and no enzyme activity has been attributed to PBP1. The degree of cross-linking of peptidoglycan was markedly reduced in strains of *S.aureus* which lack PBP4 and also in wild-type strains treated with cefoxitin such that PBP4 was effectively removed by irreversible binding of the antibiotic (3). These observations led to the conclusion that *in vivo* PBP4 acts as a secondary transpeptidase responsible for inter-peptide cross-linking.

Studies of the PBPs of a highly methicillin-resistant strain of *S.aureus* (strain MR-1) after growth in the presence of β-lactam antibiotics have confirmed the role of PBP4 as a secondary transpeptidase and have shown that either PBP2 or PBP3 remained resistant to high concentrations of methicillin (4). *In vivo* studies were not able to distinguish the physiological roles of these two PBPs and it has been suggested that they are able to compensate for each other.

S.aureus MR-1 is a strain which is highly resistant to

The Target of Penicillin

methicillin (and many other β-lactam antibiotics) but does not exhibit the characteristics of temperature-dependence or heterogeneity of expression usually associated with intrinsic methicillin-resistance. In a further attempt to clarify the roles played by PBPs 2 and 3 in peptidoglycan biosynthesis another methicillin-resistant strain, *S.aureus* 1006E, which has a very low level of PBP3 and a slightly reduced level of PBP4 when compared with *S.aureus* H, was studied. This strain is a clinical isolate and shows the typical features of heterogeneity of expression and temperature dependence (MGIC for methicillin is 640 μgml^{-1} at 30°C and 20μgml^{-1} at 37°C). The very low level of PBP3 in strain 1006E was unusual and several other methicillin-resistant strains examined showed similar PBP profiles to the sensitive strain, *S.aureus* H. The amino acid composition of the cell walls and the degree of cross-linking of the peptidoglycan of strain 1006E were similar to those of other staphylococci (4). When exponential-phase cultures of strain 1006E were treated with high concentrations (5xMGIC) of β-lactam antibiotics growth continued at the normal rate for 1-2 generations, followed by lysis and cell death. This was typical of the response of sensitive staphylococci to treatment with β-lactam antibiotics (5) and suggested that the autolytic enzymes were normal. Treatment with high concentrations of methicillin allowed continued growth for 1-2 generations and then the optical density remained constant for several hours. After further incubation for 18h the culture grew to a high optical density and the new population of organisms had an increased level of methicillin-resistance. Growth of strain 1006E for 4 generations in the presence of low concentrations of methicillin caused a large decrease in the degree of cross-linking of the peptidoglycan. These studies revealed no abnormal features in the cell walls of the organism which could be related to the very low level of PBP3.

Since the mutant *S.aureus* H28 lacks PBP1 and treatment with low levels of cefoxitin effectively removes PBP4 by irreversible binding (3) it should be possible to remove either PBP2 or 3 by selective binding of one of these PBPs to another β-lactam. The binding affinities of the PBPs of *S.aureus* H28 were measured for several β-lactam antibiotics. In most cases there were only small differences between the affinities of PBP2 and 3 for any particular β-lactam. The affinities were measured as the concentration of antibiotic that results in 50% saturation of the protein and these concentrations were used in experiments in which attempts were made to selectively bind PBP2 or 3 in *S.aureus* H28 treated with cefoxitin. Under these conditions, the organisms appeared to grow with normal morphology (as judged by phase-contrast microscopy) and with the expected reduction of peptidoglycan cross-linking resulting from inhibition of PBP4 by cefoxitin. However, the PBP profiles often showed the presence of more than one PBP and it was obviously not possible to achieve a situation where the organisms were growing with only one PBP functionally active. This is not altogether surprising since 50% saturation of any given protein may not automatically lead to 50% inhibition of its enzymic activity. Efficient functioning of a very small percentage of the total enzyme protein may be sufficient for normal biosynthesis of the cell wall.

The biosynthetic targets of penicillin in *S.aureus* have also been studied in an *in vitro* system in which the synthesis and cross-linking of peptidoglycan were measured in organisms incubated in a "wall synthesis" minimal medium. When mid-exponential phase organisms were resuspended in phosphate-buffered medium containing glucose and cell wall amino acids, either radioactive lysine, glycine or N-acetylglucosamine were incorporated into peptidoglycan. Incorporation of radioactivity was linear with time for about 3h and continued to rise slowly thereafter to reach a plateau level. To measure peptidoglycan synthesis samples were taken into 5%TCA at 0°, filtered onto glass fibre discs, washed extensively and the radioactivity counted. Further treatment at 75° in 5%TCA or at 100° in 4%SDS did not remove any radioactivity. This peptidoglycan synthesis was sensitive to inhibition by vancomycin and was apparently independent of *de novo* protein synthesis since it was unaffected by the addition of chloramphenicol. Both methicillin sensitive and resistant strains behaved similarly.

The effects of β-lactam antibiotics were then examined. Peptidoglycan synthesis could only be partially inhibited by β-lactam antibiotics both in methicillin-resistant and sensitive strains even at concentrations as high as 100xMGIC. Maximum values of 70-75% inhibition were achieved for the sensitive strain H and 50-60% inhibition for the resistant strain MR-1. This level of inhibition appeared to be independent of the MGIC. The remaining, penicillin-insensitive peptidoglycan synthesis presumably occurred by a transglycosylation mechanism in which glycan chains were elongated without further cross-linking as described in *Micrococcus luteus* (6). When exponentially growing cultures of *S.aureus* were treated with chloramphenicol and [^{3}H]-lysine, a similar peptidoglycan synthesis was observed and the inhibitory effects of β-lactam antibiotics were the same as in the wall synthesis medium. *S.aureus* rendered permeable by treatment with toluene could also synthesize peptidoglycan from nucleotide-linked precursors (3). This peptidoglycan was covalently attached to the wall and contained pentapeptide sub-units. Again synthesis could only be inhibited by 50-60% by high concentration of methicillin or benzyl penicillin. Smith and Wilkinson (7) have also reported that peptidoglycan synthesis was insensitive to high concentrations of methicillin in actively growing cells of a methicillin-resistant strain of *S.aureus* (34.2% inhibition) but was more sensitive in non-growing conditions (68.0% inhibition).In a methicillin-sensitive derivative of that strain, peptidoglycan synthesis was inhibited by 87% in both growing and non-growing cells.

The cross-linking of peptidoglycan synthesized in wall synthesis medium was first studied by oxidative deamination of the free amino groups of lysine using nitrous acid (8) and the degree of cross-linking determined as the % of the radioactivity present as [^{3}H]-lysine after separation of the hydrolysis products by paper electrophoresis. In both cases, 85-90% of the [^{3}H] was recovered as [^{3}H]-lysine. This was surprising since it had been assumed that only linear, uncross-linked peptidoglycan would have been synthesized in the presence of β-lactam antibiotics. However when this material was examined by fractionation of the products of a *Chalaropsis* muramidase digest (3) 83% of the [^{3}H]-lysine was present in the monomer fraction confirming that incorporation of peptidoglycan into the wall had occurred by a transglycosylation mechanism. In the absence of β-lactam antibiotics the radioactively labelled peptidoglycan showed the same high degree of cross-linking previously obtained for staphylococcal walls. Incorporation of [^{3}H]-lysine continued throughout the 2h incubation period and the cross-linking of the newly synthesized peptidoglycan occurred in the absence of *de novo* protein synthesis. Earlier studies (8) had shown that in actively growing cultures of *Bacillus megaterium* most rapid cross-linking of peptidoglycan occurred during the first min. of incorporation of [^{14}C] diaminopimelic acid into the cell wall and that cross-linking continued after incorporation of the subunit into the wall. Cells of *S.aureus* H containing [^{3}H]-lysine pentapeptide subunits synthesised in the presence of benzyl penicillin were re-suspended in wall synthesis medium with non-radioactive lysine and no penicillin and incubated for a further 2h. There was no evidence for redistribution of the [^{3}H] although in experiments using very short (30sec) labelling times, monomer was a precursor of cross-linked peptidoglycan (19).

Study of the *in vitro* enzymic activities of the individual PBPs of *S.aureus* has been hampered by the difficulties encountered in their purification. They are present only in very small amounts in the membrane and although they can be separated from other proteins by affinity chromatography using β-lactams as ligands, so far only PBP4 has been isolated and its *in vitro* enzymic activities characterized (10). Purification of the high molecular weight PBPs is difficult because they are similar in size and also show very similar binding affinities for most β-lactams.

However, the PBPs of *S.aureus* H have now been purified from other membrane proteins by affinity chromatography on 6-APA-Sepharose and subsequently resolved by electrophoresis on SDS-polyacrylamide gels. By using gels with a low degree of cross-linkage (10% acrylamide, 0.068% bis-acrylamide) 5 PBPs have been resolved. These proteins have then been transferred by "electrophoretic blotting" onto nitrocellulose membrane filters and their ability to bind radioactive penicillin (whilst themselves remaining attached to the nitrocellulose) has been successfully demonstrated. This technique leads to immobilised proteins which are functionally active and incubation with appropriate substrates should now allow the identification of D-alanine carboxypeptidase, transpeptidase and transglycosylase activities in those proteins.

References

1. Curtis, N.A.C., Hayes, M.V.:FEMS Microbiol Lett. *10*, 227-229 (1981).
2. Curtis, N.A.C., Hayes, M.V., Wyke, A.W., Ward, J.B.: FEMS Microbiol. Lett. *9*, 263-266 (1980).
3. Wyke, A.W., Ward, J.B., Hayes, M.V., Curtis, N.A.C.: Eur. J. Biochem. *119*, 389-393 (1981).
4. Wyke, A.W., Ward, J.B., Hayes, M.V.: Eur.J.Biochem. *127*, 553-558 (1982).
5. Rogers, H.J.: Nature *213*, 31-33 (1967).
6. Mirelman, D., Bracha, R., Sharon, N.: FEBS Lett. *39*, 105-110 (1974).
7. Smith, P.F., Wilkinson, B.J.: J.Bacteriol. *148*, 610-617 (1981).
8. Fordham, W.D., Gilvarg, C.: J.Biol. Chem. *249*, 2478-2482 (1974).
9. Tipper, D.J., Strominger, J.L.: J.Biol. Chem. *243*, 3169-3179 (1968).
10. Kozarich, J.W., Strominger, J.L.: J.Biol. Chem. *253*, 1272-1278 (1978).

PART VI

BIOSYNTHESIS OF MUREIN

PEPTIDOGLYCAN BIOSYNTHESIS : CONTROL OF PRECURSOR SYNTHESIS AND INTERMEDIATES IN ASSEMBLY

J. Barrie Ward
Glaxo Group Research Ltd., Greenford, Middlesex, UB6 OHE.

The biosynthesis of peptidoglycan has been studied extensively and although individual enzymes have been studied in detail little is known about the control and regulation of peptidoglycan biosynthesis. In this short review I will attempt to focus on two of the areas where in my view, gaps in our understanding remain. These are the control and regulation of nucleotide-linked precursor biosynthesis and the involvement of lipid-linked oligomeric intermediates in peptidoglycan assembly.

Control of nucleotide-linked precursor biosynthesis

The precursor UDP-MurAc pentapeptide is synthesized in the cytoplasm by the sequential addition of amino acids to the carboxyl group of UDP-MurAc. The sequence of the amino acids is determined by the specificity of the ligases for both the acceptor nucleotide and the amino acid or dipeptide which is being added. Various examples of feedback inhibition of these ligases and the enzymes involved in the synthesis of UDP-MurAc from UDP-*N*-acetylglucosamine have been reported and suggested as potential mechanisms for the control of cellular concentrations of peptidoglycan precursors. For example, the UDP-*N*-acetylglucosamine ; enolpyruvate transferases from *Enterobacter cloacae*, *Bacillus cereus* and *Escherichia coli* were inhibited by UDP-MurAc-pentapeptide and - tripeptide (1,2,3) whereas the same enzyme for *Staphylococcus epidermidis* was only inhibited

The Target of Penicillin

by UDP-MurAc (4). Treatment of *Staphylococcus aureus* with β-lactam antibiotics results in the accumulation of UDP-MurAc pentapeptide. Similar treatment of *E.coli*, *B.licheniformis* and *B.subtilis* has no such effect, suggesting that in these organisms control of precursor biosynthesis by feedback inhibition might occur. However incubation of the Bacilli with vancomycin (to which *E.coli* is resistant), and of all three organisms with D-cycloserine, does result in the accumulation of UDP-MurAc pentapeptide and -tripeptide respectively. Thus in *E.coli* regulation of precursor synthesis appears to be limited to pentapeptide and not tripeptide while in Bacilli the position is even less clear. It has also been reported (5) that in addition to feedback inhibition precursor biosynthesis in *E.coli* is under stringent control (ie. is regulated by intracellular ppGpp concentrations). When synthesis of UDP-MurAc pentapeptide was relaxed an approximately 3-fold increase in intracellular concentration was observed. Addition of D-cycloserine to these cultures caused marked accumulation of UDP-MurAc-tripeptide with concomitant depletion of the pentapeptide pool by ongoing peptidoglycan biosynthesis. Hence, it was argued that removal of pentapeptide, the putative feedback inhibitor, allowed accumulation of the -tripeptide precursor to occur. However, these conclusions remain difficult to reconcile with the earlier report that both penta- and tripeptide precursors inhibit the enolpyruvate transferase. Peptidases which cleave D-alanyl-D-alanine from UDP-MurAc pentapeptide have also been implicated in the control of intracellular levels of precursors (6,7). The intracellular pool of UDP-MurAc pentapeptide in *E.coli* was determined to be approximately 2×10^{-5}M and sufficient to synthesize only 0.2% of the peptidoglycan made in one generation (8). Earlier, the pool of UDP-*N*-acetylglucosamine was reported to be approximately 10-fold larger than this (9). In view of these low intracellular concentrations it appears unlikely that feedback inhibition alone is responsible for regulation of precursor biosynthesis. Moreover, control exerted only at this stage of the pathway would not regulate the overall rate of peptidoglycan biosynthesis. It could, however, protect the bacteria from accumulation of nucleotide precursors arising as a consequence of changes in the rate of polymer synthesis. The problem remains, however, of providing an explanation for the differences observed in the patterns of the precursor accumulated by different organisms. Do organisms possess unique controls and regulatory mechanisms over the cytoplasmic enzymes involved in nucleotide precursor synthesis? Does any particular organism have more than one potential regulator mechanism?

The role of lipid-linked oligomeric peptidoglycan in wall assembly

Incorporation of newly synthesized subunits into peptidoglycan requires the formation of either glyosyl- or peptide bonds and the independance of glycan polymerization from cross-linkage in many bacteria suggests that polymerization preceeds transpeptidation. Recently, lipid-linked oligomeric uncross-linked peptidoglycan was reported to be present in several organisms and in pulse-chase experiments this newly-synthesized peptidoglycan was incorporated into the wall (10,11,12). Glycan polymerization also preceeds transpeptidation in the early stages of the reversion of protoplasts and unstable L-forms of *B.licheniformis* (13). Treatment of *Micrococcus luteus*, *B.subtilis* and *B.licheniformis* with β-lactam antibiotics results in the synthesis of soluble uncross-linked peptidoglycan (14, 15). Thus polymerization by transglycosylation can occur in the absence of transpeptidation whereas the reverse situation, cross-linkage of peptide side chains in the absence of glycan chain elongation has only been demonstrated with model peptides (16).

Lipid-linked oligomeric peptidoglycan was first isolated from *B.megaterium* after extraction of whole organisms with SDS and characterized as uncross-linked peptidoglycan with a glycan chain length of 11 to 13 disaccharide units (10). Treatment with lysozyme hydrolysed the oligomer into disaccharide pentapeptide subunits and a fragment tentatively identified as undecaprenyl-P-P-disaccharide peptide. Pulse-chase experiments showed similar labelling kinetics for both UDP-MurAc pentapeptide and the lipid-linked components but a precursor-product relationship with mature peptidoglycan was not established. Lipid-linked oligomers with less than 11 disaccharide units were not detected. The possibility remains that lipid(disaccharide peptide)$_{11-13}$ could have arisen by autolytic cleavage of the nascent lipid-linked glycan chains by a specific enzyme having a substrate requirement for 11-13 units. However, in B.megaterium at least, polymerization by transglycosylation may preceed cross-linkage into the wall. In this context, it was reported that cross-linkage in B.megaterium continued for at least 25 min after initial incorporation of newly-synthesized peptidoglycan into the wall had occurred (17).

In vitro studies with M.luteus showed that low concentrations of deoxycholate appeared to selectively dirupt transpeptidation and led to accumulation of material with a glycan chain length of 40-60 disaccharides which was soluble in hot but not cold SDS (11). Lysozyme-treatment of the extracted polymer yielded disaccharide-hexapeptides and a very small amount of undecaprenyl-P-P-disaccharide hexapeptide. On removal of the detergent and reincubation with nucleotide precursors, the soluble peptidoglycan became cross-linked to the wall by a pencillin-sensitive transpeptidation. In the absence of deoxycholate, linear peptidoglycan of low molecular weight was synthesized but there was no evidence that this could be transpeptidated with the wall (18). The presence of oligomeric peptidoglycan and continued transpeptidation following incorporation of newly-synthesized material into the sacculus have also been described for E.coli (12,19). Oligomeric material could be separated into water- and detergent-soluble fractions both of which were partially cross-linked and contained a higher proportion of pentapeptide residues than did mature peptidoglycan. The peptidoglycan in both fractions was substituted with lipoprotein. Pulse-chase experiments showed that during the chase radioactivity present as 'soluble' peptidoglycan decreased and there was a concomitant increase in the amount of radioactivity incorporated into the detergent-insoluble peptidoglycan. Approximately half the TCA-insoluble peptidoglycan synthesized *in vitro* by ether-treated *E.coli* was soluble in hot -SDS and was less cross-linked than peptidoglycan isolated as detergent-insoluble material ie. the pre-existing sacculus. Pulse-chase experiments showed that during the chase period the detergent-soluble material could be covalently-linked to the sacculus by a β-lactam sensitive transpeptidation reaction.

Ether-treated *P.aeruginosa* also synthesized partially cross-linked oligomeric material but substitution (activation) of some reducing terminals of the glycan chains by undecaprenyl pyrophosphate was not detected (20). However, radioactively-labelled 'soluble' peptidoglycan was completely chased into SDS -insoluble sacculus within 20 min of incubating the ether-treated cells with unlabelled peptidoglycan precursors. Again, incorporation into the sacculus was β-lactam sensitive and presumably occurred by transpeptidation.

Although lipid-linked oligomeric peptidoglycan has been detected (and probably looked for) in only a few bacteria, quite striking differences in the structure of the isolated material are already evident. On the one hand, the oligomeric material isolated for *B.megaterium* and *M.luteus* is linear uncross-linked peptidoglycan whereas that from the Gram-negative organisms is considerably more complex with a low degree of cross-linkage, and in the case of *E.coli* substituted with lipoprotein. Notwithstanding these differences, pulse-chase experiments showed in all organisms with the possible exception of *B.megaterium*, that the oligomeric material became cross-linked into the wall suggesting it is indeed an intermediate in peptidoglycan assembly. However, the *in vivo* situation may be more complex. In *M.luteus* for example further disaccharide hexapeptide subunits may be added to extend the glycan chain and it is these subunits which become cross-linked into the wall. The situation in *E.coli* and *P.aeruginosa* is clearly different. If the proposed mechanism of glycan chain polymerization ie. elongation from the lipid-linked reducing terminal (21) is correct then extension of the glycan chains present in oligomeric peptidoglycan isolated from *P.aeruginosa*, could not occur. In this case cross-linkage of the oligomer into the pre-existing wall is presumably by transpeptidation alone, whereas incorporation by either means should be possible for the oligomer isolated from *E.coli.*

We might speculate that growth of the relatively thin sacculus of Gram-negative bacteria requires addition of more complex units than does growth of the thicker peptidoglycan layer in Gram-positive walls. In *E.coli* the lipoprotein may serve to anchor the oligomeric material to the outer membrane. Provided no cross-linkage to pre-existing peptidoglycan occurs some lateral movement of the oligomer might be possible to allow newly-snythesized polymer to be redistributed from sites of synthesis to other areas of the sacculus. Autoradiographic studies may be interpreted as providing evidence for such a redistribution (22). This aspect of peptidoglycan assembly was discussed in detail in recent papers of Koch *et al* (22,23). Alternatively, it remains possible that controlled autolysis of peptidoglycan is an essential part of wall growth and that the oligomer represents a fragment of peptidoglycan mobilized in this way. In either case the presence of lipoprotein and undecaprenyl pyrophosphate substituents could be taken to imply that the oligomer is anchored in both outer and cytoplasmic membranes.

In both types of organisms subsequent cross-linkage of oligomeric peptidoglycan into the wall suggests a requirement for multiple transpeptidases. In recent years evidence has occurred for such activities, particularly among the pencillin-binding proteins of E.coli (24,25). However, the mechanisms of peptidoglycan assembly described above may require tanspeptidation to occur not only at the wall-membrane interface but perhaps also in the 'depth' of the wall. Little evidence appears to be available suggesting the presence of such enzymes. It would be particularly interesting to see if newly-synthesized peptidoglycan has associated with it the transpeptidase necessary for the formation of cross-links to the pre-existing wall.

Thus although the structure and biosynthesis of peptidoglycan is now well established further work is obviously required to understand the control and regulation of the biosynthetic enzymes. In particular the mechanism whereby this complex macromolecule is assembled and how growth of peptidoglycan occurs as a relatively thin sacculus in Gram-negative organisms and as the thicker wall of Gram-positive bacteria await further study.

References

1. Venkateswaran, P.S., Lugtenberg, E.J.J. Wu, H.C.: Biochim. Biophys. Acta 293, 570-574 (1973).
2. Taku, A., Gunetilike, K.G., Anwar, R.A.: J. Biol. Chem. 245, 5012-5016 (1970).
3. Zemmel, R.I., Anwar , R.A.: J. Biol. Chem. 250, 3185-3192 (1975).
4. Wickus, G.G., Rubenstein, P.A., Warth, A.D., Strominger, J.L.: J. Bacteriol. 113, 291-295 (1973).
5. Ishiguro, E.E., Ramey, W.D.,: J. Bacteriol 135, 766-774 (1978).
6. Gondre, B., Flouret B., van Heijenoort, J.,: Biochimie 55, 685-691 (1973).
7. Oppenheim, B., Patchornik, A.: FEBS Letts, 48, 172-175 (1974).

8. Mengin-Lecreulx, D., Flouret, B., van Heijenoort, J.,: J. Bacteriol. 151, 1109-1117 (1982).

9. Ward, J.B., Glaser, L.,: Arch. Biochem. Biophys. 134, 612-622 (1969).

10. Fuchs-Cleveland, E., Gilvarg C.,; Proc. Natn. Acad. Sci. U.S.A. 73, 4200-4204 (1976).

11. Thorpe, S.J., Perkins, H.R.,: FEBS Letts. 105, 151-154 (1979).

12. Mett, H., Bracha, R., Mirelman, D.,: J. Biol. Chem. 255, 9884-9890 (1980).

13. Elliott, T.S.J., Ward, J.B., Rogers, H.J.,: J. Bacteriol. 124, 623-632 (1975).

14. Mirelman, D., Bracha, R., Sharon, N.,: Biochemistry 13, 5045-5053 (1974).

15. Tynecka, Z., Ward, J.B.,: Biochem. J. 146, 253-267 (1975).

16. Ghuysen, J.M., Frere, J-M., Leyh-Bouille, M., Dideberg, O. Lamotte-Brasseur, J., Perkins H.R., De Coen J-L.,: Topics in Molec. Pharmacol. 1, 63-97 (1981).

17. Fordham, W.D., Gilvarg, C.,: J. Biol. Chem., 249, 2478-2482 (1974).

18. Weston, A., Ward, J.B., Perkins, H.R.,: J. Gen. Microbiol. 99, 171-181 (1977).

19. De Pedro, M.A., Schwarz, U.,: Proc. Natl. Acad. Sci. U.S.A. 78, 5856-5860 (1981).

20. Mirelman, D., Nuchamowitz, Y.,: Europ. J. Biochem. 94, 541-548 (1979).

21. Ward, J.B., Perkins, H.R.,: Biochem. J. 135, 721-728 (1973).

22. Koch, A.L., Verwer, R.H., Nanninga, N.,: J. Gen. Microbiol. 128, 2893-2898 (1982).

23. Koch, A.L., Higgins, M.L., Doyle, R.J.,: J. Gen. Microbiol. 128, 927-945 (1982).

24. Matsuhashi, M., Ishino, F., Nakagawa, J., Mitsui, K., Nakajima-Iijima, S., Tamaki, S., Hashizumi, T.,: in β-lactam antibiotics Salton, M., Shockman, G.D. eds. pp. 169-184 (1981).

25. Ishino, F., Matsuhashi, M.,: Biochim. Biophys. Res. Commun. 101, 905-911 (1981).

CYTOPLASMIC STEPS OF PEPTIDOGLYCAN SYNTHESIS IN *E. coli* K 12.

Dominique Mengin-Lecreulx, Bernard Flouret, Claudine Parquet and Jean van Heijenoort.

E.R. n°245 du CNRS, Institut de Biochimie, Université Paris-Sud, 91405 Orsay - France.

Introduction

In *Escherichia coli*, the main peptidoglycan cytoplasmic precursors are a series of uridine nucleotides which have been characterized as UDP-GlcNac, UDP-GlcNac-enolpyruvate, UDP-MurNac and its four peptide derivatives. Their sequential formation from UDP-GlcNac, PEP, NADPH, L-alanine, D-glutamic acid, meso-diaminopimelic acid (DAP) and D-alanyl-D-alanine, is catalyzed by a set of six highly specific cytoplasmic enzymes (see 1 for references). However, there is as yet little information concerning the properties of these enzymes and their *in vivo* regulation in *E. coli*. We have undertaken investigations to determine whether specific regulatory mechanisms are involved in the control of this pathway. One method for studying the factors of regulation involved in this reaction sequence is to determine the pool levels of the different intermediates which, surprisingly, have not yet been extensively studied. We have now developed new techniques of high-pressure liquid chromatography (HPLC) for the rapid isolation of these compounds from cell extracts (1)(2)(3). Furthermore, the kinetic parameters of the six synthetase activities catalyzing the stepwise formation of UDP-MurNac-pentapeptide from UDP-GlcNac were determined and the inhibitory effect of different peptidoglycan precursors on them was also examined. Taking into consideration the data obtained, an attempt was made to compare the *in vitro* activities of these enzymes with their *in vivo* functioning.

The Target of Penicillin

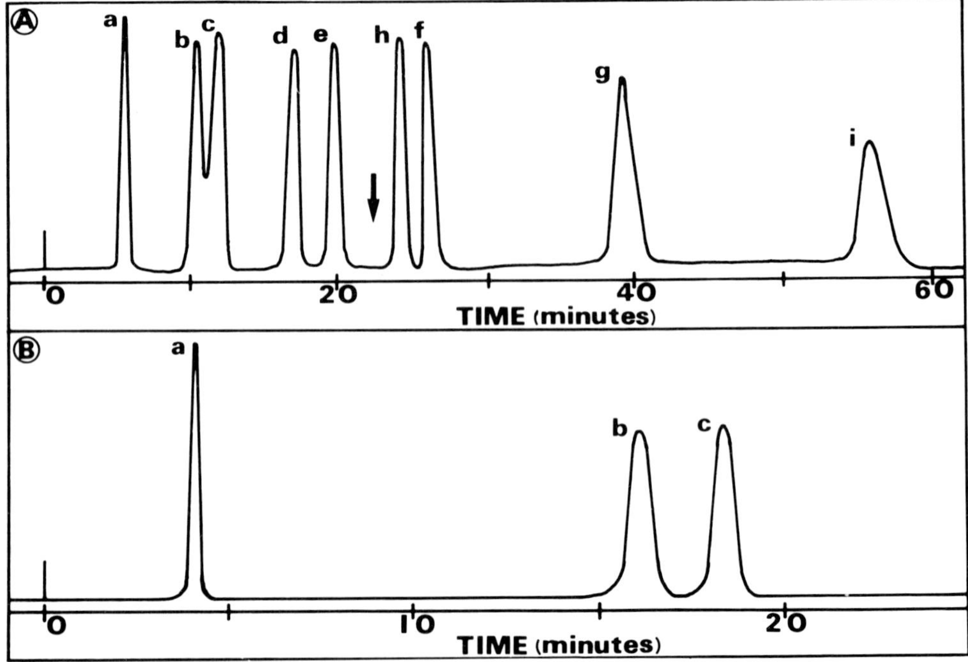

Figure 1 : Separation of nucleotide precursors of peptidoglycan by HPLC. Operating conditions : (A) µ-Bondapak C_{18} column (3.9x300 mm), elution with 0.05 M ammonium phosphate, pH 3.65, flow rate was changed from 0.5ml/min to 1.5 ml/min at the time indicated (↓). (B) µ-Bondapak NH_2 column (3.9x300 mm), elution with 0.1 M ammonium acetate, pH 4.70/methanol (3/2, v/v) at a flow rate of 2 ml/min. Precursors are indicated by letters : (a) UDP-GlcNac ; (b) UDP-GlcNac-enolpyruvate ; (c) UDP-MurNac ; (d) UDP-MurNac-tripeptide with DAP ; (e) UDP-MurNac-L-alanine ; (f) UDP-MurNac-dipeptide ; (g) UDP-MurNac-pentapeptide with DAP ; (h) UDP-MurNac-tripeptide with lysine ; (i) UDP-MurNac-pentapeptide with lysine.

Results

1) HPLC separation of uridine nucleotide precursors (2). A good separation of these compounds was routinely obtained under reverse-phase conditions on a µ-Bondapak C_{18} column by isocratic elution with a 0.05 M ammonium phosphate (or formate) buffer, pH 3.65. Figure 1A shows that the retention time of these precursors generally increases with their molecular weight. The exception observed with UDP-MurNac-L-Ala-γ-D-Glu-meso-DAP, which is eluted before UDP-MurNac-L-Ala, was most probably due to the

high polarity of the DAP residue. Furthermore, this method could be extended to precursors of bacterial strains other than E. coli, which contain lysine instead of DAP (Fig.1A). UDP-GlcNac-enolpyruvate and UDP-MurNac, which are eluted as two poorly separated peaks under such conditions, could be easily isolated by subsequent chromatography on a μ-Bondapak NH_2 column with a buffer-methanol appropriate elution (Fig.1B).

2) Pool levels of the cytoplasmic peptidoglycan precursors. A rapid three-steps procedure, including a gel filtration on fine Sephadex G25 and the two HPLC systems described above, was developed for the complete analysis of the uridine nucleotide precursors in cell extracts from E. coli K 12 (1)(3). No such preliminary purification step was necessary for the other amino acid substrates,which could be determined directly in crude cell extracts (1). The pool levels of all these precursors were determined for exponential-phase cells of E. coli grown either in glucose-minimal or rich medium (Table 1).

Table 1 : Pool levels of cytoplasmic precursors of peptidoglycan in E. coli K 12 cells.

Precursors	Pool levels (M)	
	glucose minimal medium	rich medium
UDP-GlcNac	10×10^{-5}	11×10^{-5}
UDP-GlcNac-enolpyruvate	0.2×10^{-5}	0.2×10^{-5}
UDP-MurNac	3.7×10^{-5}	0.5×10^{-5}
UDP-MurNac-L-Ala	1.1×10^{-5}	0.9×10^{-5}
UDP-MurNac-dipeptide	0.9×10^{-5}	0.9×10^{-5}
UDP-MurNac-tripeptide	0.5×10^{-5}	0.2×10^{-5}
UDP-MurNac-pentapeptide	14×10^{-5}	27×10^{-5}
D-alanine	4×10^{-4}	1×10^{-3}
L-alanine	4×10^{-3}	1×10^{-2}
D-glutamic acid	8×10^{-4}	3×10^{-3}
DAP	6×10^{-4}	2×10^{-4}
D-alanyl-D- lanine	2×10^{-4}	2×10^{-4}

It is noteworthy that UDP-GlcNac and UDP-MurNac-pentapeptide, which are the first and last nucleotide precursors of this metabolic pathway, have the highest cellular concentration (ca.: 10^{-4}M), whereas the other precursor concentrations are in the range 10^{-6} to 10^{-5} M.

3) Kinetic properties of the cytoplasmic synthetases. The kinetic parameters Km (for both substrates) and Vm were determined for each of the six enzymatic activities investigated in the 100,000 g supernatant of cell extracts from *E. coli* K 12 grown in rich medium (Table 2). It is noteworthy that the amount of L-alanine-adding enzyme in crude extract is much lower than that of the five other activities, but it is still unclear whether this low *in vitro* level is due to the instability of the enzyme or to the lack of some specific effector.

Table 2 : Km and Vm values for the substrates of the six synthetases.

Enzyme	Substrates	Km	Vmax (nmol/min/mg of protein)
PEP:UDP-GlcNac transferase	PEP UDP-GlcNac	1.8×10^{-4} 2.4×10^{-4}	1.30
UDP-GlcNac-enol-pyruvate reductase	NADPH UDP-GlcNac-enolpyruvate	4.6×10^{-6} (4) 6.5×10^{-6} (4)	1.15
L-alanine adding enzyme	L-alanine UDP-MurNac	2×10^{-4} 3×10^{-4}	0.01
D-glutamic acid adding enzyme	D-glutamic acid UDP-MurNac-L-Ala	1.3×10^{-4} 3.0×10^{-5}	2.80
meso-DAP adding enzyme	meso-DAP UDP-MurNac-dipeptide	3.6×10^{-5} 3.5×10^{-5}	1.80
D-Ala-D-Ala adding enzyme	D-Ala-D-Ala UDP-MurNac-tripeptide	3.8×10^{-5} 1.5×10^{-5}	0.45

4) Synthetase activities under pseudophysiological conditions. An attempt was made to compare the *in vitro* activities of these enzymes in extracts from exponential-phase cells with the amount of peptidoglycan material which is effectively synthesized *in vivo* by the same cells. Therefore, these activities were determined under *in vitro* conditions reflecting as closely as possible the physiological ones, in particular by fixing

Table 3 : Synthetase activities under pseudophysiological conditions.

Enzyme	Total activities (nmol of product)
PEP : UDP-GlcNac transferase	1700
UDP-GlcNac-enolpyruvate reductase	2100
L-alanine adding enzyme	7
D-glutamic acid adding enzyme	3500
meso-DAP adding enzyme	1500
D-alanyl-D-alanine adding enzyme	1100
DAP in peptidoglycan from 3.8×10^{11} cells (nmol) ..	3100

Activities were determined for a total cell extract originating from 1 liter of culture (3.8×10^{11} cells) and for one generation time (38 min).

the substrates concentrations at their pool levels. A good correlation was obtained between the *in vitro* activities of these enzymes determined under such pseudophysiological conditions and the requirements of peptidoglycan synthesis (Table 3).

5) *Inhibition of synthetase activities*: The effects of the different uridine nucleotide precursors on the six enzymatic activities were examined under the *in vitro* pseudophysiological conditions (1)(3). Most of these enzymes are greatly inhibited by their respective reaction product at a 10^{-3} M concentration and to a lesser extent by UDP-MurNac-pentapeptide. The good inhibitory effect observed with UDP-MurNac on the PEP : UDP-GlcNac transferase activity was probably due to its structural analogy with UDP-GlcNac-enolpyruvate.

Discussion

When comparing the pool levels of the various precursors with the enzymatic parameters determined for the six synthetases, it is noteworthy that the concentration of the nucleotide precursors are lower than or of the same order of magnitude as the corresponding Km values, whereas the other substrates are present in the cell at a saturating level for these activities.

Thus, high turnover values should characterize the intermediate nucleotide precursors, the very low pool of which are apparently limiting the in vivo functioning of the corresponding synthetase activities. Consequently, the inhibition of any of these cytoplasmic steps should lead to a rapid depletion of the nucleotide substrates of the subsequent steps and to an abrupt arrest of the peptidoglycan synthesis. Furthermore, the values determined for the synthetase activities under the pseudophysiological conditions seem to show that these enzymes are not in excess in the cell, but are more or less adjusted to the requirement of peptidoglycan synthesis. Finally, a few comments can be made concerning possible regulatory mechanisms. The various inhibitory effects observed on the different synthetases are probably of little physiological significance since such inhibitions seem to require much higher concentrations of effectors than those normally found in cells. Further work is now under investigation to determine the factors and sites of regulation of these cytoplasmic steps. One approach consists in studying the variations of the pool levels of precursors and of the synthetase activities under various growth conditions or after treatment with antibiotics affecting different steps of peptidoglycan synthesis. Preliminary results show that the cellular concentration of nucleotide precursors did not vary when generation times from 25 to 200 minutes were considered. Furthermore, certain of the synthetase activities, in particular the D-Glu, meso-DAP and D-alanyl-D-alanine adding enzymes,were detected always at the same level in cell extracts, whatever the growth conditions used.

References

1. Mengin-Lecreulx, D., Flouret, B., van Heijenoort, J.: J. Bacteriol. 151, 1109-1117 (1982).
2. Flouret, B., Mengin-Lecreulx, D., van Heijenoort, J.: Anal. Biochem. 114, 59-63 (1981).
3. Mengin-Lecreulx, D., Flouret, B., van Heijenoort, J.: J. Bacteriol. (submitted for publication) (1983).
4. Anwar, R.A., Vlaovic, M.: Can. J. Biochem. 57, 188-196 (1979).

EFFECT OF DRUGS ON THE FORMATION AND STRUCTURE OF INTERMEDIATES OF ESCHERICHIA COLI MUREIN BIOSYNTHESIS

Helmut Mett, Pharmaceuticals Division, Research Department, CIBA-GEIGY Limited, CH-4002 Basel, Switzerland

David Mirelman, Biophysics Department, Weizmann Institute of Science, Rehovot 76100, Israel

Introduction

Bacterial murein synthesis has been at the center of microbiological research for many years due to two basic aspects: Murein biosynthesis is a process closely related to bacterial morphogenesis, and second, murein biosynthesis represents a metabolic pathway unique to bacteria. Therefore, compounds which specifically interact with murein metabolism can be expected to exhibit little effect on the infected host, a toxicological prerequisite for any chemotherapeutic agent.
Numerous antibacterial agents have been found to inhibit overall murein synthesis. In recent studies, some of the target reactions of various inhibitors have been characterized by several techniques, including studies with partially purified enzyme preparations (1). We have investigated the effects of several antibiotics on murein biosynthesis both in intact bacterial cells and in ether-permeabilized cells (2, 3). The results of these studies support the hypothesis that a delicate balance between the different murein-synthesizing and murein-degrading enzymes is required to maintain normal bacterial growth.

Methods

We developed a system to analyze the biosynthesis of murein under conditions that are close to those found in normal, untreated bacteria. Murein synthesis is investigated in intact, growing bacterial cultures,

The Target of Penicillin

as well as in bacteria permeabilized by mild ether extraction (2, 3, 4). Experiments were done with E. coli JE5509 (dapA lysA), a diaminopimelic acid (DAP) requiring mutant obtained through the courtesy of Dr. Y. Hirota, Mishima, Japan. In experiments using intact bacteria, exponentially growing bacterial cultures were labeled usually for 20 min with radioactive DAP of high specific activity (35 Ci/mmol, CEA, Gif-sur-Yvette, France). After labeling in the presence or absence of antibiotics, bacteria were harvested, washed, and disrupted in a French Press (Aminco). Cell envelopes were collected by high speed centrifugation and subsequently solubilized by hot sodium dodecylsulfate (SDS) treatment. Three murein fractions were obtained: a) Water-soluble material (supernatant after French Press disruption), b) SDS-soluble material (obtained from envelopes pelleted after French Press disruption), c) SDS insoluble murein sacculi (pellet after SDS treatment).

Polymeric murein was separated from low-molecular weight material in the soluble fractions by gel filtration.

Ether-permeabilized bacteria were obtained from exponentially growing bacteria by mild (1 min) ether extraction. The permeabilized bacteria were able to synthesize SDS-insoluble murein from radiolabeled nucleotide-precursors, UDP-GlcNAc and UDP-MurNAc-Pentapeptide. Ether-permeabilized bacteria were especially helpful for studying the effects of antibiotics which do not penetrate through the envelopes of Gram-negative bacteria. The newly synthesized products were fractionated by trichloroacetic acid (TCA)- or SDS-extraction and subsequent filtration (0.2 μm pore size) resulting in four fractions: SDS-soluble / insoluble murein and TCA-soluble / insoluble murein.

The structural composition of murein synthesized by intact bacteria or by ether-permeabilized bacteria was analyzed by several procedures (2): Lysozyme digestion and subsequent paper chromatography revealed the degree of crosslinkage of the material and its high-molecular weight nature. Mild acid hydrolysis (0.1 N HCl, 5 min $100^{o}C$) was applied to cleave phospho-diester bonds in nucleotide- and lipid-bound murein precursors. Proteolytic digestion with trypsin was used to cleave lipoprotein from murein. Affinity chromatography on vancomycin-Sepharose allowed the separation of pentapeptide-containing murein fragments from

those with tri- and tetrapeptide side chains.
Our data suggest the formation of soluble, polymeric, slightly cross-linked murein intermediates which already contain small amounts of lipoprotein covalently attached to it (Table I).
Cefotaxime and moenomycin (= flavomycine) were gifts from Hoechst, mecillinam (FL 1060) was obtained from Leo Pharmaceuticals. Bicozamycin is a new antibacterial agent from CIBA-GEIGY.

Results

1. Effects of moenomycin on murein synthesis. Moenomycin has been suggested to inhibit the polymerization of murein glycan chains (6). The effect of moenomycin was tested only with ether-permeabilized bacteria, because the compound did not penetrate into intact bacteria (minimum inhibitory concentration (MIC) >100 ug/ml). At 0.1 ug/ml, moenomycin the inhibited formation of polymeric murein by more then 80% (Fig. 1). Ether permeabilized bacteria were unable to synthesize lipid-linked murein intermediates in the presence of moenomycin, as detected by column chromatography, paper chromatography and mild acid hydrolysis (data not shown).

Table I. Composition of murein fractions synthesized by intact *E. coli*.

Murein fraction	Degree of peptide cross-linkage (mol-%)	Pentapeptide content (% of side chains)	Covalently bound lipo-protein (% of side chains)
Water-soluble material			
Nucleotide precursors	0	67	0
Water-soluble polymers	19	24	4.6
SDS-soluble polymers	23	5.9	4.4
SDS-insoluble sacculi	27	3.1	11

2. Effects of the two ß-lactams mecillinam and cefotaxime on murein synthesis. When added to intact bacteria, both drugs led to an apparent accumulation of water-soluble precursors with a concomitant reduction of murein incorporation into the pre-existing sacculus (not shown). Structural analysis of the murein fractions synthesized by intact bacteria in the presence of the two compounds revealed that cefotaxime led to an increase in the pentapeptide content of all fractions, whereas mecillinam did not affect the pentapeptide content in the nucleotide-precursor pool (Table II). Moreover, cefotaxime inhibited cross-link formation in the insoluble murein sacculi, whereas mecillinam led to enhanced cross-linking of all polymeric murein fractions.

Essentially the same observations were made upon incubation of ether-permeabilized bacteria with these two ß-lactams (Fig. 1). Cefotaxime strongly inhibited the formation of polymeric murein and reduced its degree of cross-linkage, while mecillinam did not affect polymerization or cross-link formation to a significant extent.

Table II. Structure of murein fractions synthesized by intact bacteria in the presence of cefotaxime and mecillinam.

Murein fraction	Pentapeptide content (%)			Degree of cross-linkage (%)		
	Control	CTX[a]	MEC[b]	Control	CTX[a]	MEC[b]
Nucleotide precursors	67	81	64	--	--	--
Water-soluble polymers	12	23	19	11	10	13
SDS-soluble polymers	n.d.	n.d.	n.d.	7.7	10	13
SDS-insoluble sacculi	0.5	0.8	1.5	19	11	22

[a] CTX: cefotaxime, [b] MEC: mecillinam

3. Effect of bicozamycin on murein synthesis. Bicozamycin had been suggested to inhibit lipoprotein attachment to murein sacculi (5). We incubated E. coli JE5509 in the presence of bicozamycin (50 and 200 ug/ml equal to 1 and 4 x MIC) and radioactive DAP and analysed therafter the total soluble macromolecular murein as well as the insoluble sacculi. No difference was observed, as compared to antibiotic-free controls, with respect to the incorporation of new material into sacculi, degree of peptide cross-linkage and amount of bound lipoprotein (data not shown).

Fig. 1 Macromolecular murein synthesized by ether-permeabilized bacteria in the presence of cefotaxime, mecillinam or moenomycin.

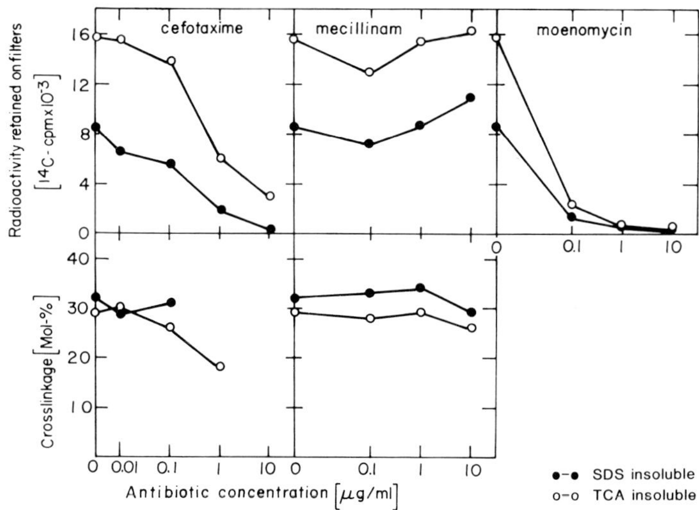

Discussion

In this study we have shown that bicozamycin does not primarily affect murein synthesis and lipoprotein insertion into the murein sacculi. This view is supported by our finding that E. coli JE5510, a lipoprotein-less mutant, is as sensitive towards bicozamycin as the isogenic, lipoprotein-containing strain JE5509 (H. Mett, unpublished). Moenomycin appears to inhibit murein synthesis already at the level of transfer of MurNAc-pentapeptide from the nucleotide to the bactoprenyl carrier lipid, since no lipid-linked intermediates were formed by ether-permeabilized bacteria.

Cefotaxime apparently inhibited the murein-polymerizing reactions, primarily via inhibition of transpeptidation. In addition, carboxypeptidase activity was reduced by cefotaxime, as evident from the increase in pentapeptide side chain content of all murein fractions.
Mecillinam did not inhibit overall cross-link formation and had only minor effects on the pentapeptide content. We therefore assume that mecillinam may interact with some regulatory mechanism of murein synthesis and bacterial cell shape generation rather than directly with murein polymerization and its insertion into the sacculus. The importance of a delicate balance of enzyme activities involved in murein metabolism has recently been demonstrated in a different approach by Spratt and coworkers (7).

References

1. Ishino, F., Matsuhashi, M.: Agric. Biol. Chem. 43, 2641-2642 (1979).
2. Mett, H., Mirelman, D.: FEMS Microbiol. Lett. 16, 39-44 (1983).
3. Mett, H., Bracha, R., Mirelman, D.: J. Biol. Chem. 255, 9884-9890 (1980)
4. Vosberg, H.P., Hoffmann-Berling, H.: J. Mol. Biol. 58, 739-753 (1971)
5. Tanaka, N.: J. Antibiotics 29, 155-167 (1976).
6. Franklin, T.J., Snow, G.A.: Biochemistry of antimicrobial action, Third edition, Chapman and Hall, London · New York 1981.
7. Markiewicz, Z., Broome-Smith, J.K, Schwarz, U., Spratt, B. G.: Nature 297, 702-704 (1982).

THE EFFECT OF ß-LACTAM ANTIBIOTICS ON MUREIN SYNTHESIS IN ETHER TREATED ESCHERICHIA COLI.

Susanne Henning, Renate Metz, Walter P. Hammes
Institut für Lebensmitteltechnologie der Universität Hohenheim, Garbenstraße 25, D-7ooo Stuttgart 7o

Introduction

Ether treated cells of bacteria (ETB) proved to be valuable tools for the study of murein synthesis (1-3). It was suggested that ETB retain normal cellular controls, involved in regulating murein synthesis, in contrast to cell-free systems, where such controls are lost as a result of mechanical disruption of the cells (4).
In these studies UDP-MurNAc-pentapeptide was used as the substrat for murein synthesis *in vitro* (X-pentapeptide system) and for determining the effect of ß-lactam antibiotics on this process.
Recently, we succeeded in establishing a murein synthesizing system from ETB, which utilizes UDP-GlcNAc and amino acids (complete system), thus allowing the study of the complete sequence of murein synthesis *in vitro* (Metz et al., manuscript in preparation).
It is the purpose of this communication to compare the effects of various ß-lactam antibiotics on murein synthesis in the X-pentapeptide system with those in the complete system. From this comparison it was expected to obtain a better understanding of the sites of the regulation of murein synthesis.

The Target of Penicillin

Material and Methods

The preparation of ETB was performed as described by Mirelman et al. (1). The following reaction mixtures were employed in a final volume of 3o µl:

Reaction mixture I (X-pentapeptide system):

o.1 mol/l of TES-buffer, pH 8.3; 2o mmol/l of $MgCl_2$; 15 mmol/l of NH_4CL; 1 mmol/l of DTT; 1 mmol/l of ATP; o.45 mmol/l of UDP-MurNAc-pentapeptide; 44 µmol/l of UDP-(^{14}C)GlcNAc (16.8 cpm/pmol) and ether treated cells (78 µg of protein).

Reaction mixture II (complete system):

o.1 mmol/l of TES-buffer, pH 8.3; 2o mmol/l of $MgCl_2$; 15 mmol/l of NH_4Cl; 1 mmol/l of DTT; 1o mmol/l of ATP; o.9 mmol/l of creatine phosphate; o.23 U of creatine kinase; o.1 mmol/l of L-alanine; 5o µmol/l of D-glutamic acid; 2.1 mmol/l of UDP-GlcNAc; o.11 mmol/l of PEP; 5 mmol/l of NADPH; 1o µmol/l of FAD; 2o mmol/l of KCl; o.11 mmol/l of (^{14}C)-diaminopimelic acid (14,4 cpm/pmol) and ether treated cells (78 µg of protein).

After incubating at 37°C for 45 min, the reation was terminated by adding 3o µl of IBS-solvent, containinq isobutyric acid/H_2O/NH_4OH (33%) at a ratio of 66/33/2 (v/v/v).

The amount of total murein was determined after chromatography on Whatman 3 MM paper in the IBS-solvent. This polymer remained on the origin of the chromatogram and was completely sensitive to lysozyme digestion. The origins were cut from the chromatogram and counted for radioactivity which was taken as a measure of the total murein synthesis. For estimation of the wall-bound murein synthesis, the paper strips were removed from the scintillation vials, dried and boiled in 1.2 ml of 4% SDS for 1o min, in order to extract that part of the murein polymer which had not undergone attachment to the cell wall. The paper strips were transferred to a Gelman filter (GN 6, pore size o.45 µm), washed three times with 3 ml of 4% SDS and three times with 3 ml of H_2O. The filter and the paper strips were dried, combined and counted for radioactivity. The

SDS-soluble murein was then calculated from the difference between the total murein and the wall-bound murein.

Results and Discussion

1) The effect of ß-lactam antibiotics on murein synthesis in the complete system (Fig.1). The synthesis of murein was sensitive to the various antibiotics at a concentration of 5 µg/ml, except for mecillinam and thienamycin, which required a higher concentration to obtain a marked inhibition. The inhibition of wall-bound murein is generally connected with an concomitant inhibition of total murein

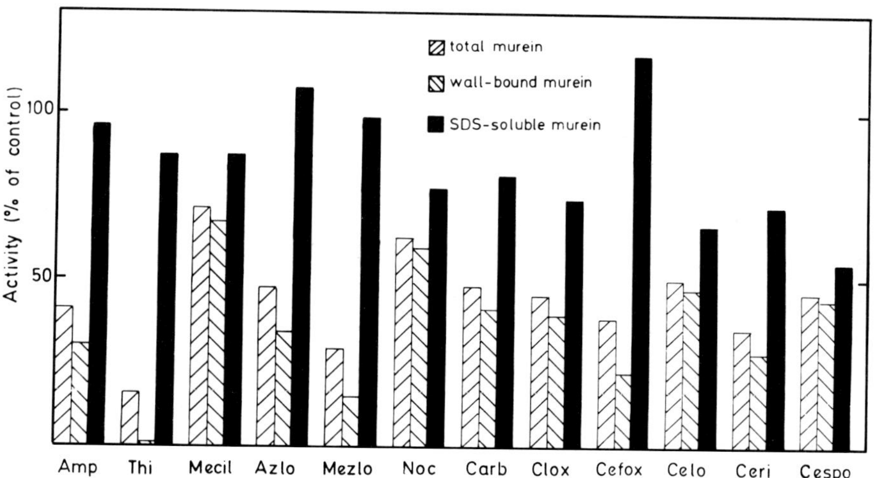

Fig.1: Effect of ß-lactam antibiotics on murein synthesis in the complete system. The following antibiotics were used at concentrations shown in parantheses: ampicillin = Amp (5µg/ml); thienamycin = Thi (5oµg/ml); azlocillin = Azlo (5µg/ml); mecillinam = Meci (5oµg/ml); mezlocillin = Mezlo (5µg/ml); nocardicin A = Noc (5µg/ml); carbenicillin = Carb (5µg/ml); cloxacillin = Clox (5µg/ml); cefoxitin = Cefox (5µg/ml); cephalothin = Celo (5µg/ml); cephaloridin = Ceri (5µg/ml); cephalosporin C = Cespo (5µg/ml). A synthesis of 1oo% corresponds to (pmol/mg protein): 1689 for total, 1399 for wall-bound and 29o for SDS-soluble murein, respectively.

synthesis. The amounts of SDS-soluble murein synthesis were not enhanced to a major extent by the various antibiotics.

2) The effect of ß-lactam antibiotics on murein synthesis in the X-pentapeptide system (Fig.2). In the X-pentapeptide system, on the other hand, an increase in soluble peptidoglycan occured in the case of the antibiotics cefoxitin and thienamycin. With the remaining antibiotics the synthesis of this type of polymer remained rather constant.

3) Comparison of the effects of ß-lactam antibiotics in the two systems. From the comparison of the Fig.1 and 2, it can be derived that, in most cases, the complete system

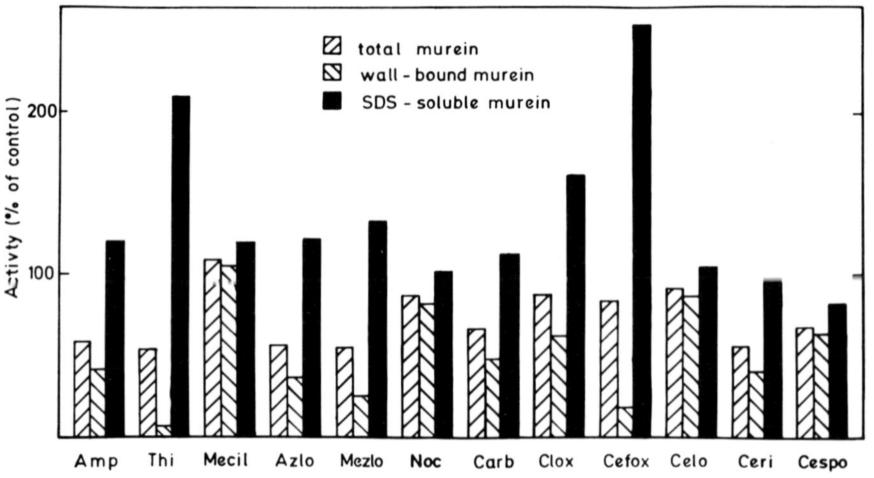

Fig.2: Effect of ß-lactam antibiotics on murein synthesis in the X-pentapeptide system. The following antibiotics were used at concentrations shown in parantheses: ampicillin = Amp (5 µg/ml); thienamycin = Thi (5oµg/ml); mecillinam = Meci (5o µg/ml); azlocillin = Azlo (5µg/ml); mezlocillin = Mezlo (5 µg/ml); nocardicin A = Noc (5µg/ml); carbenicillin = Carb (5µg/ml); cloxacillin = Clox (1oµg/ml); cefoxitin = Cefox (1o µg/ml); cephalothin = Celo (1oµg/ml); cephaloridin = Ceri (1o µg/ml); cephalosporin C = Cespo (1oµg/ml). A synthesis of 1oo% corresponds to (pmol/mg protein): 1689 for total, 1399 for wall-bound and 29o for SDS-soluble murein, respectively.

exhibits a higher sensitivity to ß-lactam antibiotics than the X-pentapeptide system. The greatest differences in the inhibition of the synthesis of wall-bound murein were caused by mecillinam and cephalothin. The degrees of inhibition at 50μg/ml of mecillinam and 5μg/ml of cephalothin were 33 and 53%, respectively, in the complete system. On the other hand, in the X-pentapeptide system the same concentration of mecillinam had no effect and 10μg/ml of cephalothin inhibited only by 14%.
In Gram positive bacteria it was observed that in the presence of ß-lactam antibiotics the synthesis of total murein remained constant, whereas the formation of cross-links is inhibited. As a result non cross-linked murein is formed, which compensates exactly for the reduced amounts of murein in the cross-linked fraction (5).
Similar results were obtained in studies of murein synthesis with the aid of particles from desintegrated *Escherichia coli* cells (6). Since in our studies with ETB, total murein synthesis was inhibited by ß-lactam antibiotics, one has to conclude that early reactions in murein synthesis are inhibited as a result of the inhibition of the transpeptidation. These reactions are catalyzed by membrane associated enzyme activities in the case of murein synthesis from UDP-MurNAc-pentapeptide. In the complete system, the reactions leading to the formation of UDP-MurNAc-pentapeptide are most likely inhibited in addition. This latter system, comprises more sites for regulation and exhibits therefore the higher ß-lactam sensitivity. It is surprising that antibiotics as thienamycin and cefoxitin cause the accumulation of SDS-soluble murein in the X-pentapeptide system, but not in the complete system. It may be concluded, that these antibiotics interfere with the regulation mechanism in such a way, that polymer synthesis continues in the absence of concomitant transpeptidation.

References

1. Mirelman, D., Yashouv-Gan, Y., Schwarz, U.: Biochem. 15, 1781-179o (1976).
2. Pelzer, H., Reuter, W.: Antimicrob. Agents Chemother. 18, 887-892 (198o).
3. Mirelman, D., Nuchamowitz, Y.: Eur. J. Biochem. 94, 541-548 (1979).
4. Ishiguro, E.E., Mirelman, D., Harkness, R.E.: FEBS Letters 12o, 175-177 (198o).
5. Hammes, W.P.: Eur. J. Biochem. 7o, 1o7-113 (1976).
6. Izaki, K., Matsuhashi, M., Strominger, J.L.: J. Biol. Chem. 243, 318o-3192 (1968).

COMPARISON OF THE EFFECTS ON MUREIN SYNTHESIS OF ß-LACTAM ANTIBIOTICS AND D-AMINO ACIDS.

Hildegard Criegee and Walter P. Hammes
Institut für Lebensmitteltechnologie der Universität Hohenheim, Garbenstraße 25, D-7ooo Stuttgart 7o

Introduction

The common effect of ß-lactam antibiotics (1) and of D-amino acids (2) on Gram negative bacteria is characterized by growth inhibition with concomitant morphological changes. For both groups of compounds it was shown that their mode of action on growth inhibition is the interference with the synthesis of murein. This polymer maintains bacterial morphology and cell integrity. _In vitro_ studies with _Gaffkya homari_ (3,4) revealed that in this organism the inhibition of the final cross-linking reaction by ß-lactam antibiotics is the indirect result of the inhibition of preceding reactions catalyzed by DD- and LD-carboxypeptidases, which create the proper substrate for the transpeptidase. Similarely, D-amino acids exert their effect by interfering with the LD-carboxypeptidase solely (5).

It was the purpose of this communication to investigate to which extent the mode of action of the two groups of inhibitors observed in _G. homari_ is also effective in a Gram negative organism.

Results and Discussion

1) The chain length of peptide subunits in murein synthesized _in vitro_. With the aid of ETB prepared from E. coli W7 as described (6), murein was synthesized _in vitro_ and the pro-

The Target of Penicillin

duct was analyzed for the chain length of the peptide subunits and for the amounts of D-alanine released. The analysis was possible by application of radioactive UDP-MurNAc-peptides as the precursors in which the peptide subunit $-Ala^{1}-DGlu^{2}-mDap^{3}-DAla^{4}-DAla^{5}$ was specifically labeled in position 3,4 or 5 (4).

As shown in Fig. 1, wall-bound and detergent soluble murein were synthesized. Both polymers contain mainly tetrapeptide subunits, but penta- and tripeptide subunits are also present.

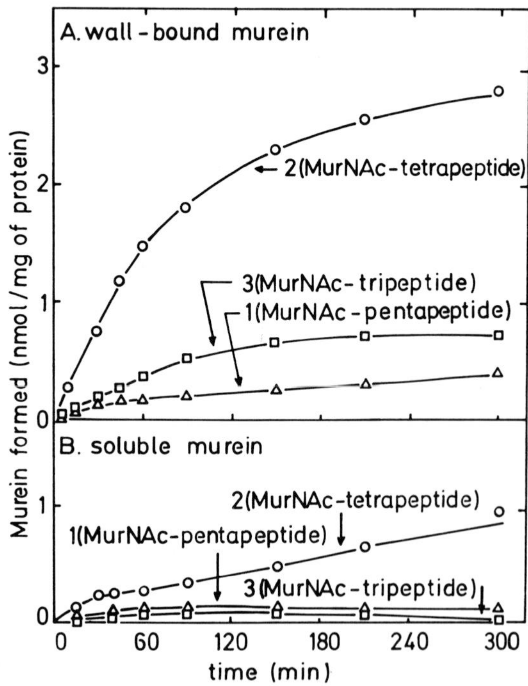

Fig. 1: Kinetics of the incorporation of muropetides into wall-bound (A) and detergent soluble (B) murein.

2) Release of D-alanine from position 4 or 5 during murein synthesis. The kinetics of the release of D-alanine residues from position 4 (Fig. 2B,C) or 5 (Fig. 2A) of the peptide subunits were studied in incubations performed in the

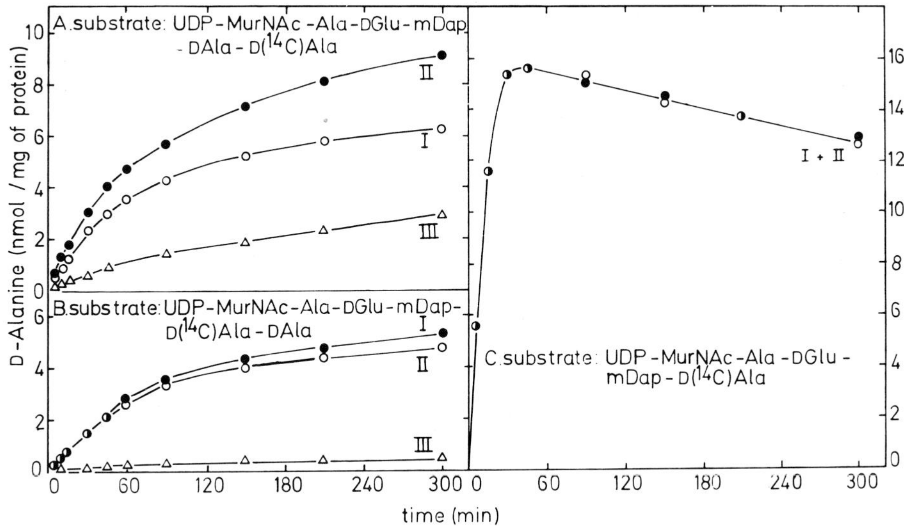

Fig. 2: Kinetics of the release of D-alanine from positions 4 and 5 of peptide subunits. Curve I: UDP-GlcNAc absent; curve II: UDP-GlcNAc present; curve III: difference from II-I.

presence (curve II) and absence (curve I) of UDP-GlcNAc. The increased amounts of D-alanine released in the presence of UDP-GlcNAc, i.e. when murein synthesis is permitted, is depicted in curve III. That amount derived from position 5 may originate from both, the action of transpeptidase and a DD-carboxypeptidase (activity II) degrading pentapeptides in the nascent murein.

In order to create tripeptide subunits, DD-carboxypeptidase activity has to precede LD-carboxypeptidase action. Therefore, only little LD-carboxypeptidase activity can be detected with UDP-MurNAc-pentapeptide as the substrate (Fig. 2B) when compared with UDP-MurNAc-tetrapeptide (Fig. 2C). The increase in D-alanine liberation in the presence of UDP-GlcNAc (Fig. 2B, curve III) indicates activity II of LD-carboxypeptidase degrading nascent murein. This activity was not detectable in incubations with UDP-MurNAc-tetrapeptide (Fig. 2C), since this precursor is a poor substrate for the synthesis of wall-bound murein and, in addition, the release of D-alanine by activity I is extremely efficient with this substrate.

3) The effect of inhibitors on murein synthesis (Tab. 1). Ampicillin inhibited the syntheses of wall-bound and of total murein from both UDP-MurNAc-pentapeptide and, at a higher concentration, from UDP-MurNAc-tetrapeptide. At 0.1 µg/ml of ampicillin the DD-carboxypeptidase degrading UDP-MurNAc-pentapeptide (activity I) was inhibited by 83%, whereas activity II of this enzyme was rather resistant at that concentration. In order to determine the effect of activity II, the following calculations were performed: Incubations with and without UDP-GlcNAc yielded a difference in D-alanine from position 5 of 1270 pmol in the absence and 1180 pmol in the presence of 0.1 µg/ml of ampicillin. At that concentration 1650 pmol of murein were synthesized with a 29% degree of cross-linkages. Thus, only 480 pmol (29% of 1650 pmol) of D-alanine were split off by transpeptidase activity and, therefore, 700 pmol (1180-480 pmol) originate from the action of activity II. The corresponding amount calculated for the incubation without inhibitor was equal to 850 pmol of D-alanine, indicating an inhibition of activity II of DD-carboxypeptidase by 18% only. LD-carboxypeptidase activity was not affected by ampicillin.

With thienamycin it was observed that murein synthesis from UDP-MurNAc-tetrapeptide was more sensitive than from -pentapeptide. Both DD-carboxypeptidase activities were sensitive to the antibiotic, whereas only activity II of LD-carboxypeptidase was inhibited. These findings are consistent with results obtained with *G. homari* (4) which revealed that LD-carboxypeptidase activity and the synthesis of wall-bound murein from UDP-MurNAc-tetrapeptide are thienamycin sensitive.

D-Alanine and D-leucine exhibited no effect on DD-carboxypeptidase activity. They caused, however, a reduction in the degree of cross-linkages and differ in that respect from

Table 1: Effect of inhibitors on murein synthesis *in vitro*

Product	Sub-strate*	Ampicillin ($\frac{\mu g}{ml}$)	Ampicillin Inhibition (% of control)	Thienamycin ($\frac{\mu g}{ml}$)	Thienamycin Inhibition (% of control)	D-Alanine ($\frac{\mu g}{ml}$)	D-Alanine Inhibition (% of control)	D-Leucine ($\frac{\mu g}{ml}$)	D-Leucine Inhibition (% of control)	Nocardicin A ($\frac{\mu g}{ml}$)	Nocardicin A Inhibition (% of control)
total murein	1	2	88	1.7 17	115 78	17	64	5	6o	5	63
wall-bound murein	1	2	5o	1.7 17	13o 6o	17	51	5	51	5	5o
cross-links**	1	o.1	27 → 29%	17	33 → 29%	17	27 → 16%	5	27 → 19%	5	27 → 27%
total murein	2	6	69	1.7 17	94 53		n.d.+		n.d.+		n.d.+
wall-bound murein	2	6	5o	1.7 17	82 36	17	153	5	95	16	5o
DD-Carboxy-peptidase	1	o.1 1o	27 18	1.7 17	12 1o	17	1oo	5	1oo	1oo	9o
	3	o.1	82	1.7 17	91 7	17	1oo	5	1oo	1oo	9o
LD-Carboxy-peptidase	2	≤5o	1o2	≤17o	1o2	5 5o	1o9 87	5	5o	1oo	7o
	3	≤1o	1o3	17	o	17	38o	5	4o	1o	o

+ not determined

* Substrat: 1= UDP-MurNAc-pentapeptide; 2= UDP-MurNAc-tetrapeptide; 3= peptide subunits in nascent murein.

** The change in the degree of cross-linkages from controls to inhibited samples is presented in the absolute numbers.

the ß-lactam antibiotics. Murein synthesis from UDP-MurNAc-tetrapeptide was activated by D-alanine, whereas D-leucine exhibited a rather inhibitory effect. A major difference was observed in the effect on LD-carboxypeptidase activity. D-Alanine activated activity II fourfold, whereas D-leucine inhibited both activities.

The effects of nocardicin A were similar to the effects of D-amino acids. Thus, the DD-carboxypeptidase activities were resistant but the LD-carboxypeptidase activity II was completely inhibited at a concentration of 10 µg/ml. The results obtained with D-amino acids and with nocardicin A are consistent with results obtained with *G. homari* (4,5). In this organism LD-carboxypeptidase was also affected by this group of inhibitors.

References

1. Duguid, J.P. Edinburgh Medical Journal 53, 401-412 (1946).
2. Trippen, B., Hammes, W.P., Schleifer, K.-H., Kandler, O.: Arch. Microbiol. 109, 247-261 (1976).
3. Hammes, W.P., Eur. J. Biochem. 70, 107 113 (1976).
4. Hammes, W.P., Seidel, H.: Eur. J. Biochem. 91, 509-515 (1978)
5. Hammes, W.P.: Eur. J. Biochem. 91, 501-507 (1978).
6. Mirelman, D., Yashouv-Gan, Y., Schwarz, U.: Biochemistry 15, 1781-1790 (1976)
7. Hammes, W.P., Seidel, H.: Eur. J. Biochem. 84, 141-147 (1978)

GLYCAN POLYMERASE WITH NO PENICILLIN-BINDING ACTIVITY IN ESCHERICHIA COLI

Hiroshi Hara, Taro Ueda and Hideho Suzuki
Department of Biology, Faculty of Science, University of Tokyo, Hongo, Tokyo, Japan

Introduction

The enzymes involved in the last stage of murein synthesis in *Escherichia coli* have been found among penicillin-binding proteins (PBPs): first PBP-1b (1, 2, 3) and subsequently PBP-1a (4) and PBP-3 (5) were identified as murein synthetases that carry out both transglycosylation and transpeptidation using the lipid-intermediate as a substrate. PBP-1b appeared to be a major transglycosylase-transpeptidase in *E. coli*.
The ability of penicillins to cause rapid cell lysis was well correlated with their binding to PBP-1b (6), and the membrane fraction of a mutant defective in PBP-1b were almost unable to support *in vitro* murein synthesis and accumulated lipid-intermediates (7, 8). Nevertheless, the mutants defective in PBP-1b grew normally and this was interpreted as a result of the murein synthesis by a detour function of PBP-1a (8). The compensatory activity of PBP-1a should be found in the membrane fraction of PBP-1b$^-$ mutants. Although the residual activity in this membrane fraction was too low to be significant so far, it was found that the residual activity was greatly increased by inclusion of glycerol in the reaction mixture.
The investigation of the glycerol-stimulated activity suggested that two or more glycan polymerases other than PBP-1b were responsible for the activity and that a glycan polymerase with no penicillin-binding ability may exist in *E. coli*.

The Target of Penicillin

Results and Discussion

Stimulation of in vitro murein synthesis by aprotic polar substance. *In vitro* murein synthesis by the membrane fraction of a PBP-1b$^-$ mutant was stimulated several times by addition of glycerol (at 25%) to the reaction mixture using UDP-MurNAc-pentapeptide and UDP-GlcNAc as substrates. Glycerol stimulated murein synthesis by about 1.3 fold in the membrane fraction of wild type and rather inhibitory to murein synthesis by the membrane fraction in which PBP-1b was overproduced due to gene amplification. The glycerol-stimulated activity was practically not increased by overproduction of PBP-1a or PBP-3. These observations suggested that the glycerol-stimulated activity may be ascribed to multiple components other than PBP-1b. Other aprotic polar substances were also effective as

Table 1. Effects of aprotic polar substances on murein-synthesizing activity of the membrane fraction defective in PBP-1b.

Addition	Murein synthesis (cpm)
None	656
Glycerol	2,990
Ethylene glycol	2,011
Propylene glycol	1,388
Polyethylene glycol 200	4,023
" 1000	4,452
" 6000	3,935
Sucrose	1,819
Dimethyl sulfoxide (DMSO)	2,983

The membrane fraction (60 µg protein) of JE5702 (PBP-1b$^-$) prepared in the presence of glycerol was incubated with 14.6 µM UDP-[^{14}C]GlcNAc (42.9 Ci/mol, Amersham) and 150 µM UDP-MurNAc-pentapeptide at 30°C for 30 min in a 40 µl reaction mixture containing 50 mM PIPES [piperazine-N,N'-bis(2-ethanesulfonic acid)] pH 6.5, 5 mM $MgCl_2$ and 0.2 mM DTT (dithiothreitol) with additions (25%) as indicated.

shown in Table 1. Stimulation of murein synthesis by aprotic polar solvents has been shown in Bacillus megaterium (9, described also on E. coli without presentation of data).

Fractionation of the active components. The solubilized membrane proteins were prepared as described previously (3), except that French press was operated at 6,000 psi and glycerol was present at 25% throughout the experimental procedure, unless specified. The membrane proteins were fractionated with ion-exchange cellulose into three fractions: the fraction adsorbed on DEAE-cellulose at 50 mM NaCl (DE-fraction), the fraction adsorbed on DEAE-cellulose without NaCl but not at 50 mM NaCl (DE°-fraction) and the fraction adsorbed on CM-cellulose (CM-fraction). When the membrane proteins of a mutant defective in PBP-1b were fractionated, 56% of the

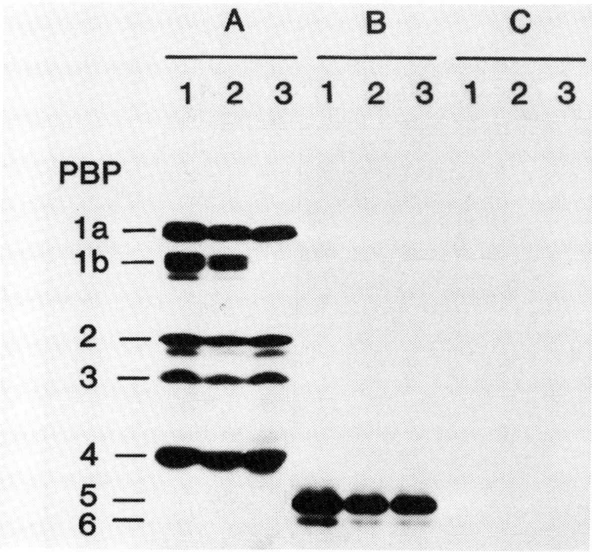

Fig. 1. Analysis of PBPs in fractions separated by ion-exchange cellulose. Portions of each fraction shown in Table 2 were analyzed for PBP by using [^{35}S]penicillinG (1.5 Ci/m mol, NEN). A, DE-fraction; B, DE°-fraction; C, CM-fraction. Lane 1 and 2, d9 (PBP-1b^{+}); lane 3, d10 (PBP-1b^{-}).

Table 2. Fractionation of activities for murein synthesis

Strain	Fraction*1	Protein*2	Polymerizing activity*3	Cross-linking		
				Dimer (cpm)	Monomer (cpm)	$\frac{D}{D+M}$ (%)
d10 (PBP-1b$^-$)	DE	61	0.54(56%)	933	2,432	27.7
	DE°	12	0.06(6%)	ND*4	ND	ND
	CM	5.2	0.37(38%)	-*5	2,333	-
d9 (PBP-1b$^+$)	DE	71	1.37(77%)	1,695	4,866	25.8
	DE°	12	0.10(6%)	ND	ND	ND
	CM	6.5	0.31(17%)	-	2,131	-
d9 (PBP-1b$^+$)	<u>DE</u>	57	0.93(70%)	2,068	5,325	28.0
	<u>DE°</u>	11	0.07(6%)	-	1,001	-
	CM	7.6	0.32(24%)	-	2,775	-

Solubilized membrane proteins were applied to a DEAE-cellulose column at pH 8 in the presence of 50 mM NaCl. The flow-through fraction was desalted by dialysis and applied again to a DEAE-cellulose column at pH 8 in the absence of NaCl. The second flow-through fraction was applied to a CM-cellulose column at pH 5.2 (Tris-maleate) in the presence of 50 mM NaCl. Adsorbed materials were eluted with 1 M (for DE) or 2 M (for DE° and CM) NaCl. Glycerol was included throughout the procedure except the fractions underlined. Finally, all the fractions were dialyzed against 20 mM Tris-HCl pH 7.4 containing 25% glycerol. ^{14}C-labeled lipid-intermediates (103 Ci/mol) accumulated by the membrane fraction defective in PBP-1b was extracted with chloroform/methanol (1:1) (3), dried under nitrogen and suspended in DMSO/10 mM DTT/n-octanol (9:1:5). A portion of each fraction was incubated with 15 µl of this radioactive substrate extract at 30 °C for 60 min in a 40 µl reaction mixture containing 50 mM PIPES pH 6.5, 5 mM $MgCl_2$, 0.2 mM DTT and 8% polyethylene glycol 1000. The reaction products were analyzed by paper chromatography with isobutyric acid/ammonia. More than 80% of synthesized murein was digested with lysozyme and separated into dimer (D) and monomer (M) by paper electrophoresis with 7% formic acid (10).

*1 DE, adsorbed to DEAE-cellulose at 50 mM NaCl.
DE°, adsorbed to DEAE-cellulose in the absence of NaCl.
CM, adsorbed to CM-cellulose.

*2 Percent of solubilized membrane protein.

*3 nmol/mg solubilized membrane protein/hr

*4 ND, not determined.

*5 -, practically no dimer detected.

recovered activity was found in DE-fraction, 6% in DE°-fraction and 38% in CM-fraction (Table 2). No detectable amount of PBP was present in CM-fraction, whereas DE-fraction contained PBP-1a, -1b, -2, -3 and -4; DE°-fraction contained mainly PBP-5 and PBP-2, -4 and -6 in small amounts (Fig. 1). Glycan polymerization was independent of penicillin-binding ability. The reaction products were digested with lysozyme and analyzed for cross-linking. Dimers were not detectable in the products by CM-fraction, while the products by DE-fraction showed about 28% cross-linking (Table 2). The activity in DE-fraction may be ascribed to PBP-1a or -3, or both. Membrane proteins were prepared from the isogenic cells of wild type with respect to PBP-1b and fractionated in the same way. With the fractions of the wild type preparation, distribution of PBPs among the fractions was essentially identical with that found in the mutant and a similar level of the polymerization activity was found in CM-fraction, although

Table 3. Effects of moenomycin and vancomycin on the murein-synthesizing activities of DE- and CM-fractions.

Antibiotic added	Conc. (ug/ml)	% inhibition		
		CM-fraction (d9, PBP-1b$^+$)	DE-fraction (d9, PBP-1b$^+$)	DE-fraction (d10, PBP-1b$^-$)
Moenomycin	0.1	28	81	73
	1	50	96	ND*
	10	62	96	96
Vancomycin	10	79	76	79
	50	92	93	87

Methods for preparation and assay were essentially as described in the legend to Table 2. Activities in the absence of antibiotics were: 4,417 cpm (d9 CM), 3,786 cpm (d9 DE) and 1,675 cpm (d10 DE) in the experiments with moenomycin; and 1,386 cpm (d9 CM), 3,701 cpm (d9 DE) and 1,675 cpm (d10 DE) in the experiments with vancomycin.

* ND, not determined.

the activity in DE-fraction was higher in this case, due to PBP-1b (Table 2). The result suggested that a mutant type of PBP-1b was not responsible for the glycan polymerization activity in CM-fraction. In all the preparations, the minor activity for glycan polymerization was found in DE°-fraction at a nearly constant level. The products by DE°-fraction were not cross-linked (Table 2). The activity of this fraction may be the same as that found in CM-fraction. The polymerase activity in the CM-fraction was inhibited by vancomycin and moenomycin, although it was less sensitive to the latter (Table 3). The glycan polymerase in CM-fraction has no transpeptidase activity and consequently should require cooperation of a cross-linking enzyme to serve sacculus synthesis. Such a cross-linking enzyme might turn out to be one of the major PBPs, or else might be found among the minor PBPs.

References

1. Nakagawa, J., Tamaki, S., Matsuhashi, M.: Agric. Biol. Chem. 43, 1379-1380 (1979).
2. Suzuki, H., Van Heijenoort, Y., Tamura, T., Mizoguchi, J., Hirota, Y., Van Heijenoort, J.: FEBS Lett. 110, 245-249 (1980).
3. Tamura, T., Suzuki, H., Nishimura, Y., Mizoguchi, J., Hirota, Y.: Proc. Natl. Acad. Sci. USA 77, 4499-4503 (1980).
4. Ishino, F., Mitsui, K., Tamaki, S., Matsuhashi, M.: Biochem. Biophys. Res. Commun. 97, 287-293 (1980).
5. Ishino, F., Matsuhashi, M.: Biochem. Biophys. Res. Commun. 101, 905-911 (1981).
6. Spratt, B. G.: Proc. Natl. Acad. Sci. USA 72, 2999-3003 (1975).
7. Tamaki, S., Nakajima, S., Matsuhashi, M.: Proc. Natl. Acad. Sci. USA 74, 5472-5476 (1977).
8. Suzuki, H., Nishimura, Y., Hirota, Y.: Proc. Natl. Acad. Sci. USA 75, 664-668 (1978).
9. Oka, T.: Antimicrob. Agents Chemother. 10, 579-591 (1976).
10. Linnett, P. E., Strominger, J. L.: J. Biol. Chem. 249, 2489-2496 (1974).

BIOSYNTHESIS OF PEPTIDOGLYCAN IN GAFFKYA HOMARI

Processing of nascent glycan by reactivated membranes[1]

Francis Neuhaus, Rabindra Sinha, Claudette Bardin[2] and Efstathia Kalomiris[3]
Department of Biochemistry, Molecular Biology and Cell Biology
Northwestern University
Evanston, Illinois

Introduction

In *Gaffkya homari* the processing of nascent peptidoglycan (PG), prior to its incorporation into the cell wall, requires the regulated actions of a variety of membrane-associated enzymes: DD-carboxypeptidase; LD-carboxypeptidase; N^{ε}-(D-Ala)-transpeptidase (1,2,3). The sequence of events catalyzed by these enzymes is only partially understood. A variety of PG-synthesizing systems containing either membranes, membrane-walls, or toluene-treated cells has been used to study these events. These membrane systems have retained many of the features of organization and processing which the bacterium requires for the assembly of its wall. In our experiments, partially purified membranes from *G. homari* reactivated by freeze-thawing have been used to study the events involved in the assembly of sodium dodecyl sulfate (SDS)-insoluble PG (4,5). The ability to reactivate these membranes in the absence of added walls for the synthesis of SDS-insoluble PG allows one to monitor stages in processing which are not influenced by wall-PG. The objectives of these experiments are directed to three questions. (1) Is the formation of cross-linked dimers correlated with the synthesis of SDS-insoluble PG? (2) Is SDS-soluble PG a precursor of SDS-insoluble PG? (3) Is cross-linking an essential feature of SDS-insoluble PG?

[1]Supported by grant AI-04615 from the National Institute of Allergy and Infectious Diseases.
[2]Present address: University of Miami, School of Medicine, Miami FL.
[3]Present address: SUNY at Stony Brook, Long Island NY.

Results

1) Time course of cross-linking. The activity of the N^{ε}-(D-Ala)-Lys transpeptidase can be estimated from the amount of bis-disaccharide peptide dimer formed. In characterizing the in vitro synthesized PG, four major fractions were resolved from the Chalaropsis sp digests by thin layer chromatography (M', M, D' and D). M' is amidated disaccharide peptide monomer, M is non-amidated disaccharide peptide monomer, D' is amidated bis-disaccharide peptide dimer and D is non-amidated bis-disaccharide peptide dimer (Sinha and Neuhaus, unpublished observations). The presence of these components is consistent with the composition of peptides isolated from the cell walls of G. homari (6).

The time courses of formation of bis-disaccharide peptide dimers, D' and D, are shown in Fig. 1A. There was essentially no lag in the formation of the amidated cross-linked dimer (D'). The formation of the non-amidated dimer (D) lagged that of amidated dimer (D'). Based on the percentage of dimers, the in vitro synthesized total PG attained a steady state cross-linkage of 19% (D' and D) by 15 min (Fig 1B).

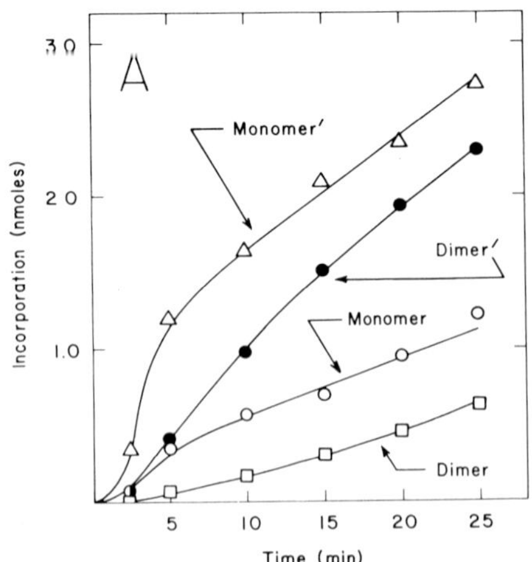

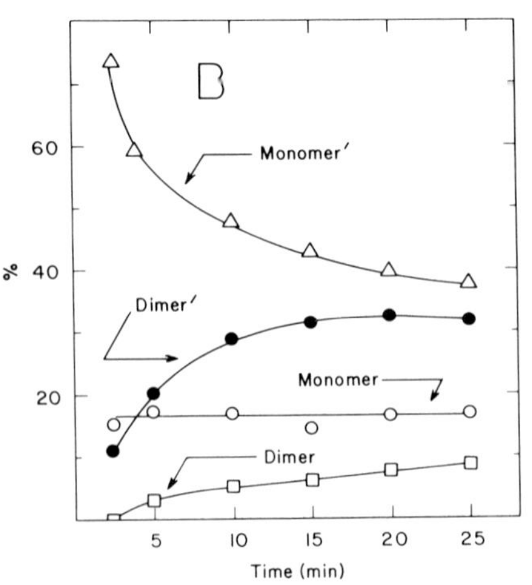

Fig. 1. Time courses of incorporation of [^{14}C]GlcNAc into M', M, D' and D of total PG synthesized by reactivated membranes (freeze-thawed six cycles). Total PG was isolated at the indicated times and digested with the Chalaropsis sp muramidase.

2) Time courses of SDS-insoluble PG formation. As described by Kalomiris *et al.*(4), the time course of formation of 10 kg SDS-insoluble PG by reactivated membranes was characterized by a 15 min lag time. However, as illustrated in Fig. 2A, the amount of SDS-insoluble PG which was formed depended on the sedimentation force applied to the sample during the isolation. For example, sedimentation at 37 kg yielded significantly larger amounts of SDS-insoluble PG than sedimentation at 10 kg. Furthermore, the apparent lag time decreased from 15 min to 5 min. In contrast, the time course of total PG synthesized was characterized by a lag of 1-2 min (Fig. 2B). These results suggested that the assay for SDS-insoluble PG is dependent in part on its aggregate or micelle size. Micelles which did not sediment at 10 kg after a 10 min incubation were easily sedimented at 37 kg (Fig. 2A). It appeared that these micelles could be further intercross-linked with time to form larger micelles which were then sedimented at 10 kg. Comparison of the results in Fig 1A (formation of D') with those in Fig 2B indicated that the 10 kg SDS-soluble PG was significantly cross-linked. The 10 kg SDS-soluble PG was the only component of the total PG fraction during the first 15 min (Fig. 2B).

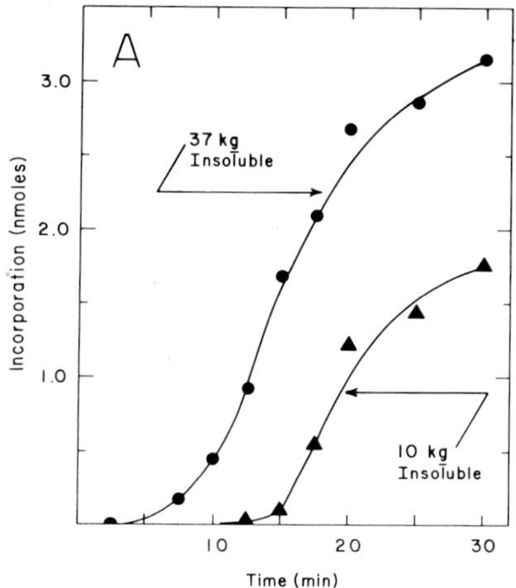

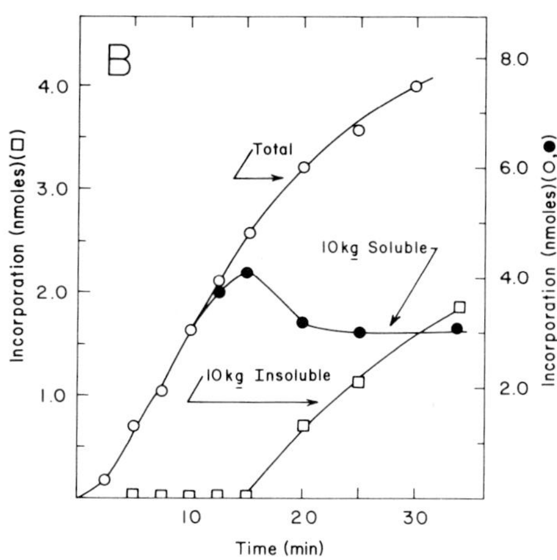

Fig. 2. Time courses of formation of 37 kg and 10 kg SDS-insoluble PG (A) and total PG, 10 kg SDS-soluble PG, and 10 kg SDS-insoluble PG (B) synthesized by reactivated membranes.

3) Utilization of SDS-soluble PG for the synthesis of SDS-insoluble PG. The above experiments suggested that SDS-soluble PG is a precursor of SDS-insoluble PG. In order to test this proposal, the time courses of syntheses of 10 kg and 37 kg SDS-insoluble PG from 37 kg SDS-soluble PG were followed in a pulse-chase experiment (Fig. 3). In this experiment reactivated membranes synthesized [^{14}C]labeled 37 kg SDS-soluble PG during a 5 min pulse (-5 min to zero). During the chase the specific activity of the UDP-[^{14}C]GlcNAc was decreased by a factor of 100. The appearance of 10 kg SDS-insoluble PG required 15 min. In contrast, only 2-3 min was required to observe 37 kg SDS-insoluble PG. These results supported the proposal that a fraction of the 37 kg SDS-soluble PG is a precursor of both the 37 kg and 10 kg SDS-insoluble PG.

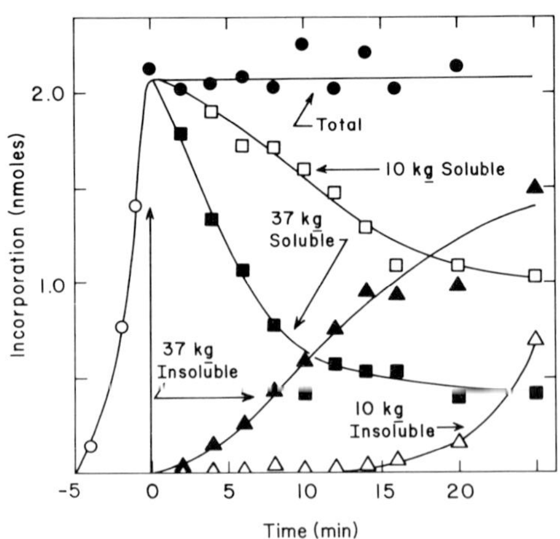

Fig. 3. Time courses of incorporation of [^{14}C]GlcNAc into 10 kg and 37 kg SDS-insoluble PG from [^{14}]GlcNAc labeled SDS-soluble PG determined in a pulse-chase experiment.

4) Utilization of SDS-soluble PG accumulated in the presence of penicillin for the synthesis of SDS-insoluble PG. The results in Fig. 3 indicated that cross-linked SDS-soluble PG is a precursor of SDS-insoluble PG. A more defined precursor-product relationship would be established if the precursor were uncross-linked SDS-soluble PG. The formation of such a precursor would allow one to separate events of processing from those involved in synthesis and elongation of PG. In order to separate these events, a two-stage incubation was designed. In the first stage,

penicillin was used to induce the accumulation of uncross-linked PG. In the second stage, penicillinase was added to degrade the penicillin allowing one or more of the inhibited enzyme(s) to reactivate. In addition, nisin was added to prevent the further elongation and synthesis of new PG (7). In Fig. 4, the time course of 37 kg SDS-insoluble PG synthesis from 37 kg SDS-soluble PG is illustrated. In this experiment PG was synthesized for 30 min (i.e. -30 to zero time) in the presence of 10 μg/ml penicillin. At zero time, penicillinase and nisin were added to the reaction mixture. Under conditions in which further synthesis and elongation were inhibited, the formation of the 37 kg SDS-insoluble PG was monitored. Its formation was correlated with a decrease in the amount of 37 kg SDS-soluble polymer. These results clearly illustrated that 37 kg SDS-soluble PG synthesized in the presence of penicillin can be utilized for the subsequent synthesis of 37 kg SDS-insoluble PG. Characterization of the SDS-insoluble PG synthesized in the first 15 min revealed only amidated and non-amidated monomers. In contrast, the SDS-insoluble PG synthesized in 30 and 45 min was cross-linked 10% and 17%, respectively.

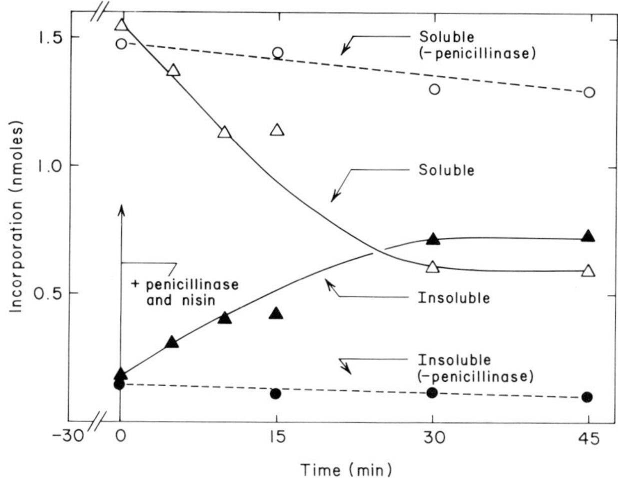

Fig. 4 Utilization of the uncross-linked SDS-soluble PG for the synthesis of SDS-insoluble PG. PG was synthesized for 30 min in reaction mixture containing 10 μg/ml of penicillin. After 30 min (zero time), penicillinase and nisin were added and the incubations were continued for the indicated times.

These experiments indicated that dimer formation is not correlated with the formation of SDS-insoluble PG.

Discussion

The assembly of SDS-insoluble PG by reactivated membranes from G. homari provided a system for studying late stage processing in the absence of pre-existing wall-PG. In this system, we defined SDS(2%)-insoluble PG as that fraction which acquired the necessary characteristics for precipitation in the absence of walls at a specific centrifugal force. Because this product may have ordered regions stabilized by intra- and inter-chain hydrogen bonds, the term micelle was used to denote the product. The designation of PG micelle was originally applied to ordered regions of wall-PG by Hammes and Kandler (2) and Formanek (8).

The formation of cross-linked dimers does not correlate with the formation of SDS-insoluble PG. In fact, SDS-soluble PG is significantly cross-linked when assembled in the absence of penicillin. Thus, cross-linking is not an essential feature correlated with SDS-insolubility. This conclusion is supported by the additional observation that SDS-insoluble PG which is assembled from uncross-linked PG (accumulated in the presence of penicillin) is not cross-linked during the first 15 min following the removal of penicillin. It would appear that micelle size is an important factor in SDS insolubility. In this system SDS-insoluble PG can be cross-linked, partially cross-linked or uncross-linked.

References

1. Hammes, W.P.: Eur. J. Biochem. 70, 107-113 (1976).
2. Hammes, W.P., Kandler, O.: Eur. J. Biochem. 70, 97-106 (1976).
3. Hammes, W.P., Seidel, H.: Eur. J. Biochem. 84, 141-147 (1978).
4. Kalomiris, E., Bardin, C., Neuhaus, F.C.: J. Bacteriol. 150, 535-544 (1982).
5. Carpenter, C.V., Goyer, S., Neuhaus, F.C.: Biochemistry 15, 3146-3152 (1976).
6. Nakel, M., Ghuysen, J.-M., Kandler, O.: Biochemistry 10, 2170-2175 (1971).
7. Reisinger, P., Seidel, H., Tschescke, H., Hammes, W.P.: Arch. Microbiol. 127, 187-193 (1982).
8. Formanek, H.: Biophys. Struct. Mechanism 4, 1-14 (1978).

THE INHIBITION OF IN VITRO SYNTHESIS OF MUREIN IN GAFFKYA HOMARI BY MONOCYCLIC ß-LACTAM ANTIBIOTICS NOCARDICIN B,E AND AZTHREONAM

Gudrun Wolf and Walter P. Hammes
Institut für Lebensmitteltechnologie der Universität Hohenheim
Garbenstr. 25, D-7000 Stuttgart 70

Introduction

Nocardicin A was the first example of a monocyclic ß-lactam antibiotic exhibiting relatively high antimicrobial potency. It is produced by *Nocardia uniformis* together with a family of derivatives named nocardicin B, C, D, E, F, and G. Nocardicin A and B are stereoisomers (Fig.1). The oxime group of nocardicin A is syn to the acylamino group, and nocardicin B is the anti-isomer. Nocardicin E is an analogue of nocardicin A, in which the D-homoserine substituent is missing (1). Previously, it was shown, that nocardicin A inhibits the peptidoglycan synthesis by exerting an effect on the LD-carboxypeptidase similar to that caused by D-amino acids (2). The aim of this study was to compare the D-amino acid analogous effect of nocardicin B with the effect of nocardicin E on the *in vitro* synthesis of murein in *Gaffkya homari*. An inhibitory effect of nocardicin E should be exerted by the ß-lactam moiety of the molecule. This moiety might exhibit an effect on murein synthesis which is general to monocyclic ß-lactams. In order to draw this conclusion the mode of action on peptidoglycan synthesis of azthreonam as an example of a synthetic monobactam antibiotic was studied (3).

The Target of Penicillin

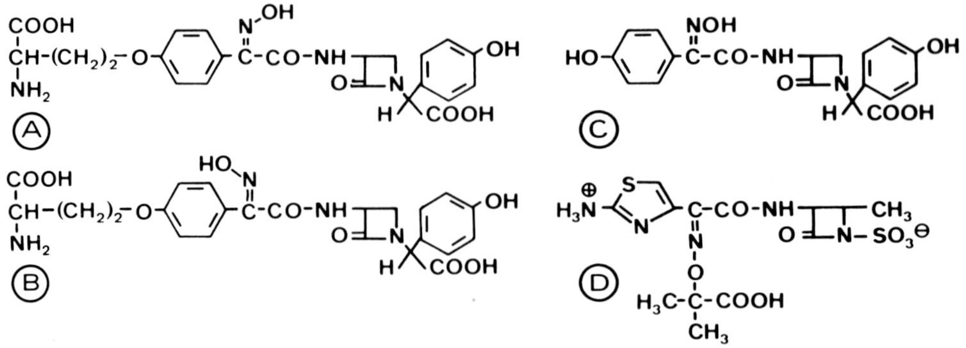

Fig.1. Structures of the monocyclic ß-lactam antibiotics nocardicin A (A), nocardicin B (B), nocardicin E (C), azthreonam (D).

Materials and Methods

Nocardicin A, B, and E were a gift from Fujisawa Pharmaceutical Co. and azthreonam from Squibb and Sons, Inc. The preparation of wall membrane enzymes from Gaffkya homari were described elsewhere (4). The protein content of the batch was 12 mg/ml according to the method of Lowry (5). The UDP-MurNAc-peptides were prepared as described (6). The standard assay of peptidoglycan synthesis from UDP-GlcNAc and UDP-MurNAc(^{14}C)peptides and the determination of the products of peptidoglycan synthesis have been described previously (4).

Results and Discussion

The effects of nocardicin B and E on the formation of peptidoglycan from UDP-MurNAc-pentapeptide (Fig.2).
The synthesis of wall-bound peptidoglycan is inhibited by 50 % at 0,6 mM/l of nocardicin B. The amount of total peptidoglycan is not affected at this concentration with nocardicin E. The ED_{50} for inhibition of wall-bound peptidoglycan is 9 mM/l.

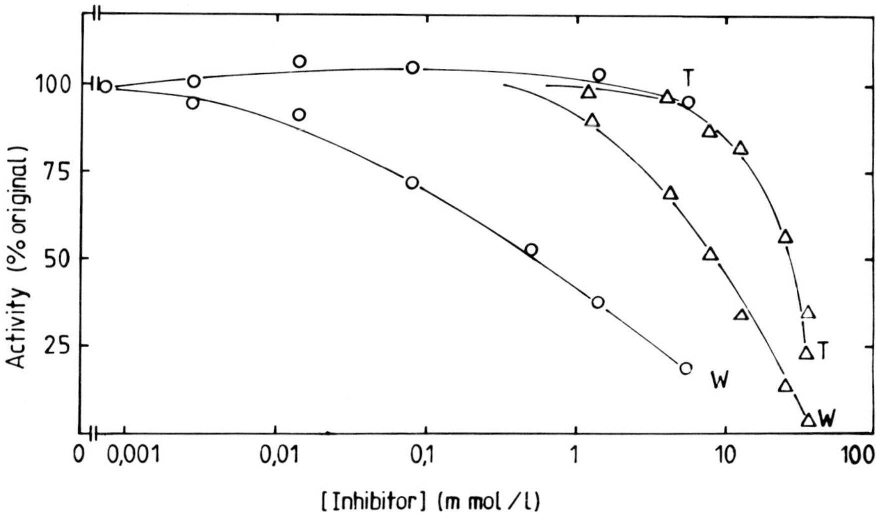

Fig.2. Effect of nocardicin B (o) and E (Δ) on the formation of peptidoglycan from UDP-MurNAc-Ala-DGlu((^{14}C)Lys-DAla-DAla). An activity of 100% corresponds to a formation of 5,1 nmol/mg protein of total peptidoglycan (T) and 4,4 nmol/mg of protein of wall-bound peptidoglycan (W).

The effect of nocardicin B and E on the formation of peptidoglycan from UDP-MurNAc-tetrapeptide (Fig.3).

A marked increase in the release of D-alanine from position 4 was observed with nocardicin B being at maximum at 1 mmolar concentration. At this concentration synthesis of wall-bound murein was inhibited by 65%. The synthesis of total murein was only poorly affected. Nocardicin E inhibited wall-bound and total murein by 50% with a concentration of 3 mmol/l. In contrast to nocardicin B, nocardicin E inhibited the release of D-alanine. The presence of a D-amino acid residue as a substituent in nocardicin B suggests an analogous mode of action as the effect described for nocardicin A (2). The enhanced release of D-alanine from position 4 of the peptide subunit is consistent with an exchange of this amino acid with the inhibiting D-amino acid or a derivative, catalyzed by the LD-carboxypeptidase. As a result less tripeptide subunits are formed in nascent peptidoglycan causing an inhibition of transpeptidation reaction.

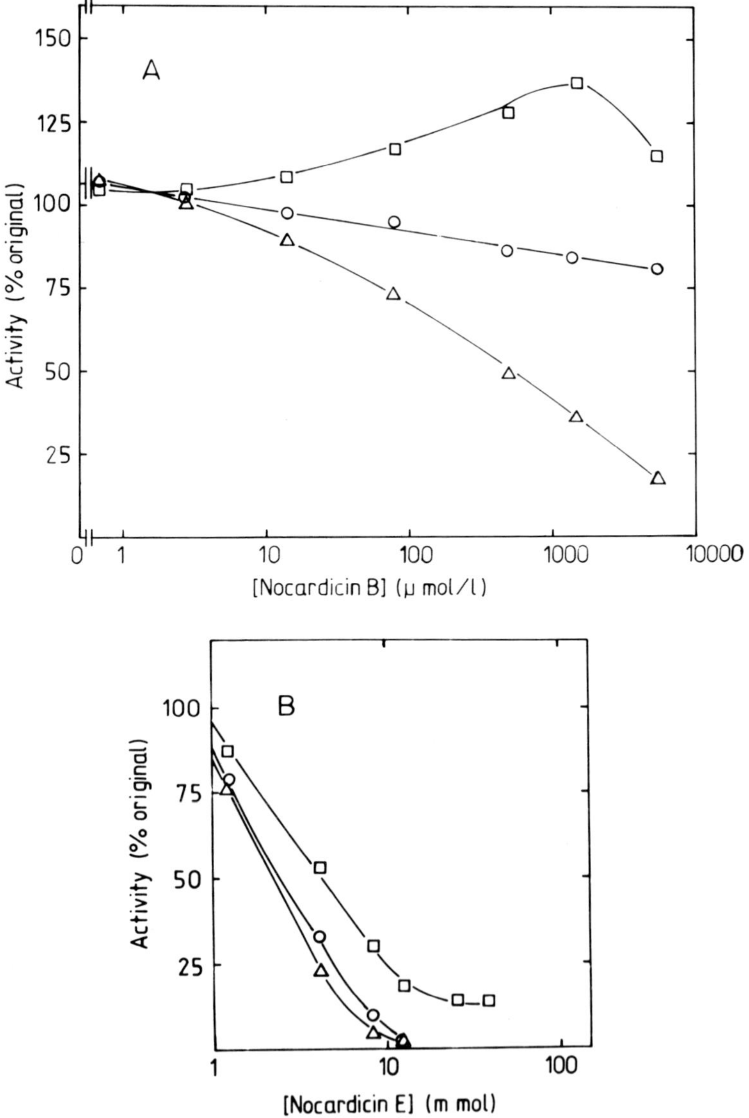

Fig.3. Effect of nocardicin B (A) and nocardicin E (B) on the formation of products from UDP-MurNAc-Ala-DGlu(Lys-D(^{14}C)Ala). An activity of 100% corresponds to a formation of 4,5 nmol/mg of protein of total peptidoglycan (o), 4,4 nmol/mg of protein of wall-bound peptidoglycan (△) and 12 nmol/mg of protein of released D-alanine (□).

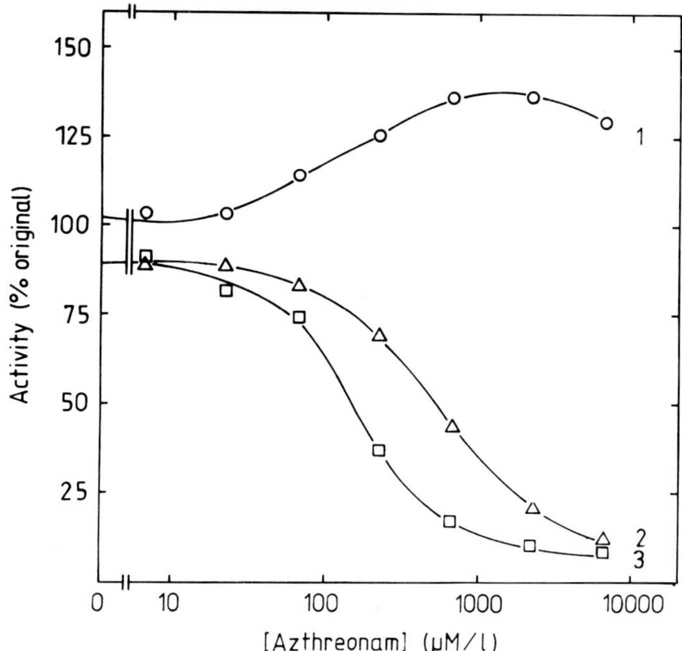

Fig.4. Effect of azthreonam on the formation of products from UDP-MurNAc-pentapeptide. The values of 100% activity (nmol/mg of protein) are shown in parantheses. Curve 1: Total peptidoglycan from UDP-MurNAc-Ala-DGlu(Lys-DAla-D(^{14}C)Ala); Curve 2: wall-bound peptidoglycan from UDP-MurNAc-Ala-DGlu(^{14}C)Lys-DAla-DAla) (5,6nmol; Curve 3: release of D-(^{14}C)alanine from UDP-MurNAc-Ala-DGlu(Lys-DAla-D(^{14}C)Ala) (1,6nmol).

The isomerism at the oxime function of nocardicin A and B does not contribute to the inhibitory effect in Gaffkya homari.

The effect of azthreonam on the formation of peptidoglycan (Fig.4).

The most sensitive reaction in peptidoglycan synthesis from UDP-MurNAc-pentapeptide is the release of D-alanine from position 5, catalyzed by the DD-carboxypeptidase. The enzyme is inhibited by 50% at 150 μM/l. As a result the incorporation of UDP-MurNAc-DGlu(Lys-DAla-D(^{14}C)Ala) into total peptidoglycan is enhanced. At slightly increased concentrations the wall-bound peptidoglycan is also inhibited (ED_{50}:540 μM/l). The biosynthe-

sis of peptidoglycan from UDP-MurNAc-tetrapeptide is virtually not affected by azthreonam (data not shown).
In summary, the effect of the monocyclic ß-lactam antibiotics can be described as follows: The mode of action and the range of inhibitory concentrations are similar for nocardicin B and D-amino acids. Nocardicin E and B differ in the inhibitory concentration, which is ten fold less with the latter compound. Further the release of D-alanine from position 4 of peptidoglycan is activated with nocardicin B, whereas nocardicin E was inhibitory. The effect of azthreonam on the *in vitro* synthesis of murein is on a qualitative basis similar to penicillin G.

Acknowledgement: We thank M. Bair for technical assistance.The work was supported by SFB 76 of the Deutsche Forschungsgemeinschaft.

References

1. Hosoda, J., Konomi, T., Tani, N., Aoki, H., Imanaka, H.: Agric. Biol. Chem. 41, 2031-2020 (1977).
2. Hammes, W.P., Seidel, H.: Eur. J. Biochem. 91, 509-515 (1978).
3. Sykes, R.B., Bonner, D.P., Bush, K., Georgopapadakou, N.H., Wells, J.S.: J. Antimicrob. Chemother. 8, Suppl. E, 1-16 (1981).
4. Hammes, W.P., Kandler, O.: Eur. J. Biochem. 70, 97-106 (1976).
5. Lowry, O.H., Rosebrough, N.J., Farr, A.J., Randall, R.J.: J. Biol. Chem. 193, 265-276 (1951).
6. Hammes, W.P., Neuhaus, F.C.: J. Biol. Chem. 249, 3140-3150 (1974).

PEPTIDOGLYCAN SYNTHESIS IN VITRO BY PARTICULATE PREPARATION FROM *Streptococcus sanguis* 34

Victor M. Reusch, Jr.

Department of Biological Sciences, University of Denver
Denver, Colorado 80208

Introduction

Dental plaque adheres tenaciously to tooth surfaces and consists primarily of oral bacteria and extracellular polysaccharides. It has been implicated as an important etiological factor in caries and periodontal disease (1). Dental plaque formation may be governed by a complex set of biological recognition processes involving cell surface receptors (2,3). *Actinomyces viscosus* strain T14V (T14V) coaggregates specifically with *Streptococcus sanguis* strain 34 (Ss34) by means of a heat-labile "lectin" (3,4) that is localized on surface fibrils of T14V (5). Coaggregates formed *in vitro* may exceed the dimensions of the participants by two orders of magnitude (Fig. 1a). The interaction is inhibited by 28 mM lactose (3,4) and does not appear to involve a specific portion of the streptococcal cell surface

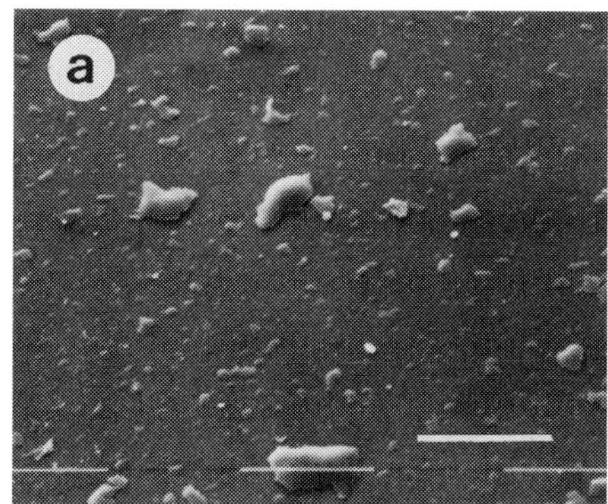

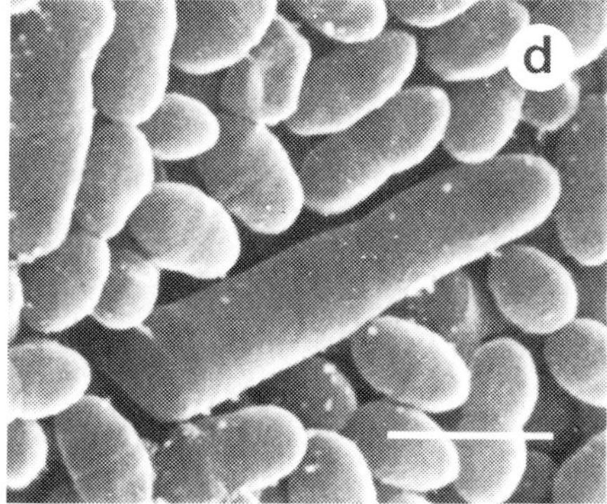

Fig. 1. Specific coaggregation of *S. sanguis* 34 with *A. viscosus* T14V. (a) The size distribution and irregular shape of coaggregates (bar=100 μm). (b) Detail of coaggregate. Note the lack of polarity evident in the association between the two cell types (bar = 1 μm).

The Target of Penicillin

Fig. 2. Thin section of packed sacculi prepared from lysozyme-treated S. sanguis 34 (bar = 1 μm).

(Fig. 1b). Our interest in specific coaggregation has focussed on the nature of attachment of the carbohydrate receptor ("site") to the streptococcus and the mechanism by which this structure is synthesized.

Cellular Localization of the Site

Sacculi (6) prepared from Ss34 by extensive extraction (90°C) of whole bacteria with 1% sodium dodecyl sulfate/0.5% 2-mercaptoethanol (SDS/ME) readily coaggregated with T14V to an extent comparable to untreated bacteria. Treatment of Ss34 with hen egg white lysozyme (HEWL) in the concentration range of 1 mg enzyme per 2.5 to 45 mg of bacteria (dry weight) released significant amounts of material from the bacteria without causing loss of the site, although the bacteria were simultaneously sensitized to bacteriolysis in the presence of 0.1% SDS (7). We reasoned that the site might be attached to an HEWL-resistant core of peptidoglycan in Ss34, analogous to that described for Micrococcus sodonensis (8). When Ss34 was grown in Todd-Hewitt broth and treated sequentially with HEWL and hot SDS/ME, DNA gel formed. The gel was sheared in a blender and the insoluble fraction collected by centrifugation (48,000 x g, 15 min). The extraction was repeated for two cycles and the insoluble fraction was subjected to SDS/polyacrylamide gel electrophoresis for 3 hr at room temperature. The insoluble fraction did not penetrate the 10% gel matrix and could be washed off the top of the gel with a stream of deionized water. After washing with 2 M NaCl and deionized water as described previously (6), the insoluble materi-

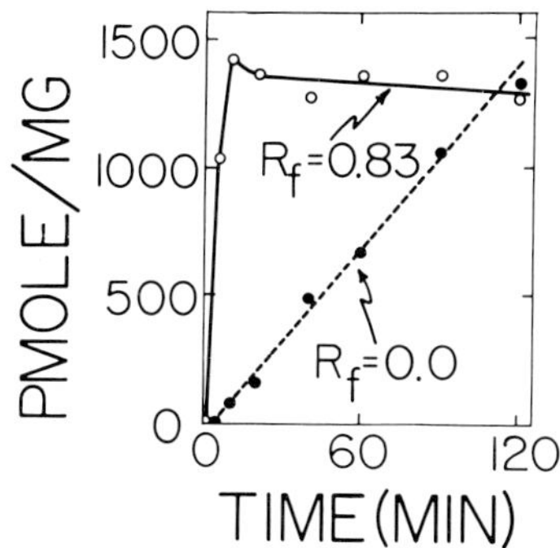

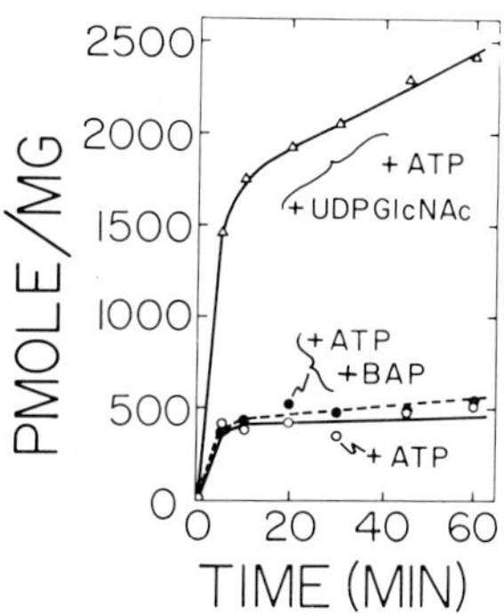

Fig. 3 (LEFT). Peptidoglycan synthesis by membrane particles. Note the lag in appearance of polymer. Reaction mixtures contained (25 µl): 50 mM Tris. HCl (pH 8.0), 25 mM $MgCl_2$, 2 mM ATP, 26 µM UDPMurNAc-L-ala-D-glu-L-lys-D-(^{14}C)ala-D-(^{14}C)ala and 1.4 mM UDPGlcNAc. The reaction mixture was incubated (37°C) for the indicated time and 1 M acetic acid (5 µl) added. The entire mixture was spotted on Whatman 3MM paper (2 x 24.5 cm) which then was developed in 1 M ammonium acetate (pH 3.8)/ethanol = 2/5, v/v. The dried chromatogram was cut into 1 cm segments and counted in nonaqueous scintillation fluid. In order to correct for losses during transfer to the paper, the results were normalized on the basis of the total radioactivity present on each chromatogram (9).

Fig. 4 (RIGHT). Incorporation from UDPMurNAc(^{14}C)pentapeptide into perchloric acid precipitate (9). Reactions were as described in Fig. 3 with the indicated modifications. BAP = *E. coli* alkaline phosphatase.

al was examined by scanning electron microscopy and found to resemble empty cloth sacks. Thin sections revealed sacculi (Fig. 2) that appeared emptier than conventional sacculi prepared from Ss34 grown in Todd-Hewitt broth (6). The sacculi possessed 4% of the absorbance units (650 nm) of the input bacteria and contained combined amino acids and carbohydrates, phosphate, nitrogen and protein. Between 93 and 116% of the coaggregating units of input bacteria were recovered as agglutinating units of sacculi. These data support the hypothesis that the site is firmly linked to a core fraction of the peptidoglycan matrix.

Peptidoglycan Synthesis In Vitro

Ss34 was disrupted by alumina grinding as described previously (9). Stimulation of the phospho-MurNAc pentapeptide (PMp) translocase (10) by ATP varied amongst preparations. Synthesis of peptidoglycan appeared normal (Fig. 3). However, incorporation of radioactivity from UDPMurNAc-(^{14}C)-pentapeptide (UMPPM(^{14}C)p) into perchloric acid precipitate was greater in the presence of UDP-N-acetyl-D-glucosamine (UDPGlcNAc) than in its absence (Fig. 4). The difference could not be accounted for as polymer synthesis (compare Fig. 3). Increases commensurate with those observed in the presence of UDPGlcNAc were not seen either in the presence of alkaline phosphatase (11) (Fig. 4) or by increasing the initial concentration of UMPPM-(^{14}C)p. UDPglucose, UDPgalactose, TDPglucose, GDPmannose and CDPglycerol did not substitute for UDPGlcNAc in evoking the synthesis of higher levels of presumptive lipid intermediates. The stimulation of additional lipid synthesis was dependent upon ATP. Products synthesized in the presence and absence of UDPGlcNAc possessed the same mobility (R_f = 0.14) on silica gel paper in $CHCl_3$/methanol/acetic acid/water = 60/40/4/5. When labelled, acid-precipitated membranes were hydrolyzed by the procedure of Anderson _et al._ (12), a single peak possessing mobility consistent with disaccharide pentapeptide was found (Fig. 5). These data suggest that the product of additional synthesis is lipid intermediate II (12) of peptidoglycan synthesis. Synthesis of these higher levels of intermediates presumably would require the utilization of additional acceptor molecules, the source of which remains obscure.

Incorporation From UDP-N-Acetyl-D-Glucosamine

In the absence of UMPPMp, incorporation from UDP(^{14}C)GlcNAc into acid-precipitable material was dependent upon ATP. Other nucleoside triphosphates substituted for ATP less efficiently, except for UTP which inhibited completely, possibly because of contaminating UDP. The requirement for ATP was satisfied by α,β-methylene ATP, but less well by β,γ-methylene ATP, suggesting that the reaction required hydrolysis of the γ-phosphoryl group of ATP. Whereas UMP, UDP, TMP and CMP were inhibitors of incorporation, only UDP reversed the reaction (Fig. 6). In the most active enzyme preparation, the plateau of synthesis was 583 pmoles per mg membrane as compared with 643 pmoles of C_{55}OPPMp or 1860 pmoles of C_{55}OPPM(GlcNAc)p per mg. When membranes were prelabelled with (^{14}C)GlcNAc, the label could not be chased

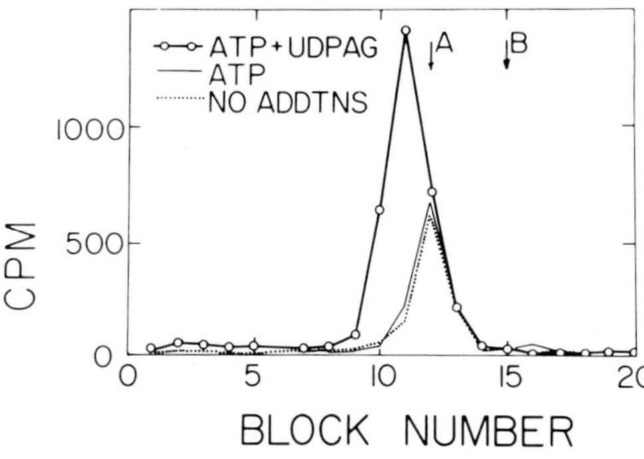

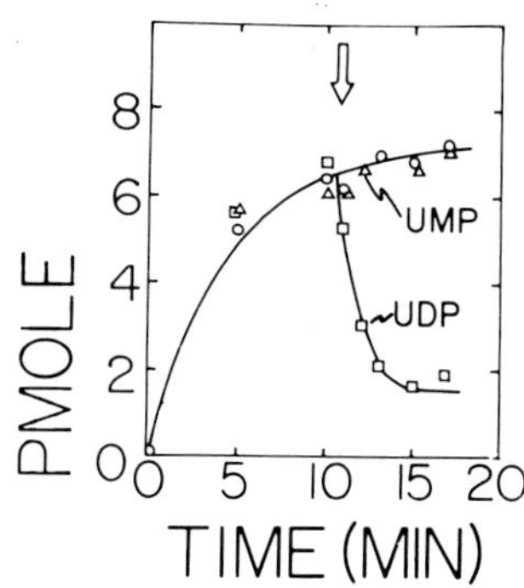

Fig. 5 (LEFT). Chromatography of hydrolysates obtained under the conditions of Anderson et al. (12). Reactions were terminated by addition of 0.1 N HCl (0°C). The pellet obtained by centrifugation was washed once and then hydrolyzed. The soluble fraction was analyzed on Whatman 3MM paper in n-propanol/con. ammonia/water = 6/3/1, v/v/v. The arrows indicate the mobilities of PMp (A) and GlcNAc (B).

Fig. 6 (RIGHT). Reversal of GlcNAc incorporation. The reaction mixture (250 µl) was similar to that described in Fig. 3 except that UMPPMp was left out and 3.6 µM UDP(^{14}C)GlcNAc was used. Aliquots (25 µl) were added to 0.3 M $HClO_4$ (0°C) and the precipitate collected on membrane filters which then were dissolved in scintillation fluid and counted. UMP or UDP was added at the time indicated by the arrow.

into other products by incubation with UMPPMp ± UDPGlcNAc ± ATP. Product was extracted into the organic phase of chloroform/methanol/water (13). It did not separate from peptidoglycan lipid intermediates on silica gel paper or thin layers in neutral, acidic or basic chloroform/methanol/water systems, and cochromatographed with them on DEAE cellulose acetate columns, suggesting that it was a pyrophosphate. When lipid was treated in 5 mM HCl in 50% n-propanol (90°C) (13), 92% of the radioactivity became aqueous-soluble in 6 minutes, but GlcNAc was not released (Fig. 7). After hydrolysis in 0.1 N HCl (12), products consistent with disaccharide pentapeptide and GlcNAc were observed. Thus, at least some of the GlcNac lipid might be peptidoglycan lipid intermediate II, but the ATP requirement for its synthesis remains unexplained. (Supported by U.S.P.H.S. grant nos. DE-05135, DE-05597 and DE-00087).

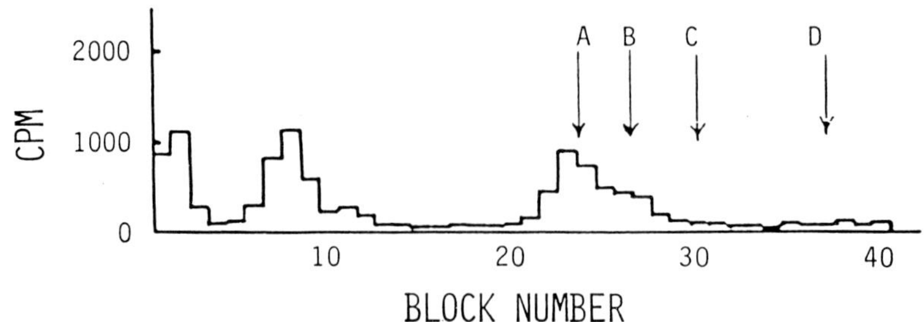

Fig. 7. Chromatography of mild acid hydrolysate (13) of GlcNAc lipid on long chromatograms (2 x 44.5 cm) as described in Fig. 5. Standards: A, PMp; B, glucosamine; C, GlcNAc; D, GlcNAc lipid untreated.

References

1. Schluger, S., Yuodelis, R.A., Page, R.C.: Periodontal disease, p. 85-106, Lea & Febiger, Philadelphia. 1977
2. Gibbons, R.J., Nygaard, M.: Arch. Oral Biol. 15: 1397-1400 (1970).
3. McIntire, F.C., Vatter, A.E., Baros, J., Arnold, J.: Infect. Immun. 21, 978-988 (1978).
4. McIntire, F.C., Crosby, L.K., Vatter, A.E.: Infect. Immun. 36, 371-378 (1982).
5. Cisar, J.O., Barsumian, E.L., Curl, S.H., Vatter, A.E., Sandberg, A.L., Siraganian, R.P.: J. Immunol. 127, 1318-1322 (1981).
6. Reusch, V.M. Jr.: J. Bacteriol. 151, 1543-1552 (1982).
7. Coleman, S.E., Van de Rijn, I., Bleiweis, A.S.: J. Dent. Res. 50, 939-943 (1971).
8. Johnson, K.G., Campbell, J.N.: Biochem. 11, 277-286 (1972).
9. Reusch, V.M. Jr., Panos, C.: J. Bacteriol. 126, 300-311 (1976).
10. Neuhaus, F.C.: Accts. Chem. Res. 4, 297-303 (1971).
11. Heydanek, M.G., Struve, W.G., Neuhaus, F.C.: Biochem. 8, 1214-1221 (1969).
12. Anderson, J.S., Matsuhashi, M., Haskin, M.A., Strominger, J.L.: J. Biol. Chem. 242, 3180-3190 (1967).
13. Bettinger, G.E., Chatterjee, A.N., Young, F.E.: J. Biol. Chem. 252, 4118-4124 (1977)

I gratefully acknowledge the collaboration of J.L.M. Foster, D.S.Haberkorn, S. McLelland, P.A. Russell, D. Barkan and R. Golub.

SYNTHESIS OF PEPTIDOGLYCAN FROM EXTERNALLY SUPPLIED PRECURSORS BY PARTLY AUTOLYSED CELLS OF Bacillus subtilis W23

Charles R. Harrington and James Baddiley

Department of Biochemistry, University of Cambridge,
Tennis Court Road, Cambridge CB2 1QW, U.K.

Introduction

Whereas the individual reactions involved in cell wall assembly in Gram-positive bacteria have been characterized in some detail, the molecular mechanism by which the substrates, synthesized as nucleotide precursors in the cytoplasm, reach their eventual location in the wall (i.e. translocation) is not understood. Recently Bertram _et al_. (1) reported that protoplasts of _B. subtilis_ W23 readily synthesized ribitol teichoic acid, plus linkage unit to peptidoglycan, when UDP-GlcNAc, CDP-glycerol and CDP-ribitol were supplied in the isotonic medium. They postulated that there is some reorientation of the biosynthetic enzyme complex in the membrane that temporarily exposes active sites to the outer surface of the membrane during wall biosynthesis. This study has now been extended to the synthesis of peptidoglycan in the same organism. While protoplasts failed to synthesize peptidoglycan when supplied with externally supplied UDP-GlcNAc and UDP-MurNAc-pentapeptide, cells with partially digested walls did synthesize peptidoglycan.

Results

1) Protoplasts

Protoplasts prepared by incubation of cells at 37°C for 1 h with lysozyme (0.5 mg/ml) in isotonic buffer (0.625M sucrose; 10 mM $MgCl_2$;

50 mM Tris-HCl, pH 7.5) and then gently washed with buffer, to remove lysozyme, did not synthesize peptidoglycan when incubated with UDP-GlcNAc and UDP-MurNAc-pentapeptide between pH 4-9. However, small amounts of polymer were synthesized from UDP-GlcNAc alone, which was optimal at pH 8.5. Peptidoglycan synthesis was taken as the radioactive polymer synthesized from UDP-$[^{14}C]$GlcNAc that remained at the origin of paper chromatograms, developed in ethanol: ammonium acetate (pH 3.8) (5:2, v/v; solvent A) for 18 h, that was dependent on the addition of UDP-MurNAc-pentapeptide. However, by separation of the reaction mixture on chromatograms developed in isobutyric acid: 0.5 M NH_3(aq) (5:3, v/v; solvent B) it was possible to detect the synthesis of material having the same mobility (R_F= 0.8) as prenyl disaccharide-pentapeptide and smaller amounts of material characteristic of free disaccharide-pentapeptide (R_F= 0.5). Synthesis of the former was optimal with 0.28 μM UDP-GlcNAc and 56 μM UDP-MurNAc-pentapeptide; at greater concentrations of the latter, the rate of synthesis decreased. Synthesis of lipid-disaccharide precursors was inhibited not only by tunicamycin but also by bacitracin (Fig. 1). This indicated that the lipid cycle had been completed and that the absence of peptidoglycan synthesis was due to traces of lysozyme, used in protoplast preparation, adhering to the membranes. Presence of the lysozyme would also account for the formation of the free disaccharide-pentapeptide.

2) Partly autolysed cells

Cells undergoing autolysis in isotonic buffer readily utilized nucleotides in the medium to synthesize muramidase-sensitivity, un-cross-linked peptidoglycan that remained only loosely associated with the cell. The optimal conditions for peptidoglycan synthesis included a 20 min-period of autolysis in isotonic buffer at 37°C and pH 7.5 prior to assay. Freshly harvested cells did not synthesize peptidoglycan. Peptidoglycan synthesis was optimal at 37°C and pH 8.5, required Mg^{2+} ions (10 mM) and continued linearly for approximately 30 min. At pH 7.5, the Km for

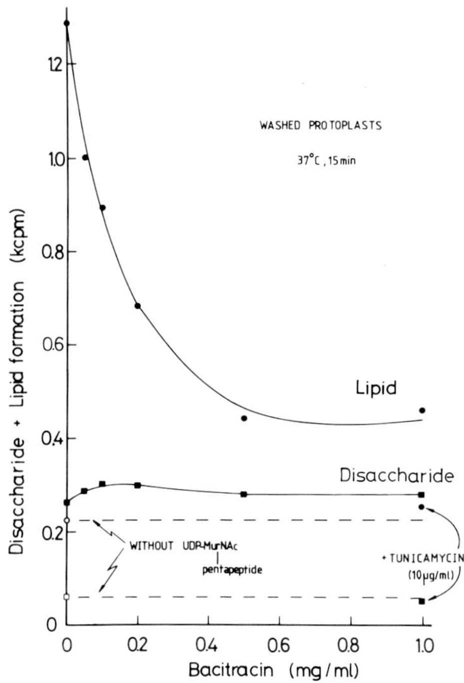

Fig. 1. Effect of bacitracin and tunicamycin on the formation of lipid-linked (●,○) and free (■,□) disaccharide-pentapeptide by washed protoplasts of B. subtilis W23

UDP-GlcNAc and UDP-MurNAc-pentapeptide were 9 µM and 16 µM, respectively. The highest specific activity obtained at pH 7.5 was 7.0 nmol $[^{14}C]$GlcNAc incorporated into peptidoglycan h^{-1} (mg of membrane protein)$^{-1}$. This value was decreased by the action of endogenous autolysins (data not shown). Only the endo-β-N-acetylglucosaminidase (pH 5-7) should decrease the length of the glycan chain whereas the muramyl-L-alanine amidase (pH 9-12) would only preclude crosslinking. Peptidoglycan was similarly formed using partly autolysed preparations of B. subtilis NCIB 3610, B. subtilis 168, B. megaterium KM and B. licheniformis ATCC 9945.

In addition, polymer was also formed from UDP-GlcNAc alone. The latter polymer was firmly cell-associated and unaffected by muramidase or by the presence of vancomycin, bacitracin or tunicamycin at concentrations which inhibited peptidoglycan formation. Its synthesis was optimal at pH 8.5 and with either Mn^{2+} or Mg^{2+} ions (10-20 mM).

3) Location of polymer synthesis

The cells remained intact during autolysis and were converted to protoplasts by incubation with lysozyme. No DNA was released from the cells during autolysis and there was no change in the rate of oxidation of NADH. Peptidoglycan synthesis by partly autolysed cells was inhibited by p-chloromercuribenzoate (pCMBS), trypsin and lysozyme thus indicating that it occurred at the outer surface of the membrane.

There was no uptake, by the cells of UMP, UTP or UDP-GlcNAc ([^{14}C]-labelled in the uridine moiety) nor of GlcNAc-1-phosphate, using a filtration method. There was some UDP-GlcNAc pyrophosphorylase activity exhibited by partly autolysed cells but the product, characterized by separation on paper chromatograms developed in solvent B for 3 days, was released on the outer surface of the cells. There was no trans-phosphorylation exchange reaction for UDP-GlcNAc with UMP although there was efficient exchange between [^{14}C] UMP and UDP-MurNAc-pentapeptide supplied externally. The translocase exchange reaction proceeded linearly during 80 min at 37°C [0.87 nmol min^{-1} (mg of membrane $protein)^{-1}$] and was extremely sensitive to tunicamycin (50% inhibition at 2.5 µg/ml). However, the exchange product did not remain associated with the cells, indicating that the supplied nucleotide precursors were not translocated by the prenyl phosphate prior to polymer synthesis.

4) Energy requirement for peptidoglycan synthesis

B. subtilis W23 grown in minimal salts + threonine + glucose readily took up GlcNAc and incorporated the sugar into highly cross-linked peptido-glycan (approximately 60% of the lysozyme-soluble product remained at the origin of paper chromatograms developed in solvent B for 3 days).

Addition of uncouplers such as 2,4-dinitrophenol (DNP) at 10 mM or carbonylcyanide *m*-chlorophenylhydrazone (CCCP) at 40 μM, although reducing sugar uptake, greatly inhibited the amount of peptidoglycan formed by exponentially growing cells pulsed with $[^{14}C]$GlcNAc (0.18 μM; 58 Ci/mol) for 10 min and then fractionated as described by Mobley *et al.* (3). The amount of lysozyme-soluble material was decreased from 56% to less than 10% in the presence of either uncoupler. There was a corresponding increase from 24% to more than 75% for the radioactive fraction soluble in cold 5% TCA. Analysis of the latter fraction showed that UDP-GlcNAc and a peptide derivative of UDP-MurNAc were both formed in the presence of uncoupler. It is not yet clear whether the inhibition of peptidoglycan formation was the result of incomplete synthesis of UDP-MurNAC-pentapeptide; due to the initiation of autolytic activity by uncoupler (2); or explained by a requirement, for peptidoglycan synthesis *in vivo*, of an energised membrane as with the case of cellulose biosynthesis in *Acetobacter xylinum* (4).

Conclusions

Four proposals have been considered to explain these results for peptidoglycan synthesis and those for teichoic acid synthesis (1).

1) Synthetic enzymes might possess active sites on both surfaces of the membrane. This seems unlikely and wasteful since nucleoside mono- or diphosphates, formed during polymer synthesis, are not released on the outer surface of the cell during growth *in vivo* and neither partly autolysed cells nor protoplasts were capable of taking in externally supplied UMP.

2) There was no evidence for translocation of substrates by prenyl lipid carriers to the inner surface of the membrane. Moreover, the transposition rates of such carriers in several membrane systems has been found to be negligible (5,6,7).

3) Partial or complete release of enzyme complexes from the membrane

might occur while the membrane remains intact. The N-acetylglucosaminyltransferase is easily removed from the membranes of toluene-treated B. megaterium by mild extraction with LiCl while the translocase and polymerase remain tightly membrane-bound (8).

4) Finally, the possibility that the teichoic acid biosynthetic complex spans the membrane and either rotates or reorientates in such a way that the polymer dissociates at the outer surface while the sites for nucleotide interaction become temporarily exposed to the external environment has been proposed (1). The necessary energy for such rotation might originate in thermal agitation of proteins or in energy generating systems. Therefore, it will be of interest to confirm whether the membrane potential plays a role in cell wall assembly in Gram-positive bacteria.

At present, the latter two proposals, or a combination of these two, are considered feasible explanations for the results thus far obtained.

References

1. Bertram, K.C., Hancock, I.C., Baddiley, J.: J. Bacteriol. 148, 406-412 (1981).
2. Jolliffe, L.K., Doyle, R.J., Streips, U.N.: Cell 25, 753-763 (1981).
3. Mobley, H.L.T., Doyle,R.J., Streips, U.N., Langemeier, S.O.: J. Bacteriol. 150, 8-15 (1982).
4. Delmer, D.P., Benziman, M., Padan, E.: Proc. Natl. Acad. Sci. U.S.A. 79, 5282-5286 (1982).
5. Weppner, W.A., Neuhaus, F.C.: J. Biol. Chem. 253, 472-478 (1978).
6. McCloskey, M.A., Troy, F.A.: Biochemistry 19, 2061-2066 (1980).
7. Hanover, J.A., Lennarz, W.J.: J. Biol. Chem. 254, 9237-9246 (1978).
8. Taku, A., Stuckey, M., Fan, D.P.: J. Biol. Chem. 257, 5018-5022 (1982).

O-ACETYLATION OF MURAMIC ACID, BIOSYNTHESIS OF MUREIN AND THE EFFECT OF PENICILLIN G IN VIVO IN SYNCHRONIZED CELLS OF PROTEUS MIRABILIS

Jobst Gmeiner and Elke Sarnow
Institut für Mikrobiologie, Technische Hochschule Darmstadt
D-6100 Darmstadt

Introduction

Murein of *Proteus mirabilis* differs from that of *Escherichia coli* in that approximately two thirds of the N-acetylmuramic acid residues are O-acetylated. These O-acetylgroups are transferred to murein subunits after their incorporation into the growing polymer. Concomitantly, an increase in murein cross-linkage is observed (1). Thus, O-acetylation, which is an easy parameter to follow, can be used as an indication of murein maturation.
In the present communication we describe experiments in which we followed murein incorporation into synchronized cells in vivo in the absence and in the presence of 6 µg/ml penicillin G. Time periods for longitudinal growth and septum synthesis were corroborated by inhibition of cell division by various antibiotics in synchronized cultures.

Results

As shown in Fig. 1, incorporation of new murein subunits into the growing murein sacculus of synchronized cells of *P. mirabilis* proceeded linearly during the first 30 min. The rate was similar in the absence and in the presence of 6 µg/ml penicillin G. At about 30 min a rate increase could be observed in

The Target of Penicillin

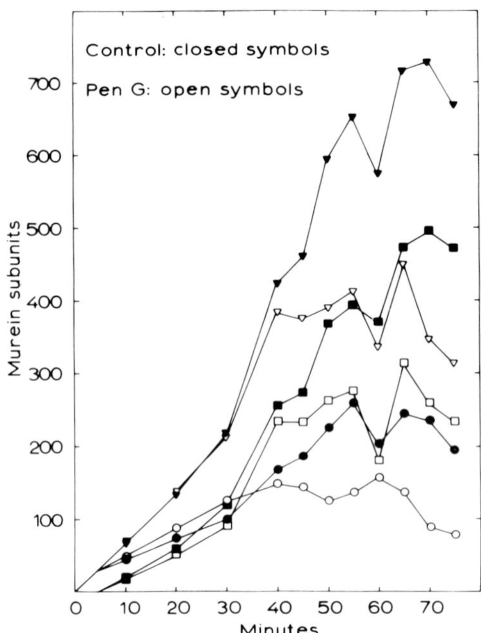

Fig. 1. Incorporation of murein subunits into polymeric murein of synchronized cells in the absence and presence of penicillin G.

Cells of P. mirabilis were synchronized by sucrose density gradient centrifugation, resuspended in diluted complex medium containing 150 µCi N-acetyl-D-[1-^{14}C]-glucosamine (specific activity 57.9 Ci/mol, Amersham-Buchler, Braunschweig) and incubated at 25°C on a rotary shaker. Penicillin G was added at time zero (final concentration 6 µg/ml). The control culture contained 1.1×10^{8} cells/ml, and the cell number increased at 50 min reaching its double amount at 65 min. The penicillin culture contained 1.9×10^{8} cells/ml and division was completely inhibited. At various times, samples were taken, cells were harvested by centrifugation and mixed with unlabelled carrier cells. Murein was isolated and the various murein subunits were quantified as described (1,2). The results are expressed as relative murein subunits incorporated into polymeric murein by 10^{10} cells. Triangles: total murein subunits incorporated; circles: non-O-acetylated subunits; squares: O-acetylated subunits.

both cases. At 40 min penicillin G inhibited further incorporation, whereas incorporation resumed after a short lag period in the control. Differentiation of the murein subunits into non-O-acetylated and O-acetylated species revealed that O-acetylation started with a delay of about 5 min in accordance with previous pulse- and pulse-chase experiments (1). During the first 30 min the rate of O-acetylation was lower in the presence of penicillin G than in the control, but similar in both cases between 30 and 40 min. These results showed that penicillin G did not inhibit overall incorporation of new subunits during the first 40 min of the cell cycle, but disturbed the process of maturation, i.e. O-acetylation during the first 30 min of the cycle.
Quantitation of the various O-acetylated fragments revealed that the amount of monomers was reduced to about 50% in penicillin-treated cells as compared with the control. At the same time, the amount of O-acetylated subunits found in the mono-O-acetylated dimer and in the di-O-acetylated dimer was slightly increased and almost the same, respectively. Accordingly, the degree of crosslinkage was higher in the penicillin-treated culture, caused mainly by the high crosslinkage of O-acetylated subunits (Fig. 2).

In order to relate these findings with the time periods of longitudinal growth and septum formation during the cell cycle, we determined the time during which various antibiotics would inhibit cell division in synchronized cultures. We used penicillin G (6 µg/ml) which blocks septum formation leading to filamentous growth, mecillinam (2 µg/ml) which interferes with longitudinal growth leading to spherical cells, and nalidixic acid (10 µg/ml) which inhibits DNA replication as precondition for septum formation. The results which will be published elsewhere are summarized in Fig. 3.
The presence of penicillin G up to 30 min inhibited the first cell division at about 55 min. Synchronized cells cultivated in the presence of penicillin G for more than two generations

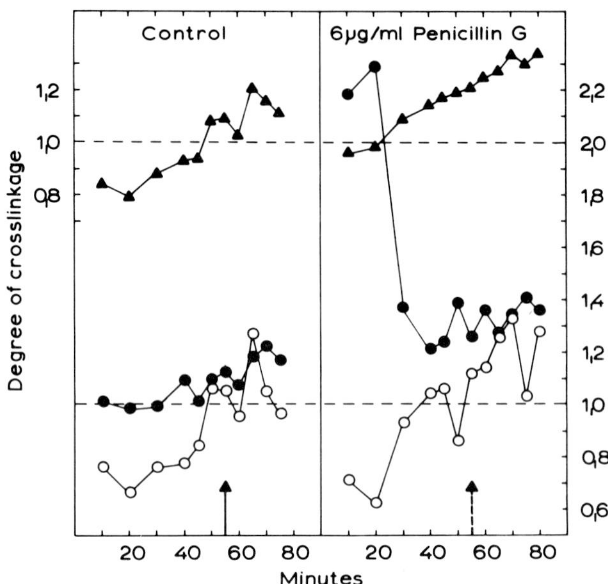

Fig. 2. Crosslinkage of isolated murein

Crosslinkage was determined as ratio of the amount of subunits in dimers and trimers versus that found as monomers. The upper part of the figure shows the degree of crosslinkage of all subunits (triangles; scale at the left margin). The lower part shows the degree of crosslinkage of non-O-acetylated subunits (open circles) and of O-acetylated subunits (closed circles; scale at the right margin). The arrow indicates the mean time of cell separation in the control.

grew out as long filaments, all of which possessed constrictions in the middle of the filaments well visible under the light microscope. Taken together with the incorporation experiments above, these observations indicate that the small newborn cells collected from sucrose gradients for synchronous growth had already initiated septum synthesis. Penicillin G interfered with the maturation of this murein portion. On the other hand, mecillinam applied at time zero or after 10 min did not affect the first cell division. Given at 20 min or later this ß-lactam even did not inhibit the second division. Hence, we have to conclude that the mecillinam-sensitive re-

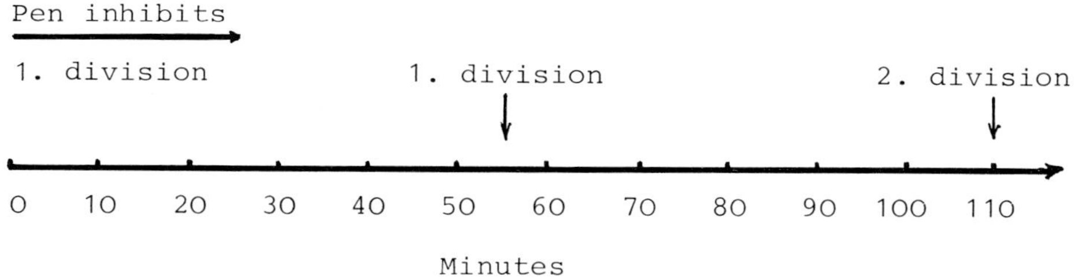

Mec inhibits

2. division

Nal inhibits

2. division

Fig. 3. Effect of various antibiotics on cell division
The figure shows the time period during which an antibiotic had to be present in order to block cell division in synchronized cultures of *P. mirabilis*. The antibiotics (penicillin G, 6 µg/ml; mecillinam, 2 µg/ml; nalidixic acid, 10 µg/ml) were given to aliquots of a synchronized culture at time zero and at intervals of 10 min during the first division cycle. From all cultures aliquots were sampled every 10 min for microscopic determination of the cell numbers in a counting chamber.

action for longitudinal growth, which is obligatory for the second division, occurred around 90-100 min before cell division. This reaction is most probably involved in the initiation of longitudinal extension of the sacculus as observed by penicillin-insensitive murein incorporation between 30 and 40 min. Finally, nalidixic acid did not affect the first cell division and, given later than 40 min after synchronization, did not affect the second division. Thymidine incorporation studies confirmed this finding since a rate increase of DNA synthesis was observed starting around 30 min and again at around 85 min after synchronization (not shown). The nalidixic acid-insensitivity of the second division after 40 min indicated that DNA replication was completed at this time, coincident with the time of penicillin-sensitive murein incorporation, i.e. initiation of septum murein synthesis for the second division (cf. Fig. 1).

Discussion

The results described in this paper show that a discrepancy exists between the rate of incorporation of new murein subunits into the sacculus of synchronized cells of *P. mirabilis*, its sensitivity towards penicillin G and the time periods where different antibiotics inhibit subsequent cell division. The most plausible explanation is that biosynthetic events for cell wall biosynthesis are not in phase with the morphologically discernible cell cycle. If we adopt the commonly accepted thesis that DNA-replication and longitudinal growth of the cells are preconditions for septum formation and cell division, we have to conclude that in the present case septum formation takes place right after collection of newborn cells and that septa are completed after about 25 min, leading to cell separation approximately 30 min later. At the time of septum completion, longitudinal growth for the next cell generation takes place while the mother cells have not yet separated. Consequently, longitudinal extension of the sacculus and septum formation for two different cell generations are observed coincidentally. The pattern of subunit incorporation into the growing sacculus and the characterization of newly synthesized murein according to degree of crosslinkage and O-acetylation indicated further that penicillin G interfered with the maturation process of septum murein and prevented initiation of new septa. A more detailed discussion of structural data obtained will be published elsewhere.

References

1. Gmeiner, J., Kroll, H.-P.: Eur. J. Biochem. 117, 171-177 (1981)
2. Essig, P., Martin, H.H., Gmeiner, J.: Arch. Microbiol. 132, 245-250 (1982).

PEPTIDOGLYCAN SYNTHESIS AND TURNOVER IN BACILLUS SUBTILIS

J.T.M.Wouters and P.D.Meyer
Laboratorium voor Microbiologie, Universiteit van Amsterdam
Nieuwe Achtergracht 127, 1018 WS Amsterdam, The Netherlands

Introduction

The activity of wall located enzymes capable of splitting bonds in peptidoglycan -autolysines- is responsible for the loss of cell wall material (peptidoglycan and covalently attached anionic polymers) during growth of *Bacillus subtilis*. This process is referred to as cell wall turnover. In order to quantitate this process in *B.subtilis* var.*niger* WM De Boer et al (1) elaborated a mathematical description of this phenomenon in terms of geometry of the cell wall and autolytic activity. Using this model one is able to calculate the turnover sensitive fraction of the wall and the rate of turnover from label-chase experiments performed with a pulse or continuous label.
Recent findings by us and others (2,3) indicate that a part of the label in *B.subtilis* is not, or to a lesser degree, subject to turnover. This is partly due to a small amount of label incorporated in non-wall polymers, but also to wall label being conserved. The consequences thereof for our calculations on cell wall turnover in *B.subtilis* will be discussed.
As turnover leads to the loss of cell wall during growth, the processes of wall synthesis and turnover have to be carefully tuned to prevent weakening of the wall during growth. This tuning was investigated in *B.subtilis* var.*niger* WM and mutants with a reduced autolytic activity. Also, under conditions of growth inhibition the balance between turnover and synthesis of peptidoglycan was investigated.

The Target of Penicillin

Materials and methods

1) Organisms and growth conditions. *B.subtilis* var.*niger* WM and two mutants with a reduced autolytic activity were used. The isolation of the mutants DM-82 and DM-263 will be described elsewhere in this volume. The strains were grown in batch or chemostat culture as detailed in (1,4). Growth rates are expressed as specific growth rate μ (h^{-1}).
2) Labeling and chasing procedures. Cultures were labeled with ^{3}H-N-acetylglucosamine (^{3}H-GlcNac) for turnover measurements and ^{14}C-GlcNac for peptidoglycan synthesis as mentioned elsewhere (4). Turnover parameters were calculated as described (1). After a lag time θ (h) label is lost with first order kinetics (rate constant λ, h^{-1}). The fraction of the wall lost per hour is called κ (h^{-1}), whereas k (%) is the percentage of wall lost per generation time. The turnover sensitive fraction of the wall is referred to as X_B (%).
3) Localisation of label. Labeled cells were suspended in 2 % (w/v) sodiumdodecylsulfate, held at 100 ^{0}C for 3 minutes, washed twice with water and suspended in 50 mM Tris/HCl buffer (pH 7.5) containing either lysozyme (1 mg/ml), trypsine (6 mg/ml), RNase (0.4 mg/ml) or DNase (0.4 mg/ml). After incubation (18 h at 37 ^{0}C) trichloroacetic acid was added to 10 % (w/v) and the total and solubilised radioactivity was assessed before and after centrifugation.

Results and discussion

The loss of a ^{3}H-GlcNac pulse label and the incorporation of ^{14}C-GlcNac were measured simultaneously in batch cultures. Pulse label was lost after a lag time with first order kinetics both in the wild type strain and in the mutant (Fig.1). The first order rate constant in the mutant was somewhat lower than in the wild type strain, whereas the lag time showed a reverse effect, being longer in the mutant. Turnover parameters and

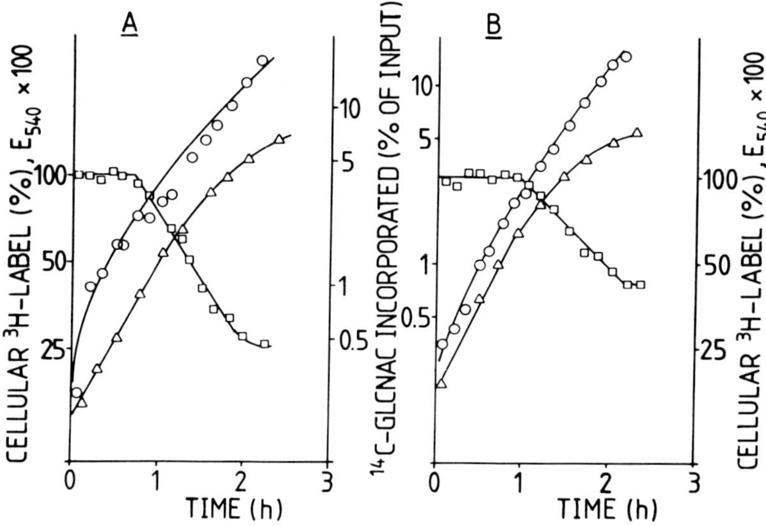

Fig.1 Wall turnover and synthesis in *B.subtilis* var.*niger* WM (A) and DM-82 (B) in enriched minimal medium. Loss of ^{3}H-label (□), incorporation of ^{14}C-GlcNac (○) and E_{540} (△) were followed as decribed in the text.

data on the incorporation of GlcNac are compiled in Table 1. As can be seen, a lower rate of turnover k is accompanied by a lower incorporation rate q^{PG}. Under these growth conditions the rate of turnover and synthesis of peptidoglycan are apparently well-balanced.

Table 1 Wall turnover parameters in *B.subtilis* var.*niger*

Strain	Growth conditions		μ	κ	k	X_B	q^{PG}
	Labeling	Chase	(h^{-1})	(h^{-1})	(%)	(%)	(†)
WM	EMM	EMM	1.34	0.27	14	25	113
DM-263	EMM	EMM	1.34	0.17	9	22	101
DM-82	EMM	EMM	1.49	0.13	6	18	90
WM	EMM	EMM +pen	-	0.14	-	-	-
WM	P-lim	MM	0.26	0.095	26	39	180
WM	P-lim	MM -P	-	0.048	-	-	-

EMM:enriched minimal medium; +pen: + penicillin G (50 μg/ml)
MM :minimal medium ; -P : no phosphate added
P-lim: phosphate limited chemostat culture
(†) nmol GlcNac incorporated per mg per generation time
- : not applicable
For a definition of the parameters, see Materials and methods

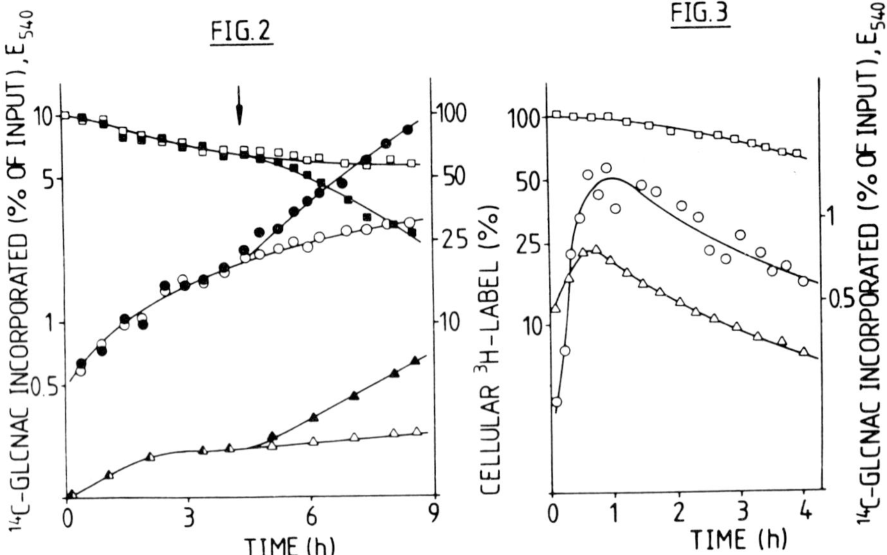

Fig.2 Wall turnover and synthesis in *B.subtilis* var.*niger* WM in minmal medium during phosphate starvation (open symbols) and after readdition of phosphate as indicated by the arrow (closed symbols). Loss of ^{3}H-label (■,□), incorporation of ^{14}C-GlcNac (●,○) and E_{540} (▲,△) were measured as described.

Fig.3 Wall turnover and synthesis in *B.subtilis* var.*niger* WM in the presence of penicillin G (50 μg/ml) in enriched minimal medium. Symbols as in Figure 1.

By means of starving a continuously labeled phosphate limited chemostat culture for phosphate we were able to study the effect of growth inhibition on peptidoglycan turnover and synthesis. The data suggested that, although turnover continued, as indicated by the release of ^{3}H-label (see κ in Table 1), the culture remained stable for many hours due to a slow peptidoglycan synthesis (Fig.2). On the readdition of phosphate,growth started without a lag, whereas turnover increased accompanied by an acceleration of peptidoglycan synthesis (Fig.2), thus maintaining the balance between these processes.

If, however, growth inhibition was accomplished by the addition of wall synthesis directed antibiotics such as penicillin G, the culture lysed as indicated by the decrease in optical density and cellular ^{3}H-label (Fig.3 and κ in Table 1).

The release of both the ^{14}C-labeled peptidoglycan precursor incorporated during the slight increase in optical density and of the ^{3}H-pulse label showed that the whole cell wall became sensitive to the hydrolytic activity of the autolysines (Fig.3) Comparable data were obtained with other peptidoglycan synthesis inhibitors, such as cycloserine and vancomycin (4).
Apparently, the inhibition of cell wall synthesis together with the ongoing action of autolysines leads to lysis of the cells, whereas during growth inhibition by phosphate starvation the tuning of peptidoglycan synthesis and breakdown is not perturbed in such a way as to cause lysis (cf. the effect of inhibiting protein synthesis by chloramphenicol, 4).
When turnover was measured during long-term experiments (i.e. more than 3-4 generations), it was found that some 25 % of the label was conserved irrespective of the labeling regime (pulse or continuous label). Enzymatic fractionation of the cells revealed that about 10 % of the label present at the start of the chase was incorporated into non-cell wall polymers (i.e. solubilised by trypsine, RNase and DNase) and this material was seemingly conserved, as this percentage remained constant throughout the chase. It appeared, however, that also a part of the lysozyme-soluble radioactivity (i.e. cell wall) was con-

Table 2 Wall turnover parameters in *B.subtilis* var.*niger* WM grown under phosphate limitation after correction for conserved label (continuous label)

Parameter		a	b	c
μ	(h^{-1})	0.25	0.25	0.25
λ	(h^{-1})	0.26	0.39	0.60
θ	(h)	1.96	2.45	3.25
κ	(h^{-1})	0.11	0.12	0.11
k	(%)	31	33	31
X_B	(%)	43	31	19

a: no corrections
b: correction for conserved non-cell wall material
c: correction for all material conserved
For a definition of the turnover parameters, see Materials and methods.

served, as even after correction for non-cell wall label, the loss of wall label leveled off after 3-4 generations. The amount of cell wall label not subject to turnover was about 10 % of the total label. Correction for label conserved leads in the first place to an increase of the first order rate constant λ (Table 2). This, in turn, causes a decrease of the turnover sensitive fraction of the wall X_B, whereas the parameters concerning autolytic activity (κ and k) are hardly influenced (Table 2). In our opinion this suggests that the turnover model, as originally put forward by De Boer et al (1), though not complete requires only a small correction to adequately describe cell wall turnover in *B.subtilis* in terms of geometry of the wall and autolytic activity.
From the data presented here no conclusions can be drawn with respect to the location of the wall not subject to turnover, nor whether turnover takes place only or mostly from the cylindrical part of the cell wall as suggested by others (2,3).

References

1. De Boer, W.R., Kruyssen, F.J., Wouters, J.T.M.: J. Bacteriol. *145*, 50-60 (1981)
2. Clarke-Sturman, A.J., Archibald, A.R.: Arch. Microbiol. *131*, 375-380 (1982)
3. Doyle, R.J., Mobley, H.L.T., Joliffe, L.K., Streips, U.N.: Curr. Microbiol. *5*, 19-22 (1981)
4. De Boer, W.R., Meyer, P.D., Jordens, C.G., Kruyssen, F.J., Wouters, J.T.M.: J. Bacteriol. *149*, 977-984 (1982)

POTENTIAL SITES FOR THE COORDINATE CONTROL OF MUREIN AND TEICHOIC ACID BIOSYNTHESIS IN S. PNEUMONIAE

Helene Fischer and Alexander Tomasz

The Rockefeller University
New York, New York 10021

Pneumococcal cell walls are made up of two major polymers, a murein and a choline-containing teichoic acid present in nearly equal amounts. Although the precise nature of many of the linkages is not yet known, murein and teichoic acid are found covalently attached to one another in the intact wall sacculus. Earlier studies have established that, in *S. pneumoniae*, nascent murein and teichoic acid polymers synthesized concurrently become covalently linked to one another at some time during their biosynthesis, forming a segregating unit that is conserved during subsequent cell divisions (1). It appears likely that some sort of control mechanism is operating, regulating and coordinating the biosynthesis of these two polymers. Our aim was to investigate this possibility, using as an experimental system the autolysis-defective pneumococci, in which preliminary experiments have indicated the production and release of cell wall polymers during treatment with penicillin (2). Two major observations will be described.

Radioactive lysine labelling of amino acid starved penicillin-treated cells permitted the detection of acid precipitable polymer released into the medium (fig. 1). The polymer was identified as uncrosslinked murein, as most was degraded upon treatment with egg white lysozyme or M1 muramidase; trypsin left the polymer intact. The chains were heterogeneous in size, with 35% having a molecular weight of greater than 80,000 as determined by gel filtration. The recent purification of this material will now permit determination of chain length, degree of pentapeptide substitution, and the existence of any additional covalent modification of the murein chain.

The Target of Penicillin

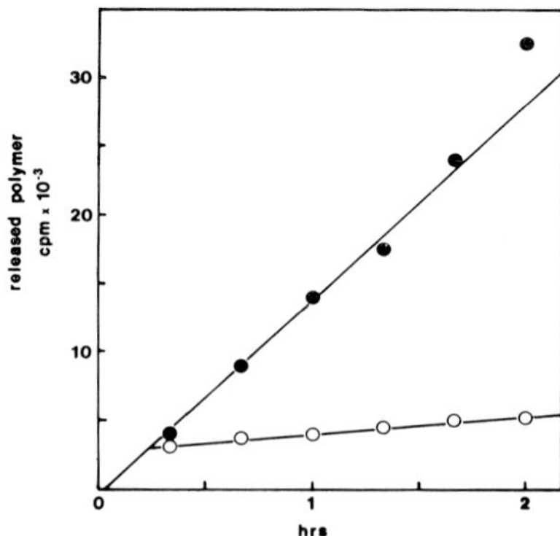

Fig. 1: Cells suspended in -leucine medium for 40 min prior to addition of 0.06 ug/ml (10X MIC) penicillin, followed 15 min later with ^{3}H-lysine. Aliquots taken at times indicated, and cell-free supernatants precipitated with 10% TCA at 4°C. Precipitates are filtered and counted.

Labelling with radioactive choline, a component unique to the teichoic acids of pneumococcus, indicated that these polymers were also released upon treatment with penicillin (2, 3). While this precipitable material included the pneumococcal Forsmann antigen (membrane lipoteichoic acid) (4), greater than 50% of the choline-containing chains could be identified as nascent wall teichoic acid on the basis of only the former to inhibit pneumococcal autolytic activity (5). The released wall teichoic acid polymers were also a heterogeneous group, with a size range similar to that of the murein chains.

Murein and teichoic acid polymers are found covalently linked to one another in the intact wall sacculus, yet our experiments indicated that when wall synthesis was interrupted with penicillin, the two polymers were released as unlinked units. The lysine-containing polymer was not precipitable with C reactive protein (directed against the C polysaccharide, or teichoic acid antigen of pneumococcus (6)), nor was it bound by TEPC-15, a myeloma protein specific for choline phosphate residues (7). However, an affinity resin of TEPC-15 bound the choline-containing polymers quantitatively, and permitted specific elution with low concentrations of free choline phosphate. This is consistent with the work of others who observed the release of murein free of teichoic acid from penicillin-treated cells (8, 9, 10), and suggests that penicillin is interfering with a reaction responsible for the linkage of these polymers both to each other and to the cell wall sacculus.

In an attempt to determine the nature of this penicillin-sensitive reaction, we compared the drug concentrations required for polymer release with those resulting in growth inhibition, using intrinsically penicillin resistant transformants and beta-lactam antibiotics with widely varying MIC values for pneumococci. Release of murein and teichoic acid chains was just detectable at the MIC value of penicillin in a sensitive strain (MIC=0.006 ug/ml); a greater percentage of the total synthesized polymer was found in the medium when the concentration of penicillin was increased. In an intrinsically resistant transformant (MIC=6.0 ug/ml), extensive release of cell wall polymers likewise required as a minimum a penicillin concentration of 6.0 ug/ml (table 1).

released polymer/total polymer	^{3}H-lysine labelled (cpm)	
penicillin cocn. (ug/ml)	strain R6 (MIC: 0.006 ug/ml)	strain pen^R (MIC: 6.0 ug/ml)
none	0.26	0.25
0.06	0.49	0.23
0.6	0.60	0.26
6.0	0.60	0.25
60.0	0.63	0.49

Table 1: Experiment described in fig. 1; total polymer determined as acid precipitable counts in cells and supernatants together. Aliquots taken at t=90 min after addition of ^{3}H-lysine.

Beta-lactam antibiotics other than benzylpenicillin, with varying MIC values, all appeared to induce release in accordance with their corresponding MIC values (fig. 2). These findings suggest that the release phenomenon results from penicillin's interference with the action of a 1° transpeptidase (11), an enzyme that is believed to be primarily responsible for the incorporation of nascent murein chains into the cell wall sacculus; other wall synthesizing enzymes are known to be inhibited at penicillin concentrations far below the MIC value (12). This lends support to previous work with other bacterial systems, suggesting that the attachment of teichoic acid to murein follows the incorporation of nascent murein into the pre-existing cell wall via the penicillin-sensitive transpeptidation reaction (10).

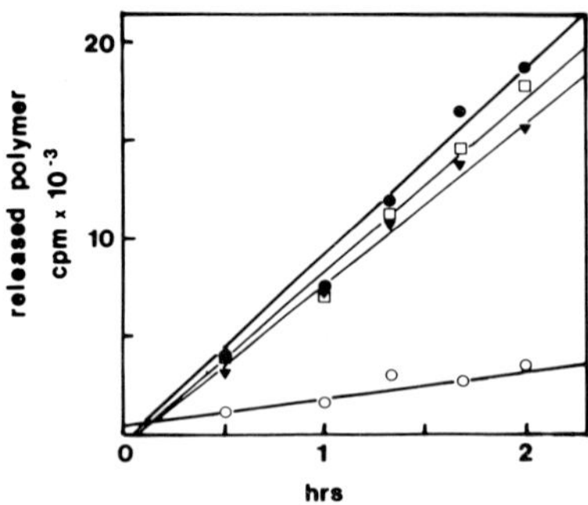

Fig. 2: Experiment described in fig. 1; all drugs tested at 20X MIC for sensitive strain. ●, penicillin (MIC=0.006 ug/ml); □, ampicillin (MIC=0.025 ug/ml); ▼, methicillin (MIC=0.8 ug/ml); O, no drug.

Although physically separable from one another, the biosyntheses of teichoic acid and murein produced by the penicillin-treated pneumococci showed additional signs of coordinate control. The presence of choline, a structural component of teichoic acid, was found to be an absolute requirement for the synthesis of murein polymers. Short term choline starvation was sufficient to completely inhibit the synthesis of murein polymer (fig. 3); readdition of choline resulted in instant, rapid resumption.

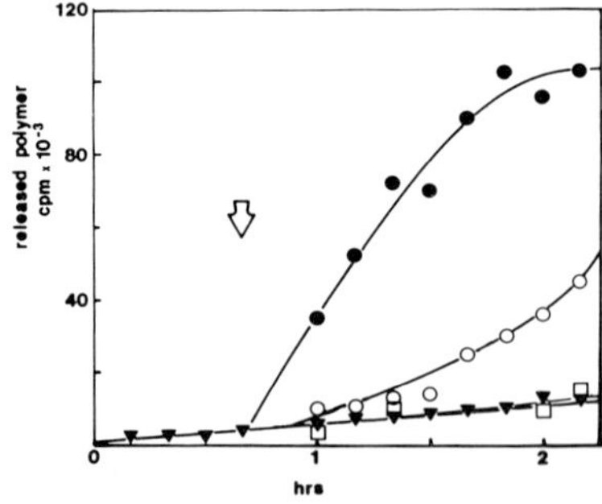

Fig. 3: Cells starved for choline 60 min prior to addition of penicillin (0.06 ug/ml) and ^{3}H-lysine. Additions made 40 min after addition of label (at arrow). ●, +15 ug/ml choline; O, +120 ug/ml ethanolamine; □, + 60 ug/ml CDP-choline; ▼, no addition.

Whereas the addition of ethanolamine also permitted renewed synthesis (albeit somewhat less efficiently than choline), CDP-choline, remaining extracellular, did not.

Our findings suggest at least two sites of coordination for the synthesis of murein and teichoic acid polymers in pneumococci. Our first experiments suggest that the transpeptidase may be a control point, acting to regulate the rate of incorporation of murein and teichoic acid polymers into the cell wall. An additional site of control operating at a step prior to polymerization is suggested by the observation that choline deprivation inhibited murein synthesis. A model to explain this finding may be based on an assumption of competition for a common bactoprenol carrier (13) (fig. 4).

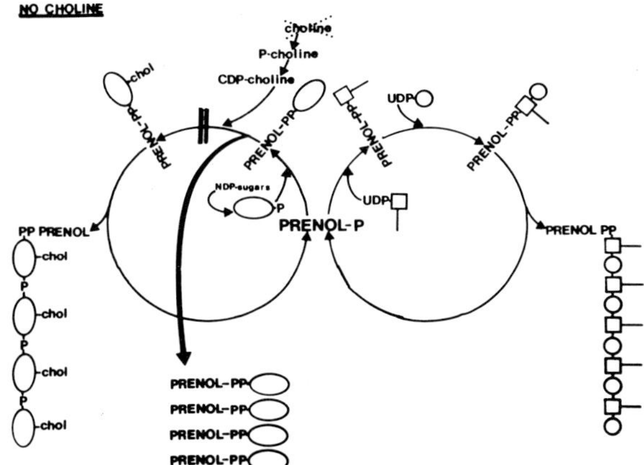

Fig. 4: See text.

We suggest that, in the absence of choline, teichoic acid subunits linked to bactoprenol but lacking the phosphoryl choline residue accumulate in the cell. These incomplete teichoic acid subunits cannot serve as substrate for the transferase catalyzing their polymerization into chains. Consequently, the pool of free bactoprenol available for muramic acid-pentapeptide binding diminishes, resulting in the inhibition of murein synthesis.

References

1. Tomasz, A., et. al.: J. Biol. Chem. 250, 337-341 (1975).
2. Waks, S., Tomasz, A.: Antimicrob. Agents and Chemotherapy 13, 293-301 (1978).
3. Hakenbeck, R., et. al.: Antimicrob. Agents and Chemotherapy 13, 302-311 (1978).
4. Goebel, W. F., et. al.: J. Biol. Chem. 148, 1 (1943).
5. Höltje, J-V., Tomasz, A.: Proc. Nat. Acad. Sci., USA 72, 1690-1694 (1975).
6. Brundish, D. E., Baddiley, J.: Bioch. J. 110, 573 (1968).
7. Potter, M., Lieberman, R.: J. Exp. Med. 132, 737-751 (1970).
8. Keglevic, D., et. al.: Eur. J. Biochem. 42, 389-400 (1974).
9. Mirelman, D., et. al: Biochemistry 13, 5043-5045 (1974).
10. Tynecka, Z., Ward, J. B.: Bioch. J. 146, 253-267 (1975).
11. Smith, P. F., Wilkinson, B. J.: J. Bact. 148, 610-617 (1981).
12. Williamson, R., et. al.: Antimicrob. Agents and Chemotherapy 18, 629-637 (1980).
13. Anderson, R. G., et. al.: Bioch. J. 127, 11-25 (1972).

MECHANISM OF STRINGENT CONTROL OF PEPTIDOGLYCAN SYNTHESIS IN *ESCHERICHIA COLI*

Edward E. Ishiguro
Department of Biochemistry and Microbiology, University of Victoria, Victoria, B.C., Canada V8W 2Y2

Introduction

Guanosine 5'-diphosphate 3'-diphosphate (ppGpp), the apparent mediator of the stringent control system, is involved in the regulation of peptidoglycan synthesis in *Escherichia coli* (1). The accumulation of ppGpp by amino acid deprived *relA*$^+$ bacteria is accompanied by the inhibition of peptidoglycan synthesis. In contrast, amino acid deprived *relA* mutants do not accumulate ppGpp, and peptidoglycan synthesis is relaxed. Both the synthesis of nucleotide-activated precursors (2) and the polymerization of peptidoglycan (3) are regulated by ppGpp, but the mechanisms involved are still uncertain. Recent *in vitro* studies (4) on the final steps of peptidoglycan synthesis indicated that the peptidoglycan biosynthetic activity of ether-permeabilized cells prepared from a culture of *relA*$^+$ bacteria decreased about 75% during the course of amino acid deprivation. This was most likely a reflection of the stringent control mechanism since peptidoglycan synthesis assayed in ether-permeabilized *relA* bacteria was virtually unaffected by amino acid deprivation. These results suggested that stringent control may not involve the allosteric inhibition of a key particulate peptidoglycan biosynthetic enzyme by ppGpp as previously thought (1) since ether treatment depletes the cells of low molecular weight compounds such as ppGpp. Thus, stringent control is apparently a manifestation of some inherent characteristic of the membranes of amino acid

The Target of Penicillin

deprived *relA*$^+$ bacteria, i.e., the expression of a functional change which occurs in membranes of *relA*$^+$ cells but not *relA* cells during amino acid deprivation. It is shown here that the mechanism may involve a dependence of peptidoglycan synthesis on on-going phospholipid synthesis.

Materials and Methods

E. coli K-12 strains LD5 (*thi lysA dapD*) and LD5456 (*thi lysA dapD relA*) have been described (1). Strain VC58 was a *gpsA* derivative of LD5456. Bacteria were grown as described previously (1) in either M9 medium (1) or, in experiments involving $^{32}P_i$ labeling, in the modified M56LP medium of Bell (5). Amino acid deprivation was imposed by omitting L-lysine from the medium. Alternatively, bacteria were starved for isoleucine by adding L-valine (500 μg/ml) to the medium. Both methods gave identical results. Peptidoglycan synthesis was determined by following the incorporation of [^{3}H]diaminopimelic acid ([^{3}H]DAP) into cold 5% trichloracetic acid-insoluble material (1). The incorporation of $^{32}P_i$ into phospholipids was determined after extraction with chloroform-methanol, 1:2 (5,6). The incorporation of [^{3}H]DAP into the peptidoglycan, nucleotide-activated precursor, and lipid intermediate fractions of cells was determined as described previously (3).

Results and Discussion

The syntheses of phospholipids (7) and peptidoglycan (1) are under stringent control, and both were therefore inhibited during amino acid deprivation in strain LD5 *relA*$^+$ (Figs. 1A, 1B) but not in strain LD5456 *relA* (Figs. 1C, 1D). Two series of experiments were performed to test the possibility that there may be a relationship between peptidoglycan synthesis

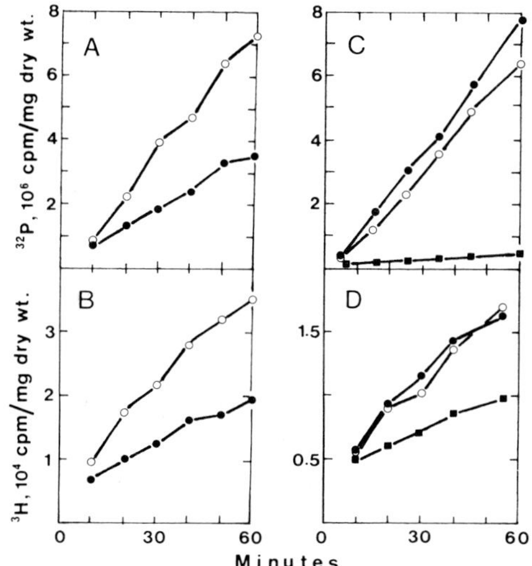

Fig. 1. Effects of amino acid deprivation and cerulenin on phospholipid and peptidoglycan synthesis. Phospholipid synthesis (A) and peptidoglycan synthesis (B) were determined in untreated control cultures (○) and isoleucine deprived cultures (●) of strain LD5 relA+. Phospholipid synthesis (C) and peptidoglycan synthesis (D) were determined in untreated control cultures (○) and isoleucine deprived cultures (●,■) of strain LD5456 relA. In this case, one of the isoleucine deprived cultures was treated with 100 μg of cerulenin per ml (■).

and phospholipid synthesis. Firstly, the effects of cerulenin on amino acid deprived cells of strain LD5456 relA were determined. Cerulenin treatment resulted not only in the expected inhibition of relaxed phospholipid synthesis (Fig. 1C) but also in the inhibition of relaxed peptidoglycan synthesis (Fig. 1D). Other experiments (data not shown) indicated that: (i) the inhibitory effects of cerulenin on both phospholipid synthesis and peptidoglycan synthesis were concentration-dependent, and (ii) reversal of the inhibitory effect of cerulenin on phospholipid synthesis by the addition of exogenous oleate and palmitate resulted in the restoration of relaxed peptidoglycan synthesis. In the second series of

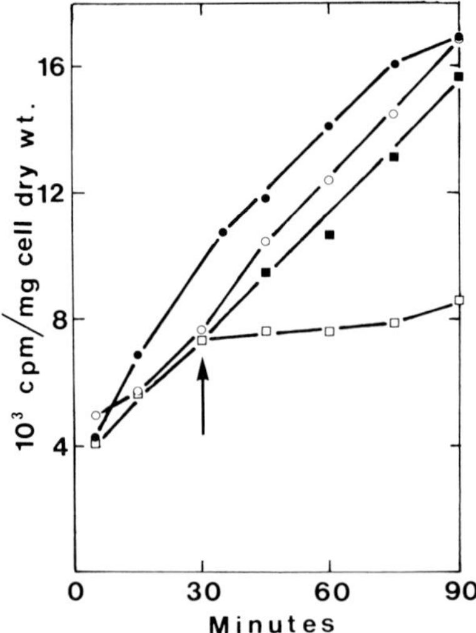

Fig. 2. Effects of amino acid deprivation and glycerol deprivation on peptidoglycan synthesis in strain VC58 relA gpsA. Symbols: ○, untreated control; ●, isoleucine deprived culture; □, culture deprived of isoleucine and glycerol; ■, portion of the isoleucine and glycerol deprived culture to which glycerol was added at 30 min (arrow).

experiments, strain VC58, a derivative of strain LD5456 relA which was auxotrophic for glycerol (gpsA), was used. When strain VC58 was simultaneously deprived of glycerol and a required amino acid, the relaxed synthesis of phospholipids was inhibited (data not shown). Fig. 2 shows that the relaxed synthesis of peptidoglycan was also inhibited during glycerol and amino acid deprivation of VC58. Furthermore, the addition of glycerol to this culture resulted in the concomitant restoration of relaxed phospholipid and peptidoglycan synthesis. These results suggest that relaxed peptidoglycan synthesis was dependent on on-going phospholipid synthesis.

The step in relaxed peptidoglycan synthesis which was apparently dependent on phospholipid synthesis was determined by comparing the amounts of [^{3}H]DAP incorporated into the various cellular fractions by the cultures shown in Figs. 1D and 2. the essential results were as follows. (i) UDP-N-acetyl-muramyl-pentapeptide accumulated over 3-fold during amino acid deprivation of relA bacteria as reported previously (2), and the inhibition of relaxed phospholipid synthesis did not interfere with this accumulation. (ii) Inhibition of phospholipid synthesis also did not inhibit the synthesis of the lipid intermediates. (iii) However, the accumulation of labelled peptidoglycan was inhibited by about 70% when phospholipid synthesis was inhibited in amino acid deprived relA cells. Thus, the terminal step in peptidoglycan synthesis apparently was dependent on phospholipid synthesis. We have recently shown that relaxed peptidoglycan synthesis is penicillin-sensitive and that the peptidoglycan synthesized by amino acid deprived relA cells is crosslinked to pre-existing cell wall peptidoglycan (submitted for publication). Thus, it is possible that the enzymatic step requiring on-going phospholipid synthesis is transpeptidation. In summary, these results suggest that in the mechanism of stringent control, ppGpp indirectly regulates peptidoglycan synthesis by acting as a negative effector in the control of phospholipid synthesis.

Acknowledgments

I thank D. Vanderwel and R.E. Harkness for their contributions to this study. This work was supported by grants from the Natural Sciences and Engineering Research Council of Canada.

References

1. Ishiguro, E.E., Ramey, W.D. J. Bacteriol. 127, 1119-1126 (1976).

2. Ishiguro, E.E., Ramey, W.D. J. Bacteriol. 135, 766-774 (1978).
3. Ramey, W.D., Ishiguro, E.E. J. Bacteriol. 135, 71-77 (1978).
4. Ishiguro, E.E., Mirelman, D., Harkness, R.E. FEBS Lett. 120, 175-178 (1980).
5. Bell, R.M. J. Bacteriol. 117, 1065-1076 (1974).
6. Bligh, E.G., Dyer, W.J. Can. J. Biochem. Physiol. 37, 911-917 (1959).
7. Sokawa, Y., Nakao, E., Kaziro, Y. Biochem. Biophys. Res. Commun. 33, 108-112 (1968).

INTERRELATIONSHIPS BETWEEN WALL AND MEMBRANE BIOSYNTHESIS

Professor Howard J Rogers and Paul F. Thurman
National Institute for Medical Research
Mill Hill, London NW7 1AA, England

Introduction

During cell growth and division of bacteria tight integration between expansion of wall and membrane is demanded, with considerable localisation either of biosynthetic enzymes or of activation of pro-enzymes. The organisation of the enzymes in the membrane may be physically controlled by the cross-linking of extending peptidoglycan strands still attached to isoprenoid pyrophosphate precursors. Disturbed membrane functions might then result from inhibiting peptidoglycan synthesis and *vice versa* apparently irrelevant membrane disturbances might disorganise the sites of peptidoglycan synthesis. One possible example of each of these situations will be given:

Inhibition of peptidoglycan synthesis affects membranes

When suitable concentrations of β-lactams or cycloserine are added to exponentially growing cultures of autolytically defective *lyt* mutants of *Bacillus subtilis* (1) the optical density of the culture approximately doubles during the first hour but thereafter remains exactly constant, with β-lactams the bacteria nevertheless die quite rapidly after about two hours, whereas little death occurs with cycloserine for six hours or more. In other words the bacteria appear truly tolerant (2,3) to cycloserine but only partially so to β-lactams. Table 1 shows the series of events which occur in these cultures. Within less than the equivalent of half-a-

The Target of Penicillin

generation of growth with either antibiotic lipid synthesis starts to increase, and with β-lactams damage to transport processes starts. By two generations, with cephalothin, for example, the intracellular metabolite pools are empty and protein and lipid synthesis have stopped; lipid has been excreted by the cells. With cycloserine lipid and protein synthesis are still only very partially inhibited even after a time equivalent to 3 - 4 generations of growth. Thus, two sets of very early membrane disturbance can be seen. 1) an increase in the rate of phospholipid synthesis, for which further work is needed to distinguish between a true increase in general rate of phospholipid synthesis or as in *S.mutans* an increase in the proportion of phosphatidyglycerol (4) with its higher rate of turnover. 2) active transport processes are damaged and the intracellular pool leaks out. Both of these events start within 0.2 - 0.3 of a generation of growth after adding the antibiotics. Process (2) becomes irreversible whilst (1) alone does not cause irreversible damage. No evidence could be obtained for formation of soluble wall polymers when β-lactams are present for up to two generations of growth, whilst uncross-linked peptidoglycan is formed and remains associated with the cells in increasing amounts up to about one generation. Ultra-structurally ferritin particles were unable to penetrate the walls of cells exposed to twice the concentration of β-lactams that were used for the above experiments for 0.5 of a generation. After 1.5 generations, wall penetration was seen in only about 1% of the cells. Results for bacteria treated with cycloserine were similar. The cross-walls of the β-lactam treated organisms appeared thickened and multi-layered presumably due to the deposition of uncross-linked material (5).

Results with the probe 1-anilinonaphalene-8-sulphonate (ANS) a molecule which fluoresces when associated with hydrophobic substances, and therefore with membranes, are more difficult to understand. No fluorescent signal is given by exponentially growing bacteria above that given by the bacterial suspension in the absence of the probe together with that of the supernatant from the suspension. After the equivalent of about 0.5 generations of treatment with either β-lactams or cycloserine fluorescence begins and increases up to about 2 - 3 generations. The intensity with β-lactams is then about 80% of that given by a culture of the same optical density completely lysed with lysozyme which in turn is about equivalent to that given by isolated membrane preparations, tested at the concentration present within the micro-organisms. Thus the situation of the probe appears to change from one in which it sees none of the membrane to one in which it sees nearly all of it. This is particularly surprising in view of other results (6) showing considerable fluorescence by the same probe at the same concentration but with suspensions of control untreated wild-type *B.subtilis*. Although the walls of autolytic-deficient mutants are less permeable that those of the wild-type (7) they are still able to allow the passage of proteins of molecular weight 70,000 Daltons. It is therefore difficult to believe that ANS is excluded from the membrane simply by the low porosity of the walls. It would seem more likely either that some layer of a hydrophilic substance exists in the envelope through which the probe cannot penetrate, or that its strong negative charge prevents wall penetration. If the latter is true it may be that membrane lipids come out through the wall to meet the probe rather than *vice versa*.

The important unanswered question in the interpretation of the above work is whether or not the small residual autolytic activity in the *lyt* mutants plays a role in either causing the membranes damage or in allowing entrance of the fluorescent probe to the membrane. All we are able to say at present is that no wall damage has been found from either chemical or ultrastructural examination at a time when reaction with the probe and membrane damage are both clearly occurring.

Partial inhibition of phospholipid synthesis affects cell morphology.

An example of the reverse phenomenon, that is of membrane disturbance affecting wall formation, would seem to involve a more sophisticated change. The effect of inhibiting lipid synthesis in *lyt* mutants with cerulenin during β-lactam treatment was examined in an attempt to understand the significance of the early increase in rate previously found. When concentrations of from 5μg to 10μg/ml of cerulenin were added to exponentially growing cultures, the growth rate was reduced but not stopped, being with 5μg/ml about 50% of that of the control culture. When cephalothin (0.04μg/ml) was also added to such a culture, growth (as measured by optical density) continued to increase slowly for at least 2 - 3 generations whereas in the absence of cerulenin it ceased after about one generation. Moreover, even with 0.5mg/ml of cephalothin the residual autolysin in the *lyt* mutant was not expressed. In the absence of the inhibitor of lipid synthesis slow lysis was fully expressed with 0.1 - 0.2 μg/ml of cephalothin. When 0.2 μg/ml of cephalothin was added to wild type *B.subtilis* cultures very rapid lysis (complete in 3-4 hours) ensued whereas in the presence of cerulenin growth stopped but only very slow lysis (20% in 5 hours) occurred. Cerulenin did not seem to prevent the lethal affects of cephalothin. Most interesting, was the effect of cerulenin on the morphology of both the *lyt* mutant and the wild type. Both organisms were changed from rods to cocci which were indistinguishable from the coccal form of *rod* mutants (8). The change took place within about two generations of growth and was not prevented by cephalothin (0.04μg/ml), the spherical cells were phase dark and viable. It would thus seem that partial inhibition of lipid synthesis a) reduces expression of autolytic activity b) apparently alters the distribution of wall growth sites so that cocci and not rods are formed. Effects of cerulenin on morphological changes in other micro-organisms have recently been recorded (9) as well as the known effect on the morphology of *Streptococcus faecalis* (10). Preliminary results suggest that there may be a considerable alteration in the rate of lipid synthesis during the morphological change of *rodB* mutants themselves. These results together with the effect of protease inhibitors previously recorded (11) may suggest that an important lesion in *rod* mutants of *B.subtilis* is in the processing of an enzyme concerned with lipid metabolism.

References

1. Fein, J.E., Rogers, H.J.:J.Bact. 127, 1427-1442 (1976).
2. Tomasz, A., Albino, A. and Zanati, E.: Nature (Lond.), 227, 138-140. (1970).
3. Tomasz, A.: Ann.Rev.Microbiol, 33, 113-137. (1979).
4. Brissette, J.L., Shockman, G.D., Pieringer, R.A.: J.Bact., 157, 838-844. (1982).
5. Rogers, H.J., Thurman, P., Burdett, I.D.J.: J.Gen.Microbiol. (in press). (1983)
6. Jolliffe, L.K., Doyle, R.J., Streips, U.N.: Cell, 25, 753-763. (1981).
7. Williamson, R., Ward, J.B.:J.Gen.Microbiol, 125, 325-334.
8. Rogers, H.J.: Adv. Microbiol. Physiol, 19, 1-62. (1979).
9. Ito, E.T., Ohlar, R.L., Inderlied, C.B.: J.Bact, 152, 880-887. (1982).
10. Higgins, M.L., Carnen, D.D., Daneo-Moore, L.: J.Bact, 143, 989-994. (1980).
11. Rogers, H.J., Thurman, P.F.: FEBS Letters, 117, 99-102, (1980).

Rapid secondary membrane effects of cell wall inhibitors acting on autolytic deficient *lyt* mutants of *Bacillus subtilis*.

Time* (min)	Generations* of growth	Observed effects	
		β-lactams	Cycloserine
15-20	0.25 - 0.3	Increase in L_R starts, pool starts to empty, uncrosslinked peptidoglycan added.	Increase in L_R starts Rapid decrease in peptidoglycan synthesis.
30-40	0.5 - 0.6	Fluorescence with probes starts, uncrosslinked peptidoglycan added.	Fluorescence with probes starts, addition of peptidoglycan stops.
50-65	0.8 - 1.0	Addition of uncrosslinked peptidoglycan ceases, Peak in L_R. P_R 60-70% inhibited, Lipid excretion starts.	Peak in L_R, P_R 30-40% inhibited.
130-195	2 - 3	L_R + P_R 80-90% inhibited, max. fluorescence with probes, pool almost empty, transport irreversibly damaged, many (50-60%) cells dead.	L_R 20-30% inhibited P_R 40-50% inhibited Max. fluoresence with probes.

* Times after adding antibiotics and equivalent generations of growth of an untreated culture

L_R = Rate of lipid synthesis P_R = Rate of protein synthesis

FUNCTIONAL DOMAINS OF COLICIN M

Ralf Dreher, Joachim-Volker Höltje[1], Volkmar Braun,
Mikrobiologie II, Universität Tübingen und [1]Max-Planck-Institut für Virusforschung, 7400 Tübingen

Introduction

Colicin M, a bactericidal protein produced by certain strains of Escherichia coli, kills sensitive cells by lysis. It inhibits murein biosynthesis both in vivo and in vitro. This penicillin-like action is unique among the colicins (3).
In an osmotically stabilizing medium (16 % sucrose) spheroplasts are formed. The rod-like shape of normal E. coli cells is converted to large round cells (Fig. 1). This suggests murein (peptidoglycan) as target of colicin M action (1).

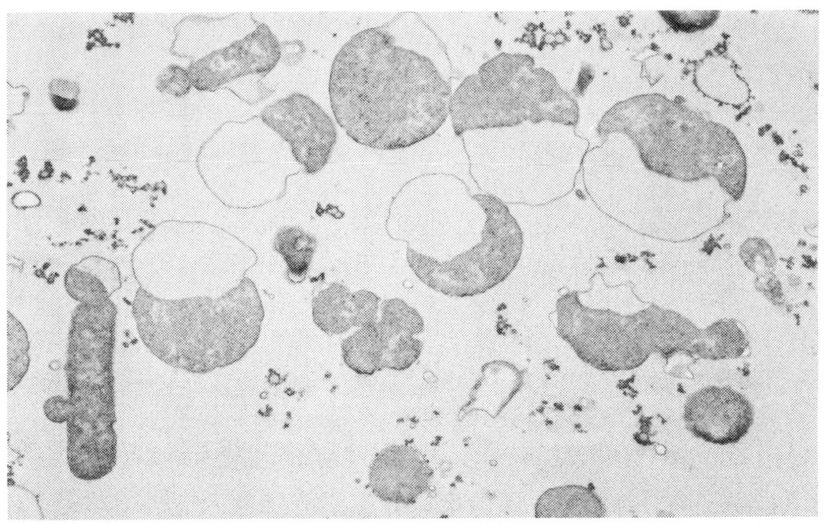

Fig. 1 Electron microscope pictures of morphological alterations of cells(E. coli K-12 ROW/V/22.1) induced by colicin M in a medium containing 16 % sucrose.

The Target of Penicillin

The mode of action of colicin M can be divided into three stages:

I. Specific binding to the receptor of the sensitive cell

II. Penetration through the cell membranes

III. Inhibition of murein biosynthesis

Results

Proteinase K splits the colicin M molecule (MW 27.000) into two fragments, A and B, with molecular weights of 3.000 and 24.000, respectively. Fragment A and B were inactive against whole cells. Characterization of these fragments revealed that fragment A is the NH_2-terminal part, while B forms the COOH-terminal part of colicin M.

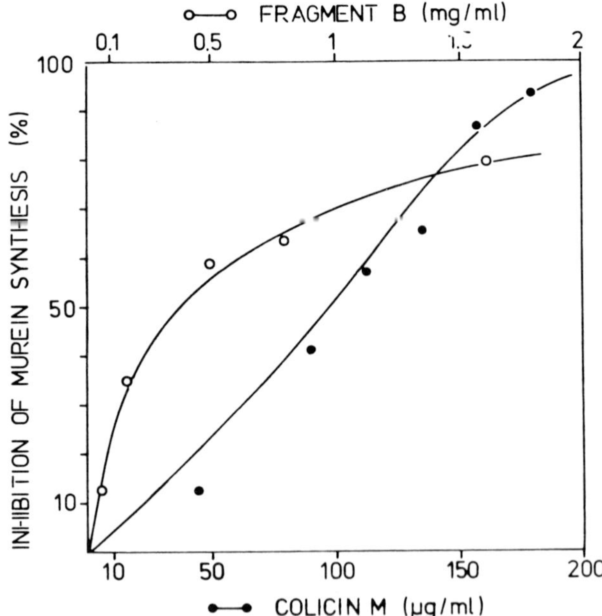

Fig. 2 Inhibition of the in vitro murein synthesis. The UDP-N-acetylmuramylpentapeptide-dependent incorporation of $[^{14}C]$UDP-N-acetylglucosamine into high-molecular-weight material was determined by the method of Izaki (2).

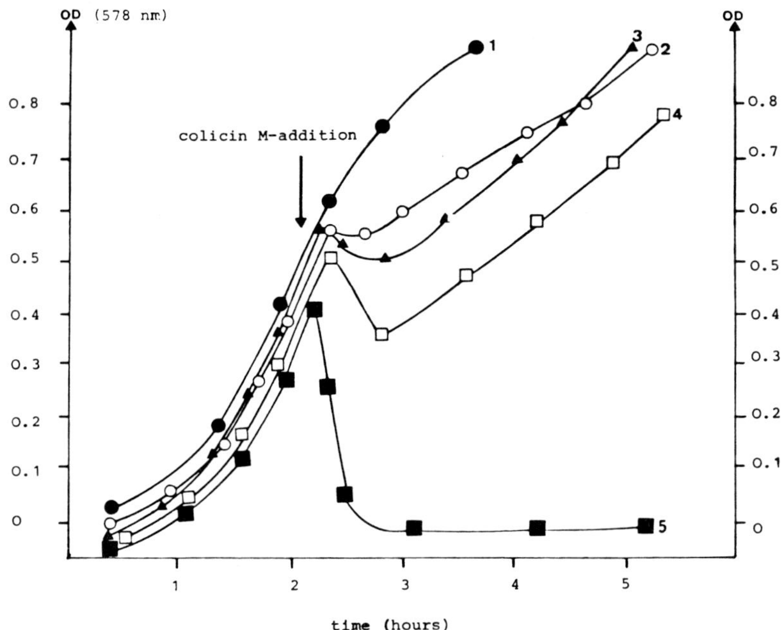

Fig. 3 Protection against colicin M action on cells of strain E. coli K-12 AB2847 by fragment B. Colicin M was added 10 Min after fragment B ($3x10^{14}$ molecules) to logarithmically growing cultures.
1 (●) : no fragment B and no colicin M
2 (○) : fragment B and colicin M ($2x10^{10}$ molec.)
3 (▲) : fragment B and colicin M ($3x10^{10}$ molec.)
4 (□) : and colicin M ($2x10^{10}$ molec.)
5 (■) : and colicin M ($3x10^{10}$ molec.)
OD_{578} of 0.5 corresponds to 5×10^8 cells/ml (8 ml).

Tab. 1 Action of fragment B and Colicin M on resistant cells under conditions of osmotic shock(4). Data are expressed as per cent of the input.

surviving cells (%)

STRAIN	SHOCK ALONE	FRAGMENT B without shock	FRAGMENT B with shock	COLICIN M with shock
P8 fhu A	51	100	0.5	2
Mo6 tol M	26	100	24	24

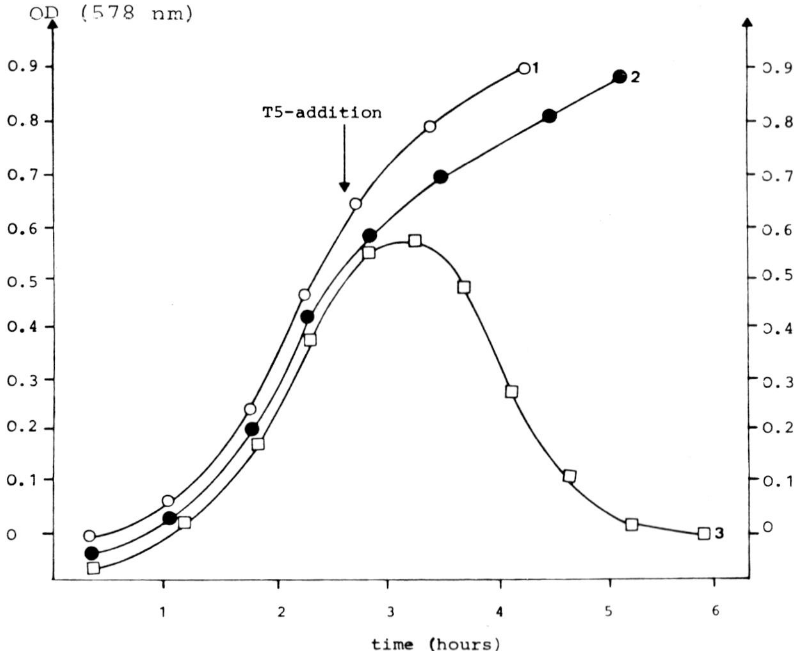

Fig. 4 Protection against the infection by phage T5 on cells of strain E. coli K-12 AB2847 by fragment B. Phage T5 was added 10 min after fragment B (3×10^{14} molecules) to logarithmically growing cultures.
1 (○) : no fragment B and no phage T5
2 (●) : fragment B and phage T5 (5×10^{9})
3 (□) : and phage T5 (5×10^{9})

1. Fragment B inhibits murein biosynthesis in an in vitro assay as has been shown for the intact colicin M (Fig. 2).
2. Fragment B could be transferred into the cells by osmotic shock treatment and killed cells (Tab. 1).
3. Fragment B adsorbed to the receptor protein at the cell surface since it prevented killing by colicin M (Fig. 3) as well as infection by phage T5 (Fig. 4).
4. These results suggested that the NH_2-terminal (fragment A) part of colicin M is required for the transport of the colicin molecule into the cell (Translocation).

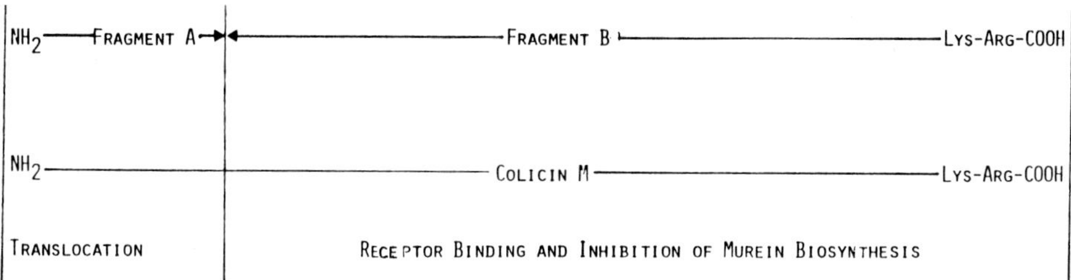

The two proteolytic fragments of colicin M were isolated in pure form. Determination of their size, amino acid composition and partial amino acid sequence revealed the position of the two fragments in the colicin M molecule. Since fragment B binds to the outer membrane receptor protein but only kills when it is shocked into the cells it is deduced that fragment A is required for colicin M translocation from the cell surface receptor to the target. Fragment B contains the inhibitory site on murein biosynthesis as is also revealed by its in vitro activity. The arrangement of the functional domains of colicin M resembles those of the E colicins and of cloacin DF13 with entirely different modes of action and receptor specificities.

References

1. Braun, V., Schaller, K., Wabl, M.R.,: Antimicrob. Agents Chemother. 5, 520-533(1974).

2. Izaki ,K.,Matsuhashi, M., Strominger, J.: J. Biol. Chem. 243, 3180-3192 (1968).

3. Schaller, K., Dreher, R., Braun, V.: J. Bacteriol. 146, 54-63 (1981).

4. Braun, V.,Frenz, I., Hantke, K., Schaller, K.: J. Bacteriol. 142, 162-168 (1980).

AUTHOR INDEX

SUBJECT INDEX

Numbers refer to the first page of the paper dealing with the subject.